Stefan Frei, Bärbel Holm, Thomas Richter, Thomas Wick, Huidong Yang (Eds.)
Fluid-Structure Interaction

Radon Series on Computational and Applied Mathematics

Volume 20

Fluid-Structure Interaction

Modeling, Adaptive Discretizations and Solvers

Edited by
Stefan Frei,
Bärbel Holm,
Thomas Richter,
Thomas Wick,
Huidong Yang

DE GRUYTER

Editors
Dr. Stefan Frei
Dept. of Mathematics
University College London
Gower Street
London
WC1E 6BT
United Kingdom
s.frei@ucl.ac.uk

Dr. Bärbel Holm
KTH Royal Institute of Technology
Computational Science & Technology
Lindstedtsvägen 5
100 44 Stockholm
Sweden
barbel@kth.se

Prof. Dr. Thomas Richter
Institut für Analysis und Numerik
Universität Magdeburg
Universitätsplatz 2
39106 Magdeburg
Germany
thomas.richter@ovgu.de

A-Prof. Dr. Thomas Wick
Centre de Mathematiques Appliquees (CMAP)
Ecole Polythechnique
Université Paris-Saclay
91128 Palaiseau cedex
France
thomas.wick@cmap.polytechnique.fr

Dr. Huidong Yang
Technische Universität Graz
Institut für Numerische Mathematik
Steyrergasse 30
8010 Graz
Austria
huidong.yang@ricam.oeaw.ac.at

ISBN 978-3-11-049527-0
e-ISBN (PDF) 978-3-11-049425-9
e-ISBN (EPUB) 978-3-11-049258-3
Set-ISBN 978-3-11-049426-6
ISSN 1865-3707

Library of Congress Cataloging-in-Publication Data
A CIP catalog record for this book has been applied for at the Library of Congress.

Bibliographic information published by the Deutsche Nationalbibliothek
The Deutsche Nationalbibliothek lists this publication in the Deutsche Nationalbibliografie; detailed bibliographic data are available on the Internet at http://dnb.dnb.de.

Typesetting: le-tex publishing services GmbH, Leipzig
Printing and binding: CPI books GmbH, Leck
♾ Printed on acid-free paper
Printed in Germany

www.degruyter.com

Preface

In January 2016, we had the pleasure to welcome a number of leading experts and many young researchers working in the field of fluid-structure interactions for a one-week international symposium including a winter school on

Modeling, Adaptive Discretizations and Solvers for Fluid-Structure Interaction

at the Johann Radon Institute for Computational and Applied Mathematics (RICAM) in the beautiful winter in Linz, Austria.

Motivated by the fascinating variety of topics presented in this symposium, we decided to collect nine of them into a book that might serve for future reference. These topics comprise both reviews of state-of-the-art fluid-structure interaction (FSI) as well as very recent research directions and are written by well-known experts in the field.

The aim of this book is a collection of various topics in order to highlight current directions in multiphysics problems. Despite the fact that a lot of progress has been made in the understanding and simulation of FSI in the last decades, FSI is still one of the most challenging topics in numerical mathematics and engineering. A number of open questions and challenges persist, e.g., in the design of efficient and robust solvers for FSI with large deformation and contact as well as regarding the numerical analysis of algorithms and discretization.

In order to facilitate the reading of this book, we classify the articles into three groups:
- Modeling and discretization;
- Solvers;
- Applications.

Despite the challenges of classical FSI on its own, there are several chapters that even go beyond FSI and a couple with further physical phenomena, e.g., thermal and fracture. Nonstandard discretization techniques such as extended ALE, high-resolution interface meshes, Eulerian FSI, and phase-field are addressed. With regard to the numerical solution, insight into monolithic and partitioned approaches is provided. Several applications in vascular flows, binary-fluid-solid interaction, and cracked solids complement the contents.

Finally, we want to thank our sponsors that supported the FSI symposium and finally enabled us to start this book: the Austrian Academy of Sciences (OEAW), Johann Radon Institute for Computational and Applied Mathematics (RICAM) Linz, the Doctoral Program DK W1214 'Computational Mathematics' at the Johannes Kepler University Linz and the graduate school HGS MathComp at the Interdisciplinary Center for Scientific Computing (IWR) in Heidelberg, Germany. We specifically acknowledge

https://doi.org/10.1515/9783110494259-201

Ulrich Langer for scientific and Annette Weihs for administrative support. We also want to mention the unique and friendly atmosphere for the organization of such events at RICAM Linz (which has also been proven by the series of Special Semesters taking place here over the last few years). Last but not least, we thank de Gruyter for their professional and friendly cooperation in producing this book.

Heidelberg, Germany — Stefan Frei
Stockholm, Sweden — Bärbel Holm
Magdeburg, Germany — Thomas Richter
Palaiseau, France; Linz, Austria — Thomas Wick
Graz, Austria; Linz, Austria — Huidong Yang

Contents

Part II: **Solvers**

Part III: **Applications**

List of contributing authors

Johan Hoffman
Department of Computational Science and Technology
School of Computer
Science and Communication
KTH Royal Institute of Technology
Stockholm, Sweden
jhoffman@kth.se

Steffen Basting
Institute of Applied Mathematics (LS III)
TU Dortmund
Dortmund, Germany
steffen.basting@math.tu-dortmund.de

Annalisa Quaini
Department of Mathematics
University of Houston
Houston TX, USA
quaini@math.uh.edu

Suncica Canic
Department of Mathematics
University of Houston
Houston TX, USA
E-mail: canic@math.uh.edu

Roland Glowinski
Department of Mathematics
University of Houston
Houston TX, USA.
roland@math.uh.edu

Philipp Birken
Lund University
Center for the Mathematical Sciences
Lund, Sweden
philipp.birken@na.lu.se

Azahar Monge
Lund University
Center for the Mathematical Sciences
Lund, Sweden
azahar.monge@na.lu.se

Ulrich Langer
Johann Radon Institute for Computational and Applied Mathematics
Linz, Austria
ulrich.langer@ricam.oeaw.ac.at

Daniel Jodlbauer
Doctoral Program Computational Mathematics
Johannes Kepler University
Linz, Austria
daniel.jodlbauer@dk-compmath.jku.at

Claudia M. Colciago
MATHICSE, Chair of Modeling and Scientific Computing
École Polytechnique Fédérale de Lausanne
Lausanne, Switzerland
claudia.colciago@epfl.ch

Simone Deparis
MATHICSE, Chair of Modeling and Scientific Computing
École Polytechnique Fédérale de Lausanne
Lausanne, Switzerland
E-mail: simone.deparis@epfl.ch

Davide Forti
MATHICSE, Chair of Modeling and Scientific Computing
École Polytechnique Fédérale de Lausanne
Lausanne, Switzerland
davide.forti@epfl.ch

E. Harald van Brummelen
Eindhoven University of Technology
Eindhoven, The Netherlands
e.h.v.brummelen@tue.nl

Mahnaz Shokrpour Roudbari
Eindhoven University of Technology
Eindhoven, The Netherlands
E-mail: m.shokrpour.roudbari@tue.nl

https://doi.org/10.1515/9783110494259-202

Görkem Şimşek
Eindhoven University of Technology
Eindhoven, The Netherlands
g.simsek@tue.nl

Kristoffer G. van der Zee
School of Mathematical Sciences
The University of Nottingham,
Nottingham, United Kingdom
kg.vanderzee@nottingham.ac.uk3

Part I: **Modeling and discretization**

Steffen Basting, Annalisa Quaini, Suncica Canic, and Roland Glowinski

1 On the implementation and benchmarking of an extended ALE method for FSI problems

Abstract: The simulation of fluid-structure interaction (FSI) problems is particularly challenging when the structural displacement is large. In this situation, certain techniques known to be robust for small structural displacement, such as arbitrary Lagrangian–Eulerian (ALE) methods, fail. Recently, we proposed an *extended ALE* method that overcomes this failure, while keeping the same mesh connectivity and enforcing mesh alignment with the structure. Our extended ALE method relies on a variational mesh optimization technique, where mesh alignment with the structure is achieved via a constraint. Once a mesh has been obtained from the constrained optimization problem, the FSI problem is solved using either a Dirichlet–Neumann algorithm, or a Robin–Neumann algorithm. Here, we provide implementation details for our extended ALE approach, with particular focus on the use of quadratic isoparametric finite elements for a more accurate representation of the fluid-structure interface, an adaptive relaxation procedure based on Aitken's acceleration to speed up the convergence of the Dirichlet–Neumann algorithm, and the structure solver. We validate the structure solver against a benchmark with an exact solution and we assess our extended ALE method through an extensive series of numerical examples involving 2D FSI problems

Keywords: Mesh optimization, Arbitrary Lagrangian–Eulerian formulation, Fluid-structure interaction, Domain decomposition methods

Classification: 74F10, 65N30, 76D05

Note: This research has been supported in part by the NSF under grants DMS-1620384 (Quaini), NIGMS DMS-1263572 (Canic and Quaini), DMS-1613757, DMS-1318763, DMS-1311709 (Canic). The research of Canic was also supported by the UH Cullen Distinguished Professor endowment.

Steffen Basting, Institute of Applied Mathematics (LS III), TU Dortmund, Vogelpothsweg 8, 44227 Dortmund, Germany, steffen.basting@math.tu-dortmund.de
Annalisa Quaini, Department of Mathematics, University of Houston, 4800 Calhoun Rd, Houston TX 77204, USA, quaini@math.uh.edu
Suncica Canic, Department of Mathematics, University of Houston, 4800 Calhoun Rd, Houston TX 77204, USA, canic@math.uh.edu
Roland Glowinski, Department of Mathematics, University of Houston, 4800 Calhoun Rd, Houston TX 77204, USA, roland@math.uh.edu

https://doi.org/10.1515/9783110494259-001

1.1 Introduction

The focus of this work is on the numerical simulation of the motion of an elastic body immersed in an incompressible, viscous fluid, with the structure undergoing large displacements. The motivation comes from fluid-structure interaction (FSI) between blood flow and heart valves. Several difficulties are associated with accurate numerical simulation of this class of fluid-structure interaction problems. Firstly, the coupled multiphysics problem is highly nonlinear and the so-called added mass effect is known to cause various numerical difficulties when the fluid and structure have comparable densities [21, 68]. Then, large changes in the fluid domain occur due to the large structural displacements. Finally, accurate approximation of the hydrodynamic force at the fluid-structure interface is needed to correctly reproduce the physics of the problem. Before outlining how we tackle the above-mentioned difficulties, let us attempt to briefly summarize the vast literature on numerical approaches for this class of problems.

To deal with the fluid domain motion associated with structural displacements, numerical methods can be classified into the methods with fixed meshes and the methods with moving meshes. The fixed mesh methods include the Immersed Boundary Method (IBM) [58–60], the Fictitious Domain Method [1, 39, 40], the level set method [23, 24, 37], and the so-called Eulerian FSI methods [28, 64, 73]. These methods rely on a fixed fluid mesh used in a fluid solver, while the presence of the structure is implemented in different ways. For example, in the IBM the fluid feels the structure through external forces acting on the fluid, where the coupling between the (fixed) fluid mesh and a (Lagrangian) structure mesh is performed via Dirac Delta functions. To get around the difficulties associated with the discretization of the Dirac Delta, and the low accuracy it causes in the calculation of the hydrodynamic force, modifications of the IBM were introduced. They include the extended IBM [71], and the Immersed Finite Element Method [74]. Conversely, in the Fictitious Domain Method, the coupling between the fluid and structure is enforced via Lagrange multipliers (imposing continuity of velocity, or the no-slip condition). This approach was applied first to problems with rigid particles and later to problems with flexible structures, where Lagrange multipliers were located along the structure surface [3, 44, 69, 70]. In all the cases discussed above, adaptive mesh refinement typically needs to be used to obtain reasonable accuracy in the calculation of the hydrodynamic force acting on the structure.

The moving mesh methods are typically based on arbitrary Lagrangian–Eulerian (ALE) approaches, introduced in [25, 47] for FSI problems discretized with the Finite Element method. Earlier works [45, 57] introduced ALE methods for the Navier–Stokes equations in moving domains discretized with the Finite Difference method. Instead of being fixed, the fluid mesh follows the motion of the elastic body via a mapping, called the ALE mapping, which is calculated based on the current location of the structure (e.g., as a harmonic extension of the current interface position onto the fluid domain).

ALE methods were proven to be accurate and robust for hemodynamics applications involving small mesh displacements (e.g., [32]). Although these methods offer many advantages provided by the explicit representation of the fluid-structure interface [7, 46, 68], problems arise whenever strong deformations or even topological changes of the interface lead to a degeneration of the computational mesh. To deal with large structural displacements, "remeshing" was introduced in [27, 52, 53]. By remeshing it is normally meant that a new mesh with different connectivity is generated from scratch when the quality of the given mesh is poor.

Mixed ALE and fictitious domain formulations have also been proposed [26, 43]. These approaches also require adaptive mesh refinement for an accurate calculation of the viscous shear stresses on the solid boundary.

For completeness, we also mention a mesh-free Lattice–Boltzmann method [20, 29, 31, 48], which has also been used for the simulation of FSI problems with large structural displacements.

In the present work we provide implementation details and additional benchmark testing for an extended ALE approach we recently proposed in [10]. Our method captures large structural displacements *without changing mesh connectivity* and it accurately approximates the fluid-structure interface and the hydrodynamic forces *without the need for adaptive mesh refinement*. The method is based on a fixed "base" mesh that is adapted to approximate the interface via an ALE-type mapping, while maintaining mesh connectivity (nodes or elements are neither inserted nor removed). The fundamental building block is a variational mesh optimization approach that does not rely on any combinatorial considerations. Alignment of the optimized mesh with the structure interface is stated as a constraint of a mesh optimization problem thanks to a level set description of the geometry.

The main features of our extended ALE approach are:

- Nondegenerate meshes of provably optimal quality;
- The alignment of the mesh with the interface, which allows for a simple definition and efficient implementation of problem-specific finite element spaces, such as spaces allowing for discontinuities across the interface;
- Fixed mesh connectivity, which makes the method easy to implement in an existing standard ALE code.

We remark that similar approaches to ours are presented in [5, 14, 22] (the fixed-mesh ALE approach) and in [35, 36] (methodology based on "universal meshes" [63]). The main benefits of our variational approach over those in [5, 14, 22, 35, 36] are: *no explicit combinatorial testing* is needed and the resulting *meshes are guaranteed to be nondegenerate*. Once a mesh has been obtained from the above-mentioned constrained optimization problem, the FSI problem is solved with classical Domain Decomposition algorithms (e.g., [62]): either the Dirichlet–Neumann (DN) method, which is combined with an Aitken's acceleration technique [49], or the Robin–Neumann method (RN). Because of the modularity of DN and RN algorithms, each physics subproblem is solved separately, with the coupling conditions enforced in an iterative fashion. We remark that the use of Robin interface conditions has led to efficient partitioned FSI algorithms when the structure lies on part of the fluid domain boundary; see, e.g., [4, 6, 18, 19, 42, 56].

We test our approach through a series of benchmark cases that involve the interaction of an inextensible, elastic beam with a 2D incompressible fluid. We show that our extended ALE method allows one to easily capture the pressure discontinuity across the interface, which coincides with the 1D elastic structure. Methods based on nonaligned fixed meshes cannot capture such a discontinuity, unless further techniques are used, such as, e.g., the enrichment of the function spaces as in X-FEM, see [34]. Moreover, thanks to the mesh alignment with the interface, the coupling conditions required by the Domain Decomposition methods are easily enforced. To treat the structure, we combine a generalized Crank–Nicolson scheme with an Uzawa-type algorithm for solving the saddle-point problem associated with an augmented Lagrangian method employed to handle the inextensibility condition. This nontrivial structure solver is described in Section 1.5.2 and validated against a benchmark with exact solution in Section 1.6.1.

The outline of the paper is as follows. In Section 1.2 we state the problem. The constrained optimization approach, which is at the core of our Extended ALE method, is explained in Section 1.3. We describe the Domain Decomposition algorithms in Section 1.4, and summarize the numerical methods that we use for the time and space discretization of the fluid and structure problems in Section 1.5. In Section 1.6, we present numerical results obtained on a series of numerical tests carefully chosen to highlight the main features of the method. Conclusions are drawn in Section 1.7.

1.2 Problem definition

Consider a time-independent domain $\Omega \subset \mathbb{R}^2$ containing an elastic beam forming a 1D manifold $\Gamma(t) \subset \Omega$ whose location depends on time. The beam is surrounded by an incompressible, viscous fluid occupying domain $\Omega_f(t) := \Omega \setminus \Gamma(t)$. See Figure 1.1.

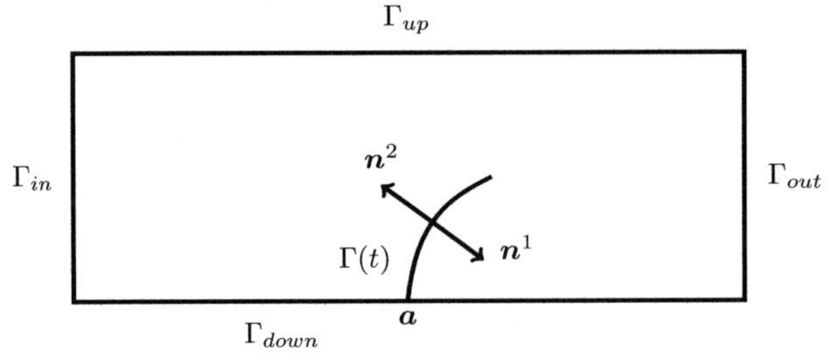

Fig. 1.1: Computational domain.

1.2.1 The fluid problem

The fluid flow is governed by the Navier–Stokes equations for an incompressible, viscous fluid:

$$\rho_{\mathrm{f}}\left(\frac{\partial \boldsymbol{u}}{\partial t}+(\boldsymbol{u}\cdot\nabla)\boldsymbol{u}\right)-\nabla\cdot\boldsymbol{\sigma}=\mathbf{0} \quad \text{in } \Omega_{\mathrm{f}}(t)\,, \tag{1.1}$$

$$\nabla\cdot\boldsymbol{u}=0 \quad \text{in } \Omega_{\mathrm{f}}(t)\,, \tag{1.2}$$

for $t \in (0, T]$, where ρ_{f} is the fluid density, $\boldsymbol{u}$ is the fluid velocity, and $\boldsymbol{\sigma}$ the Cauchy stress tensor. For Newtonian fluids $\boldsymbol{\sigma}$ has the following expression

$$\boldsymbol{\sigma}(\boldsymbol{u},p)=-p\mathbf{I}+2\mu\boldsymbol{\epsilon}(\boldsymbol{u})\,,$$

where p is the pressure, μ is the fluid dynamic viscosity, and $\boldsymbol{\epsilon}(\boldsymbol{u})=(\nabla\boldsymbol{u}+(\nabla\boldsymbol{u})^T)/2$ is the strain rate tensor. Equations (1.1)–(1.2) need to be supplemented with initial and boundary conditions.

In order to describe the evolution of the fluid domain, we begin by adopting an *Arbitrary Lagrangian–Eulerian* (ALE) approach [47]. More precisely, let $\hat{\Omega}_{\mathrm{f}} \subset \mathbb{R}^2$ be a fixed reference domain. We consider a smooth ALE mapping

$$\mathcal{A}:[0,T]\times\hat{\Omega}_{\mathrm{f}}\to\mathbb{R}^2\,,$$
$$\mathcal{A}(t,\hat{\Omega}_{\mathrm{f}})=\Omega_{\mathrm{f}}(t)\,,\quad \forall t\in[0,T]\,.$$

For each time instant $t \in [0, T]$, $\mathcal{A}$ is assumed to be a homeomorphism. The domain velocity $\mathbf{w}$ is defined as

$$\mathbf{w}(t,\,\cdot\,)=\partial_t\mathcal{A}(t,\mathcal{A}(t,\,\cdot\,)^{-1}).$$

For any sufficiently smooth function $F:[0,T]\times\mathbb{R}^2\to\mathbb{R}$, we may define the ALE time derivative of F as

$$\left.\frac{\partial F}{\partial t}\right|_{\hat{\mathbf{x}}}=\frac{\partial F}{\partial t}(t,\mathcal{A}(t,\hat{\mathbf{x}}))=\frac{\partial F}{\partial t}(t,\mathbf{x})+\mathbf{w}(t,\mathbf{x})\cdot\nabla F(t,\mathbf{x}),$$

where $\mathbf{x}=\mathcal{A}(t,\hat{\mathbf{x}})$, $\hat{\mathbf{x}}\in\hat{\Omega}$. With these definitions, we can write the incompressible Navier–Stokes equations in ALE formulation as follows:

$$\rho_{\mathrm{f}}\left.\frac{\partial \boldsymbol{u}}{\partial t}\right|_{\hat{\mathbf{x}}}+\rho_{\mathrm{f}}(\boldsymbol{u}-\mathbf{w})\cdot\nabla\boldsymbol{u}-\nabla\cdot\boldsymbol{\sigma}=\mathbf{0} \quad \text{in } \Omega_{\mathrm{f}}(t)\,, \tag{1.3}$$

$$\nabla\cdot\boldsymbol{u}=0 \quad \text{in } \Omega_{\mathrm{f}}(t)\,, \tag{1.4}$$

for $t \in (0, T]$.

1.2.2 The structure problem

For the structure problem, we consider a linearly elastic inextensible beam equation. This means that the beam is not allowed to shrink or stretch while interacting with the fluid. The motion of an inextensible beam, while well studied (e.g., [26, 38, 41] and references therein), remains a challenging problem numerically. The main difficulty stems from the nonlinearity due to the inextensibility condition.

We assume negligible torsional effects for the beam. Let us denote by ρ_s the linear density (i.e., mass per unit length), by L the length, and by EI the flexural stiffness of the beam. We will use the following notation, with s denoting arc length and t time:

$$\boldsymbol{y}' = \frac{\partial \boldsymbol{y}}{\partial s}\,, \quad \dot{\boldsymbol{y}} = \frac{\partial \boldsymbol{y}}{\partial t}\,, \quad \boldsymbol{y}'' = \frac{\partial^2 \boldsymbol{y}}{\partial s^2}\,, \quad \ddot{\boldsymbol{y}} = \frac{\partial^2 \boldsymbol{y}}{\partial t^2}\,.$$

Using the virtual work principle, the beam motion for $t \in (0, T]$ is modeled by the following. Find $\boldsymbol{x}(t) \in K$ such that:

$$\int_0^L \rho_s \ddot{\boldsymbol{x}} \cdot \boldsymbol{y}\, ds + \int_0^L EI\, \boldsymbol{x}'' \cdot \boldsymbol{y}''\, ds = \int_0^L \boldsymbol{f} \cdot \boldsymbol{y}\, ds\,, \quad \forall \boldsymbol{y} \in dK(\boldsymbol{x})\,, \tag{1.5}$$

where $\boldsymbol{x} = \boldsymbol{x}(t, s)$ is the parametric curve defining the beam position, $\boldsymbol{f}$ denotes the force acting on the beam, and

$$K = \left\{\boldsymbol{y} \in (H^2(0, L))^2, \left|\boldsymbol{y}'\right| = 1,\ \boldsymbol{y}(0) = \boldsymbol{a},\ \boldsymbol{y}'(0) = \boldsymbol{b}\right\}\,, \tag{1.6}$$

$$dK(\boldsymbol{x}) = \left\{\boldsymbol{y} \in (H^2(0, L))^2,\ \boldsymbol{x}' \cdot \boldsymbol{y}' = 0,\ \boldsymbol{y}(0) = \boldsymbol{0},\ \boldsymbol{y}'(0) = \boldsymbol{0}\right\}\,. \tag{1.7}$$

In our case, $\boldsymbol{f}$ is the hydrodynamic force, which will be specified in Section 1.2.3. For problem (1.5) we choose boundary conditions

$$\boldsymbol{x}(0) = \boldsymbol{a}\,, \quad \boldsymbol{x}'(0) = \boldsymbol{b}\,, \quad \boldsymbol{x}''(L) = \boldsymbol{x}'''(L) = 0\,. \tag{1.8}$$

The conditions at $s = 0$ are called the *essential boundary conditions*, describing a clamped beam, while the conditions at $s = L$ are called the *natural boundary conditions*. The nonlinear inextensibility condition for the beam, $|\boldsymbol{x}'| = 1$, is embedded into the set K.

1.2.3 The coupled fluid-structure interaction problem

We consider two-way coupling between the fluid and structure: the motion of the beam is driven by the contact force exerted by the fluid, while at the same time the motion of the beam influences the fluid motion. The coupling conditions are described by the following. Let us denote the interface by

$$\Gamma(t) = \{\boldsymbol{x}(t, s),\ s \in [0, L]\}\,.$$

Let $\boldsymbol{n}^1$ be the unit normal vector pointing to the "left" (left with respect to the parameterization of $\boldsymbol{x}$) and $\boldsymbol{n}^2 = -\boldsymbol{n}^1$ be the unit normal pointing to the "right", see Figure 1.1.

Notice that the fluid-structure interface coincides with the structure domain. The hydrodynamic force acting on the structure (beam) is given by the jump in the normal fluid stress across the interface $\Gamma(t)$:

$$\boldsymbol{f}_\Gamma = -\boldsymbol{\sigma}^1\boldsymbol{n}^1 - \boldsymbol{\sigma}^2\boldsymbol{n}^2 \,, \tag{1.9}$$

where $\boldsymbol{\sigma}^i(\mathbf{x}) = \lim_{\epsilon\to 0^-} \boldsymbol{\sigma}(\mathbf{x}+\epsilon\boldsymbol{n}^i)$, $\mathbf{x} \in \Gamma$, $i = 1, 2$. Using this notation, we can now state the coupling conditions. For $t \in (0, T]$, the fluid problem (1.3),(1.4) and the structure problem (1.5) are coupled by the following two conditions:

1. kinematic coupling condition (continuity of velocity, i.e., the no-slip condition)

$$\boldsymbol{u} = \dot{\boldsymbol{x}} \quad \text{on } \Gamma(t) \,; \tag{1.10}$$

2. dynamic coupling condition (balance of contact forces)

$$\boldsymbol{f}_\Gamma = \boldsymbol{f} \quad \text{on } \Gamma(t) \,, \tag{1.11}$$

   where $\boldsymbol{f}$ is given by Equation (1.5).

Here, notation $\boldsymbol{u} = \dot{\boldsymbol{x}}$ in (1.10) is used to express the relation $\boldsymbol{u}(t, \boldsymbol{x}(t, s)) = \dot{\boldsymbol{x}}(t, s)$, $s \in [0, L]$ (analogously for $\boldsymbol{f}_\Gamma$ and $\boldsymbol{f}$ in (1.11)). Since $\boldsymbol{x}$ denotes the location of structure points and not the structure displacement, both the structure and fluid are given in Eulerian coordinates.

For the purposes that will be clear below when we introduce the Robin–Neumann scheme, we remark here that the coupling conditions (1.10)–(1.11) can be written in an equivalent form by introducing the constants $\alpha_\mathrm{f} > 0$ and $\alpha_\mathrm{s} > 0$ ($\alpha_\mathrm{f} \neq \alpha_\mathrm{s}$), and writing:

$$\begin{aligned} \alpha_\mathrm{f}\boldsymbol{u} - \boldsymbol{f}_\Gamma &= \alpha_\mathrm{f}\dot{\boldsymbol{x}} - \boldsymbol{f} \quad \text{on } \Gamma(t) \,, \\ \alpha_\mathrm{s}\boldsymbol{u} + \boldsymbol{f} &= \alpha_\mathrm{s}\dot{\boldsymbol{x}} + \boldsymbol{f}_\Gamma \quad \text{on } \Gamma(t). \end{aligned} \tag{1.12}$$

1.3 Numerical representation of the geometry

The main feature of our extended ALE method is a variational mesh optimization technique combined with an additional constraint to enforce the alignment of the structure interface with the edges of the resulting triangulation. The mesh optimization in this section corresponds to a reparametrization of the ALE mapping.

1.3.1 Optimal triangulations

Let $\mathcal{T}$ be an initial, reference triangulation of the domain Ω (not necessarily approximating the structure interface at this stage). Following a variational mesh optimization technique introduced by M. Rumpf in [65], we aim at finding an "optimal" triangulation $\mathcal{T}_\mathrm{opt}$ resulting from an optimal mesh deformation χ_opt of $\mathcal{T}$, i.e., $\mathcal{T}_\mathrm{opt} = \chi_\mathrm{opt}(\mathcal{T})$.

Deformation χ_{opt} belongs to the set D of piecewise affine, orientation preserving, and globally continuous deformations:

$$D = \left\{ \chi \in \left(\mathcal{C}^0(\Omega) \right)^2 : \nabla\chi|_T \in \mathrm{GL}(2),\ \det(\nabla\chi|_T) > 0,\ \forall T \in \mathcal{T} \right\} , \tag{1.13}$$

with $\mathrm{GL}(2) = \{A \in \mathbb{R}^{2\times 2} :\ \det(A) \neq 0\}$.

Deformation $\chi_{\text{opt}} \in D$ is "optimal" in the sense that it is the argument for which a certain functional $\mathcal{F}$ attains its minimum value:

$$\mathcal{F}(\chi_{\text{opt}}) = \min_{\chi \in D} \mathcal{F}(\chi) . \tag{1.14}$$

We assume that the functional in (1.14) can be represented by a sum of weighted, element-wise contributions F_T:

$$\mathcal{F}(\chi) = \sum_{T \in \mathcal{T}} \mu_T F_T(\chi) ,$$

where $\mu_T > 0$ denotes a positive weight with $\sum_T \mu_T = 1$. Let R_T denote the linear reference mapping from a prescribed reference element T_{opt} (an equilateral simplex with customizable edge length h) to T. Under the assumptions of translational invariance, isotropy and frame indifference of the functionals, it can be shown that in two dimensions F_T may be expressed as a function of the invariants of χ: $\|\nabla R_T(\chi)\|^2$ and $\det(\nabla R_T(\chi))$ [65]. For example, given a function $\tilde{F}_T : \mathbb{R} \times \mathbb{R} \to \mathbb{R}$ we can write:

$$F_T(\chi) = \tilde{F}_T(a, d) := \tilde{F}_T(\|\nabla R_T(\chi)\|^2, \det(\nabla R_T(\chi))) .$$

Here, $\|\cdot\|$ denotes the Frobenius norm. Note that the quantity $\|\nabla R_T(\chi)\|^2$ measures the change of edge lengths with respect to the reference element, and $\det(\nabla R_T(\chi))$ measures the change in area.

In order to rule out deformations with vanishing determinant, we need

$$\lim_{\det(\nabla R_T(\chi)) \to 0} F_T(\chi) = \infty .$$

With the additional assumption that the local functional $F_T(\chi) = \tilde{F}_T(a, d)$ is polyconvex (i.e., $\tilde{F}_T(a, d)$ is convex with respect to each argument), it can be proven that an optimal deformation exists and is globally injective [65].

A classical example of such a function F_T is given by

$$F_T(\chi) = (\|\nabla R_T(\chi)\|^2 - 2)^2 + \det(\nabla R_T(\chi)) + \frac{1}{\det(\nabla R_T(\chi))} . \tag{1.15}$$

The optimally deformed simplex is obtained if $\chi_{\text{opt}}|_T = \mathbf{I}$, i.e., if

$$F_T(\chi_{\text{opt}}) = F_T(\mathbf{I}) = (2 - 2)^2 + 1 + 1 = 2 .$$

The variational mesh smoothing approach described above has several advantages:

- Minimization problem (1.14) yields triangulations that are provably optimal in the sense of the local measure (1.15).
- These triangulations can be shown to be nondegenerate, i.e., no self-intersection of elements occurs. This is the main property needed by our mesh optimization approach in order to work reliably even when the structure displacement is large.
- The element-wise representation of $\mathcal{F}$ provides built-in, local mesh quality control.

The price to pay for those advantages is that functional $\mathcal{F}$ in (1.14) is highly nonlinear, nonconvex, and global minimizers may not be unique.

1.3.2 Interface aligned mesh

We are now interested in having a triangulation that is nondegenerate, optimal (as explained in the previous subsection) *and* aligned with the beam position $\Gamma(t)$, i.e., we want the optimal triangulation edges to approximate $\Gamma(t)$. For this purpose, we introduce the following auxiliary tools:

- a "tubular box" around the structure of width δ, denoted by $\Omega_\mathrm{f}^\delta(t) \subset \Omega_\mathrm{f}(t)$, within which mesh optimization with alignment will be performed, see Figure 1.2a;
- a continuous level set function $\phi\colon\ [0, T]\times\Omega_\mathrm{f}^\delta(t) \to \mathbb{R}$ whose zero level set includes the structure position $\boldsymbol{x}$:

$$\begin{aligned} \Omega_\mathrm{f}^{\delta,1}(t) &= \left\{\mathbf{y} \in \Omega_\mathrm{f}^\delta(t)\colon\ \phi(t,\mathbf{y}) > 0\right\} , \\ \Omega_\mathrm{f}^{\delta,2}(t) &= \left\{\mathbf{y} \in \Omega_\mathrm{f}^\delta(t)\colon\ \phi(t,\mathbf{y}) < 0\right\} , \\ \Gamma^\delta(t) &= \left\{\mathbf{y} \in \Omega_\mathrm{f}^\delta(t)\colon\ \phi(t,\mathbf{y}) = 0\right\} , \end{aligned} \tag{1.16}$$

where $\Gamma^\delta(t)$ denotes a "natural" extension of $\Gamma(t)$ to the boundary of $\Omega_\mathrm{f}^\delta(t)$, and $\Omega_\mathrm{f}^{\delta,1}(t)$ and $\Omega_\mathrm{f}^{\delta,2}(t)$ denote the fluid subdomains located on the "left" and "right" side of $\Gamma^\delta(t)$, respectively. See Figure 1.2a. Notice that $\boldsymbol{n}^1$ (resp., $\boldsymbol{n}^2$) is the outward unit normal on $\Gamma^\delta(t)$ of $\Omega_\mathrm{f}^{\delta,1}(t)$ (resp., $\Omega_\mathrm{f}^{\delta,2}(t)$).

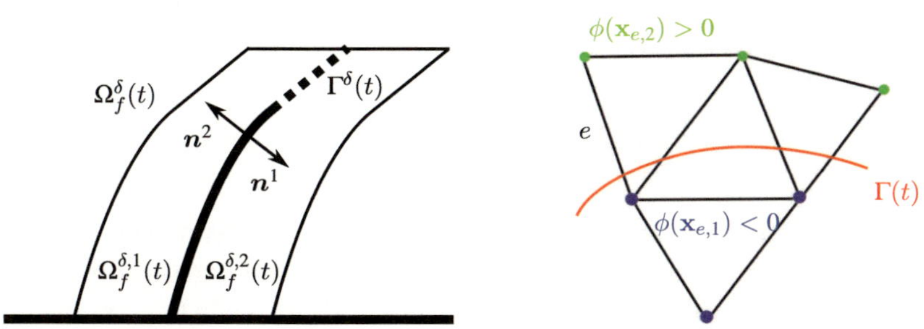

(a) Tubular box $\Omega_f^\delta(t)$ (b) Fluid mesh intersected by $\Gamma(t)$

Fig. 1.2: (a) Tubular box $\Omega_\mathrm{f}^\delta(t)$ around the structure position $\boldsymbol{x}$, zoomed in view of Figure 1.1, and (b) $\Gamma(t)$ intersecting elements of the fluid mesh.

Within $\Omega_f^\delta(t)$ we perform the following procedure. Let e be an arbitrary edge of the triangulation $\mathcal{T}$ intersected by $\Gamma^\delta(t)$ as shown in Figure 1.2b, and let $\mathbf{x}_{e,1}$ and $\mathbf{x}_{e,2}$ be its endpoints. Because of continuity of ϕ and assumption (1.16), we observe that

$$\phi(\mathbf{x}_{e,1})\phi(\mathbf{x}_{e,2}) < 0$$

if and only if e is intersected by $\Gamma^\delta(t)$, provided that the mesh size h is sufficiently small to resolve the shape of $\Gamma^\delta(t)$. We therefore define the triangulation to be *linearly aligned* with $\Gamma(t)$ if

$$\phi(\mathbf{x}_{e,1})\phi(\mathbf{x}_{e,2}) \geq 0 \quad \text{for all} \quad e \in \mathcal{T} .$$

We introduce a scalar constraint $c\colon D \to \mathbb{R}_0^+$, defined on D given by (1.13), such that:

$$c(\chi) = \sum_{e\in\chi(\mathcal{T})} \mathcal{H}(\phi(\mathbf{x}_{e,1})\phi(\mathbf{x}_{e,2})) \quad \text{where} \quad \mathcal{H}(z) = \begin{cases} > 0 & \text{if } z < 0 , \\ = 0 & \text{otherwise.} \end{cases}$$

Only deformations χ for which $c(\chi) = 0$ will give aligned triangulations. Thus, a linearly aligned triangulation of optimal quality is obtained from the following constrained minimization problem:

$$\min_{\chi\in D} \mathcal{F}(\chi) \quad \text{such that} \quad c(\chi) = 0 .$$

Given an aligned triangulation $\mathcal{T}$, we may define a linear approximation of the interface as

$$\Gamma_h = \{\text{edges } e \in \mathcal{T}\colon \phi(\mathbf{x}_{e,i}) = 0 \quad \text{and} \quad \mathbf{x}_{e,i} \in \Gamma \text{ for } i = 1, 2\} .$$

Note that condition $\mathbf{x}_{e,i} \in \Gamma$ is necessary to avoid considering points on the extension Γ^δ that do not belong to Γ.

In order to obtain a more accurate representation of the structure, we also consider piecewise quadratic approximations of $\Gamma(t)$ and make use of quadratic isoparametric finite elements. We denote by $\hat{K} = \left\{\hat{\mathbf{x}} \in \mathbb{R}^2 : \sum_{i=1}^2 \hat{x}^{(i)} \leq 1, \hat{x}^{(i)} \geq 0\right\}$ the reference simplex, and by $G_K\colon \hat{K} \to K$ the quadratic isoparametric mapping:

$$G_K(\hat{\mathbf{x}}) = \sum_{i=1}^{6} \mathbf{x}_i \varphi_i(\hat{\mathbf{x}}) , \tag{1.17}$$

where $\varphi_i, i = 1, \ldots, 6$ are the quadratic Lagrange basis functions defined on $\hat{K}$. Once a linearly aligned triangulation and the corresponding discrete interface are obtained, in order to achieve quadratic alignment we move each quadratic node (e.g., $\mathbf{x}_6$ in Figure 1.3) along the linear normal to the zero level set. Details on the numerical implementation together with an evaluation of the mesh, approximation quality, and computational costs can be found in [8, 11, 72].

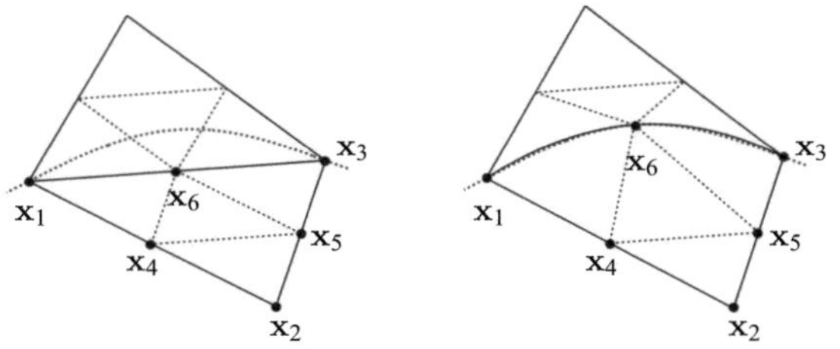

Fig. 1.3: Linearly aligned triangulation with isoparametric elements without (left) and with (right) quadratic alignment of the additional quadratic degree of freedom $\mathbf{x}_6$.

In the following, for a given structure position $\Gamma(t)$, we will denote the optimal interface aligned triangulation obtained from the strategy outlined above by $\mathcal{T}_{\text{opt}}(\Gamma(t))$. The resulting computational domain is given by:

$$\Omega_{\text{f,opt}}(t) = \bigcup_{K \in \mathcal{T}_{\text{opt}}(\Gamma(t))} K\,.$$

1.4 Partitioned methods for the fluid-structure interaction problem

The FSI problems described in Section 1.2 will be solved using two different partitioned strategies based on Domain Decomposition methods [62]: the Dirichlet–Neumann (DN) and the Robin–Neumann (RN) algorithms. Partitioned methods are appealing for solving multiphysics problems such as those discussed in this manuscript, because they allow the reuse of existing fluid and structure solvers with minimal modifications. Because of the modularity of DN and RN algorithms, each physics subproblem is solved separately, with the coupling conditions enforced in an iterative fashion. In the DN algorithm the coupling boundary condition (1.10) is imposed at the interface as a Dirichlet boundary condition for the fluid subproblem, whereas in the RN algorithm the fluid subproblem is endowed with a Robin interface condition (1.12). In both algorithms, the structure subproblem is supplemented with the Neumann "boundary condition" (1.11). Equation (1.11) is a proper Neumann boundary condition when the structure is thick; for thin structures Equation (1.11) prescribes the load on the structure.

To describe the DN and RN algorithms, we introduce the time-discretization step $\Delta t > 0$ and set $t^n = n\Delta t$, for $n = 1, \ldots, N$, with $N = T/\Delta t$. At every time t^n, the DN and RN algorithms iterate over the fluid and structure subproblems until convergence. These are Richardson (also called fixed point) iterations for the position of $\Gamma(t^n)$. Let k be the index for these subiterations.

1.4.1 The Dirichlet–Neumann method

At time t^{n+1}, subiteration $k+1$, assuming that Ω_f^n, $\boldsymbol{u}_k, p_k$, and $\boldsymbol{x}_k$ are known, the following steps are performed:

- **Step 1:** Solve the fluid subproblem for the flow variables $\boldsymbol{u}_{k+1}, p_{k+1}$ defined on Ω_f^n, with **Dirichlet boundary condition**

$$\boldsymbol{u}_{k+1} = \dot{\boldsymbol{x}}_k \quad \text{on} \quad \Gamma^n \,. \tag{1.18}$$

- **Step 2:** Solve the structure subproblem for the structure position $\boldsymbol{x}_{k+1}$, driven by the just calculated hydrodynamic force $\boldsymbol{f}_{\Gamma,k+1}$, i.e., $\boldsymbol{f}_{k+1} = \boldsymbol{f}_{\Gamma,k+1}$ on Γ^n.
- **Step 3:** Check the stopping criterion, e.g.,

$$\frac{\|\boldsymbol{x}_{k+1} - \boldsymbol{x}_k\|}{\|\boldsymbol{x}_k\|} < \epsilon_{\mathrm{FS}} \,, \tag{1.19}$$

  where ϵ_{FS} is a given stopping tolerance. If violated, repeat Steps 1–3. If satisfied, set $\boldsymbol{x}^{n+1} = \boldsymbol{x}_{k+1}$ and $p^{n+1} = p_{k+1}$.
- **Step 4:** Check the mesh quality of Ω_f^n. A mesh is considered to be “bad” if the maximum angle of the elements exceeds a certain value, for instance 130 degrees.
  - If the mesh is good: Accept and set $\tilde{\boldsymbol{u}}^{n+1} = \boldsymbol{u}_{k+1}$ and $\tilde{\Omega}_f^{n+1} = \Omega_f^n$.
  - If the mesh is bad: Apply mesh optimization to get $\Omega_{f,\mathrm{opt}}^n$, set $\tilde{\Omega}_f^{n+1} = \Omega_{f,\mathrm{opt}}^n$. Project data onto new mesh, i.e.,

$$\tilde{\boldsymbol{u}}^{n+1} = I_{\Omega_f^n \to \tilde{\Omega}_f^{n+1}}(\boldsymbol{u}_{k+1}) \,. \tag{1.20}$$

- **Step 5:** Standard ALE update: From the new structure position $\boldsymbol{x}^{n+1}$ obtain Γ^{n+1}, and from the intermediate fluid domain $\tilde{\Omega}_f^{n+1}$ obtain:

$$\Omega_f^{n+1} = E\left(\Gamma^{n+1}, \tilde{\Omega}_f^{n+1}\right)$$

  using an extension operator E (see below). Set $\boldsymbol{u}^{n+1} = \tilde{\boldsymbol{u}}^{n+1}$ and move to the next time step.

Notice that if the quality of the mesh is good, that is if the second item at Step 4 never applies, the method reduces to a standard ALE approach. We do not use a “standard” extension operator E (such as harmonic extension, or operators stemming from linear elasticity), but use the variational approach based on (1.14), (1.15). In our experience, this approach is superior to linear extension operators in terms of mesh quality.

In the “inner” loop, which corresponds to Steps 1–3 in the above iteration algorithm, the fluid domain is “frozen”, which provides important saving of computational time. Concerning Step 4, the angle-based criterion used to detect “bad” meshes is purely heuristic and may be replaced by other meaningful mesh quality criteria. The criterion should be sufficiently mild in order to prevent the reparametrization at every time step. Notice that the mesh optimization procedure presented in Subsec. 1.3.1 aims at generating triangulations made of equilateral triangles, therefore the angles after optimization are usually bounded away from, e.g., 130 degrees.

A crucial point in the above algorithm is the choice of the mesh transfer operator $I_{\Omega_f^n \to \tilde{\Omega}_f^{n+1}}$ appearing in Equation (1.20) at Step 4, needed whenever reparametrization is performed. In our case, this operator is the Lagrange interpolation operator which was also proposed in [35]. However, it is known that dynamically changing meshes may lead to spurious oscillations of the pressure for small time-step sizes [15, 17]. Indeed, we will observe those oscillations in our numerical results, as shown in Section 1.6.3. The use of special elements (e.g., divergence-free) or special variational formulations of the Navier–Stokes equations (see for instance [51]) might cure this problem which seems to be an open question.

It is well known that the convergence properties of the DN algorithm depend heavily on the added-mass effect [21]. In fact, when the structure constitutes a part of the fluid domain boundary, the number of DN iterations required to satisfy the stopping criterion (1.19) increases as the structure density approaches the fluid density. Moreover, below a certain density ratio ρ_s/ρ_f, which depends on the domain geometry, relaxation is needed for the DN algorithm to converge [21, 54, 55]. This is why we adopt a simple Aitken's acceleration technique, which is based on a relaxation approach, and is known to reduce the number of DN iterations. This strategy, introduced in [49], was proposed for a setting similar to ours in [2].

Let $\tilde{\boldsymbol{x}}_{k+1}$ be the unrelaxed structure position predicted by Step 2 of the algorithm above. Then after Step 2, we introduce a relaxation parameter ω_{k+1}, which is computed via

$$\omega_{k+1} = \frac{(\boldsymbol{x}_k - \boldsymbol{x}_{k-1}) \cdot (\boldsymbol{x}_k - \tilde{\boldsymbol{x}}_{k+1} - \boldsymbol{x}_{k-1} + \tilde{\boldsymbol{x}}_k)}{|\boldsymbol{x}_k - \tilde{\boldsymbol{x}}_{k+1} - \boldsymbol{x}_{k-1} + \tilde{\boldsymbol{x}}_k|^2} .$$

The position of the interface is then corrected via the relaxation algorithm:

$$\boldsymbol{x}_{k+1} = \omega_{k+1} \tilde{\boldsymbol{x}}_{k+1} + (1 - \omega_{k+1}) \boldsymbol{x}_k .$$

The results in [2] indicate that only a few accelerated DN subiterations are to be expected for FSI problems with an immersed structure and large added-mass effect.

It was shown in [4] that when the structure constitutes a part of the fluid domain boundary for a suitable choice of parameter α_f the RN method features excellent convergence properties: it always converges without any relaxation and its convergence is insensitive to the added-mass effect. In the next section, we present the RN method.

1.4.2 The Robin–Neumann method

At time t^{n+1}, subiteration $k + 1$, the following steps are performed:

- **Step 1:** Solve the fluid subproblem for the flow variables $\boldsymbol{u}_{k+1}, p_{k+1}$ defined on Ω_f^n, with **Robin boundary condition**

$$\alpha_f \boldsymbol{u}_{k+1} - \boldsymbol{f}_{\Gamma,k+1} = \alpha_f \dot{\boldsymbol{x}}_k - \boldsymbol{f}_k \quad \text{on} \quad \Gamma^n \,. \tag{1.21}$$

- **Step 2, 3, 4** and **5** as in Section 1.4.1

Recall that $\boldsymbol{f}_\Gamma$ denotes the jump in the normal stress across the structure, as defined in (1.9), and $\boldsymbol{f}$ stands for the right-hand side of the structure equation (1.5). Notice that the DN algorithm can be interpreted as a particular case of the RN algorithm for $\alpha_f \to \infty$.

In [4], α_f is estimated by considering a simplified structure model. Here, we follow the same approach and set the constant α_f in (1.21):

$$\alpha_f = \frac{\rho_s}{\Delta t} \,. \tag{1.22}$$

1.5 The fully discrete problem

We present the fully discrete problem for the case of the fluid problem (1.3),(1.4) with Robin boundary condition (1.21), and recall that a similar approach can be taken for the DN algorithm, since it is a particular case of the RN algorithm. We will state the problem in weak form by including only the boundary condition on $\Gamma(t)$, since those on $\partial\Omega$ are understood and do not affect the presented methodology.

1.5.1 The discrete fluid subproblem

For any given $t \in [0, T)$, we define the following spaces:

$$V(t) = \left\{\boldsymbol{v} \colon \Omega_f(t) \to \mathbb{R}^2,\ \boldsymbol{v} = \hat{\boldsymbol{v}} \circ (\mathcal{A})^{-1},\ \hat{\boldsymbol{v}} \in (H^1(\hat{\Omega}_f))^2\right\} \,,$$
$$Q(t) = \left\{q \colon \Omega_f(t) \to \mathbb{R},\ q = \hat{q} \circ (\mathcal{A})^{-1},\ \hat{q} \in L^2(\hat{\Omega}_f)\right\} \,.$$

In the following, we will use the notation $V^n := V(t^n)$ and $Q^n := Q(t^n)$ to denote the finite element spaces at the time instant t^n.

We introduce the following linear forms:

$$m(\Omega; \boldsymbol{u}, \boldsymbol{v}) = \int_\Omega (\boldsymbol{u} \cdot \boldsymbol{v})\, \mathrm{d}\Omega \,, \quad a(\Omega; \boldsymbol{u}, \boldsymbol{v}) = \int_\Omega \mu\, (\boldsymbol{\epsilon}(\boldsymbol{u}) \colon \boldsymbol{\epsilon}(\boldsymbol{v}))\, \mathrm{d}\Omega \,,$$
$$c(\Omega; \mathbf{w}; \boldsymbol{u}, \boldsymbol{v}) = \int_\Omega ((\mathbf{w} \cdot \nabla)\, \boldsymbol{u} \cdot \boldsymbol{v})\, \mathrm{d}\Omega \,, \quad b(\Omega; p, \boldsymbol{v}) = -\int_\Omega p \nabla \cdot \boldsymbol{v}\, \mathrm{d}\Omega \,.$$

The variational formulation of the fluid problem (1.3), (1.4) with boundary condition (1.12) reads: given $t \in (0, T]$, find $(\boldsymbol{u}, p) \in V(t) \times Q(t)$ such that $\forall (\boldsymbol{v}, q) \in V(t) \times Q(t)$

the following holds:

$$\rho_f m\Big(\Omega_f(t); \frac{\partial \boldsymbol{u}}{\partial t}\Big|_{\hat{\mathbf{x}}}, \boldsymbol{v}\Big) + \rho_f c(\Omega_f(t); \boldsymbol{u} - \mathbf{w}; \boldsymbol{u}, \boldsymbol{v}) + a(\Omega_f(t); \boldsymbol{u}, \boldsymbol{v}) + b(\Omega_f(t); p, \boldsymbol{v})$$
$$+ m(\Gamma(t); \alpha_f \boldsymbol{u}, \boldsymbol{v}) = m(\Gamma(t); \alpha_f \dot{\boldsymbol{x}} - \boldsymbol{f}, \boldsymbol{v}),$$
$$b(\Omega_f(t); q, \boldsymbol{u}) = 0.$$

Time and space discretization. For simplicity, the implicit Euler scheme is used to discretize the above weak formulation in time. The convective term is linearized by a first-order extrapolation formula. Notice that higher order discretization schemes and extrapolation formulas are also possible. At time t^{n+1}, and at the $(k+1)$-st RN subiteration, the time discrete linearized fluid subproblem reads as follows: Find $(\boldsymbol{u}_{k+1}, p_{k+1}) \in V^n \times Q^n$ such that

$$\rho_f m\Big(\Omega_f^n; \frac{\boldsymbol{u}_{k+1} - \boldsymbol{u}^n}{\Delta t}, \boldsymbol{v}\Big) + \rho_f c(\Omega_f^n; \boldsymbol{u}_k - \mathbf{w}^n; \boldsymbol{u}_{k+1}, \boldsymbol{v}) + a(\Omega_f^n; \boldsymbol{u}_{k+1}, \boldsymbol{v})$$
$$+ b(\Omega_f^n; p_{k+1}, \boldsymbol{v}) + m(\Gamma^n; \alpha_f \boldsymbol{u}_{k+1}, \boldsymbol{v}) = m(\Gamma^n; \alpha_f \dot{\boldsymbol{x}}_k - \boldsymbol{f}_k, \boldsymbol{v}), \tag{1.23}$$
$$b(\Omega_f^n; q, \boldsymbol{u}_{k+1}) = 0, \tag{1.24}$$

for all $(\boldsymbol{v}, q) \in V^n \times Q^n$.

For the space discretization of problems (1.23)–(1.24), we choose the inf-sup stable Taylor–Hood finite element pair $\mathbb{P}_2 - \mathbb{P}_1$. However, while the velocity field is continuous at Γ^n, the pressure space should be able to capture discontinuities across Γ^n, which are needed also for the correct evaluation of the hydrodynamic force (1.9). In order to deal with pressure discontinuities that occur at Γ^n, we introduce the following spaces which consist of piecewise continuous functions that may be discontinuous across the interface:

$$\tilde{V}_h^n = \Big\{\boldsymbol{v} \in (H^1(\Omega_f^n \backslash \Gamma^n)^2 : \ \boldsymbol{v}|_K \circ G_K \in \mathbb{P}_2(\hat{K}), \boldsymbol{v}|_{\Omega_f^n \backslash \Gamma^n} \in \big(C^0(\Omega_f^n \backslash \Gamma^n)\big)^2\Big\},$$
$$\tilde{Q}_h^n = \Big\{q \in L^2(\Omega_f^n \backslash \Gamma^n), \ q|_K \circ G_K \in \mathbb{P}_1(\hat{K}), \ q|_{\Omega_f^n \backslash \Gamma^n} \in C^0(\Omega_f^n \backslash \Gamma^n)\Big\},$$

where $\hat{K}$ is the reference simplex, and G_K is given by (1.17). The respective spaces with globally continuous functions will be denoted by V_h^n and Q_h^n. The appropriate finite element space for the unknowns in problems (1.23)–(1.24) is given by $V_h^n \times \tilde{Q}_h^n$. For the numerical implementation of our approach, we adopt a strategy called the Subspace Projection method [12, 13, 61]: we will work with spaces $\tilde{V}_h^n$ and $\tilde{Q}_h^n$, and then use an additional discrete projection to enforce continuity of the velocity on Γ^n. Note that V_h^n is a vector subspace of space $\tilde{V}_h^n$.

Let us briefly summarize how the Subspace Projection method works. We first notice that the Oseen problem (1.23) can be formally expressed as: Find $(\boldsymbol{u}_{k+1}, p_{k+1}) \in V^n \times Q^n$ such that

$$s((\boldsymbol{u}_{k+1}, p_{k+1}), (\boldsymbol{v}, q)) = g(\boldsymbol{v}, q), \quad \forall (\boldsymbol{v}, q) \in V^n \times Q^n, \tag{1.25}$$

where $s\colon (V^n \times Q^n) \times (V^n \times Q^n) \to \mathbb{R}$ is the bilinear form containing all the terms with index $k+1$, and $g\colon (V^n \times Q^n) \to \mathbb{R}$ is a linear form containing all the terms involving known quantities. Next, we define a projection operator:

$$\mathcal{P} : \tilde{V}_h^n \to V_h^n,$$

where V_h^n is a vector subspace of space $\tilde{V}_h^n$. By the Subspace Projection method, a discrete counterpart of problem (1.25) reads: Find $(\tilde{\boldsymbol{u}}_{h,k+1}, \tilde{p}_{h,k+1}) \in \tilde{V}_h^n \times \tilde{Q}_h^n$ such that

$$s((\mathcal{P}\tilde{\boldsymbol{u}}_{h,k+1}, \tilde{p}_{h,k+1}), (\mathcal{P}\tilde{\boldsymbol{v}}_h, \tilde{q}_h)) = g(\mathcal{P}\tilde{\boldsymbol{v}}_h, \tilde{q}_h)\,, \quad \forall(\tilde{\boldsymbol{v}}_h, \tilde{q}_h) \in \tilde{V}_h^n \times \tilde{Q}_h^n,$$

and then set $\boldsymbol{u}_{h,k+1} = \mathcal{P}\tilde{\boldsymbol{u}}_{h,k+1}$ and $p_{h,k+1} = \tilde{p}_{h,k+1}$. See [12, 13, 61] for a detailed description of these techniques.

The linear system resulting from linearization and discretization is solved with the direct solver PARDISO [50, 66, 67].

1.5.2 The discrete structure problem

For the time discretization of problem (1.5), we will consider a generalized Crank–Nicolson scheme [41]. At time t^{n+1}, Dirichlet–Neumann iteration $k+1$, the time discrete structure problem (1.5) is as follows: Find $\boldsymbol{x}_{k+1} \in K$ such that:

$$\int_0^L \rho_s \frac{\boldsymbol{x}_{k+1} - 2\boldsymbol{x}^n + \boldsymbol{x}^{n-1}}{\Delta t^2} \cdot \boldsymbol{y}\,ds + EI \int_0^L (\alpha \boldsymbol{x}_{k+1} + (1-2\alpha)\boldsymbol{x}^n + \alpha \boldsymbol{x}^{n-1})'' \cdot \boldsymbol{y}''\,ds$$
$$= \int_0^L (\alpha \boldsymbol{f}_{k+1} + (1-2\alpha)\boldsymbol{f}^n + \alpha \boldsymbol{f}^{n-1}) \cdot \boldsymbol{y}\,ds\,, \quad \forall \boldsymbol{y} \in dK(\boldsymbol{x}_{k+1}), \tag{1.26}$$

where K and dK are defined in (1.6) and (1.7), respectively. This scheme is known to be second-order accurate for linear problems. For the numerical results in Section 1.6, we will set $\alpha = 1/4$ since in linear cases this choice leads to an unconditionally stable scheme, which possesses a very small numerical dissipation compared to other schemes, e.g., the Houbolt method [9, 16].

Time discretization approximates problem (1.5) by a sequence of quasistatic problems that can be written as an equivalent nonconvex constrained problem:

$$\boldsymbol{x}_{k+1} = \arg\min_{\boldsymbol{y} \in K} J(\boldsymbol{y}),\ J(\boldsymbol{y}) = \frac{1}{2}\int_0^L \left(\frac{\rho_s}{\Delta t^2}|\boldsymbol{y}|^2 + EI\alpha \left|\boldsymbol{y}''\right|^2\right) ds - \int_0^L \boldsymbol{b} \cdot \boldsymbol{y}\,ds\,, \tag{1.27}$$

where $\boldsymbol{b}$ accounts for all the terms involving known quantities.

To treat the inextensibility condition $|\boldsymbol{y}'| = 1$, which is a quadratic constraint, we use an augmented Lagrangian method (e.g., [16, 33, 38, 41]). Let us introduce the following space and sets:

$$\mathcal{V} = \left\{\boldsymbol{y} \in (H^2(0,L))^2,\ \boldsymbol{y}(0) = \boldsymbol{a},\ \boldsymbol{y}'(0) = \boldsymbol{b}\right\},$$
$$\mathcal{V}_0 = \left\{\boldsymbol{y} \in (H^2(0,L))^2,\ \boldsymbol{y}(0) = \boldsymbol{0},\ \boldsymbol{y}'(0) = \boldsymbol{0}\right\},$$
$$\mathcal{Q} = \left\{\boldsymbol{q} \in (L^2(0,L))^2, |\boldsymbol{q}| = 1 \text{ a.e. on } (0,L)\right\}.$$

Problem (1.27) is equivalent to

$$\{\boldsymbol{x}_{k+1}, \boldsymbol{x}'_{k+1}\} = \arg\min_{\{\boldsymbol{y},\boldsymbol{q}\}\in W} J(\boldsymbol{y}), \quad \text{with} \quad W = \{\boldsymbol{y} \in V,\ \boldsymbol{q} \in \mathcal{Q},\ \boldsymbol{y}' - \boldsymbol{q} = \boldsymbol{0}\}.$$

With $r > 0$, we introduce the following augmented Lagrangian functional:

$$\mathcal{L}_r(\boldsymbol{y}, \boldsymbol{q}; \boldsymbol{\mu}) = J(\boldsymbol{y}) + \frac{r}{2}\int_0^L |\boldsymbol{y}' - \boldsymbol{q}|^2\, ds + \int_0^L \boldsymbol{\mu}\cdot(\boldsymbol{y}' - \boldsymbol{q})\, ds. \tag{1.28}$$

Let $\{\boldsymbol{x}, \boldsymbol{p}; \boldsymbol{\lambda}\}$ be a saddle point of $\mathcal{L}_r$ over $(\mathcal{V}\times\mathcal{Q})\times(L^2(0,L))^2$. Then a solution of problem (1.27) (and thus problem (1.26)) is $\boldsymbol{x}_{k+1} = \boldsymbol{x}$ and $\boldsymbol{p} = \boldsymbol{x}'_{k+1}$. In order to solve the above saddle-point problem, we employ the algorithm called ALG2 in, e.g., [33, 41]. As shown in, e.g., [33], this Uzawa-type algorithm is in fact a "disguised" Douglas–Rachford operator-splitting scheme:

- **Step 0**: The initial guess $\{\boldsymbol{x}_{-1}, \boldsymbol{\lambda}_0\} \in \mathcal{V}\times(L^2(0,L))^2$ is given.

Then, for $j \geq 0$, $\{\boldsymbol{x}_{j-1}, \boldsymbol{\lambda}_j,\}$ being known, proceed with:

- **Step 1**: Find $\boldsymbol{p}_j \in \mathcal{Q}$ such that:

$$\mathcal{L}_r(\boldsymbol{x}_{j-1}, \boldsymbol{p}_j; \boldsymbol{\lambda}_j) \leq \mathcal{L}_r(\boldsymbol{x}_{j-1}, \boldsymbol{q}; \boldsymbol{\lambda}_j), \quad \forall \boldsymbol{q} \in \mathcal{Q}.$$

- **Step 2**: Find $\boldsymbol{x}_j \in \mathcal{V}$ such that:

$$\mathcal{L}_r(\boldsymbol{x}_j, \boldsymbol{p}_j; \boldsymbol{\lambda}_j) \leq \mathcal{L}_r(\boldsymbol{y}, \boldsymbol{p}_j; \boldsymbol{\lambda}_j), \quad \forall \boldsymbol{y} \in \mathcal{V}_0. \tag{1.29}$$

- **Step 3**: Update the Lagrange multipliers by:

$$\boldsymbol{\lambda}_{j+1} = \boldsymbol{\lambda}_j + r((\boldsymbol{x}_j)' - \boldsymbol{p}_j).$$

To obtain $\boldsymbol{p}_j$ at step 1, we have to solve the minimization problem:

$$\min_{|\boldsymbol{q}|=1} \mathcal{L}_r(\boldsymbol{x}_{j-1}, \boldsymbol{q}; \boldsymbol{\lambda}_j), \text{ with the solution } \boldsymbol{p}_j = \frac{r(\boldsymbol{x}_{j-1})' + \boldsymbol{\lambda}_j}{|r(\boldsymbol{x}_{j-1})' + \boldsymbol{\lambda}_j|}. \tag{1.30}$$

Problem (1.29) can be stated as the equivalent problem: Find $\boldsymbol{x}_j \in \mathcal{V}$ such that for all $\boldsymbol{y} \in \mathcal{V}_0$:

$$\int_0^L \rho_s \frac{\boldsymbol{x}_j}{\Delta t^2}\cdot\boldsymbol{y}\,ds + \int_0^L EI\alpha\boldsymbol{x}''_j\cdot\boldsymbol{y}''\,ds + r\int_0^L \boldsymbol{x}'_j\cdot\boldsymbol{y}'\,ds$$
$$= \int_0^L \boldsymbol{b}\cdot\boldsymbol{y}\,ds + \int_0^L (r\boldsymbol{p}_j - \boldsymbol{\lambda}_j)\cdot\boldsymbol{y}'\,ds.$$

For the space discretization of problem (1.29) we use a third-order Hermite finite element method (e.g., [16]). For details about the discretization of $\boldsymbol{p}_j \in \mathcal{Q}$ (1.30) and $\boldsymbol{\lambda}_j \in (L^2(0,L))^2$ we refer to [41].

Steps 1, 2, and 3 are repeated until the following stopping criterion

$$\left(\int_0^L \left| \frac{\partial}{\partial s} \boldsymbol{x}^i - \boldsymbol{p}^i \right|^2 ds \right)^{1/2} \leq \epsilon \tag{1.31}$$

is satisfied for a given tolerance $\epsilon > 0$, or the number of iterations exceeds a given number.

Once (1.31) is satisfied, we set $\tilde{\boldsymbol{x}}_{k+1} = \boldsymbol{x}^i$, which defines the new structure position before relaxation.

Remark 1.1. It is known that parameter r plays a fundamental role for the convergence of algorithm ALG2, as was pointed out in [26]. We adopt the same adaptive strategy presented in [26], i.e., we start with an initial guess $r = r_0$, where r_0 is a fixed number (for instance in the range of the flexural stiffness EI). Once the Augmented Lagrangian algorithm terminates, we check if termination criterion (1.31) is met. In case (1.31) is violated, the value of r is increased (e.g., by a factor of 10) and ALG2 is repeated with the new value of r.

1.5.3 Enforcement of the coupling conditions

To describe the enforcement of the coupling conditions reported in Section 1.2.3, we first recall that at every time step the fluid mesh is aligned with the structure position. However, in general the fluid and structure meshes do *not* coincide, since they are made up of different elements: cubic Hermite elements on the structure side, and quadratic isoparametric edges on the fluid side. Because of the alignment, the fluid nodes that approximate the interface are always located on the structure mesh, as shown in Figure 1.4. In Figure 1.4, we denote by $\Gamma_h^{f,n}$ the approximation of the location of Γ^n given by the fluid mesh, and by $\Gamma_h^{s,n}$ the approximation of Γ^n by the structure mesh.

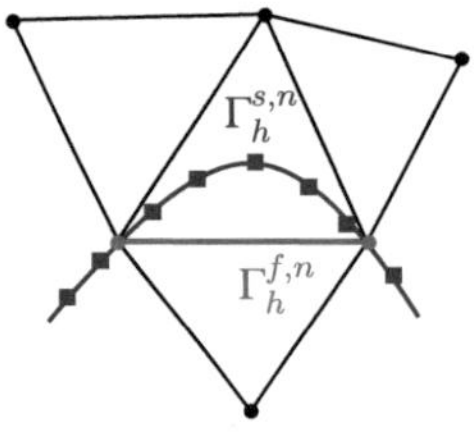

Fig. 1.4: Fluid triangulation (black) aligned with the structure mesh $\Gamma_h^{s,n}$ (blue). The fluid nodes are marked with dots, while the structure nodes are marked with squares. $\Gamma_h^{f,n}$ (red) is the approximation of the interface given by an edge of the fluid mesh.

Since Robin boundary condition (1.21) is a linear combination of the kinematic coupling condition and the dynamic coupling condition, we will only discuss the enforcement of the latter two conditions.

Enforcement of the kinematic coupling condition, i.e., the Dirichlet condition (1.18). Denote by $U_{\Gamma,k}$ and $\dot{X}_k$ the arrays of the nodal values of the corresponding fluid and structure velocities at the interface. Let B^n_{fs} be the interpolation matrix of the structure mesh at the fluid interface nodes. To impose Dirichlet condition (1.18), we set

$$U_{\Gamma,k+1} = B^n_{\mathrm{fs}} \dot{X}_k \, . \tag{1.32}$$

Enforcement of the dynamic coupling condition. The fluid load onto the structure is given by the hydrodynamic force (1.9). The computation of the hydrodynamic force (1.9) is crucial for the numerical stability and accuracy of Domain Decomposition FSI solvers [30]. In the setting considered in this paper (an immersed beam), the quality of approximation of the pressure jump across the beam is of great importance, as demonstrated by the results in Section 1.6.2.

The load exerted by the fluid onto the structure $\boldsymbol{f}_\Gamma$ can be computed as the variational residual $\mathcal{R}$ of the momentum conservation equation for the fluid, tested with test functions $\boldsymbol{v}$ that are different from zero on $\Gamma(t)$:

$$\begin{aligned} \int\limits_{\Gamma(t)} \boldsymbol{f}_\Gamma \cdot \boldsymbol{v} \, \mathrm{d}\Gamma &= -\rho_{\mathrm{f}} m\Big(\Omega_{\mathrm{f}}(t); \frac{\partial \boldsymbol{u}}{\partial t}\Big|_{\hat{\mathbf{x}}}, \boldsymbol{v}\Big) - \rho_{\mathrm{f}} c(\Omega_{\mathrm{f}}(t); \boldsymbol{u} - \mathbf{w}; \boldsymbol{u}, \boldsymbol{v}) \\ &\quad - a(\Omega_{\mathrm{f}}(t); \boldsymbol{u}, \boldsymbol{v}) - b(\Omega_{\mathrm{f}}(t); p, \boldsymbol{v}) \\ &= \mathcal{R}(\Omega_{\mathrm{f}}(t); \boldsymbol{u}, p, \boldsymbol{v}) \, . \end{aligned} \tag{1.33}$$

Let $\boldsymbol{f}^f_{\Gamma,k+1}$ denote the discrete hydrodynamic force at $\Gamma^{f,n}_h$ and subiteration $k+1$. After time and space discretization of (1.33), $\boldsymbol{f}^f_{\Gamma,k+1}$ is calculated from:

$$\int\limits_{\Gamma^{f,n}_h} \boldsymbol{f}^f_{\Gamma,k+1} \cdot \boldsymbol{v}_h \, \mathrm{d}\Gamma = \mathcal{R}(\Omega^n_{\mathrm{f}}; \boldsymbol{u}_{h,k+1}, p_{h,k+1}, \boldsymbol{v}_h) \, . \tag{1.34}$$

By using matrix notation, Equation (1.34) can be written as follows:

$$M^{n,f}_\Gamma F^f_{\Gamma,k+1} = R_{k+1}, \, , \tag{1.35}$$

where $F^f_{\Gamma,k+1}$ is the array of nodal values of $\boldsymbol{f}^f_{\Gamma,k+1}$, $M^{n,f}_\Gamma$ is the mass matrix at $\Gamma^{f,n}_h$, and R_{k+1} corresponds to the known values of the combined residuals appearing on the right-hand side of Equation (1.34). This defines the hydrodynamic force, calculated at the fluid mesh nodes along the beam.

To enforce the dynamic coupling condition (1.11), this hydrodynamic force needs to be set equal to the structural load $\boldsymbol{f}$ exerted onto the fluid. For this purpose, we need to assign the values of the hydrodynamic force to the structure mesh nodes $\Gamma^{s,n}_h$ which do not necessarily lie on the structure discretization defined by the fluid mesh, see Figure 1.4. To do that, we first project the structure mesh nodes of $\Gamma_h^{s,n}$ onto the fluid mesh interface $\Gamma_h^{f,n}$. At the fluid mesh, the fluid loading onto the structure $\boldsymbol{f}^f_{\Gamma,k+1}$ is defined by the process described above. We take those values of $\boldsymbol{f}^f_{\Gamma,k+1}$ and interpolate them first along the edges of the fluid mesh at the *projected* structural nodes, and then we assign those values back to the *original* structure nodes. More precisely, whenever the structural load $\boldsymbol{f}_{k+1}(\mathbf{x})$ needs to be evaluated for some $\mathbf{x} \in \Gamma_h^{s,n}$ (for instance at the quadrature nodes needed to evaluate the right-hand side of Equation (1.26)), we first define the projected structure node

$$\tilde{\mathbf{x}} := \arg\min_{\mathbf{y}\in\Gamma_h^{f,n}} \|\mathbf{x} - \mathbf{y}\|$$

and then let $\boldsymbol{f}_{k+1}(\mathbf{x}) = \boldsymbol{f}^f_{\Gamma,k+1}(\tilde{\mathbf{x}})$. We use the following notation to summarize this procedure:

$$F^s_{\Gamma,k+1} = B^n_{\mathrm{sf}} F^f_{\Gamma,k+1}\,, \tag{1.36}$$

where we used B^n_{sf} to denote the extrapolation of the values of the hydrodynamic quantities at the fluid nodes onto the structure nodes. This defines the hydrodynamic force at the *structure mesh nodes*, and enforces the dynamic coupling condition (1.11).

It is important to notice that in this numerical implementation of the dynamic coupling condition, the power exchanged between the fluid and structure is not perfectly balanced, i.e., at the discrete level, the energy imparted by the fluid onto the structure is not perfectly converted into the total energy of the structure, and vice versa. This is due to the nonmatching fluid and structure meshes. In the case of the DN algorithm, this mismatch can be precisely quantified as follows.

At the time t^{n+1}, after the convergence of the DN subiterations, the discrete power exchanged at the interface from the fluid side is

$$\begin{aligned} P^{f,n+1} &= \int\limits_{\Gamma_h^{f,n+1}} \boldsymbol{f}_\Gamma^{f,n+1} \cdot \boldsymbol{u}_h^{n+1}\, d\Gamma = (U_\Gamma^{n+1})^T M_\Gamma^{f,n+1} F_\Gamma^{f,n+1} \\ &= (\dot{X}^{n+1})^T (B_{\mathrm{fs}}^{n+1})^T M_\Gamma^{f,n+1} F_\Gamma^{f,n+1}\,, \end{aligned} \tag{1.37}$$

where for the last equation we used (1.32). Similarly, the discrete power exchanged at the interface from the structure side is

$$\begin{aligned} P^{s,n+1} &= \int\limits_{\Gamma_h^{s,n+1}} \boldsymbol{f}_\Gamma^{s,n+1} \cdot \dot{\boldsymbol{x}}_h^{n+1}\, d\Gamma = (\dot{X}^{n+1})^T M_\Gamma^{s,n+1} F_\Gamma^{s,n+1} \\ &= (\dot{X}^{n+1})^T M_\Gamma^{s,n+1} B_{\mathrm{sf}}^{n+1} F_\Gamma^{f,n+1}\,, \end{aligned} \tag{1.38}$$

where for the last equation we used (1.36). Thus, the power exchanged at the interface is balanced if

$$(B_{\mathrm{fs}}^{n+1})^T M_\Gamma^{f,n+1} = M_\Gamma^{s,n+1} B_{\mathrm{sf}}^{n+1}\,.$$

Since $\Gamma_h^{f,n+1}$ and $\Gamma_h^{s,n+1}$ are aligned but do not coincide ($\Gamma_h^{s,n+1}$ is a piecewise cubic globally C^1 function, and $\Gamma_h^{f,n+1}$ is a piecewise quadratic interpolation) and the fluid and structure discretizations are based on different elements, the balance equation is not necessarily fulfilled exactly. However, in Section 1.6.3 we will show that the difference between $P^{f,n+1}$ and $P^{s,n+1}$ is very small (0.01% of the power value) in our computations.

1.6 Numerical results

A series of numerical tests is presented that showcase the main features and performance of our approach. In all the tests, the fluid density of $\rho_f = 1\,\mathrm{g/cm^3}$ is considered, the structure mesh always consists of 45 nodes (with cubic Hermite elements), and the stopping tolerance in (1.19) for the partitioned schemes (either DN or RN) is always set to $\epsilon_{FS} = 10^{-8}$. We use the SI unit system, and present all the quantities in the centimeter-gram-second (CGS) units. If the units of a certain quantity are omitted for the sake of simplicity, it is implied that they are in the CGS system.

1.6.1 Validation of the structure solver

The purpose of this section is to validate the structure solver by a benchmark with an exact solution.

We consider $s \in [0, \pi/2]$ and $t \in [0, 1]$, and a family of exact solutions which is given by:

$$\boldsymbol{x}_{\mathrm{ex}}(s,t) = (\phi(t))^{-1}\left[\cos(s\phi(t)), \sin(s\phi(t))\right]^T . \tag{1.39}$$

Notice that solution (1.39) satisfies the inextensibility condition $|\boldsymbol{x}|' = 1$ pointwise for every function $\phi(t)$. We chose $\phi(t) = e^t$, for which the solution is a quarter of a circle of initial radius 1 that coils over time as its radius decreases (Figure 1.5). At $s = 0$ and $s = \pi/2$, we impose the values of $\boldsymbol{x}$ and $\boldsymbol{x}'$. The forcing term $\boldsymbol{f}_{\mathrm{ex}}$ needed to recover solution (1.39) is found by plugging $\boldsymbol{x}_{\mathrm{ex}}$ into the governing differential equations (strong form):

$$\rho_s \ddot{\boldsymbol{x}}_{\mathrm{ex}} + EI\boldsymbol{x}_{\mathrm{ex}}'''' = \boldsymbol{f}_{\mathrm{ex}} . \tag{1.40}$$

For simplicity, we set $\rho_s = 1\,\mathrm{Kg/m^3}$ and $EI = 1\,\mathrm{Kg\,m^3/s^2}$. The forcing term $\boldsymbol{f}_{\mathrm{ex}}$ is made up of two contributions: an external body force $\boldsymbol{f}_b$ and an internal force due to inextensibility $\boldsymbol{f}_{\mathrm{in}}$. To find $\boldsymbol{f}_{\mathrm{in}}$, we notice that problem (1.5) is equivalent to minimization problem $\boldsymbol{x} = \arg\min_{\boldsymbol{y}\in K} J(\boldsymbol{y})$, where the total energy of the beam can be written as:

$$J(\boldsymbol{y}) = \frac{1}{2}\int_0^L \rho_s|\ddot{\boldsymbol{y}}|^2 ds + \frac{1}{2}\int_0^L EI\left|\boldsymbol{y}''\right|^2 ds + \frac{1}{2}\int_0^L \lambda(|\boldsymbol{y}'|^2 - 1)ds - \int_0^L \boldsymbol{f}\cdot\boldsymbol{y}ds,$$

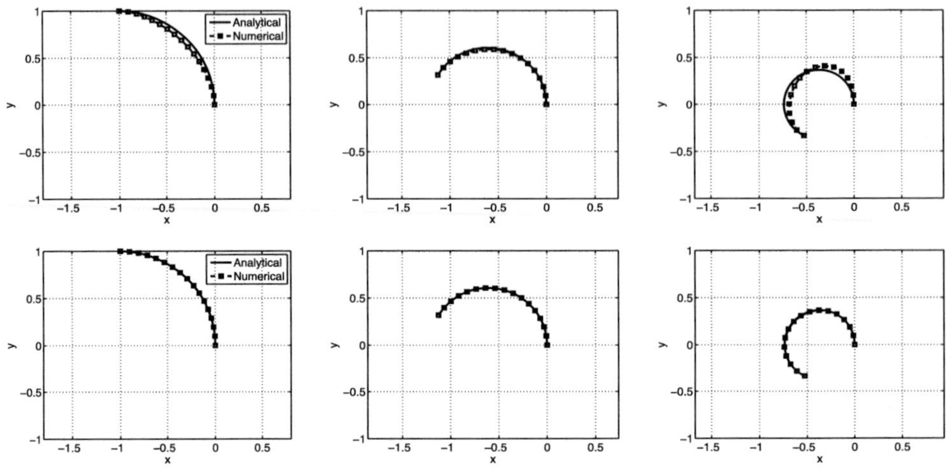

Fig. 1.5: Comparison between analytical and numerical solution at $t = 0$ s (left), $t = 0.5$ s (center), $t = 1$ s (right) for two values of stopping tolerance: $\epsilon = 10^{-1}$ (top) and $\epsilon = 10^{-5}$ (bottom). The legend in the subfigures on the left is common to all the subfigures.

and λ is a scalar function that depends on time only. If the above functional attains its minimum at $\boldsymbol{x}$, it follows that its Gâteaux derivative must be vanishing at $\boldsymbol{x}$, leading to

$$\int_0^L \rho_S \ddot{\boldsymbol{x}} \cdot \boldsymbol{y} ds + \int_0^L EI\, \boldsymbol{x}'' \cdot \boldsymbol{y}'' ds = \int_0^L \boldsymbol{f} \cdot \boldsymbol{y} ds + \int_0^L (\lambda \boldsymbol{x}')' \cdot \boldsymbol{y} ds,$$

for all $\boldsymbol{y} \in dK(\boldsymbol{x})$. The second integral on the right-hand side (equal to zero if $\boldsymbol{y} \in dK(\boldsymbol{x})$, which is not the case for the test functions used in the computations) gives the explicit contribution of $\boldsymbol{f}_{\text{in}}$.

We are going to check the convergence rates in time for the generalized Crank–Nicolson scheme in Section 1.5.2 with $\alpha = 1/4$ in two cases:

- linear case: when the forcing term is $\boldsymbol{f}_{\text{ex}}$ the inextensibility condition becomes inactive due to the fact that $\boldsymbol{f}_{\text{ex}}$ is given by (1.40) and the problem reduces to the linear beam equation;
- nonlinear case: when then forcing term is $\boldsymbol{f}_{\text{ex}} + (\lambda \boldsymbol{x}')'$, with, e.g., $\lambda = 1$, the problem becomes nonlinear and the inextensibility is treated via the augmented Lagrangian method described in Section 1.5.2.

The space resolution Δs is taken to be $\pi/240$. In the nonlinear case, for ALG2 we set stopping tolerance $\epsilon = 10^{-5}$ (1.31) and $r = 10^2$. In Figure 1.6, we plot the L^2 norm of the difference between the exact solution $\boldsymbol{x}_{\text{ex}}$ and the numerical solution $\boldsymbol{x}_h$ at $t = 1$ against time step ($\Delta t = 0.2, 0.1, 0.05, 0.025, 0.0125, 0.00625$) for the linear and non-

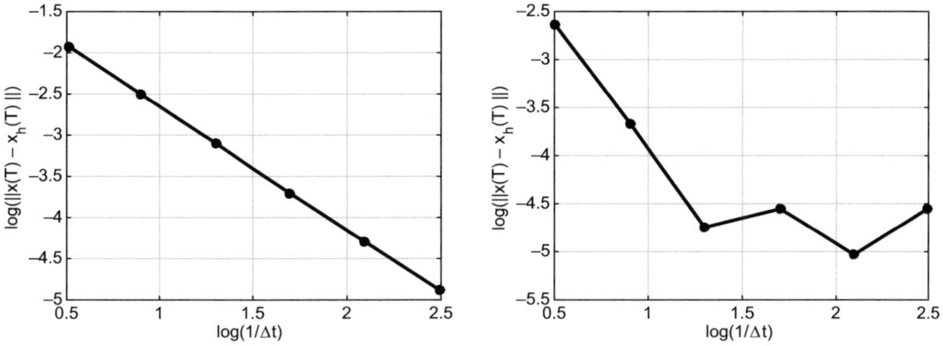

Fig. 1.6: Convergence rate in time for the generalized Crank–Nicolson scheme in the linear (left) and nonlinear/inextensible (right) case.

linear cases. The rate predicted by the theory (second order) is achieved in the linear case. In the nonlinear case, the order of convergence is even larger than 2 provided that Δt is less than a critical value for which the error reaches the stopping tolerance ϵ. If Δt is greater than that critical value, the error remains unchanged or even slightly increases.

To illustrate that the error depends on the choice of ϵ, in Figure 1.5, we compare analytical solution (1.39) with the numerical solution at $t = 0, 0.5, 1$ s and for two values of the stopping tolerance: $\epsilon = 10^{-1}$ (top) and $\epsilon = 10^{-5}$ (bottom), every other discretization parameter being the same. For $\epsilon = 10^{-1}$ the difference between analytical and numerical solution is clearly visible, while for $\epsilon = 10^{-5}$ the two solutions are almost superimposed.

Finally, in order to evaluate the dependence of the error on ϵ, we report in Figure 1.7 the convergence rates in time in the nonlinear case for different values of the stopping tolerance $\epsilon = 10^{-2}, 10^{-3}, 10^{-4}, 10^{-5}, 10^{-6}, 10^{-7}$. The values for Δt and Δs are the same as those used for the results in Figure 1.6. We see that at the critical value of Δt the curves reach a plateau for all the values of ϵ, indicating that for a given value of ϵ it does not make sense to choose a time-step size that is too small. Our computations seem to indicate that Δt should be larger than $\sqrt{\epsilon}$.

In the following sections, for the Augmented Lagrangian method in Section 1.5.2, we take $\epsilon = 10^{-4}$ in (1.31), and at the beginning of the simulation we set $r_0 = 10^{-4}$; see Remark 1.1.

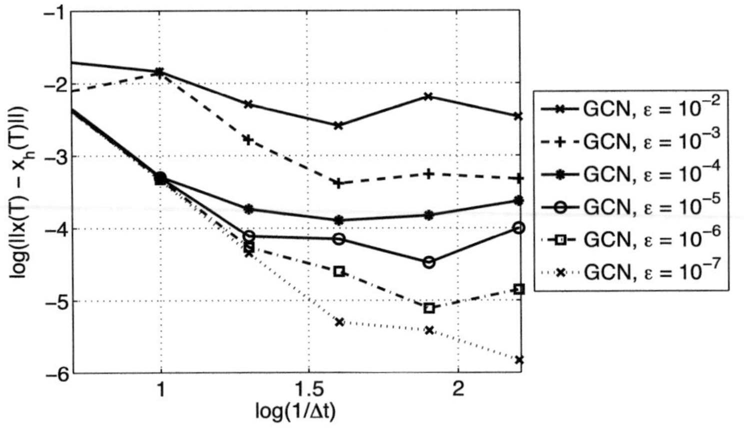

Fig. 1.7: Convergence rate in time for the generalized Crank–Nicolson scheme in the nonlinear case for different values of the stopping tolerance ϵ.

1.6.2 Continuous vs. discontinuous pressure across the interface

The goal of this preliminary test is to show the importance of capturing the pressure discontinuity across the interface. For this purpose we consider a simple test problem which involves a steady state solution.

The fluid domain is a rectangle of height 1 cm and length 6 cm: $[-3, 3]$ cm $\times[-0.5, 0.5]$ cm. The beam is clamped at the midpoint of the base, and it is 0.5 cm long. A no-slip condition is imposed on Γ_{down} (Figure 1.1), a symmetry condition is imposed on Γ_{up}, and a homogeneous Neumann condition is enforced on Γ_{out}. The steady state solution is obtained after a time-independent Poiseuille profile is imposed at Γ_{in} with maximum velocity $U = 1$ cm/s, and the initial position of the beam is a vertical straight line. We set $\rho_{\text{s}} = 10^6$ g/cm and $EI = 0.01$ g/(cm s^2). For the given inlet condition and structural parameters, the beam displacement is negligible ($\simeq 10^{-5}$ cm) during the time interval [0, 10] s. See Figure 1.9.

We consider four meshes resulting from uniform refinement of an initial coarse mesh for the fluid subdomain, with edge lengths $h_{\text{f}} = \sqrt{2}/8 \cdot 2^{-l}$ cm, $l = 0, 1, 2, 3$. We track the behavior of the hydrodynamic force along the leaflet over time to see the convergence behavior of both pressure approximations. Namely, we track, in time, the x and y components of the integrated hemodynamic force ("averaged force") over the interface:

$$F_{\Gamma,x}(t) = \int\limits_{\Gamma_h^s(t)} \boldsymbol{f}_\Gamma^f(t) \cdot [1, 0]^T \, \mathrm{d}\Gamma \,, \quad F_{\Gamma,y}(t) = \int\limits_{\Gamma_h^s(t)} \boldsymbol{f}_\Gamma^f(t) \cdot [0, 1]^T \, \mathrm{d}\Gamma \,. \tag{1.41}$$

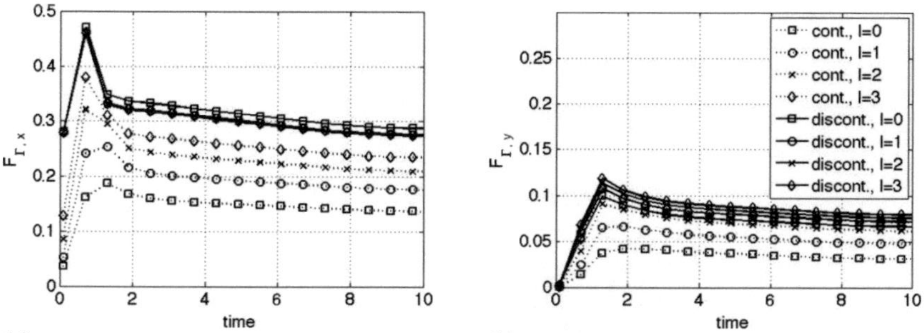

(a) x component (b) y component

Fig. 1.8: Integral over the interface of the components of the hydrodynamic force over time: (a) x component and (b) y component. The results are obtained for four different meshes ($l = 0, 1, 2, 3$) and two different pressure approximations (discontinuous and continuous). The legend in (b) is common to both subfigures.

In Figure 1.8, we plot $F_{\Gamma,x}$ and $F_{\Gamma,y}$ for all four meshes and two different pressure approximations: a discontinuous and a continuous one. The following conclusions can be reached from this study:

- In all the cases, both components reach a plateau after roughly 8 s, meaning that a steady state is well approximated by the solution at time 8 s.
- When using the discontinuous pressure approximation, mesh independence is reached in the eyeball norm at the third mesh refinement, while for the continuous mesh approximation mesh independence is not achieved even after four mesh refinements.
- At each successive refinement the solution values computed with continuous pressure get closer to the solution values computed with the discontinuous pressure on a coarse mesh. Thus, the continuous pressure approximation requires *a significantly finer mesh* than the discontinuous pressure approximation to approximate the physical solution with the same accuracy.

As a qualitative evidence, we report in Figure 1.9 the pressure at time $t = 9$ s computed on the four meshes with both the continuous and the discontinuous pressure approximation. One can see that when using the continuous pressure, there is a huge difference between the pressure computed on the coarsest mesh and the pressure computed on the finest mesh. This is not the case when using the discontinuous pressure finite element pair. Moreover, the continuous pressure computed on the finest mesh looks clearly similar to the discontinuous pressure on any mesh.

From now on, we present the results which have all been obtained with the discontinuous pressure approximation.

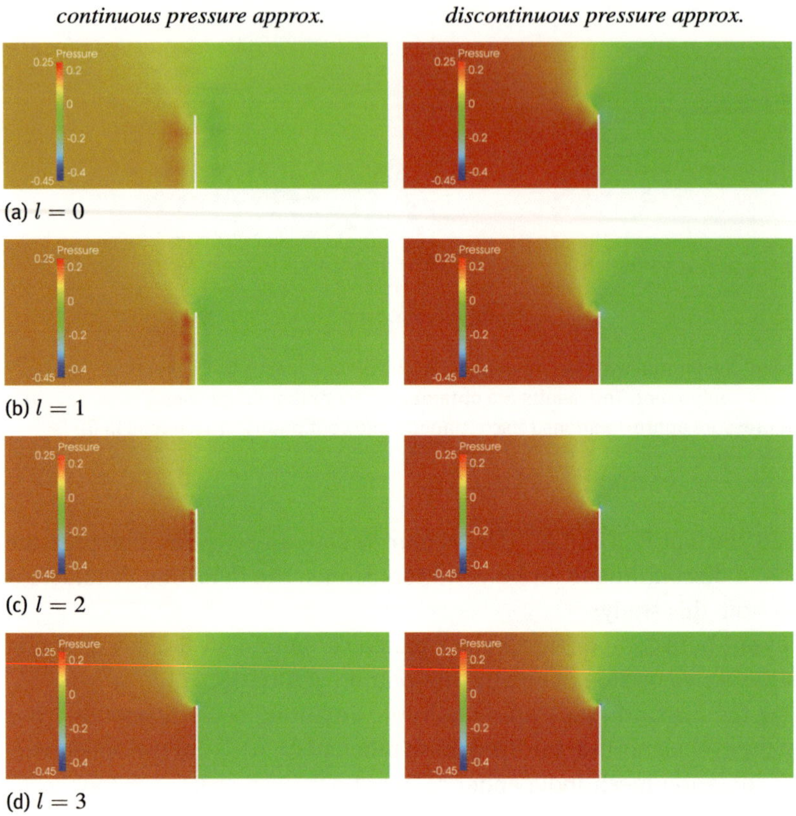

Fig. 1.9: Pressure in a portion of the computational domain at time $t = 9$ s computed with the continuous (left) and discontinuous (right) pressure finite element pair on four different refinement levels: (a) $l = 0$, (b) $l = 1$, (c) $l = 2$, and (d) $l = 3$.

1.6.3 Moderate-amplitude oscillatory motion of the beam

We consider again a channel $\Omega_f = [-3, 3] \times [-0.5, 0.5]$ filled with a viscous, incompressible fluid of viscosity $\mu = 0.01$ g/m s. A beam of length 0.5 cm, with $\rho_s = 5$ g/cm and $EI = 0.05$ g/(cm s^2), is immersed in the fluid, clamped at the mid-point $x = 0$ bottom of the channel Γ_{down}. The beam is initially straight and vertical, and the fluid is initially at rest. We consider a time-dependent FSI problem that is driven by the time-dependent inlet velocity data: a Poiseuille velocity profile, with maximum velocity

$$U(t) = \frac{1}{4}\left(1 - \cos\left(\frac{\pi}{2}t\right)\right) \text{ cm/s}$$

is prescribed at Γ_{in}, which corresponds to $x = -3$. A homogeneous Neumann condition is enforced at Γ_{out}, i.e. at $x = 3$. The no-slip condition is imposed on Γ_{down}, and a symmetry condition is imposed on Γ_{up}. See Figure 1.10. The Strouhal number for

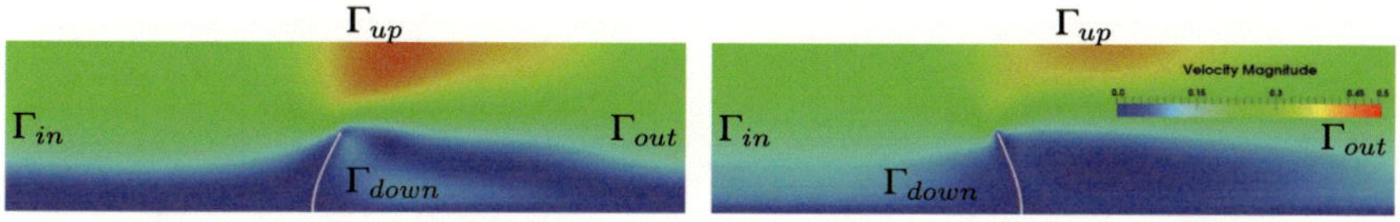

(a) $t = 7$ s (b) $t = 9$ s

Fig. 1.10: Velocity magnitude and beam position at time (a) $t = 7$ s and (b) $t = 9$ s.

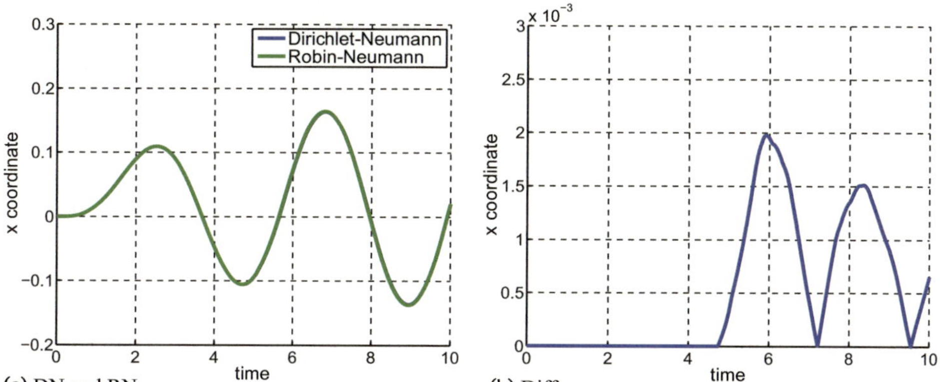

(a) DN and RN (b) Difference

Fig. 1.11: (a) Comparison of the x-component of the beam tip position computed with the Dirichlet–Neumann and Robin–Neumann algorithms and (b) difference in absolute value between the curves in (a).

this problem is 0.5. The inlet boundary condition and the structural parameters were chosen to generate a "moderate"-amplitude oscillatory motion of the beam around its initial configuration. Figure 1.10 shows two snapshots of the velocity magnitude together with the beam position at the time of maximum deflection. We will use this test case to compare the DN and RN algorithms. Thanks to the moderate amplitude of the beam oscillations we will also be able to compare our extended ALE method with a standard ALE method. Finally, we will check the power exchange at the interface.

Comparison between the DN and RN algorithms. To show that both DN and RN algorithms yield numerical results that are very close, we compare the x coordinate of the beam tip over time. The results, shown in Figure 1.11a, indicate that the two curves are almost superimposed. To zoom into the difference between the two results, we plot the absolute value of the difference between the two curves in Figure 1.11b. One can see that the maximum difference over time interval [0, 10] s is $2 \cdot 10^{-3}$, which occurs after a first ALE reparametrization at around $t = 4.8$.

In order to evaluate the performance of the DN and RN algorithms, we let the structure density vary: $\rho_s = 16, 8, 4, 2, 1$ g/cm. The convergence properties of the DN algorithm are known to depend heavily on the added-mass effect, which becomes worse as the structure density gets closer to the density of the fluid, which is $\rho_f = 1$ g/cm^3 [21].

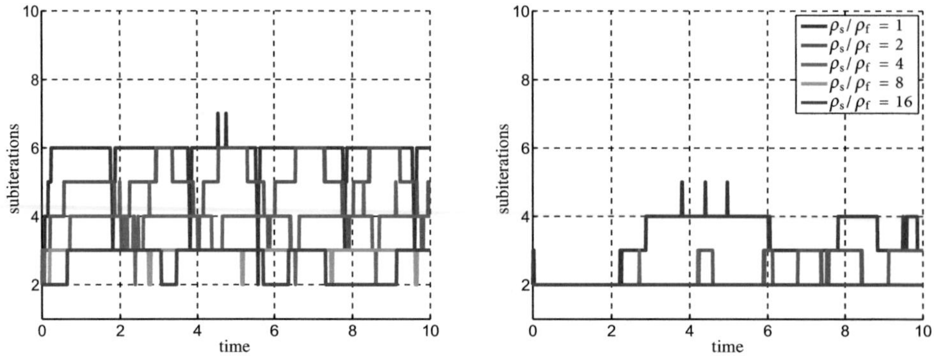

(a) DN with Aitken's acceleration

(b) RN

Fig. 1.12: Number of iterations required by (a) the DN algorithm with Aitken's acceleration method and (b) the RN method to converge over time (in s) for different values of the structure density. The legend in (b) is common to both subfloats.

Thus, we combine it with a relaxation strategy based on Aitken's acceleration technique as described in Section 1.4.1.

We compare the performance of the DN algorithm with Aitken's acceleration and the RN algorithm with no acceleration. Figure 1.12a shows the number of DN iterations within each time step over time interval $[0, 10]$, and Figure 1.12b shows the number of iterations required by the RN algorithm. The maximum number of DN iterations with Aitken's acceleration equals 7 when $\rho_s/\rho_f = 1$. We see that for all the structure densities under consideration, the RN algorithm requires less iterations to converge than the DN method with Aitken's acceleration technique. Moreover, the RN method is quite insensitive to variations in the structure density. In fact, we see in Figure 1.12b that for $\rho_s = 16, 8, 4, 2$ g/cm, the RN algorithm converges in two iterations most of the time. Even when $\rho_s = 1$ g/cm, the number of iterations required by the RN method is at most five and only for a very limited time.

Standard ALE versus extended ALE. We aim at showing that when the structure displacement is relatively small, the solution obtained using our extended ALE approach "coincides" with the solution obtained using the standard ALE approach. We fix the structure density to $\rho_s = 5$ g/cm and consider two fluid meshes of mesh size $h = 0.05 \cdot 2^j, j = 0, 1$. We compute the value of $F_{\Gamma,x}(t)$ defined in (1.41) with our extended ALE method and a standard ALE method for $t \in [0, 10]$. The comparison is reported in Figure 1.13. For the standard ALE method, we observe a smooth behavior of the force over time. Since no reparametrization takes place and no mesh transfer is performed, this is expected. For our extended ALE method we clearly observe jumps in the evaluated force at the times when the reparametrization takes place. Since the number of reparametrization steps is proportional to h^{-1}, jumps occur more frequently on the fine level. However, we also see

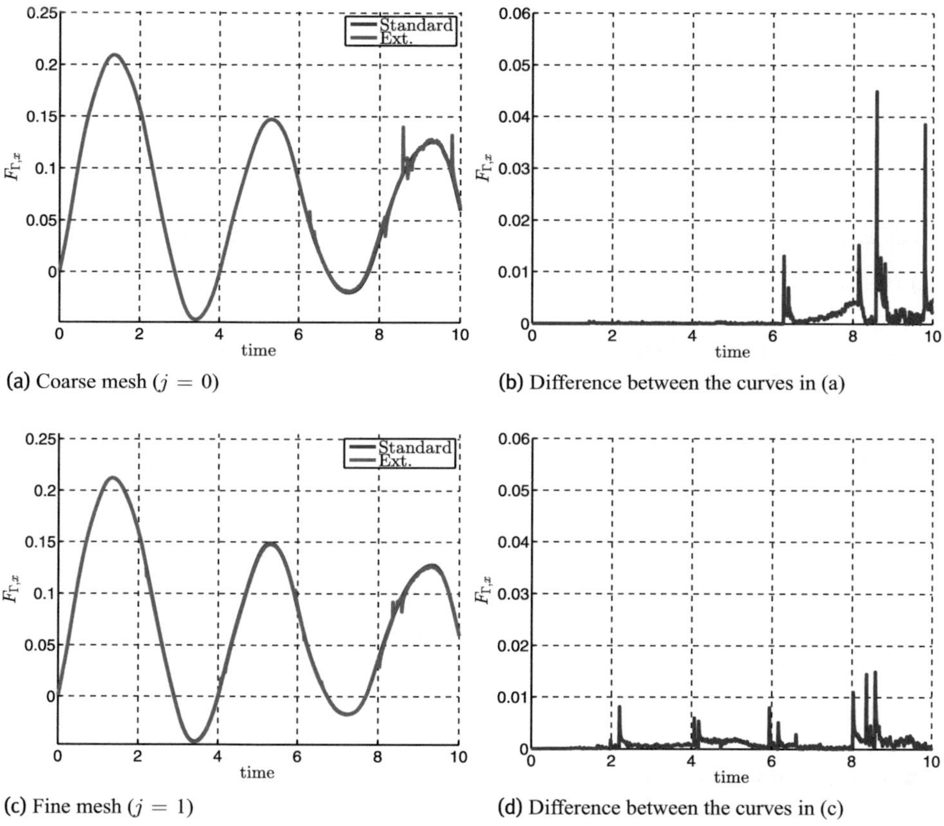

Fig. 1.13: $F_{\Gamma,x}(t)$ in (1.41) computed with a standard ALE method and our extended ALE method over the (a) coarse mesh and (c) fine mesh. The difference between the curves in subfigure (a) is reported in subfigure (b), and the difference between the curves in subfigure (c) is reported in subfigure (d).

that the jumps are local in nature and do not accumulate over time. We also observe that the absolute difference between the forces gets smaller as the mesh size gets smaller.

Power exchange at the interface. The goal is to verify how well the power exchanged at the interface is approximated. We first calculate the discrete power exchanged at the interface from the fluid side P^f, defined in (1.37), computed with two fluid meshes of mesh size $h = 0.05 \cdot 2^j$, $j = 0, 1$. Figure 1.14a shows that they are in a pretty good agreement. As for the forces in Figure 1.13, the occasional jumps in the discrete power that can be seen in Figure 1.14a occur whenever the ALE mapping is reparametrized. These jumps are to be expected for dynamically changing meshes, as pointed out in [15, 17].

Next, we quantify the unbalance in the power exchange at the interface. As explained in Section 1.5.2, at each time t^{n+1} the powers exchanged at the interface from

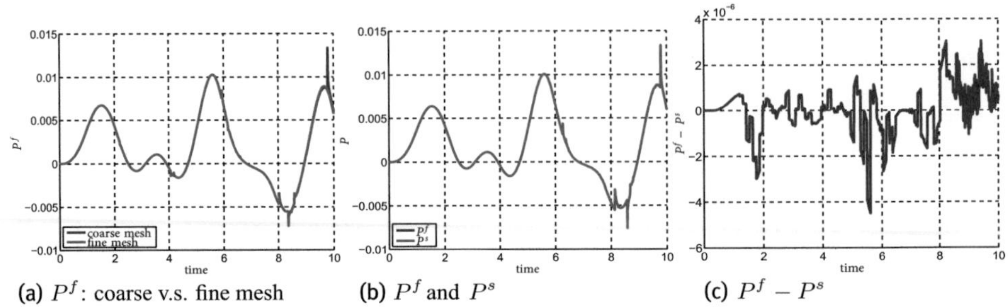

Fig. 1.14: Discrete power exchanged at the interface from the fluid side P^f, defined by (1.37), computed with two different meshes.

the fluid side $P^{f,n+1}$ and from the structure side $P^{s,n+1}$ are not necessarily equal. In Figure 1.14b, we plot the powers P^f and P^s computed with mesh $j = 1$ over the time interval under consideration, while in Figure 1.14 we show the difference $P^f - P^s$. One can see in Figure 1.14c that over a long time interval the difference between the two powers exchanged at the interface is of the order of 10^{-6} g cm/s^3. This corresponds to 0.01% of the power value, which is of the order of 10^{-2} g cm/s^3, as shown in Figure 1.14b. Such a small difference between P^f and P^s does not endanger stability.

1.6.4 Large-amplitude oscillatory motion of the beam

We consider the same fluid domain as in the previous two sections $\Omega_{\mathrm{f}} = [-3, 3] \times [-0.5, 0.5]$, but a longer immersed beam: we set the beam length to be $L = 0.95$ cm, just short of touching the top boundary Γ_{up}. See Figure 1.15a. All the physical and discretization parameters are chosen as in Section 1.6.3. The initial beam configuration is vertical, as shown in Figure 1.15a, and the initial fluid velocity is zero.

We study a time-dependent FSI problem in which the fluid flow is driven by the difference in the normal stress prescribed at the inlet and outlet, Γ_{in} and Γ_{out}. At Γ_{in} the time-dependent normal stress is given by a *time-periodic function* with period 10 s, where during each period the normal stress is given by the step function:

$$\boldsymbol{\sigma n}(t, \boldsymbol{x}) = \begin{cases} [1.5, 0]^T & \text{if } 0 \le \tilde{t} \le 3\,, \\ [0, 0]^T & \text{if } 3 < \tilde{t} < 10 \end{cases} \quad \text{for} \quad \boldsymbol{x} \in \Gamma_{\mathrm{in}}$$

where $\tilde{t} \in [0, 10)$ is $\tilde{t} = \mathrm{mod}(t, 10)$. Therefore, a constant, normal stress in the horizontal direction is applied for the first 3 s, when the stress is released and set to zero. This is repeated six times until $t = 60$ s. At Γ_{out}, the normal stress is prescribed to be zero. The no-slip condition is imposed on Γ_{down}, and a symmetry condition is imposed on Γ_{up}.

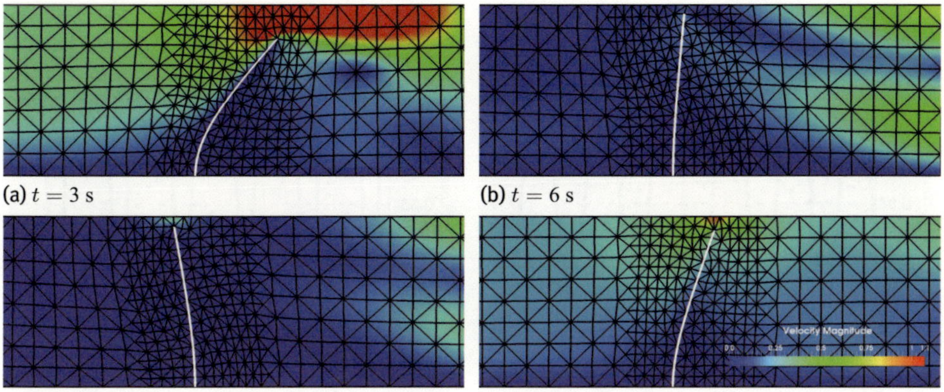

(a) $t = 3$ s (b) $t = 6$ s (c) $t = 8$ s (d) $t = 11$ s

Fig. 1.15: Beam position and velocity magnitude together with mesh deformation at (a) $t = 3$ s, (b) $t = 6$ s, (c) $t = 8$ s, and (d) $t = 11$ s.

Because of the periodic forcing of the fluid, the induced beam movement will be periodic: the beam is pushed to the right for $0 \leq \tilde{t} \leq 3$, and then bends backward due to elastic forces for $3 < \tilde{t} < 10$. This motion is repeated six times until t reaches 60 s. See Figure 1.15. Because of the symmetry condition imposed on the upper wall, this test corresponds to simulating a 1D valve which is just short of closing, i.e., near a contact. We can see in Figure 1.15 that there is only one element between the tip of the valve and Γ_{up}.

This scenario is typically associated with various problems. In particular, due to the fact that the beam tip is very close to the upper boundary of the fluid domain, the standard ALE method breaks down because the mesh associated with this simulation gets quickly severely distorted. The Extended ALE method, however, does not suffer from this problem. Indeed, Figure 1.15 shows that the mesh quality remains good, and that completing a cycle of a near-closing (regurgitant) valve is not a problem.

To show that this method does not induce spurious energy or instabilities over time, we let this simulation run for six cycles. Figure 1.16 shows the x and y coordinates of the beam tip position over time. The movement of the beam tip exhibits perfect monochromatic behavior, showing that our method does not introduce spurious energy over time and that no instabilities arise over a rather long time interval. From Figure 1.16b, we also see that the inextensibility constraint is never violated. In fact, the beam is clamped at $(0, -0.5)$ cm and its y coordinate never exceeds 0.45 cm, the beam being 0.95 cm long.

This example shows that the extended ALE method deals well with structure motion near a contact, and that it does not introduce any spurious energy or instabilities over time.

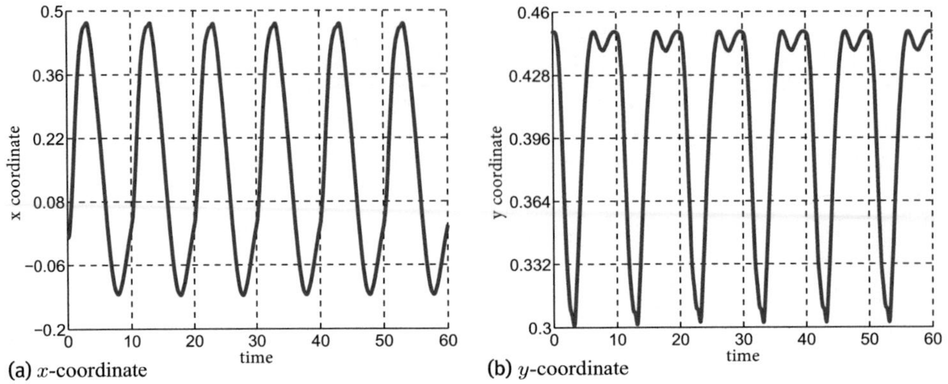

Fig. 1.16: (a) x-coordinate and (b) y-coordinate of beam tip position (in cm) over time (in s).

1.7 Conclusions

Standard ALE methods for the simulation of fluid-structure interaction problems fail when the structural displacement is large. Recently, we proposed an extended ALE method to overcome this limitation without remeshing. Our method relies on a variational mesh optimization technique with an additional constraint to enforce the alignment of the fluid-structure interface with edges of the resulting triangulation. In this manuscript, we provided a detailed description of the fundamental building blocks of our approach.

We focused on 2D FSI problems with two-way coupling and large displacements. We considered two partitioned algorithms: the classical Dirichlet–Neumann method with Aitken's acceleration and the Robin–Neumann method. Each partitioned algorithm was combined with our extended ALE method to solve the FSI problems. Because our method provides a fluid mesh aligned with the interface, nontrivial boundary conditions such as the Robin boundary conditions can be easily implemented. We showed that in problems where the structure density is close to the fluid density, the Robin–Neumann method outperforms the Dirichlet–Neumann algorithms with Aitken's acceleration.

We tested our problem on several benchmark cases involving an elastic inextensible beam interacting with an incompressible, viscous fluid in 2D. The test problems were carefully chosen to study the performance of our approach. The examples clearly show that our extended ALE method is in excellent agreement with the results obtained using a standard ALE method when structural displacement is small. For *large displacements* and in problems with *structures near a contact*, when standard ALE methods fail, we showed that the quality of the mesh given by the our approach remains high, thereby allowing full computer simulations of this class of problems.

Preliminary results using the proposed mesh optimization techniques for 3D fluid/rigid particle systems have been obtained and will be presented in a forthcoming paper. Future developments of this work include the extension to more complex structure models and 3D biomedical applications, e.g., blood interacting with heart valves.

Bibliography

[1] S. V. Apte, M. Martin, and N. A. Patankar. A numerical method for fully resolved simulation (frs) of rigid particle-flow interactions in complex flows. *Journal of Computational Physics*, 228(8):2712–2738, 2009.

[2] M. Astorino, J.-F. Gerbeau, O. Pantz, and K.-F. Traoré. Fluid-structure interaction and multi-body contact: Application to aortic valves. *Computer Methods in Applied Mechanics and Engineering*, 198(45-46):3603–3612, 2009.

[3] F. Baaijens. A fictitious domain/mortar element method for fluid-structure interaction. *Int. J. Numer. Meth. Fl.*, 35(7):743–761, 2001.

[4] S. Badia, F. Nobile, and C. Vergara. Fluid-structure partitioned procedures based on Robin transmission conditions. *Journal of Computational Physics*, 227:7027–7051, 2008.

[5] J. Baiges and R. Codina. The fixed-mesh ALE approach applied to solid mechanics and fluid–structure interaction problems. *International journal for numerical methods in engineering*, 81(12):1529–1557, 2010.

[6] J. Banks, W. Henshaw, and D. Schwendeman. An analysis of a new stable partitioned algorithm for FSI problems. part ii: Incompressible flow and structural shells. *Journal of Computational Physics*, 268:399–416, 2014.

[7] E. Bänsch, J. Paul, and A. Schmidt. An ALE finite element method for a coupled Stefan problem and Navier-Stokes equations with free capillary surface. *International Journal for Numerical Methods in Fluids*, 2012.

[8] S. Basting and R. Prignitz. An interface-fitted subspace projection method for finite element simulations of particulate flows. *Computer Methods in Applied Mechanics and Engineering*, 267(0):133–149, 2013.

[9] S. Basting, A. Quaini, R. Glowinski, and S. Canic. Comparison of time discretization schemes to simulate the motion of an inextensible beam. In *Numerical Mathematics and Advanced Applications - ENUMATH 2013. In Lecture Notes in Computational Science and Engineering*, volume 103, pages 175–183, 2015.

[10] S. Basting, A. Quaini, R. Glowinski, and S. Canic. Extended ALE Method for fluid-structure interaction problems with large structural displacements. *Journal of Computational Physics*, 331:312–336, 2017.

[11] S. Basting and M. Weismann. A hybrid level set/front tracking finite element approach for fluid-structure interaction and two-phase flow applications. *Journal of Computational Physics*, 255:228–244, 2013.

[12] S. Basting and M. Weismann. A hybrid level set/front tracking approach for finite element simulations of two-phase flows. *Journal of Computational and Applied Mathematics*, 270:471–483, 2014.

[13] K. Bäumler and E. Bänsch. A subspace projection method for the implementation of interface conditions in a single-drop flow problem. *Journal of Computational Physics*, 252:438–457, 2013.

[14] B. Bejanov, J. Guermond, and P. Minev. A grid-alignment finite element technique for incompressible multicomponent flows. *Journal of Computational Physics*, 227(13):6473–6489, 2008.

[15] M. Besier and W. Wollner. On the pressure approximation in nonstationary incompressible flow simulations on dynamically varying spatial meshes. *International Journal for Numerical Methods in Fluids*, 69(6):1045–1064, 2012.
[16] J.-F. Bourgat, M. Dumay, and R. Glowinski. Large displacement calculations of inexstensible pipelines by finite element and nonlinear programming methods. *SIAM J. Sci. Stat. Comput.*, 1:34–81, 1980.
[17] A. Brenner, E. Bänsch, and M. Bause. A priori error analysis for finite element approximations of the Stokes problem on dynamic meshes. *IMA Journal of Numerical Analysis*, page drt001, 2013.
[18] M. Bukac, S. Canic, R. Glowinski, J. Tambaca, and A. Quaini. Fluid-structure interaction in blood flow capturing non-zero longitudinal structure displacement. *Journal of Computational Physics*, 235:515–541, 2013.
[19] E. Burman and M. Fernández. Stabilization of explicit coupling in fluid-structure interaction involving fluid incompressibility. *Computer Methods in Applied Mechanics and Engineering*, 198:766–784, 2009.
[20] A. Cate, J. Derksen, L. Portella, and H. Van Den Akker. Fully resolved simulations of colliding monodisperse spheres in forced isotropic turbulence. *Journal of Fluid Mechanics*, 519:233–271, 11 2004.
[21] P. Causin, J. Gerbeau, and F. Nobile. Added-mass effect in the design of partitioned algorithms for fluid-structure problems. *Comput. Methods Appl. Mech. Engrg*, 194(42-44):4506–4527, 2005.
[22] R. Codina, G. Houzeaux, H. Coppola-Owen, and J. Baiges. The fixed-mesh ALE approach for the numerical approximation of flows in moving domains. *Journal of Computational Physics*, 228(5):1591–1611, 2009.
[23] M. Coquerelle and G.-H. Cottet. A vortex level set method for the two-way coupling of an incompressible fluid with colliding rigid bodies. *Journal of Computational Physics*, 227(21):9121–9137, 2008.
[24] G. Cottet, E. Maitre, and T. Milcent. Eulerian formulation and level set models for incompressible fluid-structure interaction. *Esaim. Math. Model. Numer. Anal.*, 42:471–492, 2008.
[25] J. Donea, S. Giuliani, and J. Halleux. An Arbitrary Lagrangian-Eulerian finite element method for transient dynamic fluid-structure interactions. *Computer Methods in Applied Mechanics and Engineering*, 33(1-3):689–723, 1982.
[26] N. D. dos Santos, J.-F. Gerbeau, and J.-F. Bourgat. A partitioned fluid-structure algorithm for elastic thin valves with contact. *Comput. Methods Appl. Mech. Engrg*, 197:1750–1761, 2008.
[27] J. Dukowicz and J. Kodis. Accurate conservative remapping (rezoning) for Arbitrary Lagrangian-Eulerian computations. *SIAM Journal on Scientific and Statistical Computing*, 8(3):305–321, 1987.
[28] T. Dunne. An Eulerian approach to fluid–structure interaction and goal-oriented mesh adaptation. *International journal for numerical methods in fluids*, 51(9-10):1017–1039, 2006.
[29] H. Fang, Z. Wang, Z. Lin, and M. Liu. Lattice Boltzmann method for simulating the viscous flow in large distensible blood vessels. *Phys. Rev. E.*, 65(5):051925–1–051925–12, 2002.
[30] C. Farhat, M. Lesoinne, and P. L. Tallec. Load and motion transfer algorithms for fluid/structure interaction problems with non-matching discrete interfaces: Momentum and energy conservation, optimal discretization and application to aeroelasticity. *Computer Methods in Applied Mechanics and Engineering*, 157(1–2):95–114, 1998.
[31] Z. Feng and E. Michaelides. The immersed boundary-lattice Boltzmann method for solving fluid-particles interaction problems. *J. Comp. Phys.*, 195(2):602–628, 2004.
[32] L. Formaggia, A. Quarteroni, and A. Veneziani. *Cardiovascular Mathematics*, volume 1 of *Modeling, Simulation and Applications*. Springer, 2009.

[33] M. Fortin and R. Glowinski. *Augmented Lagrangian Methods: Application to the Numerical Solution of Boundary Value Problem*. North-Holland, Amsterdam, 1983.
[34] T.-P. Fries and T. Belytschko. The extended/generalized finite element method: an overview of the method and its applications. *International Journal for Numerical Methods in Engineering*, 84(3):253–304, 2010.
[35] E. S. Gawlik, H. Kabaria, and A. J. Lew. High-order methods for low Reynolds number flows around moving obstacles based on universal meshes. *International Journal for Numerical Methods in Engineering*, 2015.
[36] E. S. Gawlik and A. J. Lew. High-order finite element methods for moving boundary problems with prescribed boundary evolution. *Computer Methods in Applied Mechanics and Engineering*, 278:314–346, 2014.
[37] F. Gibou and C. Min. Efficient symmetric positive definite second-order accurate monolithic solver for fluid/solid interactions. *Journal of Computational Physics*, 231(8):3246–3263, 2012.
[38] R. Glowinski and M. Holmstrom. Constrained motion problems with applications by nonlinear programming methods. *Surv. Math. Ind.*, 5:75–108, 1995.
[39] R. Glowinski, T.-W. Pan, T. Hesla, and D. Joseph. A distributed Lagrange multiplier/fictitious domain method for particulate flows. *International Journal of Multiphase Flow*, 25(5):755–794, 1999.
[40] R. Glowinski, T. Pan, and J. Périaux. A fictitious domain method for external incompressible viscous flow modelled by Navier-Stokes equations. *Methods Appl. Mech. Engrg.*, 111:133–148, 1994.
[41] R. Glowinski and P. L. Tallec. *Augmented Lagrangian and Operator-Splitting Methods in Nonlinear Mechanics*. SIAM, Philadelphia, 1988.
[42] G. Guidoboni, R. Glowinski, N. Cavallini, and S. Canic. Stable loosely-coupled-type algorithm for fluid-structure interaction in blood flow. *Journal of Computational Physics*, 228(18):6916–6937, 2009.
[43] J. D. Hart, F. Baaijens, G. Peters, and P. Schreurs. A computational fluid-structure interaction analysis of a fiber-reinforced stentless aortic valve. *Journal of Biomechanics*, 36(5):699–712, 2003.
[44] J. D. Hart, G. Peters, P. Schreurs, and F. Baaijens. A three-dimensional computational analysis of fluid-structure interaction in the aortic valve. *Journal of Biomechanics*, 36(1):103–112, 2003.
[45] C. Hirt, A. Amsden, and J. Cook. An arbitrary lagrangian-eulerian computing method for all flow speeds. *Journal of Computational Physics*, 14(3):227–253, 1974.
[46] H. Hu, N. Patankar, and M. Zhu. Direct numerical simulations of fluid-solid systems using the Arbitrary Lagrangian-Eulerian technique. *Journal of Computational Physics*, 169(2):427–462, 2001.
[47] T. J. R. Hughes, W. Liu, and T. K. Zimmermann. Lagrangian-Eulerian finite element formulation for incompressible viscous flows. *Computer Methods in Applied Mechanics and Engineering*, 29(3):329–349, 1981.
[48] M. Krafczyk, M. Cerrolaza, M. Schulz, and E. Rank. Analysis of 3D transient blood flow passing through an artificial aortic valve by Lattice-Boltzmann methods. *J. Biomech.*, 31(5):453–462, 1998.
[49] U. Küttler and W. Wall. Fixed-point fluid-structure interaction solvers with dynamic relaxation. *Computational Mechanics*, 43(1):61–72, 2008.
[50] A. Kuzmin, M. Luisier, and O. Schenk. Fast methods for computing selected elements of the Greens function in massively parallel nanoelectronic device simulations. In F. Wolf, B. Mohr, and D. Mey, editors, *Euro-Par 2013 Parallel Processing*, volume 8097 of *Lecture Notes in Computer Science*, pages 533–544. Springer Berlin Heidelberg, 2013.

[51] A. Linke, C. Merdon, and W. Wollner. Optimal L2 velocity error estimate for a modified pressure-robust Crouzeix–Raviart Stokes element. *IMA Journal of Numerical Analysis*, 2016.
[52] R. Loubere, P.-H. Maire, M. Shashkov, J. Breil, and S. Galera. Reale: A reconnection-based arbitrary-Lagrangian–Eulerian method. *Journal of Computational Physics*, 229(12):4724–4761, 2010.
[53] L. Margolin and M. Shashkov. Second-order sign-preserving conservative interpolation (remapping) on general grids. *Journal of Computational Physics*, 184(1):266–298, 2003.
[54] D. Mok and W. Wall. Partitioned analysis schemes for transient interaction of incompressible flows and nonlinear flexible structures. In *Trends in computational structural mechanics (W.A. Wall, K.U. Bletzinger and K. Schweizerhof, Eds.)*, CIMNE, Barcelona, Spain, 2001.
[55] F. Nobile. *Numerical Approximation of Fluid-Structure Interaction Problems with Application to Haemodynamics*. PhD thesis, École Polytechnique Fédérale de Lausanne, 2001.
[56] F. Nobile and C. Vergara. An effective fluid-structure interaction formulation for vascular dynamics by generalized Robin conditions. *SIAM J. Sci. Comp.*, 30(2):731–763, 2008.
[57] W. Noh. CEL: A time-dependent, two-space-dimensional, coupled Eulerian-Lagrange code. Technical report, Lawrence Radiation Lab., Univ. of California, Livermore, 1963.
[58] C. S. Peskin. Numerical analysis of blood flow in the heart. *Journal of Computational Physics*, 25(3):220–252, 1977.
[59] C. S. Peskin. The immersed boundary method. *Acta Numer.*, 11:479–517, 2002.
[60] C. Peskin and D. McQueen. A three-dimensional computational method for blood flow in the heart i. Immersed elastic fibers in a viscous incompressible fluid. *Journal of Computational Physics*, 81(2):372–405, 1989.
[61] R. Prignitz and E. Bänsch. Particulate flows with the subspace projection method. *Journal of Computational Physics*, 260(0):249–272, 2014.
[62] A. Quarteroni and A. Valli. *Domain Decomposition Methods for Partial Differential Equations*. Oxford Science Publications, 1999.
[63] R. Rangarajan and A. J. Lew. Analysis of a method to parameterize planar curves immersed in triangulations. *SIAM Journal on Numerical Analysis*, 51(3):1392–1420, 2013.
[64] T. Richter and T. Wick. Finite elements for fluid–structure interaction in ALE and fully Eulerian coordinates. *Computer Methods in Applied Mechanics and Engineering*, 199(41):2633–2642, 2010.
[65] M. Rumpf. A variational approach to optimal meshes. *Numerische Mathematik*, 72:523–540, 1996.
[66] O. Schenk, M. Bollhöfer, and R. A. Römer. On large-scale diagonalization techniques for the Anderson model of localization. *SIAM Rev.*, 50(1):91–112, February 2008.
[67] O. Schenk, A. Wächter, and M. Hagemann. Matching-based preprocessing algorithms to the solution of saddle-point problems in large-scale nonconvex interior-point optimization. *Computational Optimization and Applications*, 36(2-3):321–341, 2007.
[68] P. L. Tallec and J. Mouro. Fluid-structure interaction with large structural displacements. *Comput. Methods Appl. Mech. Engrg*, 190:3039–3067, 2001.
[69] R. van Loon, P. D. Anderson, J. de Hart, and F. P. T. Baaijens. A combined fictitious domain/adaptive meshing method for fluid-structure interaction in heart valves. *International Journal for Numerical Methods in Fluids*, 46(5):533–544, 2004.
[70] R. van Loon, P. Anderson, F. Baaijens, and F. van de Vosse. A three-dimensional fluid-structure interaction method for heart valve modelling. *Comptes Rendus Mécanique*, 333(12):856–866, 2005.
[71] X. Wang and W. Liu. Extended immersed boundary method using FEM and RKPM. *Methods Appl. Mech. Engrg.*, 193:1305–1321, 2004.

[72] M. Weismann. The hybrid level-set front-tracking approach. Master's thesis, Friedrich-Alexander University Erlangen-Nuremberg, 2012.

[73] T. Wick. Flapping and contact FSI computations with the fluid–solid interface-tracking/interface-capturing technique and mesh adaptivity. *Computational Mechanics*, 53(1):29–43, 2014.

[74] L. Zhang, A. Gerstenberger, X. Wang, and W. Liu. Immersed finite element method. *Methods Appl. Mech. Engrg.*, 193:2051–2067, 2004.

Johan Hoffman, Bärbel Holm, and Thomas Richter

2 The locally adapted parametric finite element method for interface problems on triangular meshes

Abstract: We present a locally adapted parametric finite element method for interface problems. For this adapted finite element method we show optimal convergence for elliptic interface problems with a discontinuous diffusion parameter. The method is based on the adaption of macroelements where a local basis represents the interface. The macroelements are independent of the interface and can be cut by the interface. A macroelement which is a triangle in the triangulation is divided into four subtriangles. On these subtriangles, the basis functions of the macroelement are interpreted as linear functions. The position of the vertices of these subtriangles is determined by the location of the interface in the case where a macroelement is cut by the interface. Quadrature is performed on the subtriangles via transformations to a reference element. Because of the locality of the method, its use is well suited for distributed architectures.

Keywords: Locally adapted finite element method, triangular meshes, interface problem, optimal a priori error estimates

Classification: 65N12, 65N15, 65N30, 74A50

2.1 Introduction

In this paper we assume the domain $\Omega \subset \mathbb{R}^2$ to be partitioned into two nonoverlapping parts $\Omega = \Omega_1 \cup \Gamma \cup \Omega_2$, and with the intersection $\Gamma := \partial\Omega_1 \cap \partial\Omega_2$ we denote the interface between Ω_1 and Ω_2. We consider the problem

$$-\nabla \cdot (\kappa_i \nabla u) = f \text{ in } \Omega_i \subset \mathbb{R}^2,\ i = 1, 2\,, \quad [u] = 0\,, \quad [\kappa \partial_n u] = 0 \text{ on } \Gamma\,, \tag{2.1}$$

Note: The first author gratefully acknowledges the support of the Swedish Energy Agency, project number P40435-1.

Johan Hoffman, Department of Computational Science and Technology, School of Computer Science and Communication, KTH Royal Institute of Technology, Sweden, jhoffman@kth.se
Bärbel Holm, Department of Computational Science and Technology, School of Computer Science and Communication, KTH Royal Institute of Technology, Sweden, barbel@kth.se
Thomas Richter, Universität Magdeburg, Fakultät für Mathematik, Institut für Analysis und Numerik, Universitätsplatz 2, 39106 Magdeburg, Germany, thomas.richter@ovgu.de

https://doi.org/10.1515/9783110494259-002

with positive diffusion parameters $\kappa_i > 0$ defined on Ω_i, $i = 1, 2$, where

$$[u](x) := \lim_{s\downarrow 0} u(x + sn) - \lim_{s\uparrow 0} u(x + sn) , \quad x \in \Gamma$$

denotes the jump at the interface, with n a normal vector of Γ. For a domain $\Omega \in \mathbb{R}^2$ we denote by $H^k(\Omega)$ the Sobolev space of integer order $k \geq 0$ with norm $\|\cdot\|_{H^k(\Omega)}$ and seminorm $|\cdot|_{H^k(\Omega)}$ involving only the highest derivatives. Throughout the paper, we shall use the notation of the inner product on $L^2(\Omega) = H^0(\Omega)$ given by

$$(u, \varphi)_{L^2(\Omega)} = (u, \varphi)_\Omega = (u, \varphi) = \int_\Omega u\varphi \mathrm{d}x ,$$

and the norm

$$\|u\|_{L^2(\Omega)} = \|u\|_\Omega = \|u\| = \left(\int_\Omega |u|^2 \mathrm{d}x \right)^{1/2} ,$$

induced by this inner product. We assume that both subdomains Ω_1 and Ω_2 have a boundary with sufficient regularity such that for smooth right hand sides, the solution has the regularity

$$u \in H_0^1(\Omega) \cap H^{r+1}(\Omega_1 \cup \Omega_2) ,$$

for a given $r \in \mathbb{N}$, see [2]. The variational formulation of this interface problem reads

$$\text{Find } u \in H_0^1(\Omega): \quad a(u, \varphi) := \sum_{i=1}^{2} (\kappa_i \nabla u, \nabla \varphi)_{\Omega_i} = (f, \varphi)_\Omega \quad \text{for all } \varphi \in H_0^1(\Omega) . \tag{2.2}$$

By standard arguments, the existence of solutions follows. The error between an analytical solution u and an approximation u_h by a standard finite element method with linear or higher order basis functions which do not conform to the interface will be bounded by

$$\|\nabla(u - u_h)\|_\Omega = O(h^{1/2}) ,$$

see [2, 12], and Figure 2.1, which shows the error using standard finite elements for the numerical example presented in Section 2.5.1.

To recover the optimal order of convergence for interface problems various techniques have been proposed, including so-called unfitted finite element methods that locally modify or enrich the finite element basis. Examples are the extended finite element method (XFEM) [13], the generalized finite element method [3], and the unfitted Nitsche method [10, 11]. These methods locally modify the finite element basis. Thus, the connectivity of the system matrix is changed and degrees of freedom are added or removed. In view of distributed parallel algorithms, this would lead to costly load balancing. Recent work also includes cut finite element methods (CutFEM) [5, 6]. This approach uses Nitsche's method and stabilization of the finite element method on facets close to the interface. In [8], a locally adapted (patch) finite element method is proposed for quadrilateral elements, where the mesh is locally adapted to align

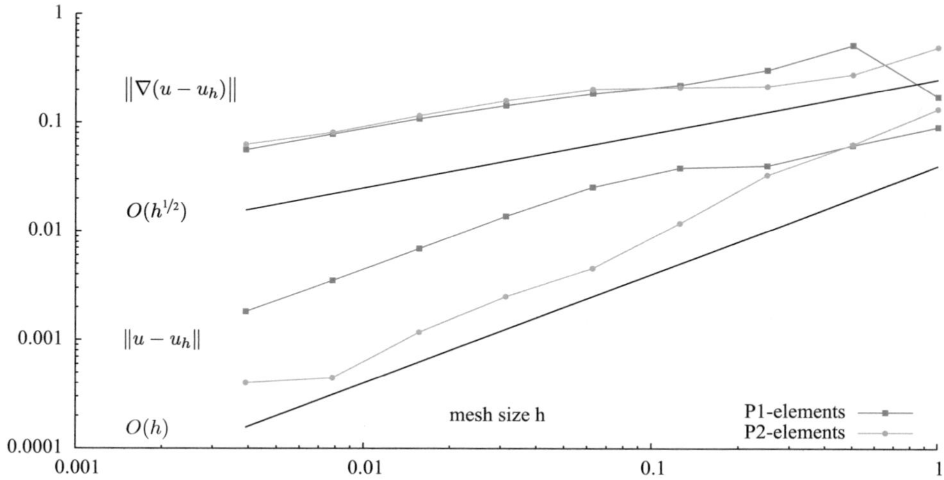

Fig. 2.1: Convergence for standard finite element methods for interface problem (2.1). Solid lines indicate a slope of $h^{1/2}$ and h, respectively.

with the interface. A similar method to the one we present here is described in [9]. In contrast to their approach, we use rectangular triangles as reference triangles and thus, get a slightly different bound for the maximum angle condition. In this paper, we describe the locally adapted finite element method for interface problems and prove a priori error estimates which are verified in numerical experiments.

The outline of this paper is as follows. In Section 2.2, we introduce the method and describe the finite element spaces on cut cells. In order to prove optimal a priori error estimates for the locally adapted finite element patch method, we show in Section 2.3 that for a certain choice of free parameters in a cell cut by the interface, a maximum angle condition is satisfied. In Section 2.4, we prove optimal a priori error bounds for the locally adapted patch finite element method for interface problems, followed by numerical examples in Section 2.5. We conclude this paper with Section 2.6.

2.2 The locally adapted patch finite element method

Let $\mathcal{T}_h$ be a shape-regular triangulation of a domain $\Omega \subset \mathbb{R}^2$ into triangles. Since we do not require that the triangulation is aligned with the interface, triangles $T \in \mathcal{T}_h$ can be cut by the interface. On these triangles we get contributions from both subproblems. To integrate those contributions we propose a locally adapted finite element method on patches of subtriangles. Note that the triangulation does not necessarily coincide with the subdivision of the domain $\Omega = \Omega_1 \cup \Gamma \cup \Omega_2$. We assume that the triangulation has a patch structure such that each triangle $T \in \mathcal{T}_h$ is divided into four smaller subtriangles T_0, T_1, T_2, T_3, see Figures 2.2a and 2.3a for the two different reference

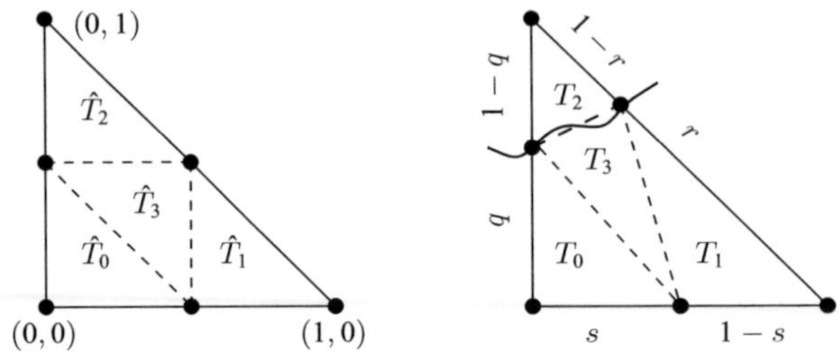

(a) Reference configuration A. (b) Cut through two edges.

Fig. 2.2: Cut through two edges, reference configuration (a) and actual, adapted configuration (b), interface shown by solid curved line, linear approximation of the interface shown by dashed thick line.

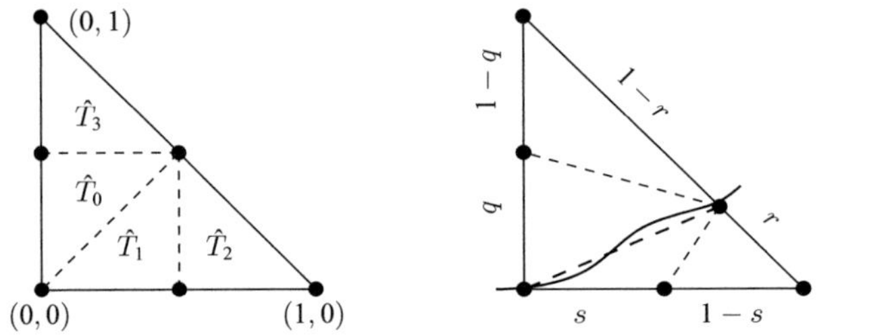

(a) Reference configuration B. (b) Cut through a vertex and an edge.

Fig. 2.3: Cut through lower left vertex and the opposite edge, reference configuration (a) and actual, adapted configuration (b), interface in solid curved line, linear approximation of the interface in dashed thick line.

configurations of a triangle $T \in \mathcal{T}_h$. By a linear transformation, the vertices close to the cut by the interface are mapped to the exact location of the cut. We will now construct a finite element method for a mesh $\mathcal{T}_h$ of locally transformed triangles T cut by the interface.

2.2.1 Definition of the finite element space on cut triangles

Before defining the finite element space on cut triangles we define when we consider a patch triangle to be cut. We allow two possible configurations which are:
1. Each (open) patch $T \in \mathcal{T}_h$ is not cut, such that $T \cap \Gamma = \emptyset$ holds or
2. a patch $T \in \mathcal{T}_h$ is cut in exactly two points on its boundary such that $T \cap \Gamma \neq \emptyset$ and $\partial T \cap \Gamma = \{x_1^T, x_2^T\}$.

If the interface cuts through two vertices of a patch, we do not consider the patch cut. We restrict our method such that if a patch is cut, the two points $\{x_1^T, x_2^T\}$ may not be inner points of the same edge. That means we do not allow a patch to be cut multiple times and the interface may not enter and leave the patch at the same edge. Using refinement of the underlying mesh, these restrictions can be avoided. The finite element space $V_h \subset H_0^1(\Omega)$ is defined as an isoparametric space on the triangulation $\mathcal{T}_h$ given as

$$V_h := \left\{\varphi \in C(\bar{\Omega}),\ \varphi \circ F_{T_i}^{-1}\Big|_{T_i} \in \hat{P} \text{ for } i = 0, \dots, 3\,, \text{ and all patches } T \in \mathcal{T}_h\right\},$$

where F_{T_i} is the mapping between the reference patch $\hat{T}$ and every patch $T \in \mathcal{T}_h$ such that

$$F_{T_i}(\hat{x}_k) = x_k, \quad k = 1, \dots, 6\,, \quad i = 0, \dots, 3\,,$$

for the six nodes $x_1^T, \dots, x_6^T$ in every patch. We choose the reference space $\hat{P}$ of polynomials as the standard space of piecewise linear functions which is given as

$$\hat{P} = \left\{\varphi \in C(\bar{T}),\ \varphi\Big|_{T_i} \in \text{span}\{1, x, y\},\ T_i \in T, i = 0, \dots, 3\right\}.$$

With $\{\hat{\varphi}_1, \dots, \hat{\varphi}_6\}$ we denote the standard Lagrange basis of $\hat{P}$ for which $\hat{\varphi}_i(\hat{x}_j^T) = \delta_{ij}$ holds. Accordingly, the transformation F_{T_i} is given by

$$F_{T_i}(\hat{x}) = \sum_{k=1}^{6} x_k^T \hat{\varphi}_k(\hat{x})\,, \quad i = 0, \dots, 3\,.$$

In the following, we will describe the transformation in more detail.

2.2.2 Quadrature on cut triangles

To define the quadrature rules on the cells that are cut by the interface, we introduce the four reference configurations that we use.

Configuration A The cell is cut through exactly two edges, see Figure 2.2 for both, the reference and an actual configuration.

Configuration B The cell is cut through the lower left vertex and the opposite edge, see Figure 2.3 for both, the reference and an actual configuration.

Configuration C The cell is cut through the right vertex and the opposite edge, see Figure 2.4a for the reference configuration.

Configuration D The cell is cut through the upper left vertex and the opposite edge, see Figure 2.4b for the reference configuration.

Since the finite element method is defined using standard polynomial basis functions, we also use standard quadrature rules on the triangles $T_0, \dots, T_3 \in T \in \mathcal{T}_h$. For the quadrature on a patch $T \in \mathcal{T}_h$, the quadrature rule is composed by a combination of standard rules on all triangles $T_0, \dots, T_3$, as is sketched in Figure 2.5.

We start from a quadrature rule for triangles on the reference triangle (Figure 2.5 on the left). This can be any quadrature rule suitable for the integration of linear polynomials on the reference triangle. Then, for each triangle $\hat{T}_i$, the quadrature points

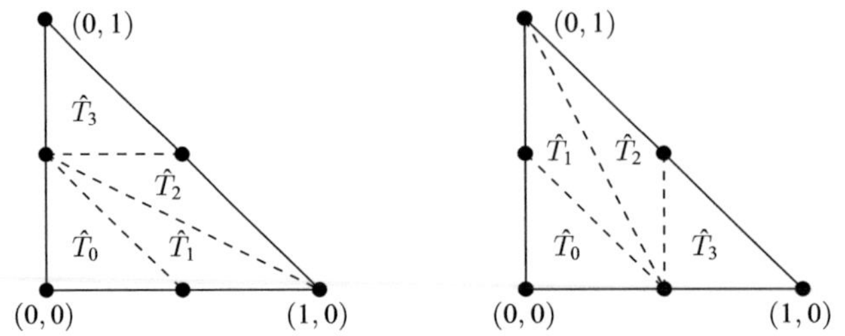

(a) Reference configuration C. (b) Reference configuration D.

Fig. 2.4: Reference configuration for the cut of the right vertex and the opposite edge (a) and for the cut of the upper left vertex and the opposite edge (b).

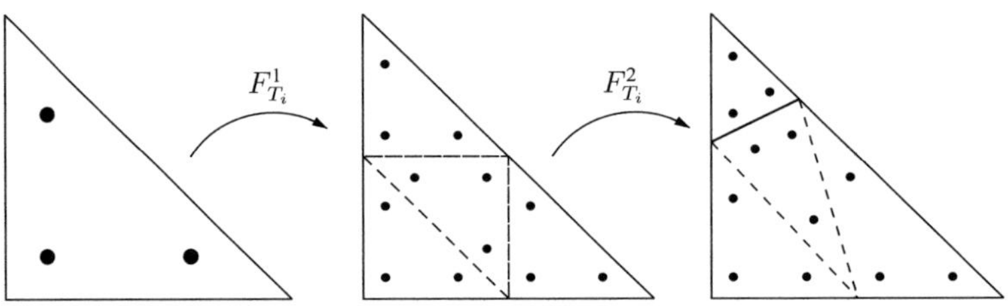

Fig. 2.5: Quadrature rule on a patch T as a composition of quadrature rules on each triangle $T_0, \dots, T_3$.

are mapped via the transformation $F^1_{T_i}$ to each of the subtriangles of one of the reference configurations A–D. With $F^2_{T_i}$, these quadrature points are then mapped to their location in real coordinates. Thus, the transformation F_{T_i} can be decomposed into

$$F_{T_i} = F^2_{T_i} \circ F^1_{T_i} \,.$$

The quadrature weights are scaled appropriately.

Example 2.1. Let us assume that the triangle with vertices $v_1 = (0, 0)$, $v_2 = (1, 0)$, and $v_3 = (0, 1)$ is cut with the parameters $q = 9/16$, $s = 1/2$, and $r = 11/16$ and further let the basis quadrature rule for integrating linear polynomials have the quadrature points

$$\hat{q}_1 = (2/3, 1/6)\,, \quad \hat{q}_2 = (1/6, 1/6), \quad \hat{q}_3 = (1/6, 2/3)\,,$$

and the weights $\hat{\omega}_i = 1/6$ for $i = 1, 2, 3$. Then, the corresponding quadrature points for a triangle based on reference configuration A are given as

$$\begin{aligned}
F_{T_0}\hat{q}_1 &= (1/3, 3/32)\,, & F_{T_0}\hat{q}_2 &= (1/12, 3/32)\,, & F_{T_0}\hat{q}_3 &= (1/12, 3/8)\,,\\
F_{T_1}\hat{q}_1 &= (77/96, 11/96)\,, & F_{T_1}\hat{q}_2 &= (53/96, 11/96), & F_{T_1}\hat{q}_3 &= (11/24, 11/24)\,,\\
F_{T_2}\hat{q}_1 &= (5/24, 23/32)\,, & F_{T_2}\hat{q}_2 &= (5/96, 21/32), & F_{T_2}\hat{q}_3 &= (5/96, 7/8)\,,\\
F_{T_3}\hat{q}_1 &= (13/96, 47/96)\,, & F_{T_3}\hat{q}_2 &= (7/24, 53/96), & F_{T_3}\hat{q}_3 &= (37/96, 5/24)\,,
\end{aligned}$$

and the weights are scaled such that

$$\omega_i = 1/24\,, \quad i = 1, 2, 3\,, \quad \text{for all } T_0, \dots, T_3\,.$$

2.2.3 Discrete variational formulation

With the definition of the discrete finite element space, we are ready to formulate the discrete counterpart of problem (2.2) as

$$\text{Find } u_h \in V_h: \quad a_h(u_h, \varphi_h) := \sum_{i=1}^{2} (\kappa_i \nabla u_h, \nabla \varphi_h)_{\mathcal{T}_{i,h}} = (f, \varphi_h)_\Omega \quad \text{for all } \varphi_h \in V_h\,. \tag{2.3}$$

Note that we do not have the standard Galerkin orthogonality property due to the fact that the value of κ differs in a small layer between the continuous interface Γ and the linear approximation Γ_h, as is shown in Figure 2.6.

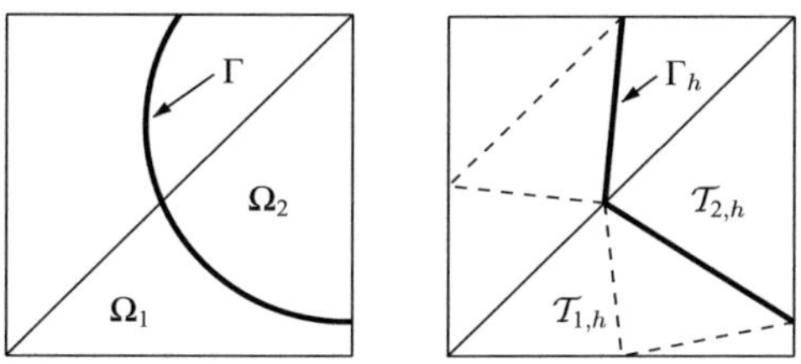

Fig. 2.6: Splitting into subdomains and sub-triangulations

2.3 Maximum angle condition

In order to prove the optimal order of convergence for the locally adapted patch finite element method, we need the Lagrangian interpolation operator

$$L_h\colon H^2(T) \cap C(\bar{T}) \to V_h$$

to satisfy

$$\|\nabla^k (v - L_h v)\|_T \le c h_{T,\max}^{2-k} \|\nabla^2 v\|_T\,, \tag{2.4}$$

where $c > 0$ is a constant and $h_{T,\max}$ is the maximum diameter of a triangle T. In [1] it is shown that a necessary condition for the above estimate to hold is the maximum angle condition.

Definition 2.2 (Maximum angle condition). There is a constant $\gamma_* < \pi$, independent of h and $T_i \in \mathcal{T}_h$ such that the maximal interior angle γ of any element T_i is bounded by γ_*:

$$\gamma \le \gamma_* < \pi\,.$$

In contrast to the corresponding finite element method on quadrilateral cells where the maximal angle condition is fulfilled by construction, see [8], the version of this method on triangles lacks this property.

There are only two possible configurations of how the interface can cut a triangle; either the interface cuts two edges (Figure 2.2b or the interface cuts one vertex and one edge, see Figure 2.3b. For each of these configurations, there are cases in which the maximum angle condition is not fulfilled as shown in Figure 2.7. Since the cut of the interface through a cell determines one of the three parameters s, r, and q in the case that the interface goes through a vertex and two parameters if the cut goes through two edges of a cell, it leaves at least one parameter per cell free to choose. In the following we discuss different possibilities to choose the free parameters and we show that by those choices the maximum angle condition will be satisfied. First, we treat the case in which only edges are cut. Later, the case of a cut through a vertex and an edge is discussed. For the estimation of the angles, we use the notation introduced in Figure 2.8.

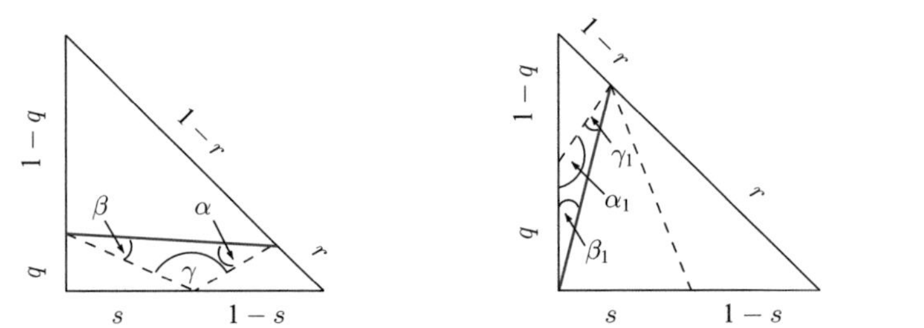

(a) Two edges are cut and if $q \to 0$ and $r \to 0$ then $\gamma \to \pi$.

(b) A vertex and an edge are cut and if $r \to 1$ then $\alpha_1 \to \pi$.

Fig. 2.7: Illustration of two cases in which one of the inner triangles becomes anisotropic and the maximum angle condition does not hold.

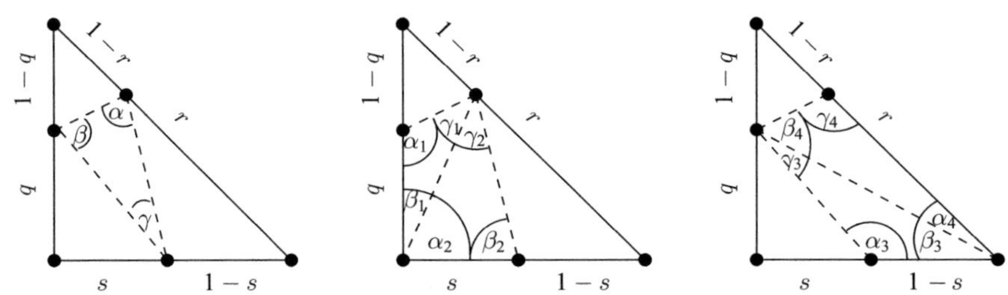

(a) Cut through two edges

(b) Cut through lower left vertex and an edge

(c) Cut through lower right vertex and an edge

Fig. 2.8: Angles to be estimated.

2.3.1 Two edges are cut

Extreme shapes of the inner triangles $T_0, \dots, T_3$ can arise. Their definition can be seen in Figure 2.2b. The discussion of how to choose the free parameters is divided into two different parts. We start with treating the case in which the interface comes arbitrarily close to an edge so that an angle will approach the value π.

The interface cuts arbitrarily close to an edge

This situation is shown in Figure 2.7a. There are three situations in which this could happen, namely if

1. $q \to 1$ and $s \to 1$, or if
2. $s \to 0$ and $r \to 1$, or if
3. $q \to 0$ and $r \to 0$.

We use that for the angle ϑ between two vectors a and b it holds

$$\cos(\vartheta) = \frac{(a, b)}{\|a\|\|b\|},$$

where $(\cdot, \cdot)$ and $\|\cdot\|$ are the Euclidean scalar product and norm, respectively. Thus, for the angles α, β and γ in Figure 2.8b it holds

$$\cos(\alpha) = \frac{(1-r)(1-r-s) + r(r-q)}{\sqrt{(q-r)^2 + (r-1)^2}\sqrt{(1-r-s)^2 + r^2}},$$
$$\cos(\beta) = \frac{s(1-r) + q(q-r)}{\sqrt{(q-r)^2 + (r-1)^2}\sqrt{s^2 + q^2}},$$
$$\cos(\gamma) = \frac{s(s-1+r) + rq}{\sqrt{(1-r-s)^2 + r^2}\sqrt{s^2 + q^2}}.$$

If $q \to 1$ and $s \to 1$ it follows that $\cos(\alpha) \to \pi$. Therefore, for $q > 1/2$ and $s > 1/2$, we recommend one of the following choices.

Choices for r

$$r := 1 - s, \qquad r := q, \qquad r := (1-s)(1-q). \tag{2.5}$$

Independent of the choice above, we can estimate such that it holds

$$1 \geq \cos(\alpha) \geq \frac{-1}{\sqrt{2}}, \qquad 1 \geq \cos(\beta) \geq \frac{1}{\sqrt{5}}, \qquad 1 \geq \cos(\gamma) \geq \frac{-1}{\sqrt{2}},$$

and therefore for all angles in these cases it follows that $\alpha, \beta, \gamma \in (0°, 135°)$ for $q, s \in (1/2, 1)$. The other two cases are treated similarly. For $s < 1/2$ and $r > 1/2$ we suggest to set the values for q as follows.

Choices for q

$$q := s\,, \qquad q := r\,, \qquad q := (1-r)s\,. \tag{2.6}$$

This means that independent of the choice above, we have that

$$1 \geq \cos(\alpha) \geq \frac{-1}{\sqrt{5}}\,, \qquad 1 \geq \cos(\beta) \geq \frac{-1}{\sqrt{2}}\,, \qquad 1 \geq \cos(\gamma) \geq \frac{-1}{\sqrt{10}}\,,$$

and therefore for all angles in these cases it holds $\alpha, \beta, \gamma \in (0°, 135°)$ for $r \in (1/2, 1)$ and $s \in (0, 1/2)$. The last case in which the interface can come arbitrarily close to an edge is when $q \to 0$ and $r \to 0$. In that case, for $q < 1/2$ and $r < 1/2$, we suggest one of the following choices.

Choices for s

$$s := 1-r\,, \qquad s := q\,, \qquad s := qr\,. \tag{2.7}$$

For symmetry reasons, we find that independent of the choice above, it follows that

$$1 \geq \cos(\alpha) \geq \frac{-1}{\sqrt{5}}\,, \qquad 1 \geq \cos(\beta) \geq \frac{-1}{\sqrt{10}}\,, \qquad 1 \geq \cos(\gamma) \geq \frac{-1}{\sqrt{2}}\,.$$

Thus, for all the angles in these cases it holds $\alpha, \beta, \gamma \in (0°, 135°)$ for q and $r \in (0, 1/2)$.

The remaining cases in which two edges are cut

In all the remaining cases in which two edges are cut, we choose the parameter which is not determined by the cut of the interface as $1/2$. For these cases it follows that

$$1 \geq \cos(\alpha) \geq \frac{-3}{\sqrt{10}}\,, \qquad 1 \geq \cos(\beta) \geq \frac{-2}{\sqrt{5}}\,, \qquad 1 \geq \cos(\gamma) \geq \frac{-2}{\sqrt{5}}\,,$$

and therefore it holds for the angles in all the cases

$$\alpha, \beta, \gamma \in (0°, 162°)\,, \quad \text{for } q, r, s \in (0, 1)\,.$$

2.3.2 A vertex and an edge are cut

In contrast to the case in which two edges are cut, only one value is determined in the case in which a vertex and an edge are cut. In the configuration shown in Figure 2.7b we are free to choose the values for s and q. The value of r is given by the cut of the interface. For symmetry reasons, it suffices to consider the cases depicted in Figure 2.8b and 2.8c to estimate the inner angles. We propose to choose the parameters as follows:

$$s = \begin{cases} 1-r & \text{if } r < 1/2\,, \\ q & \text{if } q < 1/2\,, \\ 1/2 & \text{otherwise}\,, \end{cases} \quad r = \begin{cases} q & \text{if } q > 1/2\,, \\ 1-s & \text{if } s > 1/2\,, \\ 1/2 & \text{otherwise}\,, \end{cases} \quad q = \begin{cases} r & \text{if } r > 1/2\,, \\ s & \text{if } s < 1/2\,, \\ 1/2 & \text{otherwise}\,. \end{cases} \tag{2.8}$$

For the inner angles in T_0 of the configuration in Figure 2.8b it holds

$$\cos(\alpha_1) = \frac{q-r}{\sqrt{(1-r)^2+(r-q)^2}}\,,$$
$$\cos(\beta_1) = \frac{r}{\sqrt{(1-r)^2+r^2}}\,,$$
$$\cos(\gamma_1) = \frac{(1-r)^2+r(r-q)}{\sqrt{(1-r)^2+r^2}\sqrt{(1-r)^2+(q-r)^2}}\,,$$

whereas for the inner angles in T_1, we find

$$\cos(\alpha_2) = \frac{1-r}{\sqrt{(1-r)^2+r^2}}\,,$$
$$\cos(\beta_2) = \frac{s-1+r}{\sqrt{r^2+(s-1+r)^2}}\,,$$
$$\cos(\gamma_2) = \frac{(1-r)(1-r-s)+r^2}{\sqrt{r^2+(1-r)^2}\sqrt{r^2+(1-r-s)^2}}\,.$$

Therefore, we find that

$$\begin{aligned}
&1 \geq \cos(\alpha_1) \geq \frac{-1}{\sqrt{2}} && \Rightarrow \alpha_1 \in (0^\circ, 135^\circ) && \text{for } r, q, s \in (0,1)\,,\\
&1 \geq \cos(\alpha_2) \geq 0 && \Rightarrow \alpha_2 \in (0^\circ, 90^\circ) && \text{for } r, q, s \in (0,1)\,,\\
&1 \geq \cos(\beta_1) \geq 0 && \Rightarrow \beta_1 \in (0^\circ, 90^\circ) && \text{for } r, q, s \in (0,1)\,,\\
&1 \geq \cos(\beta_2) \geq \frac{-1}{\sqrt{2}} && \Rightarrow \beta_2 \in (0^\circ, 135^\circ) && \text{for } r, q, s \in (0,1)\,,\\
&1 \geq \cos(\gamma_1) \geq 0 && \Rightarrow \gamma_1 \in (0^\circ, 90^\circ) && \text{for } r, q, s \in (0,1)\,,\\
&1 \geq \cos(\gamma_2) \geq 0 && \Rightarrow \gamma_2 \in (0^\circ, 90^\circ) && \text{for } r, q, s \in (0,1)\,.
\end{aligned}$$

If the cut goes through the lower right vertex and an edge as shown in Figure 2.8c, we arrive at

$$\begin{aligned}
\cos(\alpha_3) &= \frac{-s}{\sqrt{s^2+q^2}}, & \cos(\alpha_4) &= \frac{1+q}{\sqrt{2}\sqrt{1+q^2}}\,,\\
\cos(\beta_3) &= \frac{1}{\sqrt{1+q^2}}, & \cos(\beta_4) &= \frac{(1-r)+q(r-q)}{\sqrt{1+q^2}\sqrt{(1-r)^2+(r-q)^2}}\,,\\
\cos(\gamma_3) &= \frac{s+q^2}{\sqrt{1+q^2}\sqrt{s^2+q^2}}, & \cos(\gamma_4) &= \frac{2r-1-q}{\sqrt{2}\sqrt{(1-r)^2+(r-q)^2}}
\end{aligned}$$

for the remaining angles to be estimated. For the angles, we find that it holds

$$
\begin{aligned}
1 \geq \cos(\alpha_3) \geq \frac{-1}{\sqrt{2}} \quad &\Rightarrow \quad \alpha_1 \in (0^\circ, 135^\circ) \quad \text{for } r, q, s \in (0,1)\,, \\
1 \geq \cos(\alpha_4) \geq \frac{1}{\sqrt{2}} \quad &\Rightarrow \quad \alpha_2 \in (0^\circ, 45^\circ) \quad \text{for } r, q, s \in (0,1)\,, \\
1 \geq \cos(\beta_3) \geq \frac{1}{\sqrt{5}} \quad &\Rightarrow \quad \beta_3 \in (0^\circ, 64^\circ) \quad \text{for } r, q, s \in (0,1)\,, \\
1 \geq \cos(\beta_4) \geq 0 \quad &\Rightarrow \quad \beta_4 \in (0^\circ, 90^\circ) \quad \text{for } r, q, s \in (0,1)\,, \\
1 \geq \cos(\gamma_3) \geq \frac{1}{\sqrt{2}} \quad &\Rightarrow \quad \gamma_3 \in (0^\circ, 45^\circ) \quad \text{for } r, q, s \in (0,1)\,, \\
1 \geq \cos(\gamma_4) \geq \frac{-3}{\sqrt{10}} \quad &\Rightarrow \quad \gamma_4 \in (0^\circ, 162^\circ) \quad \text{for } r, q, s \in (0,1)\,.
\end{aligned}
$$

For the cut through the upper vertex and the opposite edge, the estimates for the inner angles follow for symmetry reasons. The findings are collected in the following Lemma.

Lemma 2.3. *With the choice of parameters in* (2.5), (2.6), (2.7), *and* (2.8), *all the interior angles in the triangles that can occur through a cut of an interface are bounded by* 162° *independent of* $r, q, s \in (0, 1)$.

With this result we are in the position to analyze the a priori error of this locally adapted finite element patch method.

Remark 2.4. Because of the adjustments of the parameters which are not determined by the location of the interface, we have to ensure continuity across edges. That means that the parameter for the neighboring element across an edge has to be set to the same value.

Remark 2.5. Note that the analysis of the angles also includes the case that two of the parameters r, q and s are equal $1/2$ and the remaining parameter takes any value between 0 and 1. This means that if a parameter is set in order to ensure continuity, we do not have to adjust another parameter in those cells. In particular, we only adjust those cells which are direct neighbors of a cut cell.

The case that two neighboring cells are cut by the interface can be circumvented by refinement of the mesh.

2.4 A priori error analysis

With the maximum angle condition satisfied, we can define a robust Lagrangian interpolation operator $L_h\colon H^2(T) \cap C(\bar{T}) \to V_h$ such that

$$\|\nabla^k(v - L_h v)\|_T \leq c h_{T,\max}^{2-k} \|\nabla^2 v\|_T$$

holds with c a positive constant, see [1]. In order to derive a priori error estimates we have to take into account that the partitioning of the mesh into submeshes $\mathcal{T}_h = \mathcal{T}_{1,h} \cup \mathcal{T}_{2,h}$ connected to κ_1 and κ_2 does not coincide with the partitioning of the domain $\Omega = \Omega_1 \cup \Gamma \cup \Omega_2$ which means that $\mathcal{T}_{i,h}$ does not necessarily cover Ω_i, see Figure 2.9. This would only be possible if the interface Γ is a polygon.

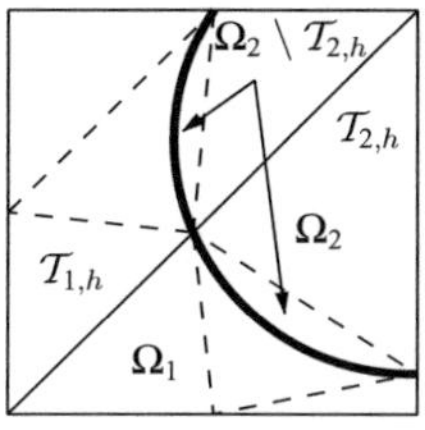

Fig. 2.9: Region between Ω_i and $\mathcal{T}_{i,h}$

Later, we will use the following auxiliary result which is an estimate for functions in the region between Ω_i and $\mathcal{T}_{i,h}$.

Lemma 2.6. *Let $v \in H^1(\Omega)$. For the convex region between Ω_i and $\mathcal{T}_{i,h}$ it holds*

$$\|v\|_{\Omega_i \setminus \mathcal{T}_{i,h}} \le ch\|v\|_{H^1(\Omega)} , \quad i = 1, 2 .$$

For $v_h \in V_h$ it holds

$$\|v_h\|_{H^s(\Omega_i \setminus \mathcal{T}_{i,h})} \le ch^{1/2}\|v_h\|_{H^s(\Omega)} , \quad i = 1, 2, \ s = 0, 1 .$$

Proof. For a cell $T \in \mathcal{T}_h$ and v_h a linear polynomial on T, the inverse inequality

$$\|v_h\|_{\partial T} \le ch^{-1/2}\|v_h\|_T$$

holds, see [7]. Therefore, we can estimate the norm over the discrete interface Γ_h as

$$\|v_h\|_{\Gamma_h} \le ch^{-1/2}\|v_h\|_\Omega .$$

For the convex region between Ω_i and $\mathcal{T}_{i,h}$, the relations

$$\|v\|^2_{\Omega_i \setminus \mathcal{T}_{i,h}} \le c(h^2\|v\|^2_\Gamma + h^4\|\nabla v\|^2_\Omega) ,$$
$$\|v_h\|^2_{\Omega_i \setminus \mathcal{T}_{i,h}} \le c(h^2\|v\|^2_{\Gamma_h} + h^4\|\nabla v_h\|^2_\Omega)$$

are proven in [4]. With the inverse inequality and the discrete trace inequality, the estimate follows for $v_h \in V_h$. The global trace inequality for $v \in H^1(\Omega)$

$$\|v\|_\Gamma \le c\|v\|_{H^1(\Omega)}$$

yields the remaining inequality to be proven. □

We cannot directly apply the interpolation estimate (2.4) to all cells in the triangulation $\mathcal{T}_h$ as the solution u is not smooth enough on cells that are cut by the interface, i.e., for cells $T \in \mathcal{T}_h$ with $T \cap \Gamma \neq \emptyset$. Therefore, we define the set of all elements $T \in \mathcal{T}_h$ that are cut by the interface

$$S_h = \{T \in \mathcal{T}_h \mid T \cap \Gamma \neq \emptyset\} . \tag{2.9}$$

An interpolation estimate on these cells of the triangulation is proven first.

Lemma 2.7. *For the cells $T \in S_h$, the interpolation estimate*

$$\|\nabla(L_h u - u)\|_{S_h} \leq ch\|u\|_{H^2(\Omega_1 \cup \Omega_2)} \tag{2.10}$$

holds with a positive constant c.

Proof. We divide the estimation of the interpolation error into the cells that are effected by the interface and those that are not effected. On those cells not cut by the interface, we use the standard interpolation estimate and extend the domain to the complete domain again:

$$\begin{aligned} \|\nabla(u - L_h u)\|_\Omega^2 &= \|\nabla(u - L_h u)\|_{\Omega \setminus S_h}^2 + \|\nabla(u - L_h u)\|_{S_h}^2 \\ &\leq ch^2\|\nabla^2 u\|_{\Omega_1 \cup \Omega_2}^2 + \|\nabla(u - L_h u)\|_{S_h}^2 . \end{aligned} \tag{2.11}$$

For the last term we introduce a continuous extension that allows us to use the interpolation estimate. Let $\tilde{u}_i \in H^2(\Omega)$ be a continuous extension of $u \in H^2(\Omega_i)$ to the complete domain Ω. For this extension it holds

$$\|\tilde{u}_i - u\|_{H^2(\Omega_i)} = 0 , \quad \|\tilde{u}_i\|_{H^2(\Omega)} \leq c\|u\|_{H^2(\Omega_i)} , \quad i = 1, 2 , \tag{2.12}$$

if the interface Γ is smooth enough, see [14]. For the remaining term we add and subtract the continuous extension $\tilde{u}$ and derive

$$\begin{aligned} \|\nabla(u - L_h u)\|_{S_h} &\leq \|\nabla(u - \tilde{u})\|_{S_h} + \|\nabla(\tilde{u} - L_h u)\|_{S_h} , \\ &= \|\nabla(u - \tilde{u})\|_{S_h} + \|\nabla(\tilde{u} - L_h \tilde{u})\|_{S_h} \end{aligned} \tag{2.13}$$

as for the nodal interpolant on S_h it holds $L_h u = L_h \tilde{u}$. The continuous extension $\tilde{u}$ has enough regularity to apply (2.4) which gives us

$$\|\nabla(\tilde{u} - L_h \tilde{u})\|_{S_h} \leq ch\|\nabla^2 \tilde{u}\|_{S_h} \leq ch\|\nabla^2 \tilde{u}_i\|_\Omega \leq ch\|u\|_{H^2(\Omega_1 \cup \Omega_2)} . \tag{2.14}$$

Here, we enlarged the domain from S_i to Ω and used the continuity of the extension (2.12). For the first term in (2.13), we have

$$\|\nabla(u - \tilde{u})\|_{S_h} \leq \|\tilde{u}_i - u\|_{H^2(\Omega)} = 0 ,$$

which completes the proof. □

Theorem 2.8. *Let $\Omega \in \mathbb{R}^2$ be a domain with a convex polygonal boundary. We assume that the interface Γ admits a C^2-parameterization and that it splits the domain into $\Omega = \Omega_1 \cup \Gamma \cup \Omega_2$ such that the solution $u \in H_0^1(\Omega)$ satisfies a stability estimate*

$$u \in H_0^1(\Omega) \cap H^2(\Omega_1 \cup \Omega_2)\,, \quad \|u\|_{H^2(\Omega_1 \cup \Omega_2)} \leq c_S \|f\|\,.$$

Then the estimate for the adapted finite element solution $u_h \in V_h$

$$\|\nabla(u - u_h)\|_\Omega \leq Ch\|f\|\,, \quad \|u - u_h\|_\Omega \leq Ch^2\|f\|$$

holds.

Proof. 1. We prove the first inequality $\|\nabla(u - u_h)\| \leq Ch\|f\|$:
For the error $e_h = u - u_h$ and for all $\varphi_h \in V_h$ it holds

$$\begin{aligned}(\kappa\nabla e_h, \nabla\varphi_h)_\Omega &= \sum_{i=1}^{2}(\kappa_i\nabla e_h, \nabla\varphi_h)_{\Omega_i}\\ &= \sum_{i=1}^{2}\{(\kappa_i\nabla u, \nabla\varphi_h)_{\Omega_i} - (\kappa_i\nabla u_h, \nabla\varphi_h)_{\Omega_i}\}\,.\end{aligned}$$

Using $\Omega_i = (\mathcal{T}_{i,h} \setminus (\mathcal{T}_{i,h} \setminus \Omega_i)) \cup (\Omega_i \setminus \mathcal{T}_{i,h})$ and the relations

$$\Omega_1 \setminus \mathcal{T}_{1,h} = \mathcal{T}_{2,h} \setminus \Omega_2\,, \qquad \Omega_2 \setminus \mathcal{T}_{2,h} = \mathcal{T}_{1,h} \setminus \Omega_1\,,$$

results in

$$\sum_{i=1}^{2}(\kappa_i\nabla u_h, \nabla\varphi_h)_{\Omega_i} = \sum_{i=1}^{2}(\kappa_i\nabla u_h, \nabla\varphi_h)_{\mathcal{T}_{i,h}} + \sum_{i=1}^{2}(\delta\kappa_i\nabla u_h, \nabla\varphi_h)_{\Omega_i\setminus\mathcal{T}_{i,h}}\,,$$

with

$$\delta\kappa_i = \begin{cases}\kappa_1 - \kappa_2, & i = 1\,,\\ \kappa_2 - \kappa_1, & i = 2\,.\end{cases}$$

Taking (2.2) and (2.3) into account, it follows that

$$\sum_{i=1}^{2}(\kappa_i\nabla u, \nabla\varphi_h)_{\Omega_i} = \sum_{i=1}^{2}(\kappa_i\nabla u_h, \nabla\varphi_h)_{\mathcal{T}_{i,h}}$$

and thus, a perturbed Galerkin orthogonality

$$(\kappa\nabla e_h, \nabla\varphi_h)_\Omega = \sum_{i=1}^{2}(\delta\kappa_i\nabla u_h, \nabla\varphi_h)_{\Omega_i\setminus\mathcal{T}_{i,h}}\,, \tag{2.15}$$

holds. Estimating

$$\|\nabla e_h\|^2 \leq (\kappa\nabla e_h, \nabla e_h) = \big(\kappa\nabla e_h, \nabla(u - \varphi_h)\big) + \big(\kappa\nabla e_h, \nabla(\varphi_h - u_h)\big)\,,$$

and picking the Lagrangian interpolant $\varphi_h = L_h u \in V_h$ yields

$$\|\nabla e_h\|^2 \le c\|\nabla e_h\|\|\nabla(u - L_h u)\| + \sum_{i=1}^{2} \|\delta\kappa_i \nabla u_h\|_{\Omega_i \setminus \mathcal{T}_{i,h}} \|\nabla(L_h u - u_h)\|_{\Omega_i \setminus \mathcal{T}_{i,h}} .$$

By Lemma 2.6, we can bound the terms on the convex remainders as

$$\|\delta\kappa_i \nabla u_h\|_{\Omega_i \setminus \mathcal{T}_{i,h}} \le c_\kappa h^{1/2} \|\nabla u_h\|_\Omega ,$$
$$\|\nabla(L_h u - u_h)\|_{\Omega_i \setminus \mathcal{T}_{i,h}} \le c h^{1/2} \|\nabla(L_h u - u_h)\|_\Omega ,$$

and arrive at

$$\|\nabla e_h\|^2 \le c\|\nabla e_h\|\|\nabla(u - L_h u)\| + ch\|\nabla u_h\|\|\nabla(u_h - L_h u)\| . \tag{2.16}$$

Applying Young's inequality $ab \le a^2/(2\varepsilon) + (\varepsilon b^2)/2$ and adding and subtracting u results in

$$\|\nabla e_h\| \le c\|\nabla(u - L_h u)\| + ch\|\nabla u\| .$$

That leaves us to estimate the interpolation error on the whole domain Ω. We divide the estimation into the cells that are affected by the interface and those that are not affected. By Lemma 2.7, we have an estimate for cells $T \in S_h$:

$$\|\nabla(L_h u - u)\|_{S_h} \le ch\|u\|_{H^2(\Omega_1 \cup \Omega_2)} .$$

On those cells not cut by the interface, we use the standard interpolation estimate and extend the domain to the complete domain again:

$$\begin{aligned}\|\nabla(u - L_h u)\|_\Omega^2 &= \|\nabla(u - L_h u)\|_{\Omega \setminus S_h}^2 + \|\nabla(u - L_h u)\|_{S_h}^2 \\ &\le ch^2\|u\|_{H^2(\Omega_1 \cup \Omega_2)}^2 .\end{aligned} \tag{2.17}$$

With the stability estimate, the first estimate

$$\|\nabla(u - u_h)\|_\Omega \le Ch\|f\|_\Omega \tag{2.18}$$

is proven.

2. We prove the second inequality $\|u - u_h\| \le Ch^2\|f\|$:
To show the estimate for the L^2-error, we apply a standard duality argument. Therefore, let $z \in H_0^1(\Omega)$ be the solution of the adjoint problem

$$\sum_{i=1}^{2} (\kappa_i \nabla\varphi, \nabla z) = (e_h, \varphi)\|e_h\|^{-1}$$

for all $\varphi \in H_0^1(\Omega)$. For the dual solution it holds $z \in H_0^1(\Omega) \cup H^2(\Omega_1 \cup \Omega_2)$ and $\|z\|_{H^2(\Omega_1 \cup \Omega_2)} \le c_S$. Using the perturbed Galerkin orthogonality (2.15), we obtain

$$\begin{aligned}\|e_h\| &= (e_h, e_h)\|e_h\|^{-1} = (\kappa\nabla e_h, \nabla z) \\ &= (\kappa\nabla e_h, \nabla(z - L_h z)) + \sum_{i=1}^{2} (\delta\kappa_i \nabla u_h, \nabla L_h z)_{\Omega_i \setminus \mathcal{T}_{i,h}} ,\end{aligned}$$

and we estimate

$$\|e_h\| \le c\|\nabla e_h\|\|\nabla(z - L_h z)\| + c \sum_{i=1}^{2} \|\delta\kappa_i \nabla u_h\|_{\Omega_i \setminus \mathcal{T}_{i,h}} \|\nabla L_h z\|_{\Omega_i \setminus \mathcal{T}_{i,h}} . \tag{2.19}$$

By adding and subtracting u and $L_h u$, we find that

$$\|\nabla u_h\|_{\Omega_i \setminus \mathcal{T}_{i,h}} \le \|\nabla u\|_{\Omega_i \setminus \mathcal{T}_{i,h}} + \|\nabla(u - L_h u)\|_{\Omega_i \setminus \mathcal{T}_{i,h}} + \|\nabla(L_h u - u_h)\|_{\Omega_i \setminus \mathcal{T}_{i,h}} .$$

With Lemma 2.6, it holds

$$\|\nabla u\|_{\Omega_i \setminus \mathcal{T}_{i,h}} \le ch\|u\|_{H^2(\Omega_1 \cup \Omega_2)} ,$$

and thus, using Lemma 2.6 and (2.17), it follows that

$$\|\nabla u_h\|_{\Omega_i \setminus \mathcal{T}_{i,h}} \le ch\|u\|_{H^2(\Omega_1 \cup \Omega_2)} + ch^{1/2}\|\nabla(L_h u - u_h)\|_\Omega .$$

After adding and subtracting u again, and employing (2.17) and (2.18), we derive

$$\begin{aligned}\|\nabla u_h\|_{\Omega_i \setminus \mathcal{T}_{i,h}} &\le ch\|u\|_{H^2(\Omega_1 \cup \Omega_2)} + ch^{1/2}\|\nabla(u - L_h u)\|_\Omega + ch^{1/2}\|\nabla(e_h)\|_\Omega ,\\ &\le ch\|f\|_\Omega .\end{aligned}$$

Similarly, by adding and subtracting the dual solution z, we find that for the interpolation of the dual solution it holds

$$\|\nabla L_h z\|_{\Omega_i \setminus \mathcal{T}_{i,h}} \le \|\nabla z\|_{\Omega_i \setminus \mathcal{T}_{i,h}} + \|\nabla(z - L_h z)\|_{\Omega_i \setminus \mathcal{T}_{i,h}} \le ch\|z\|_{H^2(\Omega_1 \cup \Omega_2)} .$$

Using (2.17), (2.18), the stability estimate $\|z\|_{H^2(\Omega_1 \cup \Omega_2)} \le c_S$, and (2.19) we deduce the estimate

$$\|u - u_h\|_\Omega \le Ch^2\|f\|_\Omega .$$

□

2.5 Numerical examples

In this section we present three different numerical test cases which are chosen to numerically also show the analytically proven convergence. All tests are taken from [8] and results can directly be compared.

We also test the behavior of the presented method dependent on how we choose the free parameters r, q, s in the cells that are cut by the interface. Note that we proposed different strategies to choose the free parameter only for the case where two edges are cut by the interface. We will show results for three different choices of parameters, namely:

Strategy 1: Choose the free parameters to be $1/2$, also if a vertex is cut.
Strategy 2: Choose the free parameters to be $r = 1 - s$, $q = s$, and $s = 1 - r$.
Strategy 3: Choose the free parameters to be $r = (1-s)(1-q)$, $q = (1-r)s$, and $s = qr$.

All computations in this paper are done with the initial mesh shown in Figure 2.10a.

(a) Initial mesh used for all computations.

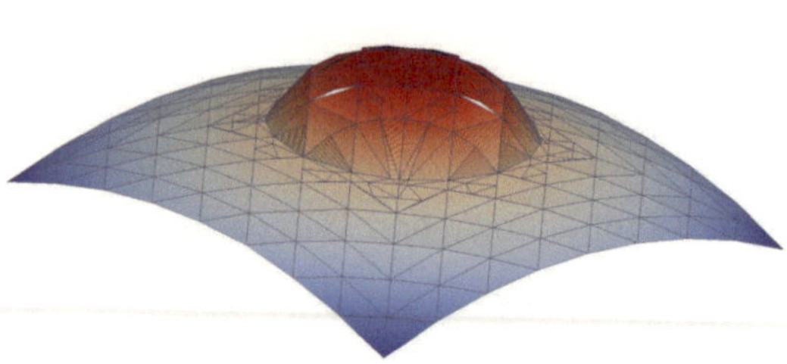

(b) Finite element approximation of (5.1) with the presented method after several global refinements of the initial mesh.

Fig. 2.10: Initial mesh (a) and approximated solution for a circular interface (b).

2.5.1 Circular interface

This example was used to compute the results for Figure 2.1 in Section 2.1. We choose an analytical solution to the interface problem (2.1) as

$$u(x) = \begin{cases} -2\kappa_2 \|x\|^4 , & x \in \Omega_1 , \\ -\kappa_1 \|x\|^2 + \frac{1}{4}\kappa_1 - \frac{1}{8}\kappa_2 , & x \in \Omega_2 \end{cases} \tag{2.20}$$

and compute the right hand side and boundary conditions accordingly. The domains are given as

$$\Omega_1 = \{x \in \mathbb{R}^2 : \|x\| < 1/4\} ,$$
$$\Omega_2 = (-1, 1)^2 \setminus \Omega_1 ,$$

and the diffusion coefficient is defined as

$$\kappa = \begin{cases} 0.1, & x \in \Omega_1 \\ 1, & x \in \Omega_2 \end{cases} .$$

A sketch of a finite element approximation with the locally adapted patch method is given in Figure 2.10b. We note that in contrast to [8], we only split the cut cells into subtriangles for visualization. Thus, there are hanging nodes in the visualization. However, the solution is continuous at these edges. For the presented adapted finite element method we plot the error in the L^2- and in the H^1-norm for several levels of global refinement of the mesh in Figure 2.11. We recover the optimal quadratic convergence for the error in the L^2-norm and linear convergence in the H^1-norm as proven in Theorem 2.8.

In Figure 2.12, we present three meshes which arise in the approximation of (2.20). The subtriangles of the cells that are cut by the interface are shown. In the mesh in Figure 2.12a, the free parameters which are not determined by the cut of the interface are chosen to be 1/2. Therefore, the maximum angle condition could be violated. For

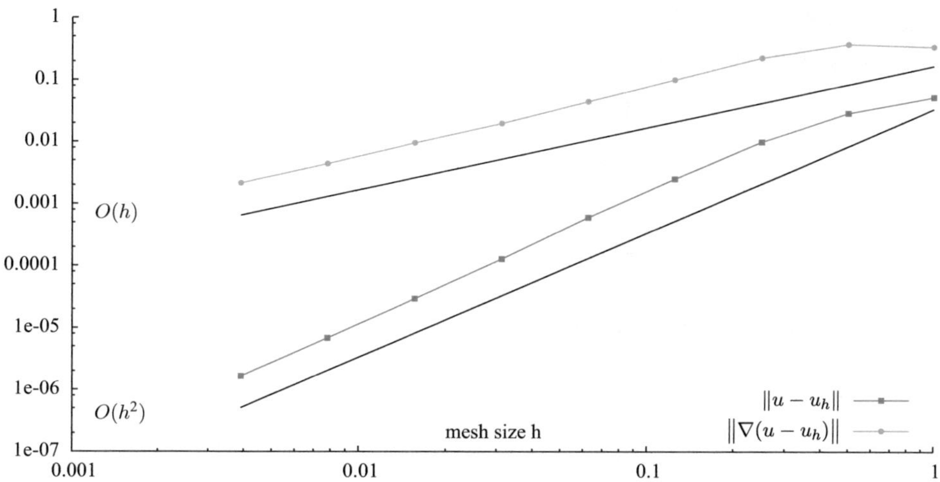

Fig. 2.11: Optimal convergence recovered with the adapted finite element method. Solid lines indicate a slope of h and h^2, respectively.

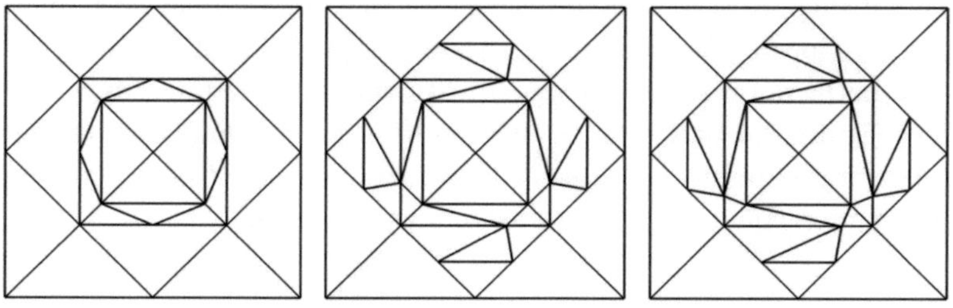

(a) Strategy 1. (b) Strategy 2. (c) Strategy 3.

Fig. 2.12: Cut cells for the approximation of (2.20) after one level of global refinement of the initial mesh in Figure 2.10a, free parameters chosen after different strategies.

the other two meshes in Figures 2.12b and 2.12c, we apply two different strategies for choosing the free parameters. In those meshes, the maximum angle condition will be satisfied.

2.5.2 Horizontal cut

In this example we study the behavior of the locally adapted patch finite element method for patch triangles which get very anisotropic. Let us assume that the domain $\Omega = (-1, 1)^2$ is cut horizontally into

$$\Omega_1(\varepsilon) = \{x \in \Omega | x_2 < \varepsilon h\} \quad \text{and} \quad \Omega_2(\varepsilon) = \{x \in \Omega | x_2 > \varepsilon h\} \,,$$

where h is the maximum edge length of a triangle T. The analytical solution to the interface problem (2.1) is chosen as

$$u(x) = \begin{cases} \frac{\kappa_2}{\kappa_1}(x_2 - \varepsilon h) - (x_2 - \varepsilon h)^2, & x \in \Omega_1 , \\ (x_2 - \varepsilon h) - (x_2 - \varepsilon h)^2, & x \in \Omega_2 . \end{cases}$$

The right hand side and the boundary conditions are computed accordingly. The vertical edge length in a patch triangulation will vary due to the value of ε. Vertical edge lengths between εh and $(1 - \varepsilon)h$ will occur for $\varepsilon \in [0, 1]$. Here, h denotes the maximum edge length of a triangle T.

We plot the error in the L^2- and H^1-norm for different values of $\varepsilon \in [0, 1]$ in Figures 2.13 to 2.18. As expected and already reported in [8], we observe the smallest errors for $\varepsilon = 0$, $\varepsilon = 1/2$, and $\varepsilon = 1$, as then the cut is resolved by the mesh.

The largest errors arise for the cases in which $\varepsilon \to 0$ and $\varepsilon \to 1$ since then, the anisotropies for the patch cells become maximal. However, the errors stay bounded. This behavior is similar on finer meshes but the variations get smaller.

The errors are smaller if we choose the free parameters such that the maximum angle condition is satisfied. However, the error behaves in a quite similar fashion for all three strategies for choosing free parameters. There is no significant difference between the error in the L^2-norm shown in Figures 2.14 and 2.15 and the error in the H^1-norm shown in Figures 2.17 and 2.18. For this example, we conclude that it is important to choose the free parameter to satisfy a maximum angle condition. However, it does not seem to matter which of the suggested values is chosen.

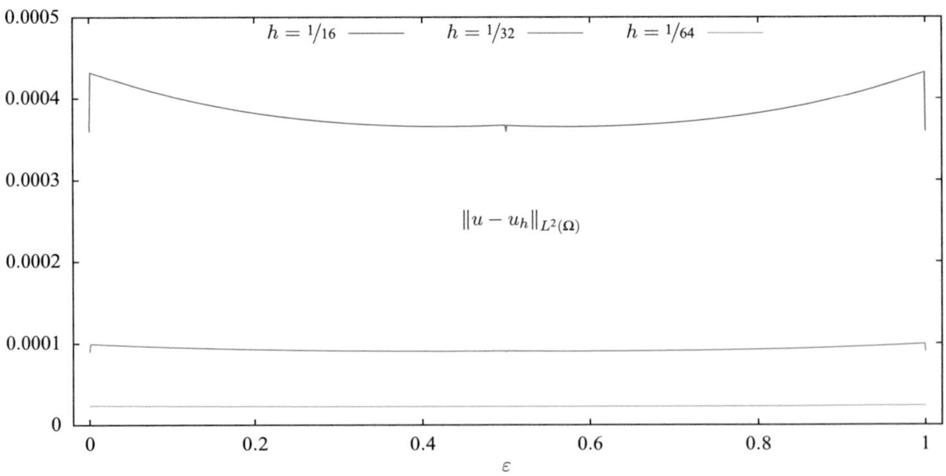

Fig. 2.13: Error in the L^2-norm depending on ε, strategy 1.

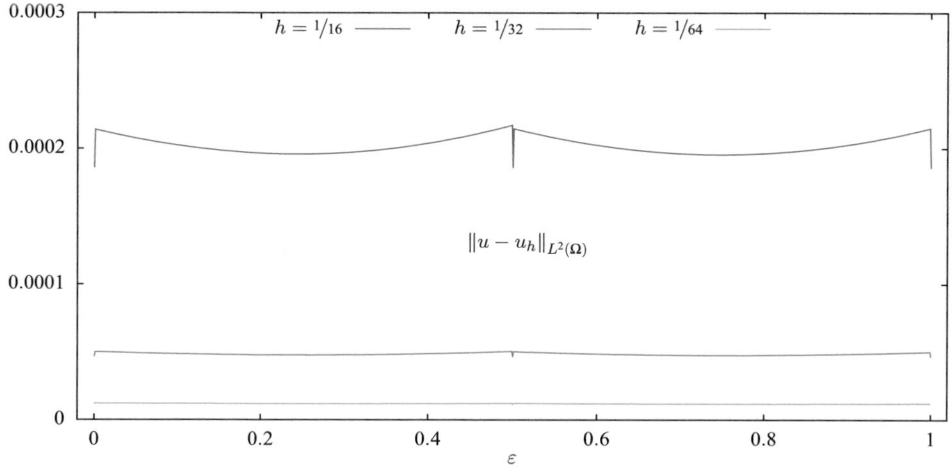

Fig. 2.14: Error in the L^2-norm depending on ε, strategy 2.

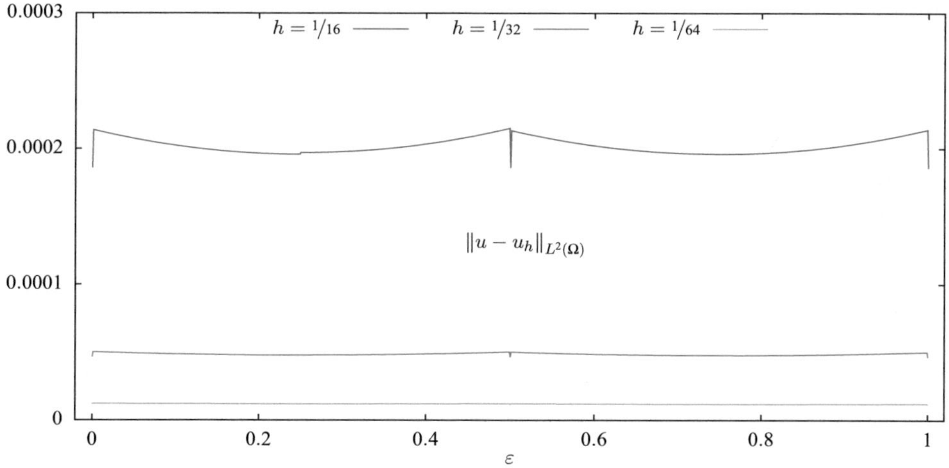

Fig. 2.15: Error in the L^2-norm depending on ε, strategy 3.

2.5.3 Tilted interface line

In this numerical example, we consider a straight interface that cuts the domain $\Omega = (-1, 1)^2$ into two subdomains

$$\Omega_1(\alpha) = \{x \in \Omega | \cos(\alpha)x_2 < \sin(\alpha)x_1\} ,$$

and

$$\Omega_2(\alpha) = \{x \in \Omega | \cos(\alpha)x_2 > \sin(\alpha)x_1\} .$$

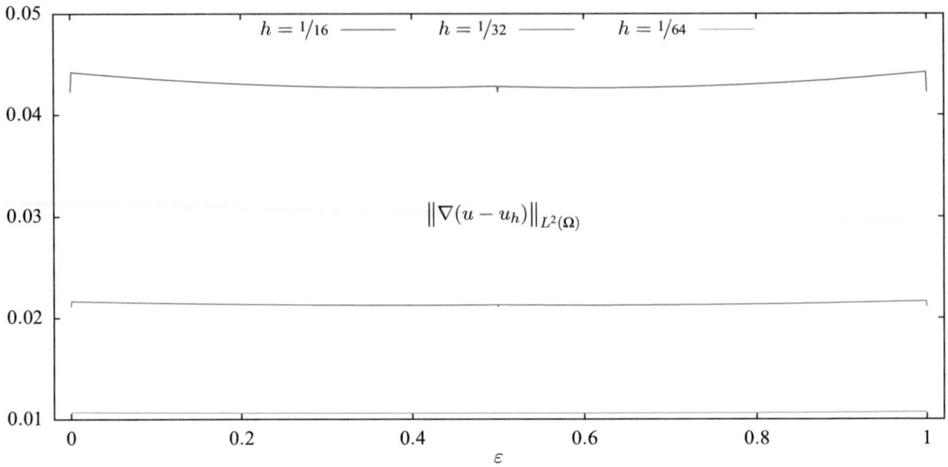

Fig. 2.16: Error in the H^1-norm depending on ε, strategy 1.

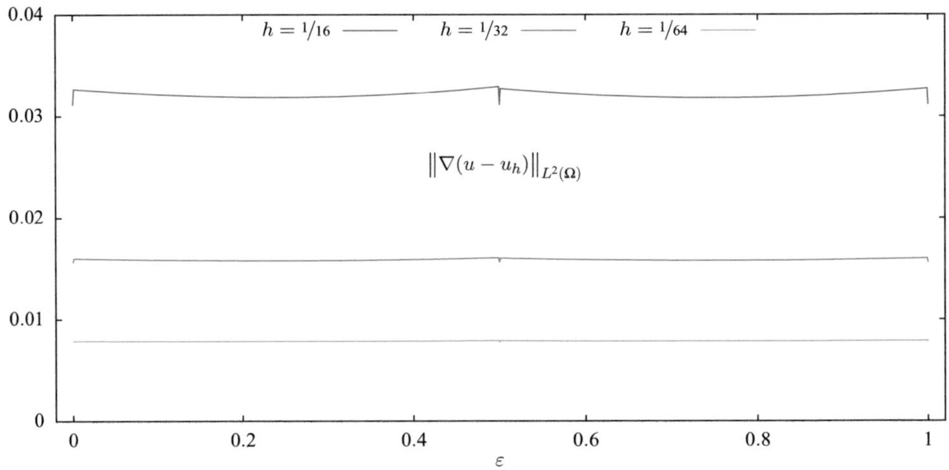

Fig. 2.17: Error in the H^1-norm depending on ε, strategy 2.

Depending on the value of α, the straight interface is defined by $\cos(\alpha)x_2 = \sin(\alpha)x_1$. We choose an analytical solution to problem (2.1) as

$$u(x) = \begin{cases} \sin\left(\frac{\kappa_2}{\kappa_1}(\cos(\alpha)x_2 - \sin(\alpha)x_1)\right), & x \in \Omega_1 , \\ \sin(\cos(\alpha)x_2 - \sin(\alpha)x_1), & x \in \Omega_2 . \end{cases}$$

From this exact solution, the right hand side and boundary conditions are computed accordingly. We plot the error in the L^2- and H^1-norm for different values of $\alpha \in [0, \pi]$ in Figures 2.19 to 2.24.

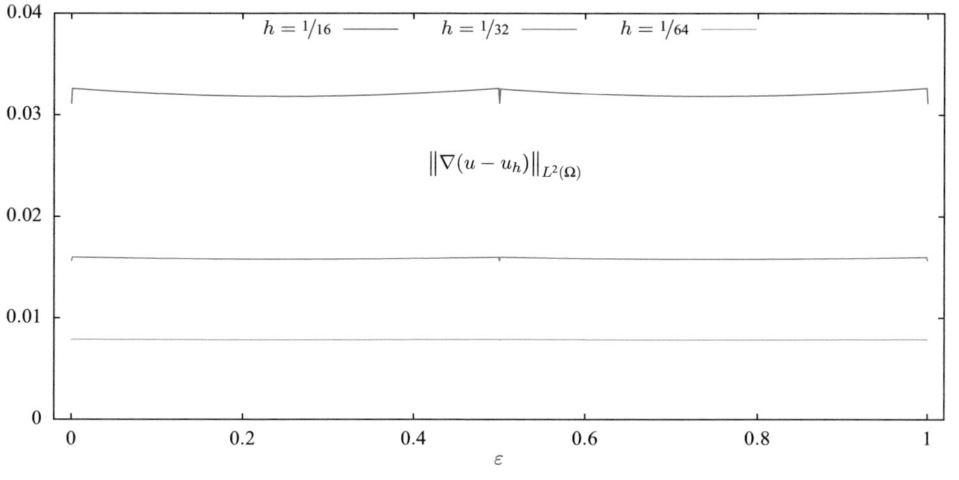

Fig. 2.18: Error in the H^1-norm depending on ε, strategy 3.

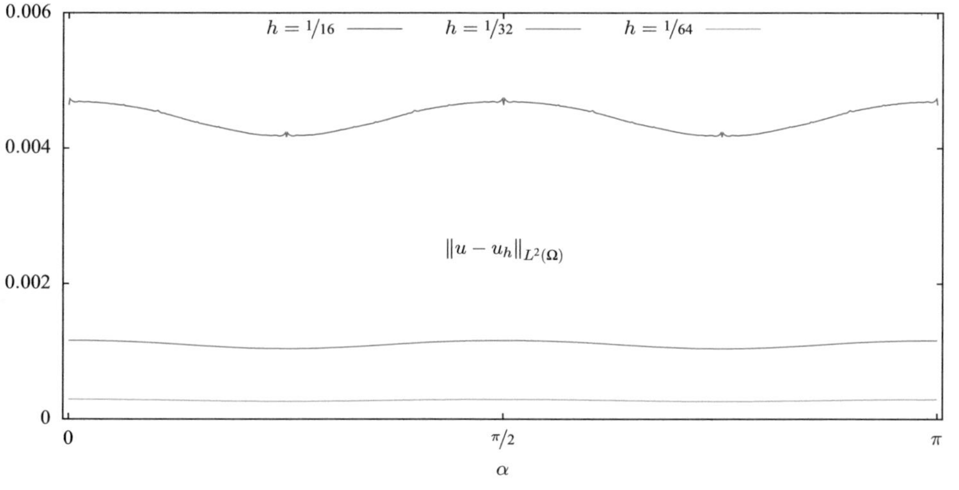

Fig. 2.19: Error in the L^2-norm depending on α, strategy 1.

In this example, the free parameter choice does not make any difference. In contrast to the previous example we even get the same error behavior if the free parameters are chosen to be $1/2$.

2.6 Conclusions

We presented an extension of the locally adapted patch finite element method to triangles. In order to resolve a solution at an interface, patch triangles that are cut by the

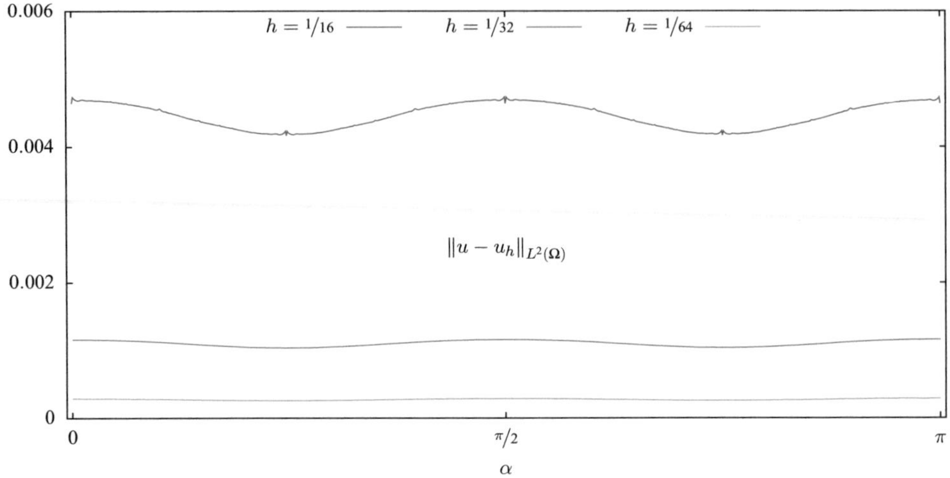

Fig. 2.20: Error in the L^2-norm depending on α, strategy 2.

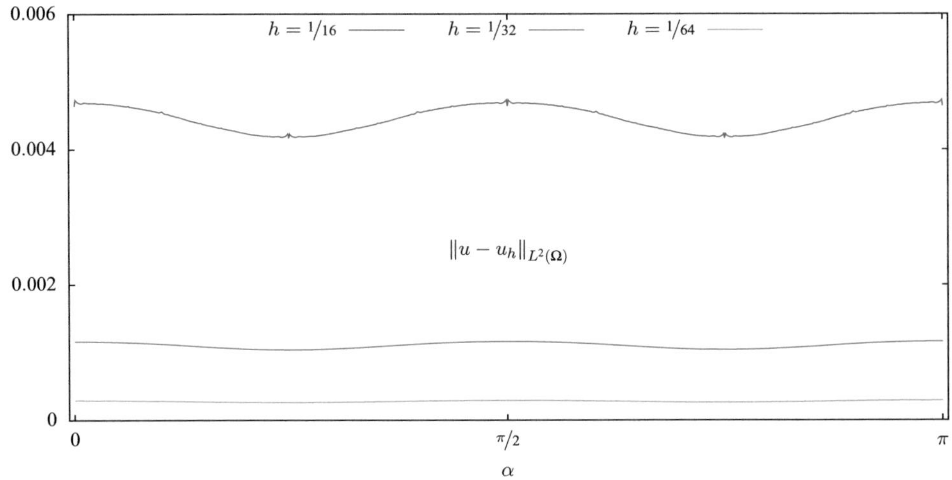

Fig. 2.21: Error in the L^2-norm depending on α, strategy 3.

interface are divided into subtriangles. We apply linear polynomials on these subtriangles together with a local quadrature rule.

Thanks to linear transformations of the subtriangles to the exact location of the cut in the patch triangle, no degrees of freedom have to be moved, inserted, or deleted. Thus, this method is perfectly suitable for parallel computing on distributed architectures.

A key ingredient for the a priori error analysis of the locally adapted finite element patch method was the maximum angle condition. In contrast to the locally adapted

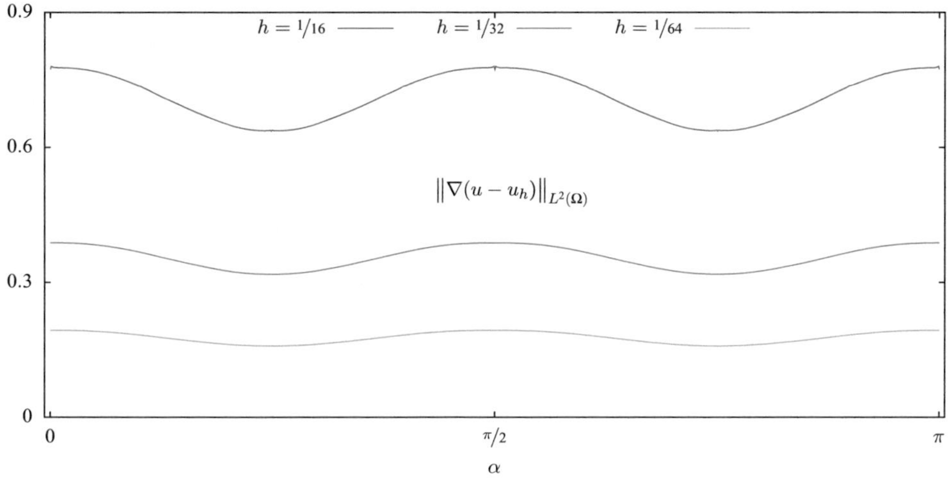

Fig. 2.22: Error in the H^1-norm depending on α, strategy 1.

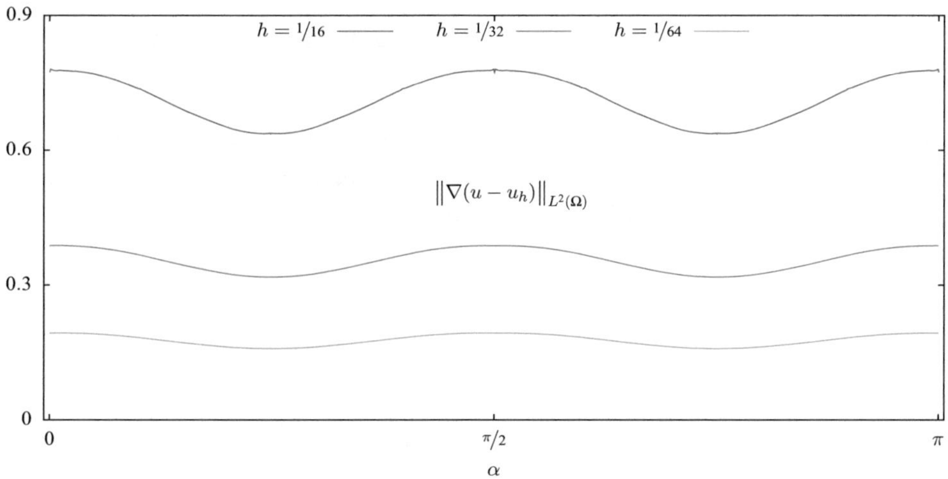

Fig. 2.23: Error in the H^1-norm depending on α, strategy 2.

patch finite element method on quadrilaterals, the maximum angle condition is not automatically satisfied on triangular meshes. We suggested various possibilities to overcome this shortcoming. With these adjustments, the optimal order of convergence for the error in the L^2- and the H^1-norm are recovered. This is also shown in the numerical experiments. Furthermore, we observe in the numerical tests that these adjustments do not necessarily lead to improved error behavior.

In future research, we plan to parallelize the method and to extend it to three-dimensional tetrahedral meshes. Furthermore, we would like to apply our method to

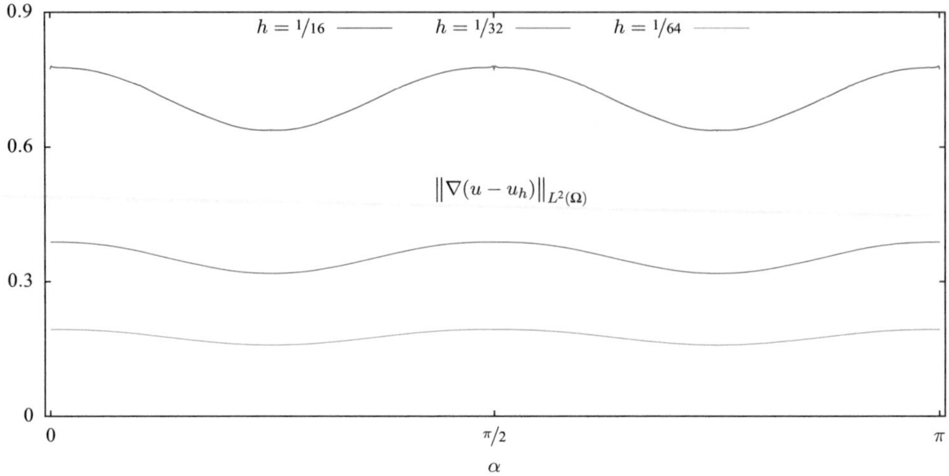

Fig. 2.24: Error in the H^1-norm depending on α, strategy 3.

time-dependent and more complex problems such as fluid-structure interaction problems.

Acknowledgment: The authors would like to thank Aurélien Larcher for the implementation of higher order finite elements in the high-performance computing branch of DOLFIN. Without his contribution, this work would not have been possible.

Bibliography

[1] T. Apel. *Anisotropic Finite Elements: Local Estimates and Applications*. Advances in Numerical Mathematics. Teubner, Stuttgart, 1999.

[2] I. Babuška. The finite element method for elliptic equations with discontinuous coefficients. *Computing*, 5(3):207–213, 1970.

[3] I. Babuška, U. Banerjee, and J. E. Osborn. Generalized finite element methods – Main ideas, results and perspective. *International Journal of Computational Methods*, 01(01):67–103, 2004.

[4] J. H. Bramble and J. T. King. A robust finite element method for nonhomogeneous dirichlet problems in domains with curved boundaries. *Mathematics of Computation*, 63(207):1–17, 1994.

[5] E. Burman, S. Claus, P. Hansbo, M. G. Larson, and A. Massing. CutFEM: Discretizing geometry and partial differential equations. *International Journal for Numerical Methods in Engineering*, 104(7):472–501, 2015.

[6] E. Burman, P. Hansbo, M. G. Larson, and S. Zahedi. Cut finite element methods for coupled bulk–surface problems. *Numerische Mathematik*, pages 1–29, 2015.

[7] P. Ciarlet. *The Finite Element Method for Elliptic Problems*. Studies in Mathematics and its Applications. Elsevier Science, 1978.

[8] S. Frei and T. Richter. A Locally Modified Parametric Finite Element Method for Interface Problems. *SIAM J. Numerical Analysis*, 52(5):2315 – 2334, 2014.

[9] P. Gangl and U. Langer. A Local Mesh Modification Strategy for Interface Problems with Application to Shape and Topology Optimization. *ArXiv e-prints*, 2016.

[10] A. Hansbo and P. Hansbo. An unfitted finite element method, based on Nitsche's method, for elliptic interface problems. *Computer Methods in Applied Mechanics and Engineering*, 191(47 – 48):5537–5552, 2002.

[11] A. Hansbo and P. Hansbo. A finite element method for the simulation of strong and weak discontinuities in solid mechanics. *Computer Methods in Applied Mechanics and Engineering*, 193(33 – 35):3523–3540, 2004.

[12] R. J. Mackinnon and G. F. Carey. Treatment of material discontinuities in finite element computations. *International Journal for Numerical Methods in Engineering*, 24(2):393–417, 1987.

[13] N. Moës, J. Dolbow, and T. Belytschko. A finite element method for crack growth without remeshing. *International Journal for Numerical Methods in Engineering*, 46(1):131–150, 1999.

[14] J. Wloka. *Partial Differential Equations*. Cambridge University Press, 1987.

Stefan Frei and Thomas Richter

3 An accurate Eulerian approach for fluid-structure interactions

Abstract: The Fully Eulerian method is a monolithic approach for fluid-structure interactions, where the two fields are coupled in a fully variational setting. The striking property of the Fully Eulerian model is its capability to cope with arbitrary large deformations and contact while keeping the coupling monolithic.

We study the Fully Eulerian approach as it was introduced by Dunne and Rannacher in 2006. For the first time, this manuscript gives a thorough discussion of accurate numerical methods for the spatial and temporal discretization. The characteristic difficulty in the Fully Eulerian model is the interface – the coupling boundary between fluid and solid – that, in a Eulerian description moves freely within the domain. On fixed Eulerian background meshes, standard finite elements cannot properly resolve the interface, where the solution might have kinks with limited regularity (for the velocity) or even discontinuities (pressure). As the interface moves with time and will change its position on a Eulerian background mesh, certain nodal points might change their affiliation from one side of the interface to the other within one time-step. Regularity is limited in the temporal direction and accuracy will be lost, if standard time-stepping schemes are used.

Another problem arises due to the Eulerian description of the solid equations, which brings along convective terms. In particular close to the interface, robust and stable discretization is difficult and asks for special stabilization tools.

The main part of this chapter introduces and analyses novel and high-order discretization schemes for interface problems with moving interfaces. For the first time in a fully Eulerian setting, we are able to derive numerical schemes of second order in space and time and show reliable results for different numerical test-cases.

Keywords: fluid-structure interactions, finite elements, monolithic coupling, Fully Eulerian, interface-problems, contact problems

Classification: 65M60, 65N30, 74F10, 35Q30, 35Q35, 35Q74

3.1 Introduction

The interaction of a fluid flow with a moving or deformable structure plays an important role in many physical processes in nature, biology and in engineering applica-

Stefan Frei, Stefan Frei, Dept. of Mathematics, University College London, Gower Street, London, WC1E 6BT, s.frei@ucl.ac.uk

Thomas Richter, Thomas Richter, Universität Magdeburg, Universitätsplatz 2, 39106 Magdeburg, Germany, thomas.richter@ovgu.de

https://doi.org/10.1515/9783110494259-003

tions. Prominent examples include the interaction of planes, vehicles or wind turbines with the surrounding air in aerodynamics as well as blood flow with the surrounding vessels or the heart valves in hemodynamics. These problems have in common that fluid forces cause a deformation or movement of the structure. This, conversely, might have a considerable influence on the surrounding fluid flow. In this article, we are especially interested in two-way-coupled problems, meaning that both subproblems have a strong influence upon each other. Fluid-structure interaction problems pose a variety of numerical challenges:

- While the solid equations are traditionally formulated in the particle-centered Lagrangian coordinate framework, for the fluid equations spatial-centered Eulerian coordinates are usually used. Consequently, the numerical methods that have been developed for the subproblems in the past decades are based on different coordinate frameworks. Thus, it is not straight-forward to combine them into one method to tackle the whole problem.
- The type of equation changes from the solid to the fluid domain. While the fluid equations have a parabolic-like character, the solid equations behave similarly to hyperbolic equations. Among other difficulties, e.g., regarding time discretization, this leads to a regularity gap between fluid and solid velocity at the interface.
- The fluid-structure interface moves with time such that the subdomains of fluid and solid are variable. A fixed point $x \in \mathbb{R}$ might belong to the fluid at time t_1 and to the solid at time t_2. This has to be carefully taken into account in the design of a numerical algorithm, especially when it comes to discretization.
- If solid and fluid densities are of the same magnitude, the coupling between the subproblems is typically so strong that partitioned algorithms that iterate between solvers for the subproblems trying to fulfil the coupling conditions in the limit converge very slowly or show no convergence at all (the *added-mass instability*, e.g., Causin et al. [20]). In this case monolithic algorithms that tackle the whole problem at once are preferable.
- In many applications the structure undergoes large movements or deformations. Even contact of different structures or a structure with a wall is something one often has to deal with. This leads to topology changes in the fluid domain which is a problem for many of the numerical approaches that are typically used to simulate fluid-structure interaction problems.

Fluid-structure interactions have been studied numerically at least since the early 1970s (e.g., Peskin [78]). A lot of advances have been made since then and a multitude of different approaches have been proposed to tackle these problems numerically. For an overview, we refer to the literature, e.g., van Loon et al. [94], Felippa et al. [35] or Frei [37] (in particular Chapter 3.2).

This article is devoted to the *Fully Eulerian* approach, which goes back to Dunne [29]. The idea is to formulate both subproblems, fluid and solid, in a fully Eulerian coordinate system. While this is the standard approach in fluid dynamics, it is not common in solid dynamics. The usual objection to this approach is the loss of conservation properties. Accurate discretization techniques resolving the interface are necessary to avoid this issue. Furthermore, we will see, that a Eulerian framework will cause substantial difficulties for discretization schemes. Conversely, the *Fully Eulerian* approach will easily be able to handle problems of very large deformation and contact in a completely *monolithic and variational* way. This makes the *Fully Eulerian* approach unique.

The *Fully Eulerian* formulation is a relatively young approach that was introduced by Dunne [29–31] in 2006. Independently, Cottet and co-workers [24, 25] derived a very similar formulation in the context of finite difference discretizations. Further contributions were made by Rannacher & Richter [81], Richter & Wick [85], and Richter [82] and Wick [98]. A survey of the fundamentals as well as recent developments can be found in [37, 40], an overview of several applications in [41]. The *Fully Eulerian* approach has been applied to simulate heart valve dynamics (Wick [99]) and to study solid growth and clogging in a vessel up to contact of the outer vessel walls [42]. Furthermore, it has been used to study a coupled chemical fluid-structure interaction problem arising in cardiac cells by Laadhari et al. [66]. Sugiyama et al. applied a *Fully Eulerian* method with a finite difference discretization to different fluid-structure interaction problems using a St. Venant–Kirchhoff, a neo-Hookean and a Mooney–Rivlin material law for the structure [90, 91]. Later on, Ii et al. used the method for simulating deformable blood cells and platelets in capillary vessels [63]. Recently, Pironneau [80] introduced a realization of the *Fully Eulerian* approach with accurate discretization based on the *Galerkin-Characteristics* finite element method.

The realizations of a Eulerian formulation for fluid-structure interactions mostly differ in specific details related to the discretization and implementation. As a common difficulty, all of them must deal with moving domains and moving interfaces. Therefore, Eulerian fluid-structure interaction problems belong to the class of *interface problems*. Two problems must be tackled: First, the moving interface must be captured. Second, the moving interface must be resolved sufficiently by the discretization to guarantee accuracy. Various approaches exist for *interface capturing*. The *level set method* is the most prominent one, see Sethian [88]. Here, the interface is captured by one scalar function, that is transported with the interface velocity and that indicates the distance to the nearest interface point. In terms of Eulerian solid mechanics and to assemble stresses, it is necessary to reconstruct the deformation field. For this purpose, multiple level set functions can be used [25, 55]. Another approach for a Eulerian representation of the solid deformation is the *initial point set method*, also denoted as *backward characteristics*, where the entire deformation field is captured on a Eulerian domain, see [29, 82] or [25].

Once the location of the interface between solid and fluid is known, it must be resolved by the discretization, as the solution typically has limited smoothness across the internal domain boundary. Finite element schemes that do not consider interfaces have limited approximation properties, see Babuška [2]. This problem is common to multiphase flows, where varying densities will lead to reduced interface regularity.

Here, a common approach is an averaging of the material parameters close to the interface [77] such that the proper order of convergence can be recovered. For fluid-structure interactions, an averaging of completely different types of materials, i.e., an elastic solid and a viscous fluid is not recommendable.

Instead, techniques for an accurate discretization at the interface are required. In principle, two different approaches are available. First, the mesh can be adjusted to the interface by a remeshing in every time-step [80]. This comes at considerable cost, in particular, as the projection of solutions from mesh to mesh is expensive, if no hierarchical structure can be used. Instead of a full remeshing, a local adaption of the mesh is possible. The second approach is to use a fixed Eulerian mesh and adapt the finite element space, such that an accurate discretization is possible. Most prominent examples are the *generalized finite element methods* [3, 43] or the *extended finite element method (XFEM)* [21, 53]. The technique that we propose falls in between both approaches. It can be regarded as a local modification of the mesh, we realize it however as a local modification of the finite element spaces.

In the following section, we introduce the *Fully Eulerian* model for fluid-structure interactions. In Section 3.3, we give an overview over alternative and related approaches.

Section 3.4 of this contribution gives details on our realization of the *Fully Eulerian* formulation of fluid-structure interactions. In particular, we will address details concerning the spatial and temporal discretization of interface problems, as this is the dominant difficulty in Eulerian approaches. We have already discussed, that an accurate spatial resolution of the interface is essential. In Section 3.4.1, we describe and analyze the spatial discretization scheme, which can be seen as a combination of locally adapted remeshing and a locally adapted generalized finite element approach.

As we consider the coupling to the incompressible Navier–Stokes equations, we need robust numerical schemes for saddle-point problems. In Section 3.4.2, we describe a new stabilization scheme for locally modified finite elements that takes into account the anisotropy of the subcells that might arise.

For dynamic problems with moving interfaces, similar accuracy issues must be handled in time: if the interface moves, the solution has limited regularity in time. Common time-stepping techniques will fail or will give a reduced convergence order. Up to now, this issue has had little attention with very few contributions that are both numerically efficient and analytically understood. Section 3.4.3 is devoted to time discretization. Problems of solid mechanics are usually formulated in Lagrangian coordinates. A Eulerian representation is not natural and brings along convective terms. This results in a loss of conservation principles and gives rise to possible numerical stability problems. In Section 3.4.4, we present a novel stabilization scheme for the structure equations in Eulerian coordinates.

Finally, we present different applications for a validation of the proposed methods. We start in Section 3.5 by analyzing a well-known benchmark problem, introduced by Hron and Turek [60]. This problem describes the coupled dynamics of an incompressible flow interacting with an elastic beam attached to a rigid obstacle. The resulting deformations are moderate, and the Reynolds number is sufficiently large, such that moderate boundary layers appear at the moving beam. For both reasons, the *Fully Eulerian* approach is not the optimal approach for this situation. Nevertheless, we will present accurate results for different benchmark configurations. To our knowledge, this is the first time that accurate quantitative results are given for a *Fully Eulerian* approach. Next, in Section 3.6, we describe the problem of an elastic ball, that is falling in a viscous fluid. The ball touches the boundary of the computational domain and bounces back due to elasticity. Modeling and discretization of the contact problem is the key focus of this application. Most of the standard monolithic approaches, e.g., the *arbitrary Lagrangian–Eulerian* framework, are not able to deal with contact.

3.2 Modeling

In this section we will briefly derive the constitutive equations in Eulerian coordinates and show how the equations and coupling conditions can easily be included in a monolithic variational formulation.

3.2.1 Continuum mechanics

We begin by introducing our notation concerning different coordinate systems.

Notation and coordinate systems

Traditionally there are two coordinate systems of interest in fluid and solid mechanics: spatial-centered Eulerian coordinates and particle-centered Lagrangian coordinates. In Figure 3.1, we sketch a moving volume $V(t)$ and its stress-free initial configuration $\hat{V}$. Throughout this chapter we will use the "hat" notation for Lagrangian quantities. If the movement $\hat{T}\colon \hat{V} \times I \to V(t)$ is known, we can relate Eulerian quantities f to their Lagrangian counterpart $\hat{f}$ by the relation

$$f(x, t) = f(\hat{T}(\hat{x}, t), t) = \hat{f}(\hat{x}, t) . \tag{3.1}$$

Lagrangian and Eulerian coordinates are related by the displacement $\hat{u}$

$$x = \hat{T}(\hat{x}, t) = \hat{x} + \hat{u}(\hat{x}, t) . \tag{3.2}$$

In fluid-structure interactions, we are typically interested in flows within a certain spatial domain, e.g., the flow around an obstacle. In this case it is practically impossible to formulate the fluid dynamics in Lagrangian coordinates as the majority of the fluid particles that are within the frame of interest at time $t \gg 0$ were far away from it at time $t = 0$. Conversely, solid equations can be formulated in both Lagrangian and

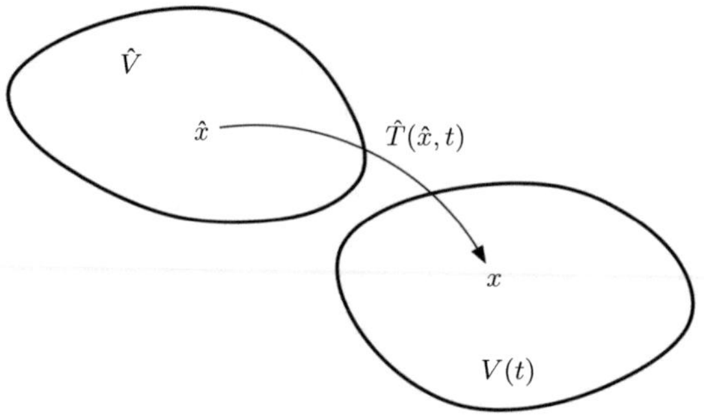

Fig. 3.1: Moving volume $V(t)$ and stress-free reference state $\hat{V}$. Quantities in Lagrangian coordinates are given on the domain $\hat{V}$ (left), while Eulerian quantities are given in the moving volume $V(t)$ (right).

Eulerian coordinates; the former one being the approach that is traditionally used in literature.

By $T\colon \bigcup_{t\in I}(t, V(t)) \to \hat{V}$ we denote the inverse mapping to $\hat{T}$ defined by

$$T(\hat{T}(\hat{x}, t), t) = \hat{x} . \tag{3.3}$$

Due to (3.2), this mapping is related to the Eulerian displacement u (defined by $u(x, t) = \hat{u}(\hat{x}, t)$) by

$$T(x, t) = x - u(x, t) . \tag{3.4}$$

Furthermore, we define the Lagrangian and the Eulerian deformation gradient by

$$\hat{F} = \hat{\nabla}\hat{T} = I + \hat{\nabla}\hat{u} , \qquad F = \nabla T = I - \nabla u$$

where I stands for the identity matrix. Their respective determinants are denoted by

$$\hat{J} = \det \hat{F}, \quad J = \det F .$$

Differentiating (3.3) by $\hat{x}$ yields with the help of the chain rule

$$F(x)\hat{F}(\hat{x}) = I \quad \Rightarrow \quad F(x) = \hat{F}^{-1}(\hat{x}) . \tag{3.5}$$

Conservation principles

The conservation principles we will exploit are conservation of mass and momentum. Both of them are typically derived in Eulerian coordinates. Conservation of mass reads

$$\partial_t \rho + \operatorname{div}(\rho v) = 0 \quad \text{in } \Omega(t) , \tag{3.6}$$

and conservation of momentum

$$\partial_t(\rho v) + \operatorname{div}(\rho v \times v) - \operatorname{div} \sigma = \rho f \quad \text{in } \Omega(t) .$$

Here, v stands for velocity, σ is the Cauchy stress tensor, ρ is the density, f a volume force and $\times$ denotes the external product of two vectors. By using (3.6), the conservation of momentum simplifies further to

$$\rho\partial_t v + \rho(v\cdot\nabla)v - \operatorname{div}\sigma = f \quad \text{in } \Omega(t)\,. \tag{3.7}$$

These principles are assumed both in the solid and in the fluid part of the domain.

Material laws

To close the system of equations, we have to specify material laws, i.e., the way the Cauchy stress tensors σ depend on the primal variables displacement u, velocity v and pressure p. While the derivation of (3.7) was based on physical laws, material laws vary strongly for different materials and can only be considered as approximation of the reality.

In the fluid part, we will consider incompressible Newtonian fluids, i.e.,

$$\operatorname{div} v = 0 \quad \text{in } \Omega_{\mathrm{f}}(t)\,, \tag{3.8}$$

$$\sigma_{\mathrm{f}} = \rho_{\mathrm{f}}\nu_{\mathrm{f}}(\nabla v + \nabla v^T) - p_{\mathrm{f}} I\,. \tag{3.9}$$

Here, $\nu_{\mathrm{f}} > 0$ is the kinematic viscosity of the fluid, ρ_{f} its density and p_{f} stands for pressure.

While this model is a good approximation for a wide range of fluids, there is a whole zoo of solid material laws available, each of them describing specific materials more or less appropriately. Here, we will consider compressible materials and include the conservation of mass in the following way. Let $\rho^0 = \rho(0)$ be the density at initial time $t = 0$. Conservation of mass means that for all times t and an arbitrary volume element $V(t)$, it holds that

$$\int\limits_{V(0)} \rho^0\,dx = \int\limits_{V(t)} \rho(x,t)\,dx\,. \tag{3.10}$$

By transforming the integral on the left-hand side to the corresponding current volume element $V(t)$ and assuming continuity of the integrands, we obtain the pointwise relation

$$\rho(x,t) = J\rho^0(T(x,t),t)\,, \tag{3.11}$$

which will be exploited in the following.

The results presented in this chapter will be based on the St. Venant–Kirchhoff material law, which reads in Lagrangian coordinates

$$\hat{\Sigma}_{\mathrm{s}} = 2\mu_{\mathrm{s}}\hat{E}_{\mathrm{s}} + \lambda_{\mathrm{s}}\operatorname{tr}(\hat{E}_{\mathrm{s}})\,, \tag{3.12}$$

where $\hat{\Sigma}_{\mathrm{s}}$ denotes the second Piola–Kirchhoff stress tensor, $\mu_{\mathrm{s}}, \lambda_{\mathrm{s}} > 0$ the Lamé constants and the Green–Lagrange strain tensor $\hat{E}_{\mathrm{s}}$ is given by

$$\hat{E}_{\mathrm{s}} = \frac{1}{2}\left(\hat{F}_{\mathrm{s}}^T\hat{F}_{\mathrm{s}} - I\right)\,. \tag{3.13}$$

In contrast to the Cauchy stress tensor σ_S, Σ_S relates directions in the Lagrangian framework to forces in Lagrangian coordinates. In Eulerian coordinates, relation (3.12) reads

$$\Sigma_S = 2\mu_S E_S + \lambda_S \operatorname{tr}(E_S) \tag{3.14}$$

where the Eulerian version of the second Piola–Kirchhoff stress tensor Σ_S and the Green–Lagrange strain tensor E_S are related to their Lagrangian counterparts by (3.1). By relation (3.5) for the deformation gradients, the Eulerian Green–Lagrange strain tensor writes

$$E_S = \frac{1}{2}\left(F_S^{-T} F_S^{-1} - I\right) .$$

Finally, we use the physical relation between the Cauchy stress tensor and Piola–Kirchhoff stress tensor (e.g., Holzapfel [59], Richter [84]) to obtain the St. Venant–Kirchhoff material law in Eulerian coordinates

$$\sigma_S = J_S F_S^{-1} \Sigma_S F_S^{-T} = J_S F_S^{-1} \left(2\mu_S E_S + \lambda_S \operatorname{tr}(E_S)\right) F_S^{-T} . \tag{3.15}$$

We remark that the choice of the material law is exemplarily and that the presented methodology can be applied directly to different material laws, e.g., the (incompressible) neo-Hookean law (e.g., Holzapfel [59], Dunne [30]).

Coupling conditions for fluid-structure interactions

We assume that an overall domain $\Omega(t)$ is split into subdomains $\Omega_f(t)$ and $\Omega_S(t)$ with a lower-dimensional interface $\Gamma_i(t)$ such that

$$\Omega(t) = \Omega_f(t) \cup \Gamma_i(t) \cup \Omega_S(t) .$$

Solid and fluid equations are coupled by means of physical principles. The first principle we exploit is called the *kinematic condition* and stems from the observation that a viscid fluid follows the motion of the neighboring solid at the fluid-structure interface. This implies the continuity of fluid and solid velocity at the interface Γ_i

$$v_S = v_f \quad \text{on } \Gamma_i(t) . \tag{3.16}$$

The second physical principle we will exploit is the third Newton law *"actio = reactio"*. We will refer to this as the *dynamic condition* . It says that fluid forces and solid forces in normal direction are equal at the interface, i.e.,

$$\sigma_f n = \sigma_S n \quad \text{on } \Gamma_i(t) . \tag{3.17}$$

Here, n might denote either the outer normal n_f of the fluid domain or the outer normal $n_S = -n_f$ of the solid domain.

Complete system of equations

In addition to the equations derived in the previous sections, we make use of the physical relation between solid velocity and solid displacement

$$v_s = d_t T(x(t), t) = d_t u_s(x(t), t) = \partial_t u_s + v_s \cdot \nabla u_s \,,$$

where d_t denotes the total (Lagrangian) time derivative. The complete set of equations is given by

$$\left.\begin{aligned} \rho_f \partial_t v_f + \rho_f (v_f \cdot \nabla) v_f - \operatorname{div} \sigma_f &= \rho_f f \\ \operatorname{div} v_f &= 0 \end{aligned}\right\} \quad \text{in } \Omega_f(t),$$

$$\left.\begin{aligned} J\rho_s^0 (\partial_t v_s + v_s \cdot \nabla v_s) - \operatorname{div} \sigma_s &= J\rho_s^0 f \\ \partial_t u_s + v_s \cdot \nabla u_s - v_s &= 0 \end{aligned}\right\} \quad \text{in } \Omega_s(t)\,, \tag{3.18}$$

$$\left.\begin{aligned} v_f &= v_s \\ \sigma_f n &= \sigma_s n \end{aligned}\right\} \quad \text{on } \Gamma_i(t),$$

with the Cauchy stress tensor specified in (3.9) and (3.15). For the sake of completeness, we define further boundary conditions on the outer solid boundary Γ_s and the outer fluid boundary Γ_f that will be used in the following

$$\begin{aligned} v_f &= v_f^d \quad \text{on } \Gamma_f^d, & \rho_f \nu_f \partial_n v_f - p_f n &= 0 \quad \text{on } \Gamma_f \setminus \Gamma_f^d\,, \\ u_s &= u_s^d \quad \text{on } \Gamma_s^d\,, & \sigma_s n &= 0 \quad \text{on } \Gamma_s \setminus \Gamma_s^d\,. \end{aligned} \tag{3.19}$$

While for the solid part Γ_s, Γ_s^d or $\Gamma_s \setminus \Gamma_s^d$ might be empty, we assume that Γ_f^d and $\Gamma_f \setminus \Gamma_f^d$ are nonempty. Finally, the system of equations is supplemented with suitable initial values

$$v(x, 0) = v^0(x) \quad \text{in } \Omega(t)\,, \qquad u(x, 0) = u^0(x) \quad \text{in } \Omega_s(t)\,.$$

3.2.2 A monolithic fully Eulerian approach

The complete system of equations is formulated in one monolithic variational formulation in Eulerian coordinates, see also Dunne [29]. The coupling conditions are included by means of variational principles in the following way: The *kinematic condition* is included by the definition of a global ansatz space for solid and fluid velocity

$$v \in v^d + \mathcal{V}, \quad \mathcal{V} = [H_0^1(\Omega(t); \Gamma_f^d)]^2\,,$$

where $v|_{\Omega_f} = v_f$ and $v|_{\Omega_s} = v_s$. By means of the trace theorem a function $v \in \mathcal{V}$ fulfils

$$v_f = v_s \;\text{ a.e. in } \Gamma_i(t)\,.$$

The *dynamic condition* is included by adding the fluid and solid momentum equations and using the global space $\mathcal{V}$ also as a test space. To see this, we use integration by parts for the fluid and solid stress terms and choose a test function $\phi \in H_0^1(\Omega(t))$. This yields

$$(\sigma_\mathrm{f}, \nabla\phi)_{\Omega_\mathrm{f}(t)} + (\sigma_\mathrm{s}, \nabla\phi)_{\Omega_\mathrm{s}(t)} = -(\operatorname{div}\sigma_\mathrm{f}, \phi)_{\Omega_\mathrm{f}(t)} - (\operatorname{div}\sigma_\mathrm{s}, \phi)_{\Omega_\mathrm{s}(t)} + (\sigma_\mathrm{f} n_\mathrm{f}, \phi)_{\Gamma_i(t)} + (\sigma_\mathrm{s} n_\mathrm{s}, \phi)_{\Gamma_i(t)} \,.$$

By the fundamental theorem of variational calculus, we obtain the *dynamic condition*

$$\sigma_\mathrm{f} n = \sigma_\mathrm{s} n \quad \text{a.e. in } \Gamma_i(t) \,.$$

Altogether the monolithic variational formulation in fully Eulerian coordinates reads:

Find the global velocity $v \in v^d + \mathcal{V}$, the solid displacement $u_s \in u_s^d + \mathcal{W}_s$ and the fluid pressure $p_f \in \mathcal{L}_f$ such that

$$\begin{aligned}
(\rho(\partial_t v + v\cdot\nabla v), \phi)_\Omega + (\sigma, \nabla\phi)_\Omega - (\rho_\mathrm{f}\nu_\mathrm{f}\nabla v^T n_\mathrm{f}, \phi)_{\Gamma_\mathrm{f}\setminus\Gamma_\mathrm{f}^d} &= (\rho f, \phi)_\Omega && \forall\phi \in \mathcal{V} \,,\\
(\partial_t u_\mathrm{s} + v\cdot\nabla u_\mathrm{s} - v, \psi_\mathrm{s})_{\Omega_\mathrm{s}(t)} &= 0 && \forall\psi_\mathrm{s} \in \mathcal{W}_\mathrm{s} \,,\\
(\operatorname{div} v, \xi_\mathrm{f})_{\Omega_\mathrm{f}(t)} &= 0 && \forall\xi_\mathrm{f} \in \mathcal{L}_\mathrm{f} \,.
\end{aligned}$$

Here we have used the abbreviations $\rho|_{\Omega_\mathrm{s}} = \rho_\mathrm{s} = J\rho_\mathrm{s}^0$ and $\rho|_{\Omega_\mathrm{f}} = \rho_\mathrm{f}$. Furthermore, we have defined $\sigma|_{\Omega_\mathrm{f}} = \sigma_\mathrm{f}$ and $\sigma|_{\Omega_\mathrm{s}} = \sigma_\mathrm{s}$ and analogously for the right-hand side f. The function spaces are given by

$$\mathcal{V} = [H_0^1(\Omega(t); \Gamma_\mathrm{f}^d \cup \Gamma_\mathrm{s}^d)]^2 \,, \quad \mathcal{W}_\mathrm{s} := H_0^1(\Omega_\mathrm{s}(t); \Gamma_\mathrm{s}^d), \quad \mathcal{L}_\mathrm{f} := L^2(\Omega_\mathrm{f}(t)) \,.$$

The fluid boundary term on the left-hand side stems from the fact that the full symmetric stress tensor σ_f enters the fluid equations while the *do-nothing* condition on $\Gamma_\mathrm{f}\setminus\Gamma_\mathrm{f}^d$ includes the reduced stress tensor $\sigma_\mathrm{f}^{\mathrm{red}} = \rho_\mathrm{f}\nu_\mathrm{f}\nabla v_\mathrm{f} - p_\mathrm{f} I$.

Note that the variational formulation contains the solid displacement as primal variable. This is in contrast to the works of e.g., Laadhari et al. [66] and Sugiyama et al. [90], where the solid equations are formulated only in terms of the velocity. The displacement-based formulation has the advantage that tracking the interface is straightforward by means of the *initial point set method.*

Initial point set method

To close the system of equations, we need to be able to determine the domain affiliation in a specific point $x \in \Omega$, i.e., we need to be able to find out whether it lies in the fluid or solid domain. Therefore, Dunne & Rannacher [31] introduced the *Initial point set function.* Independently, Cottet and co-workers [24] proposed a very similar way to capture the interface. The idea is to trace back a point x to its initial position in $\Omega(0)$. This is trivial within the solid domain $\Omega_\mathrm{s}(t)$, as the solid displacement u_s as a primal variable is known in each time step. In the fluid domain, an extension of u_s is used with the only restriction that points in the fluid domain do not cross the interface. The *Initial point set function* reads

$$\Phi_{IPS}(x,t) := \begin{cases} x - u_s(x,t) & x \in \Omega_s(t)\,, \\ x - \mathrm{ext}(u_s)(x,t) & x \in \Omega_f(t)\,. \end{cases} \tag{3.20}$$

The domain affiliation is then given by

$$\begin{aligned} \Phi_{IPS}(x,t) \in \Omega_s(0) \quad &\Leftrightarrow \quad x \in \Omega_s(t)\,, \\ \Phi_{IPS}(x,t) \notin \Omega_s(0) \quad &\Leftrightarrow \quad x \in \Omega_f(t)\,. \end{aligned} \tag{3.21}$$

It has been noted by Richter [82] that an extension by one layer of mesh cells into the fluid domain is already sufficient if the time-step size is sufficiently small (such that the interface does not move over more than one mesh cell within one time step). As we do not require any regularity of the map Φ_{IPS}, the *Fully Eulerian* method is able to handle arbitrary displacements up to contact of different structures. In particular we note, that the extension of the initial point set to the fluid domain $\Omega_f(t)$ does not necessarily define a map between the initial condition $\Omega_f(0)$ and the current configuration $\Omega_f(t)$. It is sufficient, that Φ_{IPS} maps out of the solid reference configuration $\Omega_s(0)$ for $x \in \Omega_f(t)$, see (3.21)

Alternatively, the domain affiliation can be determined by *Level-set functions*, as in Ladhaari et al. [66] and Sugiyama et al. [90]. This, however, requires one to solve at least one additional equation per time step and errors in the approximation of the level set function might lead to poor mass conservation in the solid.

3.3 Comparison with other approaches

A large number of different approaches have been used for fluid-structure interaction simulations in the past decades. In this section we try to give a brief overview over some important classes of methods, especially those that are in some sense related to the *Fully Eulerian* approach.

We will focus particularly on methods that allow for a monolithic formulation. Monolithic formulations have the advantage of being able to deal with *strongly coupled* problems, e.g., due to the *added-mass effect* (Causin et al. [20]) when fluid and structure density are of the same order. Furthermore, monolithic schemes allow in a straight-forward way for error control, e.g., by means of an a posteriori error analysis, and for gradient-based optimization algorithms. Partitioned schemes, conversely, have the advantage that existent fluid and structure solvers might be used.

Recently some authors succeeded in using the advantages of partitioned solvers within monolithic approaches (e.g., Brummelen et al. [15], Heil et al. [56], Gee et al. [46], Richter [83] and the chapters by Wick & Jodlbauer [65] and Langer & Yang [67] in this book). Here, partitioned solvers are used as preconditioners within a monolithic Krylov space solver or as smoothers within a monolithic multigrid iteration. This combines the advantages of partitioned and monolithic schemes.

Although different criteria are possible, we categorize the approaches based on the coordinate systems that are used to formulate the subproblems. The fluid problem might be formulated in Eulerian or in *arbitrary Lagrangian–Eulerian* (ALE) coordinates, while for the solid problem the Eulerian or Lagrangian coordinate framework is possible.

We begin in Section 3.3.1 with the popular *arbitrary Lagrangian–Eulerian* (ALE) method that uses ALE coordinates for the fluid and Lagrangian coordinates for the solid equations (cf. the chapters by Basting et al. [6], Colciago et al. [23] and Wick [100] in this book). Next, in Section 3.3.2, we describe two important classes of methods that use Eulerian coordinates for the fluid and Lagrangian coordinates for the solid equations. We describe approaches that are based on the *Fictitious domain method* using Lagrange multipliers for the coupling and on relatively new methods using Nitsche's method for the weak incorporation of interface conditions.

Finally, in Section 3.3.3, we describe *Immersed boundary methods*. In these partitioned approaches, Eulerian coordinates are used for the fluid equations as well as for an artificial extension of the fluid equations to the solid domain. Then, solid forces are evaluated in a Lagrangian framework and enter the right-hand side of the Eulerian system of equations. This means that the treatment of the solid includes both the Eulerian and Lagrangian coordinate framework, and, typically, an interpolation between them. These methods are mostly used for rigid body motions, low-dimensional structures or structures occupying no or only a very small volume.

3.3.1 Arbitrary Lagrangian–Eulerian method

The most popular monolithic coupling method is the *arbitrary Lagrangian–Eulerian method* (ALE). Its first ideas within a Finite Volume or Finite Difference discretization date back to the 1960s in the work of Noh [75] (1964) and Franck & Lazarus [36] (1964). In the finite element context, the ALE method was first applied by Donea et al. [26] (1977), Belytschko et al. [10] (1980) and Hughes et al. [62] (1981). A detailed survey can be found in Donea et al. [27]. A very similar approach in a space-time Galerkin framework called *deforming spatial domain/space time (DSD/ST)* has been proposed by Tezduyar and co-workers [7, 92].

Here, the solid equations are solved in the Lagrangian coordinate framework. For the fluid part, we define a domain $\hat{\Omega}_f$ whose boundary matches the boundary $\hat{\Omega}_s$ at the interface and a map $\hat{T}_f\colon \hat{\Omega}_f \times I \to \Omega_f(t)$. A suitable choice for $\hat{\Omega}_f$ is e.g., the initial fluid domain $\Omega_f(0)$. The map $\hat{T}_f$ is *arbitrary* inside $\hat{\Omega}_f$ with the condition to match the solid boundary at the interface).

As in Section 3.2.1, we relate Eulerian quantities f to ALE quantities $\hat{f}$ by the relation

$$\hat{f}(\hat{x}, t) = f(\hat{T}_f(x, t), t) .$$

If the map $\hat{T}_f$ is a $C^{1,1}$-diffeomorphism, an equivalent variational formulation of the Navier–Stokes equations on the fixed domain $\hat{\Omega}_f$ is given by

$$\begin{aligned} \hat{J}_f\rho_f\partial_t\hat{v}_f + \hat{J}_f\rho_f\hat{F}_f^{-1}(\hat{v}_f - \partial_t\hat{T}_f)\cdot\hat{\nabla}\hat{v}_f - \widehat{\operatorname{div}}\,(\hat{J}_f\hat{\sigma}_f\hat{F}_f^T) &= \hat{J}_f\rho_f\hat{f} \\ \widehat{\operatorname{div}}(\hat{J}_f\hat{F}_f^{-1}\hat{v}_f) &= 0 \quad \text{in } \hat{\Omega}_f , \end{aligned}$$

where $\hat{F}_f = \hat{\nabla}\hat{T}_f$ and $\hat{J}_f = \det\hat{F}_f$ (e.g., Dunne et al.[32], Richter[84]). As in the *Fully Eulerian* approach the coupling conditions can be included by choosing a global ansatz and test space $\mathcal{V}$. To define a suitable map $\hat{T}_f$, a harmonic extension $\hat{u}_f$ of $\hat{u}_s$ to the fluid domain might be used to define $\hat{T}_f = \mathrm{id} + \hat{u}_f$.

The key advantage of the ALE method is that for all times $t > 0$ the system of equations is defined on fixed domains $\hat{\Omega}$, $\hat{\Omega}_f$ and $\hat{\Omega}_s$ that are independent of the movement of the physical domains. In a discrete context, this is especially advantageous for the dynamics in the interface region; if the interface is resolved accurately by mesh lines at time $t = 0$, it will be resolved accurately for all times (up to discretization errors). Thus, both interface conditions are fulfilled accurately on the interface $\hat{\Gamma}_i$. Furthermore, in contrast to *fixed-grid* methods no interface-capturing has to be applied, but the interface is tracked naturally by evolving mesh lines.

The disadvantage of the ALE approach lies in the regularity requirement for the ALE map $\hat{T}_f$. This requirement rules out changes of topology, e.g., contact of the solid body with a wall. Furthermore, the ALE map might lose its regularity in the case of large solid displacements. Here, it may help to use a linear elastic or biharmonic extension (Stein et al. [89], Wick [97]) of the solid deformation. However, in the case of large structural displacements, all of these techniques will fail at some point. As an example consider the rotation of a beam (Figure 3.2) where all of the mentioned extensions lead to degenerate mesh cells at some time t before the beam completes a quarter of a full rotation.

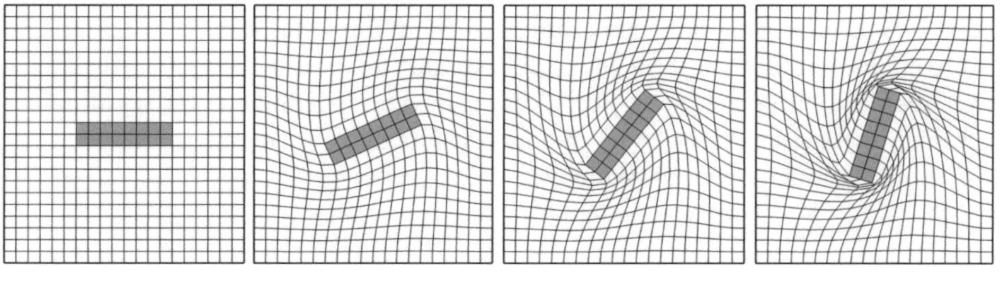

Fig. 3.2: Rotating beam with the ALE method using a biharmonic extension in the fluid domain. Because of degeneration of mesh cells near the beam, the simulation breaks down before the beam completed a quarter of a full rotation.

In this example, the limitations of the ALE method might be overcome by remeshing techniques. The difficulty, especially in a three-dimensional setting, lies in the construction of a new mesh that resolves the interface. Furthermore, the projection has to be chosen carefully, as hereby a further error contribution might be introduced. A simple linear projection might decrease the accuracy considerably.

To overcome the difficulty of remeshing Codina and co-workers [5, 22] defined an interesting variant of the ALE method called the *Fixed-mesh ALE* method. As in the ALE method they use a mesh-moving technique in each time step that follows the interface. Afterwards, however, they project the solution back to a fixed background mesh. In this way they are able to deal with large movements of the structure. We will come back to this method in Section 3.4.3 where we will derive a second-order accurate time-stepping scheme for a Eulerian approach and show its equivalence to a certain space-time variant of the *Fixed-mesh ALE* method.

3.3.2 Euler–Lagrangian methods

There is a big variety of different partitioned methods that solve the fluid equations in Eulerian coordinates and the solid equations in Lagrangian coordinates. It is beyond the scope of this chapter to give a comprehensive overview of all of them. Instead, we want to concentrate on two particular classes of methods that allow for a monolithic formulation.

Coupling via Lagrange multipliers / fictitious domain methods

The fundamental problem with Euler–Lagrangian methods is that the fluid domain $\Omega_f(t)$ and the solid domain $\hat{\Omega}_s$ do not match, which makes the imposition of interface conditions rather challenging. A promising way to tackle this problem is to introduce Lagrange multipliers that are defined at the interfaces $\Gamma_i(t)$ and $\hat{\Gamma}_i$, respectively. This approach stems from the work by Glowinski et al. [50].

Here, we follow the works of Baaijens [1] and Legay, Chessa and Belytschko [68] who applied *Fictitious domain methods* to fluid-structure interaction problems. The idea is to formulate the kinematic interface condition (3.16) weakly in an integral sense but with separate test functions defined only on the interface $\Gamma_i(t)$. To obtain a well-posed system of equations that fulfils the dynamic condition (3.17), we introduce a further variable λ at the interface that has the function of a Lagrange multiplier. The variational formulation reads as follows:

Find the fluid velocity $v_f \in v_f^d + \mathcal{V}_f$, fluid pressure $p_f \in \mathcal{L}_f$, solid velocity $\hat{v}_s \in \mathcal{V}_s$, solid displacement $\hat{u}_s \in u_s^d + \mathcal{W}_s$ and a Lagrange multiplier $\lambda \in \mathcal{Z}_i$ such that

$$\begin{aligned} &(\rho_f(\partial_t v_f + v_f \cdot \nabla v_f), \phi)_{\Omega_f(t)} + (\sigma_f, \nabla\phi)_{\Omega_f(t)} && \\ &-(\rho_f \nu_f \nabla v^T n_f, \phi)_{\Gamma_f \setminus \Gamma_f^d} - (\lambda, \phi)_{\Gamma_i(t)} = (\rho_f f, \phi)_{\Omega_f(t)} && \forall \phi \in \mathcal{V}_f\,, \\ &(\operatorname{div} v, \xi_f)_{\Omega_f(t)} = 0 && \forall \xi_f \in \mathcal{L}_f\,, \\ &(v_f - v_s, \eta_i)_{\Gamma_i(t)} = 0 && \forall \eta_i \in \mathcal{Z}_i\,. \\ &(\rho \partial_t \hat{v}_s, \hat{\phi})_{\hat{\Omega}_s} + (\hat{J}\hat{\sigma}\hat{F}^T, \hat{\nabla}\hat{\phi})_{\hat{\Omega}_s} + (\hat{\lambda}, \hat{\phi})_{\hat{\Gamma}_i} = (\hat{\rho}_s^0 \hat{f}, \hat{\phi})_{\hat{\Omega}_s} && \forall \hat{\phi} \in \mathcal{V}_s\,, \\ &(d_t \hat{u}_s - \hat{v}_s, \hat{\psi})_{\hat{\Omega}_s} = 0 && \forall \hat{\psi} \in \mathcal{W}_s\,. \end{aligned} \tag{3.22}$$

Note that the Lagrange multiplier appears in the solid equation $\hat{\lambda}$ and in the fluid equation as λ. However, $\hat{\lambda}$ is not an additional variable but the Lagrangian counterpart of λ defined by $\hat{\lambda}(\hat{x}, t) = \lambda(\hat{T}(\hat{x}, t), t)$, where $\hat{T}$ denotes the map from Lagrangian to Eulerian coordinates defined in Section 3.2.1. Hence, the dynamic condition (3.17) follows after integration by parts and a change of the coordinate system

$$\sigma_f n_f = \lambda\,, \quad \hat{\lambda} = -\hat{J}\hat{\sigma}_s \hat{F}^T \hat{n}_s \qquad \Rightarrow \qquad \sigma_f n = \sigma_s n\,.$$

A similar approach introducing further variables has been proposed by Park et al. [76] and Gerstenberger & Wall [48].

The following side ingredients are necessary to obtain an efficient and accurate numerical algorithm:

- A strategy to capture or track the interface has to be added to close the system of equations. Possible choices are a level-set approach (Legay et al. [68]) or a Lagrangian tracking technique to evolve the interface from time step to time step (Gerstenberger & Wall [48]).
- In a finite element discretization, an *inf-sup* condition has to be ensured for both fluid and solid subproblems. A popular choice for the Lagrange multipliers is to use polynomials of one degree lower than the velocity ansatz space, see e.g., Gerstenberger & Wall [48].
- As in the *Fully Eulerian* method, the fluid domain $\Omega_f(t)$ changes with time. The idea of the *Fictitious domain* method is to use a fixed background mesh including the solid domain for all times. Here, however, the interface is not aligned with mesh cells. For an accurate representation of the interface forces, a suitable enrichment of the finite element space (e.g., XFEM [73]) can be used [48].

Coupling via Nitsche's method

Nitsche's method [74] was originally introduced to incorporate Dirichlet boundary conditions for elliptic problems weakly into the variational formulation. Hansbo & Hansbo extended this concept to interface conditions for elliptic problems with discontinuous coefficients [52]. Recently, this method has also been used for imposing interface conditions between nonmatching grids for fluid-structure interaction problems (e.g., Hansbo & Hermansson [54], Burman & Fernández [16], Massing et al. [71]).

As in the previous subsection, the idea is to use a Eulerian framework and a fixed grid for the fluid equations and a Lagrangian framework for the solid equations. Again, the system of equations can be formulated into one common monolithic formulation by transforming the respective boundary conditions between the two coordinate systems. In contrast to the *Fully Eulerian* and the ALE method as formulated in Section 3.2.2 and Section 3.3.1, it is important that no global test or trial spaces are used for any of the variables and the spaces for fluid and solid variables remain separated.

Applying Nitsche's method to the kinematic interface condition, the momentum equation reads:

$$\begin{aligned}
&(\rho_f(\partial_t v_f + v_f \cdot \nabla v_f), \phi_f)_{\Omega_f(t)} + (\sigma_f, \nabla\phi_f)_{\Omega_f(t)} - (\rho_f \nu_f \nabla v^T n_f, \phi)_{\Gamma_f \setminus \Gamma_f^d} \\
&\quad + (\rho_s \partial_t \hat{v}_s, \hat{\phi}_s)_{\hat{\Omega}_s} + (\hat{J}\hat{\sigma}_s \hat{F}^T, \hat{\nabla}\hat{\phi}_s)_{\hat{\Omega}_s} - (\sigma_f n_f, \phi_f - \phi_s)_{\Gamma_i(t)} \\
&\quad - (v_f - v_s, S_f(\phi_f, \xi_f) n_f)_{\Gamma_i(t)} + \frac{\gamma\nu}{h}(v_f - v_s, \phi_f - \phi_s)_{\Gamma_i(t)} \\
&\qquad = (\rho_f f, \phi_f)_{\Omega_f(t)} + (\hat{\rho}_s^0 \hat{f}, \hat{\phi}_s)_{\hat{\Omega}_s} \qquad \forall \phi_f \in \mathcal{V}_f, \hat{\phi}_s \in \mathcal{V}_s \,. \qquad (3.23)
\end{aligned}$$

Here, $\gamma > 0$ denotes the Nitsche penalty parameter, the test function ϕ_s on $\Gamma_i(t)$ is defined by $\phi_s(x) = \hat{\phi}_s(T(x,t),t)$ and the operator

$$S_f(\phi_f, \xi_f) = \frac{1}{2}\rho_f \nu_f (\nabla\phi_f + \nabla\phi_f^T) - \xi_f I$$

is the Cauchy stress operator applied to the fluid test functions. Because of the dynamic condition (3.17), σ_s and the solid stress operator $S_s(\phi_s)$ as well as a combination of fluid and solid stresses could be used instead of σ_f and $S_f(\phi_f, \xi_f)$ in the Nitsche interface terms. The interface conditions are again included in the variational formulation, which can be seen after integration by parts.

One advantage of this formulation is a solid mathematical foundation. Burman & Fernández showed in a theoretical analysis for a linear FSI problem both the well-posedness of the complete system of equations as well as convergence of discrete finite element solutions [16]. For discretization, a Lagrangian (moving) mesh is used for the solid and a Eulerian (fixed) mesh for the fluid part. The fixed mesh covers either the complete domain $\Omega(t)$ or might be reduced in each time step by excluding those cells that lie completely within the solid domain. In any case, fluid and solid meshes overlap and fluid cells are cut by the interface in arbitrary portions. To ensure discrete coercivity, a stabilization term $g_h(v_f, \phi_f)$ is added to the fluid part. The authors claim that the system matrix stays bounded independently of the position of the cut.

The approach can be used both in a monolithic and in a partitioned way (Hansbo & Hermansson [54], Burman & Fernández [16]). In order to impose the interface conditions accurately, the Nitsche approach might be combined with an enrichment of the finite element basis [16].

3.3.3 Immersed boundary methods

The *Immersed boundary method* introduced by Peskin [78] was originally designed for fiber-like or low-dimensional structures occupying no or only negligibly small volume. The idea is to formulate the whole system of equations as if it would be a flow problem and use the Navier–Stokes equations in the complete domain $\Omega(t)$. In order to simplify notation, let us assume that the source terms f on the right-hand side of fluid and solid equations are zero. The equations are then written in the following form

$$\begin{aligned} \rho_f \partial_t v + \rho_f (v \cdot \nabla) v - \operatorname{div} \sigma_f &= g_s \\ \operatorname{div} v &= 0 \quad \text{in } \Omega(t) \, . \end{aligned} \tag{3.24}$$

The coupling is included by means of a force term g_s on the right-hand side that is zero in the fluid domain and that contains all the solid contributions subtracted by the fluid terms that appear on the left-hand side in the solid domain. To be precise g_s is defined by

$$g_s(x,t) = \begin{cases} (\rho_f - \rho_s)(\partial_t v + v \cdot \nabla v) + \operatorname{div}(\sigma_f - \sigma_s) & \text{in } \Omega_s(t) \\ 0 & \text{in } \Omega_f(t) \, . \end{cases} \tag{3.25}$$

This force term can be equivalently represented with a Dirac distribution δ

$$g_s(x,t) = \int_{\Omega_s(t)} g_s(y,t) \delta(x-y) dy \, . \tag{3.26}$$

This representation will be used for discretization. The *Immersed boundary method* was originally used with finite difference discretizations. The solid forces g_s are evaluated on a Lagrangian solid mesh that is typically much finer than the fluid mesh and that is moved with the fluid velocity. On this mesh, a discrete smooth Dirac delta function δ_h is used as approximation to δ. A lot of effort has been spent in the literature (e.g., Wang & Liu [95], Gil et al. [49]) on defining appropriate and accurate discrete approximations of the Dirac function. To the knowledge of the authors, however, a theoretical analysis on how these approximations affect the convergence of discrete solutions is missing.

Later on, the method was used as an *Immersed finite element method* and with structures occupying a finite volume (Wang & Liu [95], Zhang et al. [101], Boffi et al. [12]). In this case it is possible to formulate the whole system of equations without the delta function by means of a variational formulation, see Boffi & Gastaldi [11].

While the continuous formulation might be formulated as a variational coupling approach [11], the *Immersed boundary method* is usually used as a partitioned approach, where the solid forces are evaluated on a separate Lagrangian mesh and interpolated to the fluid mesh before the fluid problem is solved. In fact, in most works, the method is applied in situations, where no solid equations have to be solved at all, but the solid forces under consideration can be directly calculated from interface quantities.

3.4 Discretization

In this section we present discretization techniques in space and time for Eulerian approaches that take the low regularity of the solution across the interface into account. Furthermore, we introduce a pressure stabilization technique that is capable of dealing with anisotropic elements and a stabilization for the structure equations that is able to deal with some inherent instabilities.

3.4.1 Space discretization

Using interface-capturing methods, the fluid-structure interface typically moves over a fixed background mesh. The mesh cells that are cut by the interface lie partly in the fluid domain, partly in the solid domain. Without further adjustment of the mesh or enrichment of the local finite element space, this leads to a poor approximation in the interface cells and may give rise to stability problems. To avoid these problems, we use a modified finite element scheme [38] in the interface cells. Our approach fits both into the frameworks of a fitted finite element approach and an enrichment method.

The key idea is to use a fixed background mesh consisting of patches $P \in \mathcal{T}_{2h}$ that is independent of the interface location and a further subdivision into triangular or quadrilateral cells that resolves the interface. In this article, we use quadrilateral patch cells. Corresponding approaches on triangular patches have been developed by Gangl & Langer [45] and by Hoffman et al. [58].

Let us assume that $\mathcal{T}_{2h}$ is a form and shape-regular triangulation of the domain $D \subset \mathbb{R}^2$ into open quadrilaterals and make the assumption that a patch is either not cut or cut in exactly two edges. If the interface Γ_i is smooth, this can be achieved after sufficient refinement.

We split each patch into four quadrilaterals, if it is not cut by Γ_i. If Γ_i goes through a patch P, we split it into eight triangles so that the boundary is resolved in a linear approximation. To achieve this, we place degrees of freedom to the points of intersection $e_j \in \partial P \cap \Gamma_i, j = 1, 2$, see Figure 3.3 and Figure 3.4.

We define the finite element trial space V_h as an iso-parametric space on the triangulation $\mathcal{T}_{2h}$:

$$V_h = \left\{ \phi \in C(\bar{D}),\ \phi \circ T_P^{-1}\Big|_P \in \hat{Q}_P \quad \text{for all patches } P \in \mathcal{T}_{2h} \right\},$$

where $T_P \in [\hat{Q}_P]^2$ is the mapping between the reference patch $\hat{P} = (0,1)^2$ and the patch $P \in \mathcal{T}_{2h}$, such that $T_P(\hat{x}_i) = x_i^P$, $(i = 1, \dots, 9)$ for the nine nodes $x_1^P, \dots, x_9^P$ of the patch, see Figure 3.4. The local space $\hat{Q}_P$ is defined depending on whether it is cut by the interface or not. In the first case, we use a piecewise linear space

$$\hat{Q}_P = \hat{Q}_{\text{mod}} := \left\{ \phi \in C(\bar{P}),\ \phi\Big|_{T_i} \in \text{span}\{1, x, y\},\ T_1, \dots, T_8 \subset P \right\},$$

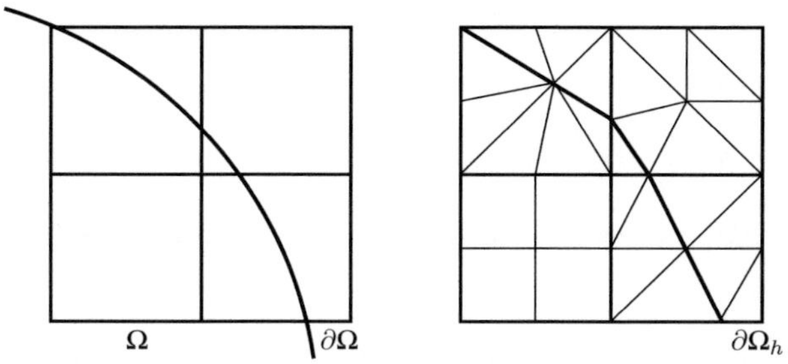

Fig. 3.3: Left: Triangulation $\mathcal{T}_{2h}$ of a domain D that contains Ω. Right: Subdivision of the patches $P \in \mathcal{T}_{2h}$ such that the boundary $\partial\Omega$ is resolved in a linear approximation by the discrete boundary $\partial\Omega_h$.

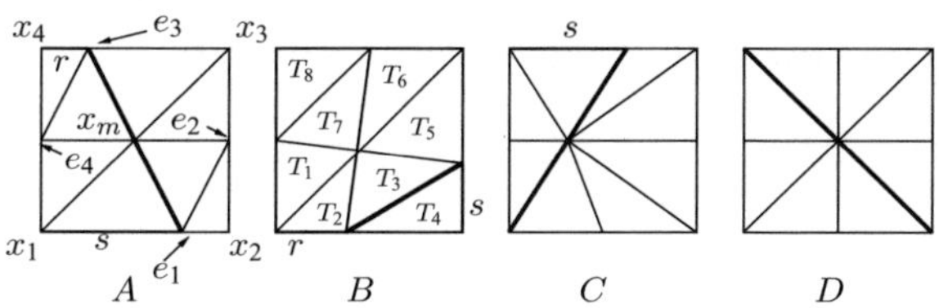

Fig. 3.4: Different types of cut patches. The subdivision can be anisotropic with $r, s \in (0, 1)$ arbitrary.

in the second case a piecewise bilinear one:

$$\hat{Q}_P = \hat{Q} := \left\{ \phi \in C(\bar{P}),\ \phi\big|_{K_i} \in \operatorname{span}\{1, x, y, xy\},\ K_1, \dots, K_4 \subset P \right\}.$$

Note that here and in our implementation the reference elements are entire patches, not the subcells. Depending on the position of the boundary $\partial\Omega$ in the patch P, three different reference elements are considered. For the details, we refer to [38]. The functions in $\hat{Q}$ and $\hat{Q}_{\text{mod}}$ are all piecewise linear on the edges ∂P, such that mixing different element types does not affect the continuity of the global finite element space.

As the cut of the elements can be arbitrary with $r, s \to 0$ or $r, s \to 1$, the triangle's aspect ratio can be very large, considering $h \to 0$ it is not necessarily bounded. We can however guarantee that the maximum angles in all triangles will be well bounded away from 180°:

Lemma 3.1 (Maximum angle condition). *All interior angles of the triangles shown in Figure 3.4 are bounded by* 144° *independent of* $r, s \in (0, 1)$.

Proof. See Frei & Richter [38]. □

Although we formulate the approach as a parametric approach on the patch mesh $\mathcal{T}_{2h}$, it is obvious that the discretization is equivalent to a mixed linear-bilinear discretization on a finer mesh $\mathcal{T}_h$ that consists of the subtriangles and subquadrilaterals. With the help of the maximum angle condition, the following interpolation estimate is well known for the nodal interpolant $I_h u$

$$\|u - I_h u\|_{L^2(\Omega)} + h_P \|u - I_h u\|_{H^1(\Omega)} \leq C h_P^2 \|\nabla^2 u\|_\Omega .$$

On the basis of this result, we can show the following theorem for the elliptic interface problem

$$-\operatorname{div}(\kappa \nabla u) = f \text{ in } \Omega , \quad [\kappa \partial_n u] = 0, \ [u] = 0 \text{ on } \Gamma_i \tag{3.27}$$

where $\Omega = \Omega_1 \cup \Gamma_i \cup \Omega_2$ and the coefficient κ might be discontinuous across Γ_i.

Theorem 3.2 (A priori estimate). *Let $\Omega \subset \mathbb{R}^2$ be a domain with convex polygonal boundary, split into $\Omega = \Omega_1 \cup \Gamma_i \cup \Omega_2$, where Γ_i is a smooth interface with C^2-parametrization. We assume that Γ_i divides Ω in such a way that the solution $u \in H_0^1(\Omega)$ of (3.27) satisfies the stability estimate*

$$u \in H_0^1(\Omega) \cap H^2(\Omega_1 \cup \Omega_2), \quad \|u\|_{H^2(\Omega_1 \cup \Omega_2)} \leq c_S \|f\| .$$

For the corresponding modified finite element solution $u_h \in V_h$, it holds that

$$\|\nabla(u - u_h)\|_\Omega \leq C h_P \|f\|, \quad \|u - u_h\|_\Omega \leq C h_P^2 \|f\| .$$

Proof. Frei & Richter [38]. □

Condition number

The drawback of this approach is that, even for this simple elliptic interface problem, the condition number of the corresponding system matrix might blow up for certain positions of the interface, that lead to very anisotropic cells, that do not fulfil a minimum angle condition. A solution to this issue is to use a particular finite element basis for V_h. Therefore, we split the space in a hierarchical manner

$$V_h = V_{2h} + V_b, \quad N := \dim(V_h) = \dim(V_{2h}) + \dim(V_b) =: N_{2h} + N_b .$$

The space V_{2h} is the standard space of piecewise bilinear or linear functions on the patches $P \in \Omega_{2h}$, equipped with the usual nodal Lagrange basis $V_{2h} = \operatorname{span}\{\phi_{2h}^1, \dots, \phi_{2h}^{N_{2h}}\}$. Patches cut by the interface are split into two large triangles.

The space $V_b = V_h \setminus V_{2h}$ collects all functions that are needed to enrich V_{2h} to V_h. This space is spanned by the nodal Lagrangian basis corresponding to the five interior nodes of a patch, see Figure 3.5 for an example of the splitting. The basis of this space is denoted by $V_b = \operatorname{span}\{\phi_b^1, \dots, \phi_b^{N_b}\}$. The finite element space V_{2h} is fully isotropic and a minimum angle condition holds. Functions in V_{2h} do not resolve the interface, while the basis functions $\phi_b^i \in V_b$ will depend on the interface location if $\Gamma_i \subset \operatorname{supp} \phi_b^i$.

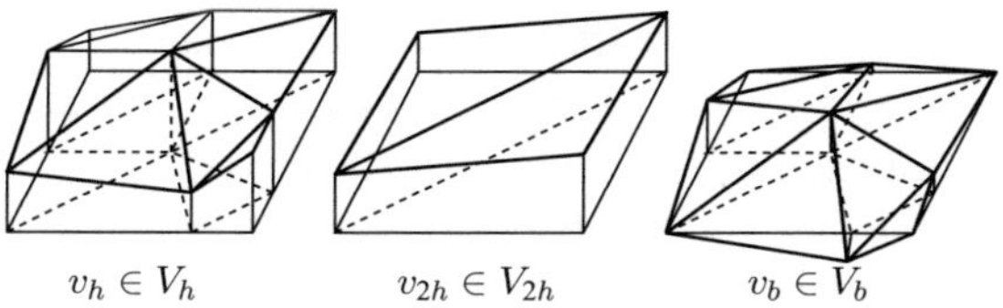

Fig. 3.5: Example for a hierarchical splitting of a function $v_h \in V_h$ into coarse mesh part $v_{2h} \in V_{2h}$ and fine mesh fluctuation $v_b \in V_b$.

Remark 3.3. By using the splitting $V_h = V_{2h} + V_b$, the proposed method can also be seen as an enrichment method (e.g., an extended finite element method [73]). Therefore, we consider the space V_{2h} as the basic finite element space and the space V_b as the enrichment to capture the *weak discontinuity* at the interface.

The system matrix is now built up by combining the usual Lagrangian basis functions of both the coarse space V_{2h} and the fine space V_b. We can guarantee the usual bound for the system matrix after an appropriate scaling of the basis functions, such that

$$C^{-1} \le \|\nabla \phi_h^i\| \le C, \quad i = 1, \dots, N_b \,. \tag{3.28}$$

In practice, this can be achieved by a simple row- and columnwise scaling with the diagonal elements of the matrix. For this modified basis, we show the following result:

Theorem 3.4 (Condition number). *Using the hierarchical finite element basis consisting of the functions $\phi_b^1, \dots, \phi_b^{N_b}, \phi_{2h}^1, \dots, \phi_{2h}^{N_{2h}}$ for the construction of the system matrix corresponding to Problem (3.27), the following bound for the condition number holds with a constant $C > 0$ not depending on the interface location*

$$\mathrm{cond}_2(\mathbf{A}) \le C h_P^{-2} \,.$$

For the technical details, we refer to [38]. Note that only the basis of the finite element space has been changed. The resulting space V_h is the same and hence the a priori error estimates of Theorem 3.2 remain valid.

3.4.2 Pressure stabilization

To apply this discretization technique to the full FSI system, we will use *equal-order* locally modified finite elements for all the components of the solution. It is well known that this choice for velocity and pressure violates the *inf-sup* condition in the fluid problem that is necessary to ensure the well-posedness of the discrete system of equations. To cope with this, we will add stabilization terms to the discrete formulation.

The challenge for pressure stabilization for the locally modified FEM lies in the anisotropies that are present in the interface region including abrupt changes of anisotropy between neighboring cells. Typically, the stabilization methods used on anisotropic meshes require an assumption on the change of anisotropy between neighboring cells that can not be guaranteed here (e.g., Braack & Richter [13]).

As the discretization is based on a background patch mesh, one could think of a stabilization based on *local projections* (LPS) onto the patch grid (Becker & Braack [8]). However, there are patches that lie only partly in the fluid domain Ω_{f}. Therefore, an application of local projections needs an artificial extension of the fluid pressure to the entire interface patches, which couples back to the fluid equations by means of the stabilization terms.

To avoid this, we will use a variant of the *Continuous Interior Penalty* (CIP) stabilization technique introduced by Burman & Hansbo for convection-diffusion-reaction problems [18] and later on for pressure stabilization within the Stokes [19] and the Navier–Stokes equations [17]. The original CIP technique is based on penalizing jumps of the gradient over element edges weighted by a factor of $\mathcal{O}(h^s)$ for $s = 2$ or $s = 3$. This is not applicable for the case of abrupt changes of anisotropy, however, as the cell size of the two neighboring cells can be very different. Hence, we have to modify this technique in the interface patches using a weighted average of the pressure gradient instead of the jump terms.

To define the stabilization, we split the set of edges into two parts: By $\mathcal{E}_h^0$, we denote all edges that lie between two quadrilateral cells $K_1, K_2 \subset \Omega_h^f$. By $\mathcal{E}_h^i$ we denote the edges that are edges of at least one triangular element $K \subset \Omega_h^f$. In other words, this means that $\mathcal{E}_h^i$ contains all edges that are part of patches cut by the interface.

We define the stabilization term by

$$S(p_h, \psi_h) := \gamma h_P^2 \Big(\sum_{e \in \mathcal{E}_h^i} \int_e \{h_{\mathrm{n}} \nabla p_h \cdot \nabla \psi_h\}_e \, do + \sum_{e \in \mathcal{E}_h^0} \int_e h_{\mathrm{n}} [\nabla p_h]_e \cdot [\nabla \psi_h]_e \, do \Big) ,$$

where $\gamma > 0$ is a constant, h_{n} is the cell size in the direction normal to e, $[\,\cdot\,]_e$ denotes the jump term across the edge e and

$$\{v_h\}_e := \frac{1}{2} \left(v_{h|K_1} + v_{h|K_2} \right)$$

is the mean value of the two cells K_1, K_2 sharing the edge e. In the case that one of the cells K_1, K_2 does not lie in Ω_h^f, we set the respective contribution to zero. This term is then added to the divergence equation.

Using this stabilization, error estimates of optimal order could be shown for the locally modified finite element method applied to the Stokes problem: *Find* $v_h \in \mathcal{V}_h := (V_h \cap H_0^1(\Omega_h^f))^2$, $p_h \in \mathcal{L}_h := V_h$ *such that*

$$A_h(v_h, p_h)(\phi_h, \psi_h) + S(p_h, \psi_h) = (f, \phi_h)_\Omega \quad \forall \phi_h \in \mathcal{V}_h, \psi_h \in \mathcal{L}_h , \tag{3.29}$$

where

$$A_h(v_h, p_h)(\phi_h, \psi_h) : = \nu(\nabla v_h, \nabla \phi_h)_{\Omega_h} - (p_h, \operatorname{div} \phi_h)_{\Omega_h} + (\operatorname{div} v_h, \psi_h)_{\Omega_h} .$$

Theorem 3.5. *Let (v_h, p_h) be the solution of the stabilized discrete system (3.29). The following stability estimate holds with a constant C that is independent of the interface position*

$$\begin{aligned}\|\nabla v_h\|_{\Omega_h} + \|p_h\|_{\Omega_h} + h_P\|\nabla p_h\|_{\Omega_h} \\ \le C \sup_{(\phi_h,\psi_h)\in\mathcal{V}_h\times\mathcal{L}_h} \frac{A_h(v_h,p_h)(\phi_h,\psi_h) + S(p_h,\psi_h)}{\|\nabla\phi_h\|_{\Omega_h} + \|\psi_h\|_{\Omega_h} + h_P\|\nabla\psi_h\|_{\Omega_h}}\,.\end{aligned}$$

Furthermore, let (v, p) be the solution of the corresponding Stokes problem on the continuous level. For $v \in H^2(\Omega) \cap H_0^1(\Omega)$ and $p \in H^1(\Omega)$, it holds that

$$\|\nabla(v - v_h)\|_\Omega + \|p - p_h\|_\Omega \le C h_P \left(\|\nabla^2 v\|_\Omega + \|\nabla p\|_\Omega\right)\,.$$

and

$$\|v - v_h\|_\Omega \le C h_P^2 \left(\|\nabla^2 v\|_\Omega + \|\nabla p\|_\Omega\right)\,.$$

For the details and the technical proof, we refer to Frei [37]. Finally, we remark that in our practical implementation, we also used the smaller stabilization term

$$\begin{aligned}S_2(p_h,\psi_h) := \gamma\Big(&\sum_{e\in\mathcal{E}_h^i}\int_e \{h_{\mathrm{n}}(h_{\mathrm{n}}^2\partial_n p_h\partial_n\psi_h + h_\tau^2\partial_\tau p_h\partial_\tau\psi_h)\}_e\,do \\ &+ \sum_{e\in\mathcal{E}_h^0}\int_e h_{\mathrm{n}}^3[\nabla p_h]_e\cdot[\nabla\psi_h]_e\,do\Big)\end{aligned} \tag{3.30}$$

that uses the local cell sizes h_{n} in direction normal to the edge and the edge length h_τ instead of the patch size h_P. This case is, however, not covered by the stability and convergence analysis leading to Theorem 3.5.

3.4.3 Temporal discretization

Concerning time discretization, we have to deal with similar problems as in space. The global velocity v is continuous across the interface, but not differentiable, neither in space nor in time. Applying a standard time-stepping scheme will thus lead to a reduced order of convergence.

Surprisingly, only very few works can be found in the literature on this issue. Among these works, we mention the work of Zunino [102], who derived a complete error analysis for the backward Euler method applied to a parabolic interface problem. A second-order scheme based on a space-time dG(1) approach was presented by Lehrenfeld and Reusken [69] including error analysis in space and time. For a Crank–Nicolson-like approach, Fries and Zilian found a reduced convergence order of 1.5 [44]. A modified Crank–Nicolson-type scheme and a complete convergence analysis of second order was introduced by the authors recently [39].

To illustrate the difficulties, we define the space-time domain

$$Q := \{(t, \Omega(t)),\ t \in I := [0, T]\} \subset \mathbb{R}^{d+1}$$

and consider the simple parabolic interface problem

$$\begin{aligned} \partial_t u_i - \operatorname{div}(\kappa_i \nabla u_i) &= f_i && \text{in } Q_i,\ i = 1, 2\,, \\ u_1 = u_2, \quad \vec{n} \cdot \kappa_1 \nabla u_1 &= \vec{n} \cdot \kappa_2 \nabla u_2 && \text{on } \Gamma(t)\,, \\ u(\cdot, 0) = u^0 \text{on } \omega(0), \quad u(\cdot, t) &= 0 && \text{on } \partial\Omega(t)\,, \end{aligned} \tag{3.31}$$

where the diffusion coefficient $\kappa\colon Q \to \mathbb{R}$ takes two values $\kappa_1, \kappa_2 \in \mathbb{R}$ in the subdomains Q_1, Q_2.

We discretize the domain in time and apply Rothe's method with a simple one-step method for $\theta \in [0, 1]$

$$\begin{aligned} \frac{1}{t_m - t_{m-1}}(u^m - u^{m-1}, \phi) + \theta(\kappa(t_m)\nabla u^m, \nabla\phi) + (1-\theta)(\kappa(t_{m-1}\nabla u^{m-1}, \nabla\phi) \\ = (\theta f(t_m) + (1-\theta) f(t_{m-1}, \phi) \quad \forall \phi \in V(t_m)\,. \end{aligned} \tag{3.32}$$

Here, $u^{m-1} \in V(t_{m-1})$ is the solution at time t_{m-1} and $u_m \in V(t_m)$ is the sought solution at time t_m. In the case of moving domains, it holds $\Omega(t_{m-1}) \neq \Omega(t_m)$ and therefore $V(t_{m-1}) \neq V(t_m)$. The problem comes to the fore, if one considers the role of the scalar product $(u^m - u^{m-1}, \phi) = \int_\Omega (u^m - u^{m-1})\phi\, \mathrm{d}x$. Whether we choose $\Omega(t_{m-1})$ or $\Omega(t_m)$ as the domain for integration, the integral is not defined for one of the solutions u^m or u^{m-1}.

In many applications, it is sufficient to consider interface problems on a fixed domain $Q := I \times \Omega$ where only the interior interface moves, but where the outer boundary is fixed. Here, the problem looks less severe. Equation (3.32) is well defined. However, consider a point $x \in \Omega$ with $x \in \Omega_1(t_{m-1})$ and $x \in \Omega_2(t_m)$ close to the interface. Then, the difference quotient $(u^m(x) - u^{m-1}(x))/(t_m - t_{m-1})$ in (3.32) is not a reasonable approximation to the time-derivative u', as u is not differentiable across the interface.

Galerkin time discretization

The most important variants of the one-step methods (3.32) are the backward Euler method ($\theta = 1$) and the Crank–Nicolson scheme ($\theta = 0.5$). It is well known that for linear problems with smooth coefficients, the Crank–Nicolson scheme is equivalent to a continuous Galerkin approach using linear polynomials (cG(1)) in time and the backward Euler scheme is equivalent to a discontinuous Galerkin scheme using piecewise constants (dG(0)) up to numerical quadrature. Hence, the backward Euler scheme is first-order, the Crank–Nicolson scheme second-order convergent. If the coefficient κ is discontinuous across the interface, however, a reduced order of convergence is observed. We will analyze this effect for the corresponding Galerkin approaches.

The continuous variational formulation of (3.31) reads: Find $u \in X$ such that

$$\begin{aligned} B(u, \phi) &= (f, \phi)_Q + (u^0, \phi(0))_{\Omega(0)} \quad \forall \phi \in X\,, \\ B(u, \phi) &:= (\partial_t u, \phi)_Q + (\kappa \nabla u, \nabla \phi)_Q + (u(0), \phi(0))_{\Omega(0)}\,, \end{aligned} \tag{3.33}$$

where $(f, g)_Q := \int_0^T (f(t), g(t))_{\Omega(t)}\, dt$ and

$$X := W(0, T) = \left\{ v \colon Q \to \mathbb{R} \;\middle|\; v \in L^2(I, H_0^1(\Omega(t))),\ \partial_t v \in L^2(I, H^{-1}(\Omega(t))) \right\} .$$

The well-posedness of this variational problem for moving outer boundaries has been studied by Dziri & Zolésio [33].

In order to derive a time-stepping scheme, we split the time interval into M discrete subintervals $I_j = (t_{j-1}, t_j], j = 1 \ldots M$ and denote the resulting space-time slabs by

$$Q^j := \{(x, t) \mid t \in I_j,\ x \in \Omega(t)\} .$$

Let us first assume that the outer boundary $\partial\Omega(t)$ is fixed such that $\Omega(t) = \Omega$ for all times t. The standard Crank–Nicolson scheme ($\theta = 0.5$) is equivalent to a space-time variational formulation with the following cG(1) Galerkin ansatz and trial spaces

$$\begin{aligned} u_k \in \tilde{X}_k^1 &= \left\{ v \in C(\bar{I}, H_0^1(\Omega)) \;\middle|\; v_{|I_m} \in P_1(I_m, H_0^1(\Omega)), v(0) \in H_0^1(\Omega) \right\} \\ \phi_k \in \tilde{X}_k^0 &= \left\{ v \in L^2(I, H_0^1(\Omega)) \;\middle|\; v_{|I_m} \in P_0(I_m, H_0^1(\Omega)), v(0) \in H_0^1(\Omega) \right\} . \end{aligned} \tag{3.34}$$

For the backward Euler scheme, we choose for the trial space $u_k \in \tilde{X}_k^0$ as for the test space.

If the coefficient κ was continuous across the interface (in our case $\kappa_1 = \kappa_2$), second-order convergence for the time discretization error would result for the cG(1) scheme. This is not the case for a discontinuous coefficient, however, as the scheme does not account for the (moving) discontinuity of κ, f and ∇u at the space-time interface G. Instead the functions $u_k \in \tilde{X}_k^1$ are polynomial on lines ζ that cross the interface (e.g., the arrow crossing the interface in Figure 3.6), which means in particular $u_k \in C^\infty(\zeta)$. As for the spatial discretization, it follows that, in general, there is no second-order in time interpolant within the space $\tilde{X}_k^1$ and we can only expect a reduced order of convergence. In the next subsection, we will therefore modify these spaces in order to show optimal-order convergence estimates.

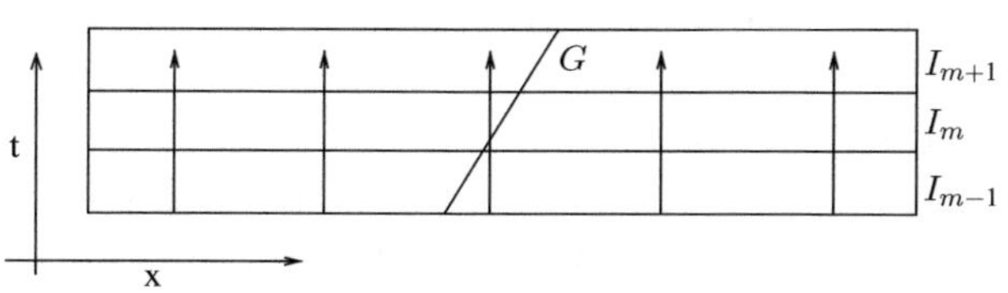

Fig. 3.6: Space-time domain for a fixed outer domain $\Omega(t) = \Omega$. Functions $v_k \in \tilde{X}_k^0, \tilde{X}_k^1$ are polynomial on vertical lines (e.g., the indicated arrows).

For the dG(0) scheme, first-order convergence can still be shown for the standard formulation. Nevertheless, we will derive a modified dG(0) scheme as well, which will show better stability properties in practice. Furthermore, the modified schemes will be applicable on moving domains $\Omega(t)$ in a natural way.

Modified Galerkin approaches

Motivated by the previous considerations, we define a Galerkin space of functions that are polynomial on trajectories that stay within the subdomains and are aligned to the space-time boundary and the interface. The construction of a second-order interpolant in time will be straight-forward within this space.

In principle, it would be convenient to introduce smooth global trajectories in the whole time-interval I. However, in the case of large movements of the interface, this might be a challenging task. Furthermore, in fluid-structure interaction problems, the interface movement depends on the solution itself and is therefore only known from time step to time step. Therefore, we define the trajectories piecewise in each time interval I_m (Figure 3.7).

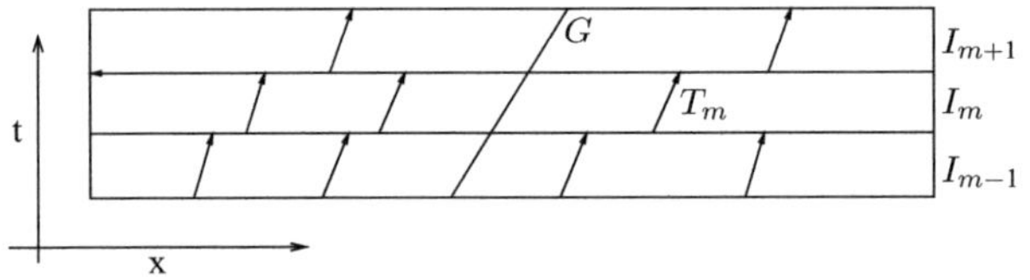

Fig. 3.7: Illustration of the modified Galerkin ansatz spaces X_k^0, X_k^1. The functions $v_k \in X_k^0, X_k^1$ are polynomial on trajectories that stay within each subdomain Q_i, $i = 1, 2$.

Specifically, we define the following (semidiscrete) test and trial spaces for the cG(1) approach

$$\begin{aligned} u_k \in X_k^1 &= \left\{ v \in C(\bar{I}, H_0^1(\Omega)) \;\middle|\; (v \circ T_m)|_{I_m} \in P_1(I_m, H_0^1(\Omega)),\ v(0) \in H_0^1(\Omega) \right\} \\ \phi_k \in X_k^0 &= \left\{ v \in L^2(I, H_0^1(\Omega)) \;\middle|\; (v \circ T_m)|_{I_m} \in P_0(I_m, H_0^1(\Omega)),\ v(0) \in H_0^1(\Omega) \right\}. \end{aligned} \quad (3.35)$$

By T_m we denote an arbitrary transformation from a reference domain $\hat{\Omega}^m$ to the space-time domain Q^m that maps $\hat{\Gamma}^m$ to $\Gamma(t)$, $\hat{\Omega}_1^m$ onto $\Omega_1(t)$ and $\hat{\Omega}_2^m$ onto $\Omega_2(t)$. It is important that a point $x \in \hat{\Omega}_j^m$ will never be mapped to $\Omega_k(t)$ for $j \neq k$. Again, the dG(0) scheme results by choosing X_k^0 for both trial and test functions.

In this work, we choose the domain at the new time step $\hat{\Omega}^m = \Omega(t_m)$ as the reference domain. Other choices, e.g., $\hat{\Omega}^m = \Omega(t_{m-1})$ would be possible, as well. The space-time slabs in the reference system will be denoted by $\hat{Q}^j := \hat{\Omega}^j \times I_j$, $j = 1, \dots, M$.

Now the discrete formulation in the ansatz and trial space defined in (3.35) reads: Find $u_k \in X_k^1$ (resp. X_k^0) such that

$$B(u_k, \phi_k) = (f, \phi_k)_Q + (u^0, \phi(0))_{\Omega(0)} \quad \forall \phi_k \in X_k^0 . \tag{3.36}$$

Time-stepping scheme

Our goal is to derive an efficient time-stepping scheme from the Galerkin approach. Therefore, note that we can formulate the bilinear form on the reference domain $\hat{\Omega}^m$ in I_m using the local maps T_m, cf. Figure 3.8:

$$\begin{aligned} B^m(u, \phi) : &= (\partial_t u, \phi)_{Q^m} + (\kappa \nabla u, \nabla \phi)_{Q^m} \\ &= \left(J_m \partial_t \hat{u} - \partial_t T_m (JF^{-T})_m \hat{\nabla}_m \hat{u}, \hat{\phi}\right)_{\hat{Q}^m} \\ &\quad + \left(\hat{\kappa} (JF^{-T})_m \hat{\nabla}_m \hat{u}, F_m^{-T} \hat{\nabla}_m \hat{\phi}\right)_{\hat{Q}^m} , \end{aligned} \tag{3.37}$$

where $F_m = \nabla T_m$ denotes the spatial derivative of the transformation and $J_m = \det(F_m)$ its determinant. The notation $\hat{\nabla}_m$ denotes the gradient with respect to the coordinates $\hat{x}_m$ on the domain $\hat{\Omega}^m$. For better readability, we will skip the subscripts m in the following.

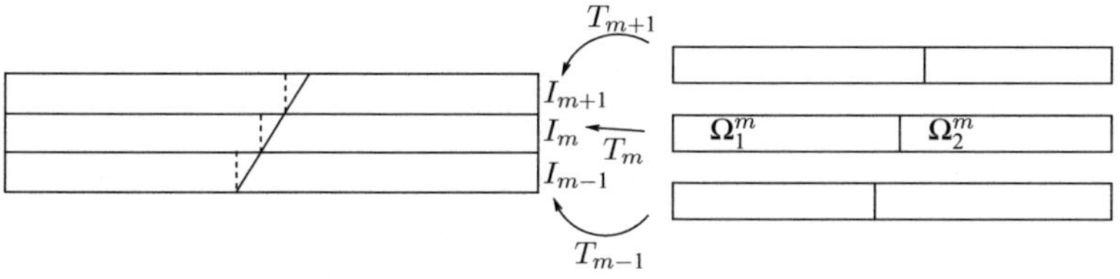

Fig. 3.8: Piecewise definition of maps T_m. The reference domain (right sketch) corresponds to the *new* domain Ω^m and changes in each time step.

Now problem (3.36) splits into the time-stepping scheme

$$B^m(u_k, \phi_k) = \left(J\hat{f}, \hat{\phi}_k\right)_{\hat{Q}^m} , \tag{3.38}$$

where

$$\begin{aligned} B^m(u_k, \phi_k) &= \frac{1}{k} \left(J(\hat{u}_k^m - \hat{u}_k^{m-1}), \hat{\phi}^m\right)_{\hat{Q}^m} \\ &\quad - \left(\partial_t T\, JF^{-T} \left(\frac{t - t_{m-1}}{k} \hat{\nabla} \hat{u}_k^m + \frac{t_m - t}{k} \hat{\nabla} \hat{u}_k^{m-1}\right), \hat{\phi}^m\right)_{\hat{Q}^m} \\ &\quad + \left(\hat{\kappa}\, JF^{-T} \left(\frac{t - t_{m-1}}{k} \hat{\nabla} \hat{u}_k^m + \frac{t_m - t}{k} \hat{\nabla} \hat{u}_k^{m-1}\right), F^{-T} \hat{\nabla} \hat{\phi}^m\right)_{\hat{Q}^m} \end{aligned} \tag{3.39}$$

and $\hat{u}_k^m$ and $\hat{u}_k^{m-1}$ denote the solution at time t_m and t_{m-1}. Each of the time steps is formulated on a new reference domain, see Figure 3.8.

In practice, the interface and boundary movement is often implicitly defined by the solution variables and might thus be available only at the time points t_{m-1} and t_m. Therefore, we will use a further simplification of (3.36). For the modified cG(1) scheme, we use approximations of the form

$$\bar{J} = \frac{1}{2}\left(J(t_m) + J(t_{m-1})\right)$$

and analogously $\overline{JF^{-T}}$, $\overline{F^{-T}}$ and $\overline{\partial_t T}$. The final bilinear form for $u_k \in X_k^1$ and $\phi_k \in X_k^0$ is given by

$$\begin{aligned} B_k^m(u_k, \phi_k) = &\frac{1}{k}\left(\bar{J}\left(\hat{u}_k^m - \hat{u}_k^{m-1}\right), \hat{\phi}_k^m\right)_{\hat{Q}^m} \\ &- \frac{1}{2}\left(\overline{\partial_t T}\,\overline{JF^{-T}}\hat{\nabla}\left(\hat{u}_k^m + \hat{u}_k^{m-1}\right), \hat{\phi}_k^m\right)_{\hat{Q}^m} \\ &+ \frac{1}{2}\left(\hat{\kappa}\overline{JF^{-T}}\hat{\nabla}\left(\hat{u}_k^m + \hat{u}_k^{m-1}\right), \overline{F^{-T}}\hat{\nabla}\hat{\phi}_k^m\right)_{\hat{Q}^m}, \end{aligned} \tag{3.40}$$

and the discrete variational formulation reads: Find $u_k \in X_k^1$ such that

$$B_k(u_k, \phi_k) = (f, \phi_k)_Q + (u^0, \phi_k(0))_{\Omega(0)} \quad \forall \phi_k \in X_k^0\,. \tag{3.41}$$

For the dG(0) scheme, we use the approximation $\bar{J} = J(t_m)$, $\bar{F} = F(t_m)$ and $\overline{\partial_t T} = \overline{\partial_t T}(t_m)$. As we have chosen, the new domain $\Omega(t_m)$ as the reference domain, we have $\bar{J} = 1$ and $\bar{F} = I$, such that only $\overline{\partial_t T}$ appears in the equation.

Remark 3.6 (Fixed-mesh ALE). The resulting time-stepping scheme can be considered a variant of the Fixed-mesh ALE method introduced by Codina and co-workers in [22] in combination with a projection scheme that is based on projecting residuals. There are two main differences: The first one lies in the approximation of the quantities J, F and $\partial_t T$ related to the transformation. Secondly, to fit into their framework, we would have to define $\hat{u}_k^{m-1}$ in (3.40) by a residual-based projection from the old to the new reference domain. To the best of our knowledge such a projection has not been used before within the *Fixed-mesh ALE* framework. Furthermore, in our practical implementation, we do not have to calculate such a projection, but a direct evaluation of the terms in (3.40) is sufficient, see [39].

Remark 3.7 (Characteristic-Galerkin method). The proposed method shares similarities with the Characteristic-Galerkin method (Douglas & Russell [64], Pironneau [79]). The Characteristic-Galerkin method has been applied recently for FSI in Eulerian coordinates by Pironneau [80]. The main difference is that the trajectories introduced here are arbitrary. While they follow the interface motion as in the Characteristic-Galerkin method, they might be very different outside the interface region. In our practical implementation, we typically define them as the identity $T_m(x) = x$ for points x above a certain distance to the interface.

For the time-discrete problem (3.41), we were able to show the following second-order error estimates. To state the result, we define the Bochner spaces

$$\mathcal{H}^{k,l} := H^k\left(I, H^l(\Omega_1(t))\right) \cap H^k\left(I, H^l(\Omega_2(t))\right) .$$

Theorem 3.8. *Let $u \in X$ be the solution of (3.33) and $u_k \in X_k^1$ the time discrete solution of (3.41). For $f \in \mathcal{H}^{1,0} \cap \mathcal{H}^{0,2}$, $u^0 \in H^4(\Omega_1(0) \cup \Omega_2(0))$ and sufficiently regular mappings T_m, it holds that*

$$\|u - u_k\|_Q + \|u(T) - u_k(T)\|_{\Omega(T)} \leq ck^2 \left(\|f\|_{1,0} + \|f\|_{0,2} + \|u^0\|_{H^4(\Omega_1(0)\cup\Omega_2(0))}\right) .$$

Proof. Frei [37]. □

Finally, we would like to mention that besides a gain in accuracy, we also observe a strong advantage in robustness and stability of the modified schemes, especially when applied to complex fluid-structure interaction problems. For the application to full FSI problems and the practical calculation of the domain maps T_m, we refer to Frei [37].

3.4.4 Stabilization of the solid equation

The full fluid-structure interaction system (3.18) contains a regularity problem at the interface $\Gamma_i(t)$. The natural trial space for the solid velocity in the variational formulation (Section 3.2.2) is $L^2(\Omega_s)$, which means that a trace on the interface Γ_i is not well defined. The trace is however needed for the *kinematic* interface condition (3.16). Furthermore, numerical tests show that the solid velocity is sensitive to stability problems caused by perturbations or discretization errors, especially in the interface region (see Chapter 4 in [37]).

To analyze this problem, we will first of all focus on the simplest form of a solid equation, a linear wave equation, and study different techniques to handle the aforementioned problems. In mixed variational formulation, the system of equations is given by: *Find $u \in \mathcal{W}, v \in \mathcal{V}$ such that*

$$\begin{aligned} (\partial_t v, \phi) + \lambda(\nabla u, \nabla\phi) &= 0 \quad \forall\phi \in \mathcal{W} , \\ (\partial_t u, \psi) - (v, \psi) &= 0 \quad \forall\psi \in \mathcal{V} , \end{aligned} \tag{3.42}$$

with a positive parameter $\lambda > 0$. It is well known that the homogeneous wave equation is energy-conserving in the following sense (e.g., Großmann & Roos [51])

$$\lambda\|\nabla u(t)\|_{\Omega_s}^2 + \|v(t)\|_{\Omega_s}^2 = \lambda\|\nabla u(0)\|_{\Omega_s}^2 + \|v(0)\|_{\Omega_s}^2 . \tag{3.43}$$

Hence, any kind of perturbations will not be damped, but is in some sense “conserved” and may accumulate over time. Furthermore, (3.43) gives neither control over derivatives of v nor over the trace of v on the boundary of Ω. We will see that this may give rise to severe stability issues, already for this simple set of equations, when the domain Ω is moving.

To deal with these instabilities, we propose a simple, but nonstandard stabilization technique, for which we will show stability and error estimates.

A stabilized formulation for the solid equation

We consider the following stabilized formulation for the linear wave equation: *Find $u_h \in \mathcal{W}_h, v_h \in \mathcal{V}_h$ such that*

$$\begin{aligned}(\partial_t v_h, \phi_h) + \lambda(\nabla u_h, \nabla \phi_h) &= 0 \quad \forall \phi_h \in \mathcal{W}_h\,,\\ (\partial_t u_h, \psi_h) - (v_h, \psi_h) - \alpha h^s(\nabla v_h, \nabla \psi_h) &= 0 \quad \forall \psi_h \in \mathcal{V}_h\,.\end{aligned} \tag{3.44}$$

Different choices for the exponent $s > 0$ will be discussed below. To motivate why this formulation increases stability, we consider a standard time discretization with the backward Euler method, multiply the second equation by -1 and write (3.44) in matrix form

$$\begin{pmatrix} \lambda A_h & \frac{1}{k}M_h \\ -\frac{1}{k}M_h & M_h + \alpha h^s A_h \end{pmatrix} \begin{pmatrix} u_h^{m+1} \\ v_h^{m+1} \end{pmatrix} = \begin{pmatrix} \frac{1}{k}v_h^m \\ -\frac{1}{k}u_h^m \end{pmatrix}. \tag{3.45}$$

Here, M_h is the discrete mass matrix and A_h the discrete Laplacian. We see that the stabilization term increases the diagonal part of the matrix considerably, especially for $s \le 2$. It can be interpreted as adding artificial diffusion to the diagonal part of the system matrix. We have the following conservation property:

Lemma 3.9. *Let $(u_h, v_h) \in \mathcal{W}_h \times \mathcal{V}_h$ be the semidiscrete finite element solution of (3.44). For $t > 0$, energy is conserved in the following sense*

$$\begin{aligned}&\lambda\|\nabla u_h(t)\|_\Omega^2 + \|v_h(t)\|_\Omega^2 + \alpha h^s\|\nabla v_h(t)\|_\Omega^2\\ &\qquad = \lambda\|\nabla u_h(0)\|_\Omega^2 + \|v_h(0)\|_\Omega^2 + \alpha h^s\|\nabla v_h(0)\|_\Omega^2\,.\end{aligned} \tag{3.46}$$

Proof. The proof is analogous to a standard proof that can be found in textbooks, e.g., Großmann & Roos [51]. For the details, see also Frei [37]. □

Lemma 3.9 gives us control over the first derivative of the velocity and thus (by the trace lemma) over the trace of the velocity on $\partial\Omega$.

We define the discrete spaces $\mathcal{V}_h = \mathcal{Q}_1 \cap H^1(\Omega_s)$ and $\mathcal{W}_h = \mathcal{Q}_1 \cap H_0^1(\Omega_s)$, where $\mathcal{Q}_1$ denotes the space of piecewise bilinear finite element functions on a triangulation $\mathcal{T}_h$. Furthermore, we use the Ritz projection R_h and the L^2-projection P_h to define the discrete initial values by $u_h(0) = R_h u(0)$ and $v_h(0) = P_h v(0)$. We can prove the following a priori error estimate for the solution of the stabilized formulation (3.44).

Lemma 3.10. *Let $(u, v) \in \mathcal{W} \times \mathcal{V}$ be the solution of (3.42) and $(u_h, v_h) \in \mathcal{W}_h \times \mathcal{V}_h$ the semidiscrete finite element solution of the stabilized formulation (3.44). Furthermore, let $I = [0, T]$, $Q = \Omega \times I$ and $u \in H^2(I, H^2(\Omega))$. For $s > 0$ and $t \in I$, it holds that*

$$\begin{aligned}\|u - u_h\|_Q + \|v - v_h\|_Q + \|(u - u_h)(t)\|_\Omega + \|(v - v_h)(t)\|_\Omega &\\ \le c(t)h^{\min(2,s-1/2)}\|u\|_{H^2(I,H^2(\Omega))}&\\ \|\nabla(u - u_h)\|_Q + \|\nabla(u - u_h)(t)\|_\Omega \le c(t)h^{\min(1,s-1/2)}\|u\|_{H^2(I,H^2(\Omega))}&\,.\end{aligned}$$

Proof. Frei [37] □

The result holds true in the case $s = \infty$, which corresponds to the nonstabilized system of equations. Lemma 3.10 suggests the choice of $s = 3/2$ for optimal convergence in the H^1-norm and $s = 5/2$ for optimal convergence in the L^2-norms. Nevertheless, in view of Lemma 3.9, a smaller choice of s might be necessary to guarantee the stabilizing effect of the approach. To demonstrate the effect of the stabilization and to study possible choices for s, we present a numerical example next.

Numerical example

As a numerical test, we consider a moving circle that is moved to the right by a spatially constant force $g = \min\{0.1t, 0.1\}$ which is applied as Neumann condition $\partial_n u = g$ on the boundary. We use a circle of radius 0.4: $\Omega_s(t = 0) = B_{0.4}(0)$ as the initial domain and $\lambda = 10$.

As for fluid-structure interactions, we prescribe the movement of the domain by the solid deformation itself. To be precise, the domain $\Omega(t)$ at time $t > 0$ is defined by the map

$$\Omega(t) = T(\Omega(0), t)\,, \quad T(x, t) = \begin{pmatrix} x_1 + u(x, t) \\ x_2 \end{pmatrix}.$$

For spatial discretization, we use the *locally modified finite element scheme*, see Section 3.4.1. We discretize a domain $\Omega_b = [-1, 1.2] \times [-1, 1]$ that contains the moving domain $\Omega(t)$ for all times $t \in [0, 1.6]$. For convenience, we extend deformation and velocity by a harmonic extension to the domain $\Omega_b \setminus \Omega(t)$. An illustration of the movement of the circle as well as the spatial discretization is given in Figure 3.9. In order to capture the interface, we use the *Initial Point Set function*; see (3.20).

For the temporal discretization, we considered both a standard backward Euler scheme as well as a locally modified $dG(0)$ time discretization as defined in Section 3.4.3 with a small time-step size $k = 10^{-2}$. The results we obtained differed only marginally for the two approaches. Thus, we will only present results based on the standard implicit Euler scheme here.

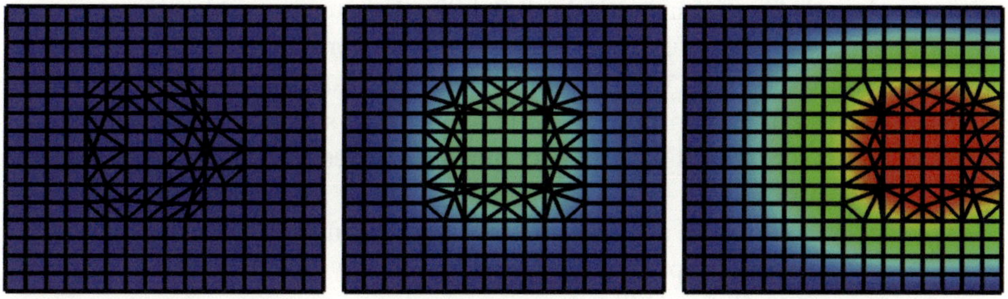

Fig. 3.9: Moving circle at start time $t = 0$, at $t = 0.8$ and at the end time $t = 1.6$. The color illustrates the horizontal velocity of the solution and its harmonic extension to the exterior domain.

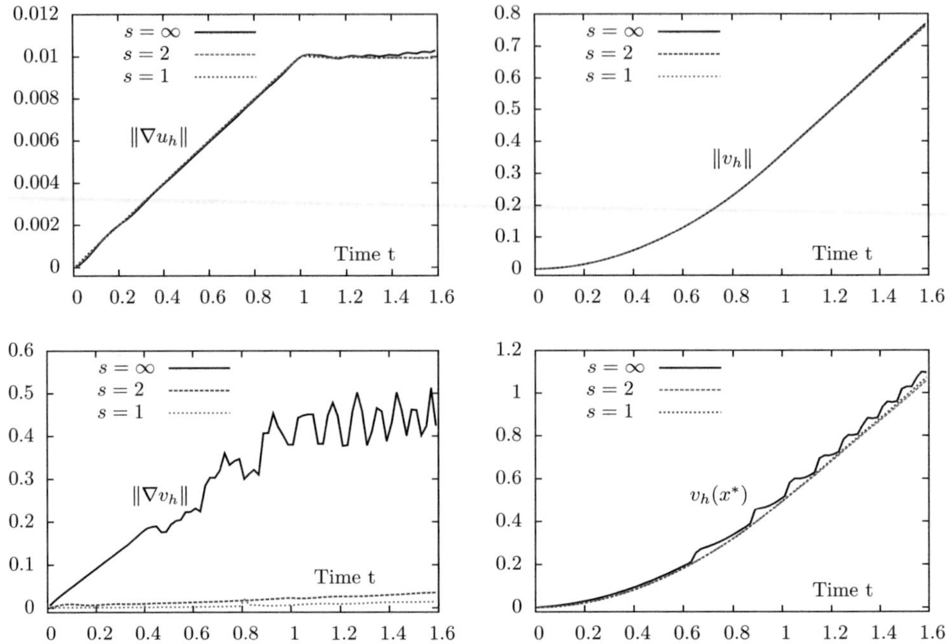

Fig. 3.10: Functionals over time for the wave equation on a moving domain. We observe strong oscillations for the solution of the nonstabilized system of equations.

In Figure 3.10, we plot the values of the three norms estimated in Lemma 3.10 as well as the point value $v_h(x^*)$ at the (moving) point x^* on the very right of the circle over time. We observe very strong oscillations for the solution of the nonstabilized system of equations in the H^1-norm of the velocity, and also considerable oscillations for the point functional $v_h(x^*)$. The instabilities for the point functional occur every time the point (and thus the interface) jumps over a patch edge. Conversely, it was sufficient to use the stabilization with $s = 2$ to dampen (almost) all visible oscillations. The solutions of the stabilized system for $s = 1$ and $s = 2$ show very good agreement in all functionals. For the L^2-norm of the velocity (and also of the deformation), the values for the solutions with and without stabilization are almost identical.

In Table 3.1, we show functional values of the L^2-norm of v_h as well as the point value $v_h(x^*)$ at time $t = 1$. The L^2-norm of v_h (as well as the L^2- and H^1-norm of u that are not shown here) converge well with and without stabilization terms. The convergence orders for the stabilized version are slightly reduced in comparison to the nonstabilized version.

Conversely, we observe a big difference when looking at the point value $v_h(x^*)$. Here, the nonstabilized approach converges very slowly due to the oscillations observed in Figure 3.10. With stabilization, the values converge with approximately second order. These results indicate that in the context of full FSI problems, where the

Tab. 3.1: Functional values of the L^2-norm of v_h and the point value $|v_h(x^*)|$ at time $t = 1$. Furthermore, we give an extrapolated value and estimated convergence rates. For the nonstabilized discretization ($s = \infty$), the point functional converges very slowly .

	$\|v_h\|_{L^2}$			$\|v_h(x^*)\|$		
#nodes	$s = 1$	$s = 2$	$s = \infty$	$s = 1$	$s = 2$	$s = \infty$
289	0.366	0.365	0.365	0.525	0.523	0.522
1,089	0.360	0.359	0.359	0.502	0.499	0.515
4,225	0.358	0.357	0.358	0.496	0.494	0.510
Extr.	0.357	0.357	0.358	0.493	0.493	0.498
Conv.	1.96	1.69	2.29	2.43	1.79	0.48

fluid dynamics might be significantly driven by the velocity at the interface, we may expect severe stability issues if the solid equations are not stabilized.

Alternative methods

Alternative approaches that have been chosen in the literature for similar problems include a so-called damping of the solid equations and the streamline diffusion method (SUPG), which might be needed anyway because of the convective term in (3.18).

For the case of the linear wave equation, weak and strong damping are defined in the following way

$$\begin{aligned} (\partial_t v, \phi) + \lambda(\nabla u, \nabla\phi) + \gamma_s(\nabla v, \nabla\phi) + \gamma_w(v, \phi) &= 0 \quad \forall\phi \in \mathcal{W}\,, \\ (\partial_t u, \psi) - (v, \psi) &= 0 \quad \forall\psi \in \mathcal{V}\,, \end{aligned} \tag{3.47}$$

with damping parameters $\gamma_s, \gamma_w \geq 0$. Such a damping might in some situations also be physically motivated (Wick [96]). If we write this formulation in the matrix form (3.45), the additional terms appear in the off-diagonal entry of the first column.

For the SUPG method, the situation is similar. The discrete system of equations reads: *Find* $v_h \in \mathcal{V}_h$, $u_h \in \mathcal{W}_h$ *such that*

$$\begin{aligned} (\partial_t v_h, \phi_h) + \lambda(\nabla u_h, \nabla\phi_h) + S_h^{\mathrm{SUPG}}(v_h, \phi_h) &= 0 \quad \forall\phi_h \in \mathcal{W}_h\,, \\ (\partial_t u_h, \psi_h) - (v_h, \psi_h) + S_h^{\mathrm{SUPG}}(u_h, \psi_h) &= 0 \quad \forall\psi_h \in \mathcal{V}_h\,. \end{aligned} \tag{3.48}$$

In matrix-vector form, this means

$$\begin{pmatrix} \lambda A_h & \frac{1}{k}M_h + \delta_s h_P^2 A_h \\ -\frac{1}{k}M_h - \delta_s h_P^2 A_h & M_h \end{pmatrix} \begin{pmatrix} u_h^{m+1} \\ v_h^{m+1} \end{pmatrix} = \begin{pmatrix} \frac{1}{k}v_h^m \\ -\frac{1}{k}u_h^m \end{pmatrix}.$$

We observe that both methods do not increase the diagonal part of the matrix. Applying these methods to the numerical example studied above, neither was able to reduce the oscillations significantly. For the details, we refer to Frei [37].

3.5 Validation by means of benchmark problems

In this section we apply the fully Eulerian approach and the discretization techniques derived in the previous sections to two FSI benchmark problems proposed by Hron & Turek [61]. These benchmark problems are widely used to test approaches and software for fluid-structure interaction. A comparison of different results has been published by Turek et al. [93].

3.5.1 Setting of the benchmark problems

In these benchmark problems, a two-dimensional elastic beam is attached to a fixed and rigid cylinder, see Figure 3.11. Because of a nonsymmetry in their vertical position, a flow field causes a movement of the beam as well as elastic deformations. If the Reynolds number of the flow configuration is large enough, we observe self-induced oscillations of the beam.

The system of equations has been summarized in Section 3.2.2. The boundary conditions for the fluid are a parabolic inflow profile on Γ_{in} on the left side, a *no-slip* condition on the upper and lower boundary and the *do-nothing* outflow condition $\rho_f \nu_f \partial_n v_f - p_f n = 0$ on the right boundary Γ_{out}. As suggested by Hron & Turek [61], we start the simulation with zero initial velocity and increase the inflow velocity gradually.

The authors specified three different sets of material parameters, see [61]. Here, we will study their first and third set of parameters. While the first set (FSI1) leads to a movement of the beam in the vertical direction that converges to a stationary state, the third set (FSI3) results in oscillatory movements. The authors proposed four different functionals to compare the results: horizontal and vertical deflection of the beam at the tip (point $A(t)$ in Figure 3.11) $u_x(A(t))$ and $u_y(A(t))$ and the drag and lift mean values at the boundary of the cylinder and the beam.

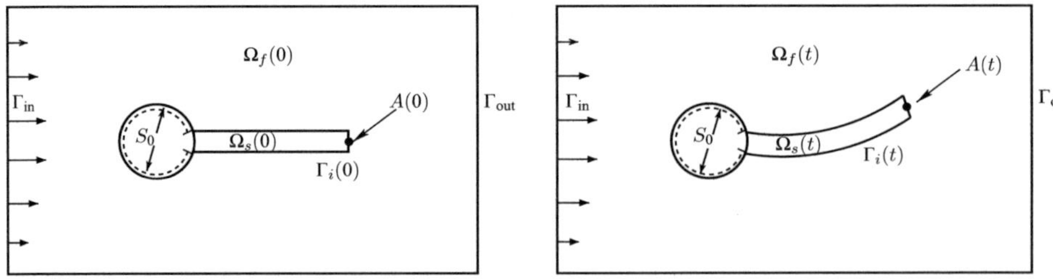

Fig. 3.11: FSI benchmark configuration (Hron and Turek [61]): The vertical position of the tip of the beam is initially $A(0) = 0.2$, while the total height of the channel is $H = 0.41$. Because of this nonsymmetric setting the beam starts to move when a parabolic inflow is applied at Γ_{in}. We measure the deformation at the tip $A(t)$ and drag and lift mean values at the FSI boundary $\Gamma_i(t)$ and the (fixed) boundary of the cylinder S_0. Right: initial configuration, *left:* current system at time $t > 0$.

Most of the methods applied to study the benchmark problem are based on the *arbitrary Lagrangian–Eulerian* method. As the displacements are rather moderate, it is possible to use the same reference domain for all time steps without any remeshing. Hence, for these problems, we cannot expect that the *Fully Eulerian* approach is able to fully compete with an ALE approach. The *Fully Eulerian* approach is designed for problems where the ALE method has problems due to large movement of the structure. We will study such an example in the next section. Here, however, we want to show that the results obtained with the *Fully Eulerian* approach converge to known reference values both with respect to time and space discretization.

3.5.2 Practical implementation

All results shown in this and the next section have been obtained with the finite element library *Gascoigne 3D* [9]. To solve the nonlinear system of equations, *Gascoigne 3D* uses Newton's method. The resulting linear systems have been solved with a direct solver. Therefore, we use the *Trilinos* [86] interface of the *SuperLU_DIST* solver [70].

Discretization

We use the discretization and stabilization techniques that have been presented in Section 3.4. Here, we use the $dG(0)$ variant of the modified time-stepping scheme which is more robust than both the $cG(1)$ variant of the modified scheme and the standard backward Euler time-stepping scheme.

For spatial discretization, we use the *locally modified finite element* scheme introduced in Section 3.4.1, with one particular modification. In the present application, the fluid-structure interface Γ_i is not smooth at the corners of the beam. Hence, using the standard linear approximation of the interface described in Section 3.4.1 may lead to an unsatisfactory approximation in these regions. Conversely, we have one additional freedom that we have not exploited yet, i.e., the position of the patch midpoint. Here, we set the midpoint of the patch that contains the corner to the corner itself (Figure 3.12). In some specific situations, we have to move some further nodes in order to ensure a maximum angle condition.

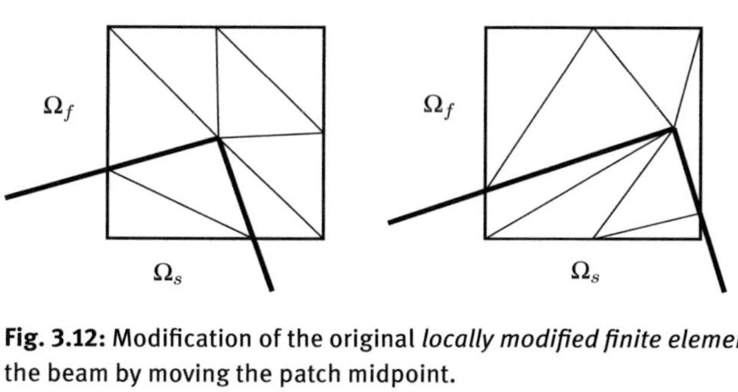

Fig. 3.12: Modification of the original *locally modified finite element method* to resolve the corners of the beam by moving the patch midpoint.

In addition to the pressure stabilization presented in Section 3.4.4, we add the additional stabilization term

$$S_p^m(p, \psi) = \gamma_p h_P (p^m - p^{m-1}, \psi^m)_{\Gamma_i} \tag{3.49}$$

at the fluid-structure interface Γ_i. This term penalizes nonphysical pressure oscillations in the interface region.

Evaluation of surface integrals

We evaluate the drag and lift functionals using the *Babuška–Miller trick* [4]. Using the divergence theorem and the strong formulation of the momentum equation, one can show that the drag and lift functionals can be expressed as

$$(F_{\text{drag}}, F_{\text{lift}}) = (\sigma_{\text{f}}, \nabla\phi)_{\Omega_{\text{f}}} + (\rho_{\text{f}}(\partial_t v_{\text{f}} + v_{\text{f}} \cdot \nabla v_{\text{f}}) - f, \phi)_{\Omega_{\text{f}}} := a(v_{\text{f}}, p_{\text{f}}, \phi)\,,$$

where $\phi \in [H^1(\Omega(t))]^2$ is 1 on $S_0 \cup \Gamma_i(t)$ and zero on the remaining fluid boundaries $\partial\Omega \setminus S_0$. This is exactly the bilinear form that we have to evaluate when we set up the right-hand side and the system matrix. Furthermore, this way of evaluating the surface forces is much more accurate than the direct evaluation of the boundary integrals, as has been shown by Braack & Richter [14]. In the present case of linear/bilinear finite elements, we expect an accuracy of $\mathcal{O}(h^2)$ compared to $\mathcal{O}(h^{1/2})$ for a direct evaluation of the surface force.

3.5.3 FSI1 benchmark

Because of the low Reynolds number $Re = 20$, the FSI1 configuration converges to a stationary state which is reached after a slight vertical movement of the tip of the beam towards the top. In Figure 3.13, we show the horizontal and vertical deflection as well as the drag and lift functional over time. After a short settling time, a stationary state is attained after approximately 8 s.

For pressure stabilization, we use the parameters $\gamma = 5 \cdot 10^{-4}$ and $\gamma_p = 2 \cdot 10^{-3}$ in (3.30) and (3.49). For the FSI1 configuration it is not necessary to add stabilization terms to the solid equation as the solid velocity is relatively small and vanishes in the stationary state. We use a time-step size of $k = 10^{-2}$.

In Table 3.2, we show values for the four functionals on three different grids with 4128, 16,192 and 64,128 nodes, respectively. The functional values show a very good convergence behavior in all the functionals. While we observe approximately linear convergence in the horizontal deflection $u_x(A)$ and the drag functional, the vertical deflection $u_y(A)$ and the lift functional converge much faster. The values on the finest grid are in excellent agreement with the reference values obtained by Turek et al. on a grid with more than 19 million degrees of freedom with an ALE approach [93]. We observe the biggest deviation from the reference values in the horizontal deflection $u_x(A)$ showing a relative error below 0.1%.

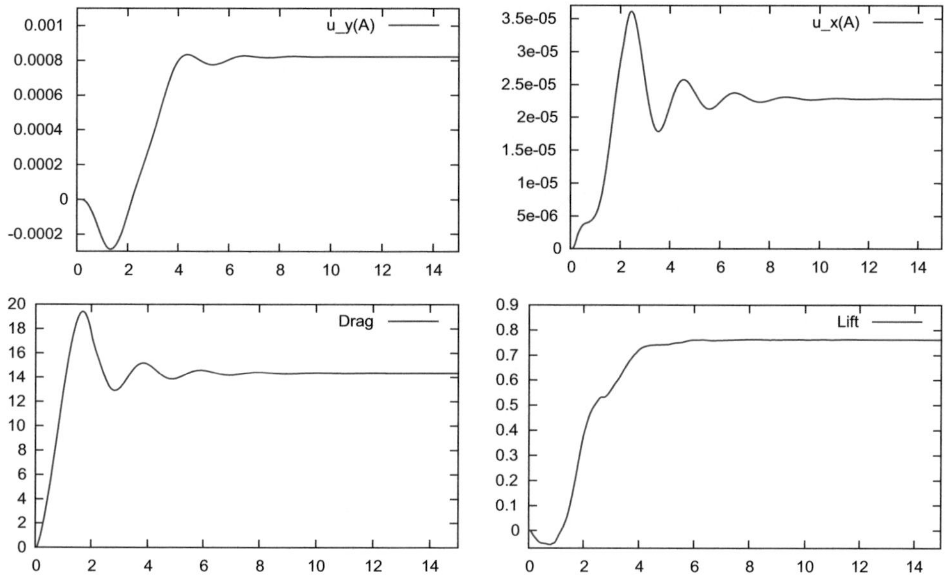

Fig. 3.13: Functional values for the FSI1 benchmark over time.

Tab. 3.2: Results for the FSI1 benchmark for four different functionals. We observe a very good convergence behavior. Furthermore, the results on the finest grid are in excellent agreement with the reference values.

#nodes	$u_x(A)$	$u_y(A)$	F_{drag}	F_{lift}
4128	$2.278 \cdot 10^{-5}$	$8.455 \cdot 10^{-4}$	14.328	0.7721
16,192	$2.274 \cdot 10^{-5}$	$8.263 \cdot 10^{-4}$	14.305	0.7656
64,128	$2.272 \cdot 10^{-5}$	$8.211 \cdot 10^{-4}$	14.298	0.7637
Ref [93]	$2.270 \cdot 10^{-5}$	$8.209 \cdot 10^{-4}$	14.294	0.7637

When we compare these values with the results of other approaches presented in [93], we observe that many of them are far less accurate even on much finer meshes. This is especially true for most of the approaches that use a partitioned coupling of different fluid and solid solvers.

3.5.4 FSI3 benchmark

The FSI3 benchmark has a Reynolds number of $Re = 200$ which, at first, causes a slightly larger displacement of the beam towards the top compared to the FSI1 benchmark. Then, after some time, the beam starts to oscillate periodically. This induces vortices in the fluid domain behind the beam. Some snapshots of the solution are given in Figure 3.14. In Figure 3.15, we show a visualization of a grid obtained with the *locally*

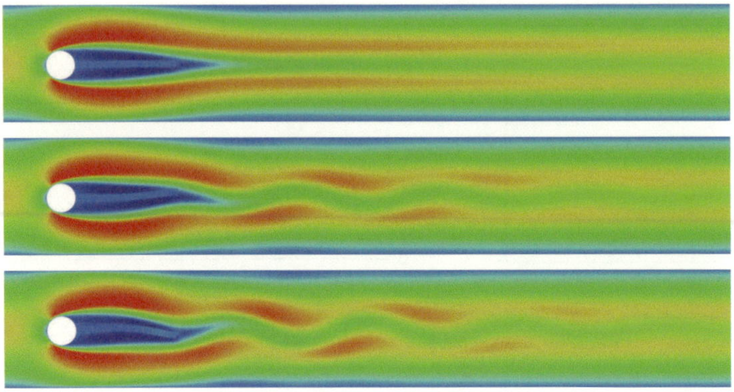

Fig. 3.14: Snapshots of the FSI3 benchmark at three different instants of time. First, we see a slight movement of the beam towards the top (top). Then, the beam starts to oscillate (middle) and a periodic movement can be observed (bottom). The color illustrates the Euclidean norm of the velocity field.

modified finite element method. For better illustration, we show only the mesh for the fluid part and plot the solid domain in black.

For pressure stabilization, we choose exactly the same parameters as for the FSI1 benchmark. In addition, here, it is necessary to stabilize the solid equation. We use the parameter $\alpha = 10$ and $s = 1$, i.e.,

$$S_v(v, \phi) = 10 h_P (\nabla v, \nabla \phi)_{\Omega_s(t)} .$$

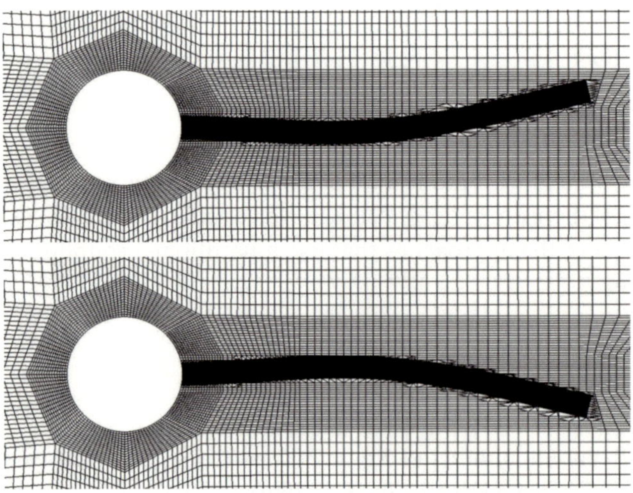

Fig. 3.15: Illustration of the grid obtained with the locally modified finite element discretization with 13,104 mesh points at the time of minimum and maximum displacement. For better visualization of the subdomains, the grid in the solid part is not shown here, but filled with black color.

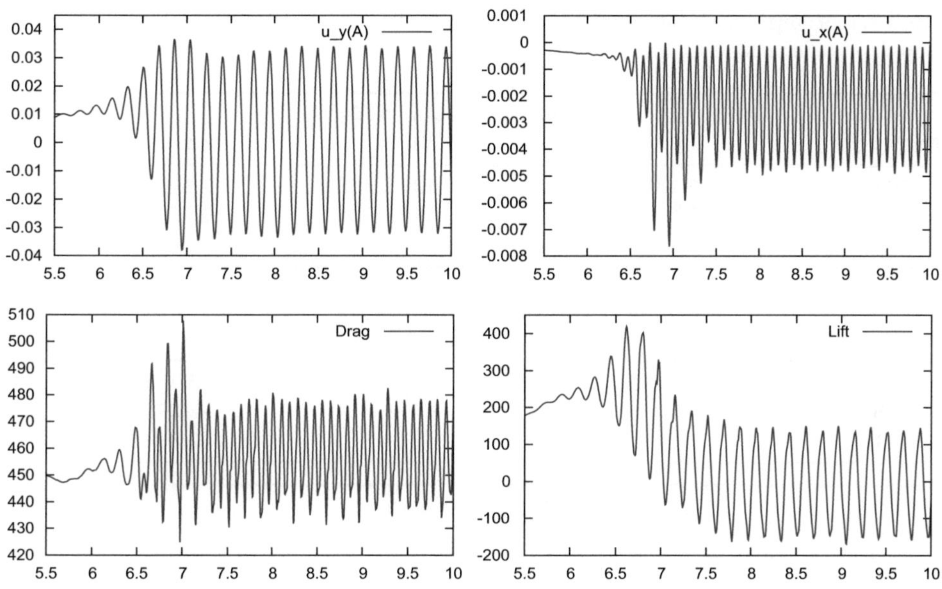

Fig. 3.16: Functional values for the FSI3 benchmark over time. A state of uniform oscillations is reached after approximately 8 *s*.

In Figure 3.16, we show the course of the four functionals over time. A uniform oscillatory movement is observed after approximately $t = 8\,s$. Before, we observe small overshoots in both the vertical and horizontal displacement. This is in agreement with results obtained with ALE calculations, but in contrast to the findings of Dunne with the Fully Eulerian approach in its early version who observed a uniform increase of the oscillations [30].

For $t > 8\,s$, the horizontal and vertical displacements show nearly perfectly uniform oscillations. For the horizontal displacements, we have to remark that two maxima and minima are attained within one period, as the horizontal displacement becomes minimal both when the beam reaches its uppermost and its lowermost position. The same holds true for the drag functional. The oscillations in the drag and lift functional are not exactly uniform. Having a closer look at the times of minimum and maximum deviation, we observe small instabilities in the pressure around the corners of the beam. These instabilities can be reduced by further increasing the pressure stabilization. In this case, however, all four functionals show considerably smaller values as the pressure stabilization dampens too much.

In Table 3.3, we give our results of the mean value and the amplitude of the four functionals on two different meshes and for three different time-step sizes after the system reached a state of uniform oscillations. To obtain precise values, we calculated the mean value of ten minima and maxima for all functionals. In all functionals, we observe a reasonable convergence behavior under refinement in space. In time, we

Tab. 3.3: Functional values $u_y(A)$, $u_x(A)$, F_{drag} and F_{lift} for the FSI3 benchmark with different time-step sizes and on different grids.

	Vertical deflection $u_y(A)$ in $10^{-2}m$		
#nodes \ k	10^{-3}	$5 \cdot 10^{-4}$	$2.5 \cdot 10^{-4}$
13,104	0.135 ± 2.847	0.130 ± 3.039	0.190 ± 3.086
51,808	0.039 ± 3.167	0.114 ± 3.288	0.055 ± 3.214
Ref [93]			0.147 ± 3.499

	Horizontal deflection $u_x(A)$ in $10^{-3}m$		
#nodes \ k	10^{-3}	$5 \cdot 10^{-4}$	$2.5 \cdot 10^{-4}$
13,104	-1.87 ± 1.81	-2.15 ± 2.07	-2.23 ± 2.16
51,808	-2.29 ± 2.16	-2.49 ± 2.38	-2.29 ± 2.19
Ref [93]			-2.88 ± 2.72

	Drag surface force		
#nodes \ k	10^{-3}	$5 \cdot 10^{-4}$	$2.5 \cdot 10^{-4}$
13,104	466.9 ± 13.8	469.4 ± 16.0	468.9 ± 22.8
51,808	452.6 ± 13.4	456.3 ± 22.5	453.1 ± 24.7
Ref [93]			460.5 ± 27.7

	Lift surface force		
#nodes \ k	10^{-3}	$5 \cdot 10^{-4}$	$2.5 \cdot 10^{-4}$
13,104	-4.6 ± 164.8	-1.4 ± 176.9	-9.4 ± 193.6
51,808	-5.0 ± 130.6	-10.2 ± 148.4	-27.9 ± 162.6
Ref [93]			2.5 ± 153.9

observe a strong improvement of the values when decreasing the time-step size from $k = 10^{-3}$ to $k = 5 \cdot 10^{-4}$. Decreasing the time step once more by a factor of two has only positive impact on the drag force and on horizontal and vertical deflections on the coarser grid. Here, the spatial discretization error seems to dominate the temporal error. We suppose that this is mainly caused by the rather large pressure stabilization term which was used to ensure stability.

Comparing these values to the results that have been presented in [93], we observe -as expected- that we cannot fully compete with monolithic ALE discretizations. To reach the same level of accuracy, we would need a much finer grid. The relative errors in the amplitude, that we get for the best values, range between 3.6% for the lift functional and 13.3% for the drag functional. The monolithic ALE approach submitted by Rannacher & Dunne in contrast shows relative errors of 0.8 to 5.6%. Similar results have been contributed by Schäfer who used an implicitly coupled partitioned approach using an ALE finite volume technique for the fluid and a Lagrangian finite element solver for the structure. All the other contributed results show either accuracies that are comparable to our results or are significantly worse.

To summarize, we have shown in this section that the *Fully Eulerian* method with the presented discretization techniques yields results that converge against known reference values in both space and time. Our results are slightly less accurate than results obtained with a monolithic ALE approach. This was expected, as the advantage of the *Fully Eulerian* approach lies in the ability to deal with large solid movements and contact. We will present such an application in the next section.

3.6 A simple contact problem

In this final section, we study contact problems of an elastic ball that falls down to the ground in a viscous fluid. At the time of contact the ball is compressed and bounces off. We study the vertical fall of a ball towards a horizontal wall and its vertical rebound in Section 3.6.1. In Section 3.6.3, we consider a more complex situation where a ball drops onto an inclined plane and bounces three stairs down afterwards.

As mentioned previously, problems of this kind cannot be simulated with standard ALE approaches as we have to deal with large structural displacements, and, in the case of "real contact" (meaning that no fluid layer remains between ball and ground), with topology changes in the fluid domain. In this case a monolithic Eulerian approach is a promising alternative.

We remark that from a modeling point of view, it is questionable if the incompressible Navier–Stokes equations are a valid model in the case of real contact, see Hillairet [57], Feireisl [34] or Gérart-Varet et al. [47] for analytical results regarding this question. For this case, we introduce a simple contact model in Section 3.6.2 that prevents the solid from touching the ground. As the model is based on an artificial contact force, we investigate its influence on the contact dynamics in detail.

As in the previous section the equations under consideration are given by the nonstationary FSI equations (3.18). Here, we apply a nontrivial volume force $f_s = (0, -1)$ that represents gravity.

3.6.1 Example 1: Vertical fall

A sketch of our first test problem is given in Figure 3.17 on the left. We consider a ball of radius $r = 0.4$ whose midpoint is initially located at the origin. Because of gravity it falls down towards a horizontal wall $\Gamma_w = \{(x, y) \in \mathbb{R}^2 \mid y = -1\}$. As a boundary condition, we impose a homogeneous velocity on Γ_w. In combination with the *kinematic condition* and the velocity-displacement relation $d_t u_s = v_s$ this ensures that the solid cannot pass "through the wall". Note that in this section, we do not use any contact algorithm.

On the remaining boundaries, we use the *do-nothing* outflow condition. The same example but with different material parameters has been studied by Richter [82]. To

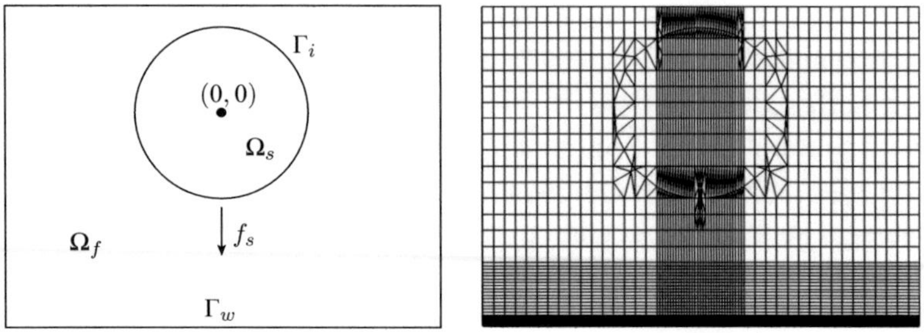

Fig. 3.17: Sketch of the configuration of the first test case and the initial mesh. To simulate the contact dynamics accurately, fine mesh cells are used in the contact region.

simulate the interval of contact accurately, we use an anisotropic mesh with fine cells around the lower boundary (Figure 3.17, right).

For the first test, we use the Lamé parameters $\mu_s = 2 \cdot 10^5$ and $\lambda_s = 8 \cdot 10^5$ and the fluid viscosity $\nu_f = 10^{-3}$. The fluid and solid density are $\rho_s = \rho_f = 10^3$. Before we discuss the numerical parameters, we describe the temporal dynamics by means of some simulation results.

In Figure 3.18, we show the falling ball at six different times. The coloring illustrates the vertical velocity. First, the ball is accelerated by gravity and falls down. At time $t \approx 1.6$, the bottom is almost reached and the ball slows down due to a high fluid pressure. It comes closest to the ground at time $t \approx 1.8$, where the minimal distance is $d \approx 1.2 \cdot 10^{-3}$. At this time the ball is significantly compressed at its bottom.

The discretization at this point is illustrated in Figure 3.19 (top) for the coarsest mesh we used. The interface shows a domed shape due to a high fluid pressure in the middle and the minimal distance is not attained in the center but left and right of it. In this configuration, there is no real contact but a small layer of fluid remains between ball and ground. Nevertheless, here and in the following we will call this period the "contact time" or "contact interval" for simplicity.

Then, due to the compression at the bottom the ball is accelerated upwards. The ball reaches its highest elevation at a maximum distance $d \approx 8.3 \cdot 10^{-2}$ from the ground at time $t \approx 2.4$ and falls down again. After a smaller second bounce with distance $d \approx 4.6 \cdot 10^{-3}$, it comes to rest, being in real contact with the ground at time $t \approx 4.4$, see Figure 3.19 (bottom) for an illustration of the mesh at the time of real contact.

In Figure 3.20, we plot the minimal distance between the ball and the ground including a zoom-in of the contact and rebound interval in the upper row. Furthermore, we show the distance between the top and the bottom of the ball and an averaged vertical velocity of the solid in the lower row.

In the lower left plot, we observe that the distance between the top and the bottom of the ball attains minima at the two contact times due to the compression. After the rebounds, we observe oscillations that become smaller over time. These oscillations are physical: First, the ball is maximally compressed at the bottom at the contact

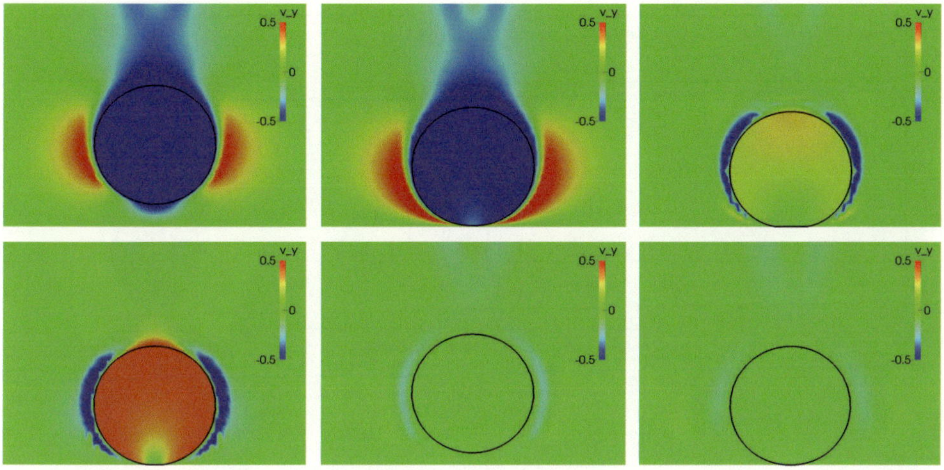

Fig. 3.18: Illustration of the free fall of an elastic ball, its contact with the ground and the subsequent rebound at six different times. The ball falls down due to gravity (top left, $t \approx 1.3$), is almost in contact with the ground (top middle: before contact, $t \approx 1.6$, top right: smallest distance, $t \approx 1.8$, bottom left: after contact, $t \approx 1.81$) and bounces off. It reaches its highest elevation again at $t \approx 2.4$ (bottom middle). After a second bounce, it comes to rest at $t \approx 4.4$ (bottom right) being in real contact with the ground. The color illustrates the vertical velocity v_y and the black contour line is the discrete interface.

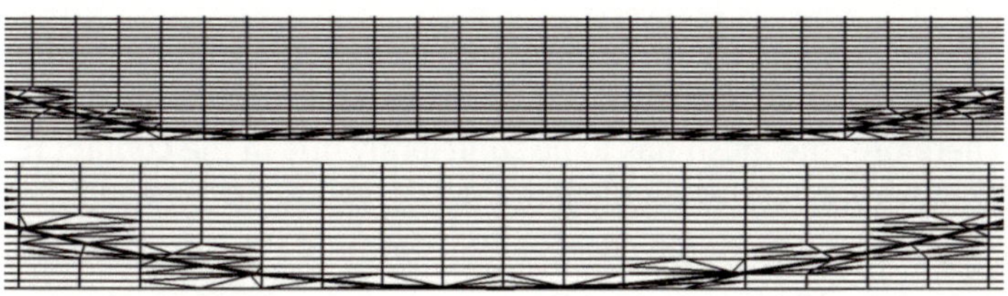

Fig. 3.19: Illustration of a coarse mesh during the first rebound (top sketch) and at the end time when the ball is at rest (lower sketch). During the first rebound, a small layer of fluid remains between ball and ground.

time. Once the ball bounces off again, the deformation is relaxed. The ball is even overstretched at some point and starts to oscillate between an expanded and a compressed state periodically.

Numerical parameters

We are using the discretization and stabilization techniques described in Section 3.4 with a few modifications. In this example, we do not use any structural stabilization. Using the stabilization proposed in Section 3.4.4 might be dangerous here as it alters the velocity-displacement relation $d_t u_s = v_s$. This relation is crucial as it ensures that the ball cannot "pass" through the ground.

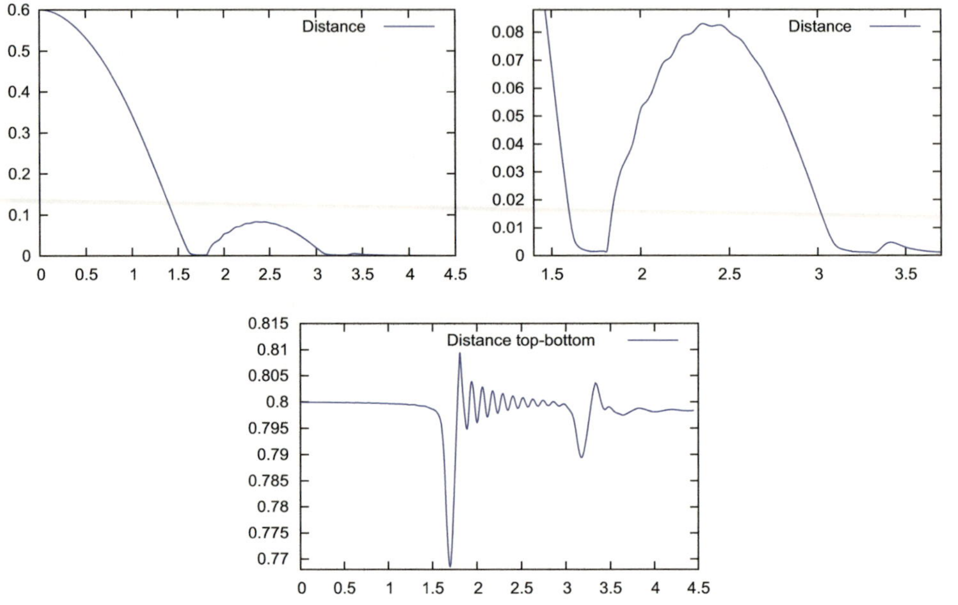

Fig. 3.20: Top: Minimal distance between ball and ground over time and a zoom-in at the interval of contact and rebound. Bottom: Distance between top and bottom of the ball.

At the time of contact the fluid pressure shows a high peak in the contact region, see Figure 3.21. Resolving this peak accurately is important as it prevents the ball from touching the ground. Therefore, the mesh has to be sufficiently fine in the contact region and the pressure stabilization has to be sufficiently small. We choose two different stabilization parameters $\gamma_i = 2 \cdot 10^{-5}$ for the edges $e \in \mathcal{E}_h^{f,i}$ (around the interface) and $\gamma_0 = 10^{-3}$ for the edges $e \in \mathcal{E}_h^{f,0}$. The first parameter γ_i has been chosen small to alter the pressure profile in the interface region as little as possible at the contact time.

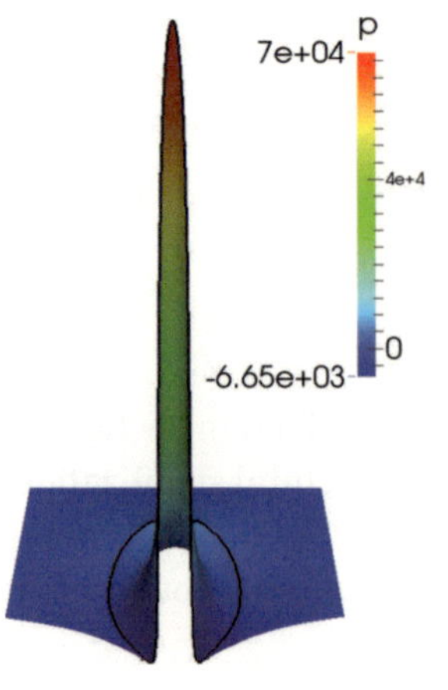

Fig. 3.21: Illustration of the pressure peak during the contact between ball and ground.

For time discretization we use the modified dG(0) scheme. This scheme is able to handle the case that two structures come arbitrarily close, but requires a modification when it comes to real contact. In this case, we set $T_m = \mathrm{id}$ in the contact region which results in the standard backward Euler scheme there.

Both the time-step size k in and around the contact interval and the spatial discretization in the contact region have to be sufficiently fine in order to capture the contact dynamics accurately. We will see that a coarse resolution might change the results considerably.

Convergence studies

We study the test configuration on different grids and for different time-step sizes. The coarsest mesh we use is shown in Figure 3.17, right sketch. It consists of patches of size 0.1×0.15 in the upper right and upper left part and of size $2.5 \cdot 10^{-2} \times 3.1 \cdot 10^{-3}$ in the contact region. Furthermore, we show the results on two finer meshes that are obtained from this coarse mesh by global refinement.

In Table 3.4, we show the minimal distance during the first contact, the maximum elevation after the first rebound and the error in mass conservation on these meshes for three different time-step sizes. Note that the calculation with the largest time-step size $k = 2 \cdot 10^{-3}$ on the finest mesh broke down before the contact, as the restriction that the interface may not pass over more than one patch per time step, that we imposed in our implementation, was violated.

We observe that both the minimal distance during the contact and the maximal distance after the rebound are significantly smaller on the coarsest mesh. Conversely, the results on the finer meshes show good agreement. On the finest mesh, the min-

Tab. 3.4: Top: Minimal distance between ball and ground during the first contact interval and maximal distance after the first rebound. Bottom: Relative error in mass conservation at time $t = 3$. The three functionals are calculated for three different time-step sizes and on three different meshes.

	First contact: Minimum distance			**First bounce: Maximum distance**		
#nodes \ k	$2\cdot10^{-3}$	$1\cdot10^{-3}$	$5\cdot10^{-4}$	$2\cdot10^{-3}$	$1\cdot10^{-3}$	$5\cdot10^{-4}$
4225	$6.38\cdot10^{-4}$	$6.49\cdot10^{-4}$	$6.26\cdot10^{-4}$	$4.71\cdot10^{-2}$	$5.19\cdot10^{-2}$	$5.20\cdot10^{-2}$
16,641	$1.22\cdot10^{-3}$	$1.24\cdot10^{-3}$	$1.24\cdot10^{-3}$	$7.41\cdot10^{-2}$	$8.29\cdot10^{-2}$	$8.49\cdot10^{-2}$
66,049	–	$1.27\cdot10^{-3}$	$1.27\cdot10^{-3}$	–	$8.71\cdot10^{-2}$	$8.96\cdot10^{-2}$

	Relative mass conservation error		
#nodes \ k	$2\cdot10^{-3}$	$1\cdot10^{-3}$	$5\cdot10^{-4}$
4225	$8.96\cdot10^{-3}$	$8.94\cdot10^{-3}$	$8.94\cdot10^{-3}$
16,641	$2.54\cdot10^{-3}$	$2.36\cdot10^{-3}$	$2.31\cdot10^{-3}$
66,049	–	$5.07\cdot10^{-4}$	$5.05\cdot10^{-4}$

imum distance, i.e., the minimum height of the fluid layer, is a little less than two patches.

We conclude that the resolution of the contact region on the coarse mesh was not fine enough to resolve the contact dynamics accurately.

The relative error in mass conservation is defined by

$$j_{\text{mass}} = \frac{|\int_{\Omega_s} J\rho_s^0 \,\mathrm{dx} - \pi r^2 \rho_s^0|}{\pi r^2 \rho_s^0}.$$

Here, we observe a good convergence behavior in both space and time, even on the coarsest mesh. The spatial discretization error is dominating and decreases with order $\mathcal{O}(h_P^2)$. This convergence behavior was expected, as it is the approximation error of the interface, see Section 3.4.1. Furthermore, this result indicates that the Eulerian approach proposed in this article, shows very good mass conservation properties.

Influence of material parameters

Next, we want to address the question of whether a small layer of fluid is maintained between ball and ground or if it comes to real contact. Therefore, we increase and decrease the solid Lamé parameters. Keeping the ratio between the Lamé parameters μ_s and λ_s constant, $\lambda_s = 4\mu_s$ (which corresponds to a Poisson ration of $\nu_s = 0.4$), we vary λ_s and μ_s simultaneously.

For a set of parameters ranging from $\mu_s = 5 \cdot 10^4$ to $3.2 \cdot 10^6$, we plot the distances between ball and ground over time in Figure 3.22. For the stiffest material ($\mu_s = 3.2 \cdot 10^6$) no fluid layer remains during the first contact interval. Ball and ground are in real contact and the ball does not bounce off at all. The same happens for the second stiffest material ($\mu_s = 3.2 \cdot 10^6$) at the second contact time. It is, however, questionable whether this corresponds to the physical situation. Instead the contact might be

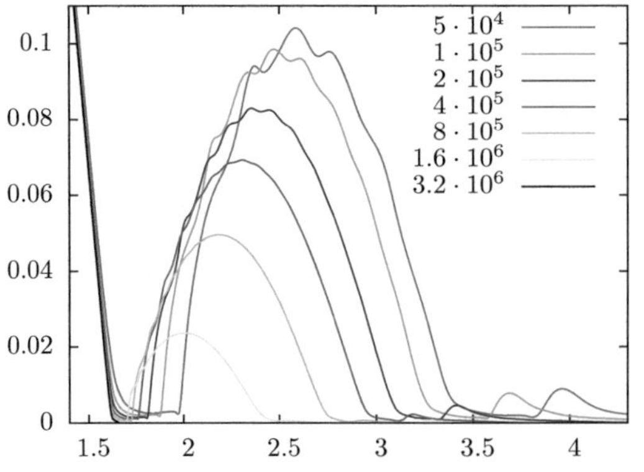

Fig. 3.22: Distance between ball and ground over time for different solid parameters μ_s during the contact and rebound interval. The first Lamé parameter is chosen as $\lambda_s = 4\mu_s$. The rebound height is higher for softer solids.

caused by numerical errors due to a too large time step or an insufficient resolution in the contact region. Conversely, a sufficiently fine resolution might cause considerable computational costs. Once the ball is in contact with the ground, the *no-slip* condition on Γ_{w} prevents it from bouncing. We give details below on contact algorithms that enable us to handle such situations.

For the remaining parameters, the solid bounces twice. Because of a higher compression during contact, the first and the second rebound heights are bigger, the softer the solid is.

3.6.2 A simple contact algorithm

In the case where no fluid layer is maintained between the ball and the ground, the application of a contact algorithm is necessary. A simple contact algorithm was used by Sathe & Tezduyar [87]. Their strategy is to add an artificial contact force g_{c} on the interface to the balance of momentum if the ball comes very close to the ground. The force depends on the distance to the ground and goes to infinity as the distance tends to zero. In this way, contact becomes impossible if the time-step size is sufficiently small. The interface condition becomes

$$(\sigma_{\mathrm{f}} - g_{\mathrm{c}} I) n_{\mathrm{f}} = \sigma_{\mathrm{s}} n_{\mathrm{f}} ,$$

where the contact force is defined by

$$g_{\mathrm{c}}(x) = \begin{cases} 0 & \mathrm{dist}(x, \Gamma_{\mathrm{w}}) \geq \mathrm{dist}_0 , \\ \gamma_{\mathrm{c}} \frac{\mathrm{dist}(x,\Gamma_{\mathrm{w}}) - \mathrm{dist}_0}{\mathrm{dist}(x,\Gamma_{\mathrm{w}})} & \mathrm{dist}(x, \Gamma_{\mathrm{w}}) < \mathrm{dist}_0 , \end{cases}$$

on Γ_i with a contact parameter γ_{c}.

One additional advantage of such a contact force is that the usage of the structure stabilization techniques and a larger pressure stabilization are possible again. However, the size of the parameters dist_0 and γ_{c} has to be chosen carefully. If we choose a force that is too big, the results will no longer be physical. If they are too small, the contact might not be prevented.

More involved contact strategies are based on variational inequalities (e.g., Diniz dos Santos et al. [28], Mayer et al. [72], Pironneau [80]) that impose the constraint

$$\mathrm{dist}(x, \Gamma_{\mathrm{w}}) \geq 0 \quad \text{on } \Gamma_i .$$

To ensure the well-posedness of the system of equations, a Lagrange multiplier is added to the balance of momentum that acts in a similar way to the contact force g_{c} when the constraint is active. Because of the additional computational complexity of numerical algorithms for variational inequalities, we stick in this work as a first step to the prior simple contact algorithm.

Influence of the contact force

As this simple contact algorithm is not physically motivated, but is based on an artificial force, we have to analyze its effect on the contact dynamics. Therefore, we consider the situation from above with the stiffest material parameters $\mu_s = 3.2 \cdot 10^6$ and $\lambda_s = 1.28 \cdot 10^7$ where a contact algorithm is necessary. We use $\text{dist}_0 = 10^{-3}$, $5 \cdot 10^{-3}$ and 10^{-2} and choose contact parameters ranging from $\gamma_c = 400$ to 1600.

For $\text{dist}_0 = 10^{-3}$, even the biggest contact parameter $\gamma_c = 1600$ was not large enough to prevent the contact. For $\text{dist}_0 = 5 \cdot 10^{-3}$, the contact was only prevented for $\gamma_c = 1600$, while for $\text{dist}_0 = 10^{-2}$ the parameter $\gamma_c = 800$ was sufficient.

In Figure 3.23, we plot the distances to the ground over time for $\text{dist}_0 = 5 \cdot 10^{-3}$ and 10^{-2}, and $\gamma_c = 800$ and 1600. The plot shows significant differences. For the largest parameters $\text{dist}_0 = 10^{-2}$ and $\gamma_c = 1600$, the rebound height is 35 percent bigger than for $\gamma_c = 800$. Furthermore, the ball stays at rest at a distance of $3.18 \cdot 10^{-3}$ from the ground which is rather large compared to $d_{rest} \approx 1.77 \cdot 10^{-3}$ for $\gamma_c = 800$ and $d_{rest} \approx 1.21 \cdot 10^{-3}$ for $\text{dist}_0 = 5 \cdot 10^{-3}$. Here it is obvious that this distance depends more on the artificial contact force than on physical effects.

We conclude that the contact parameters have to be chosen carefully and their influence must be taken into account when interpreting the results.

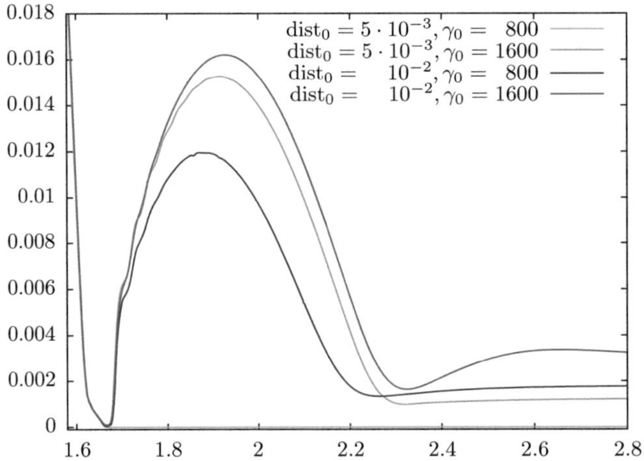

Fig. 3.23: Influence of different contact parameters γ_c and dist_0 for the Lamé parameters $\mu_s = 3.2 \cdot 10^6$, $\lambda_s = 1.28 \cdot 10^7$. The contact force with parameters $\gamma_c = 800$ and $\text{dist}_0 = 5 \cdot 10^{-3}$ as well as for $\gamma_c \leq 400$ or $\text{dist}_0 \leq 10^{-3}$ was not large enough to prevent the contact.

3.6.3 Example 2: Bouncing down the stairs

Finally, we will study a more complex numerical example, i.e., an elastic ball bouncing down some stairs. We give a sketch of the geometry under consideration in Figure 3.24, left sketch. In order to get the desired direction, we let the ball bounce on an inclined plane first. Afterwards, it bounces down three stairs. Depending on the material parameters it can bounce once or several times on a stair or just roll over it. We consider the lower, left and right walls as rigid and impose a homogeneous Dirichlet condition for the velocity there. On the top Γ_{top}, we use again a *do-nothing* boundary condition. We use the same material parameters as in Section 3.6.1 and vary only the fluid density to $\rho_\text{f} = 100, 150, 300$ and 1000.

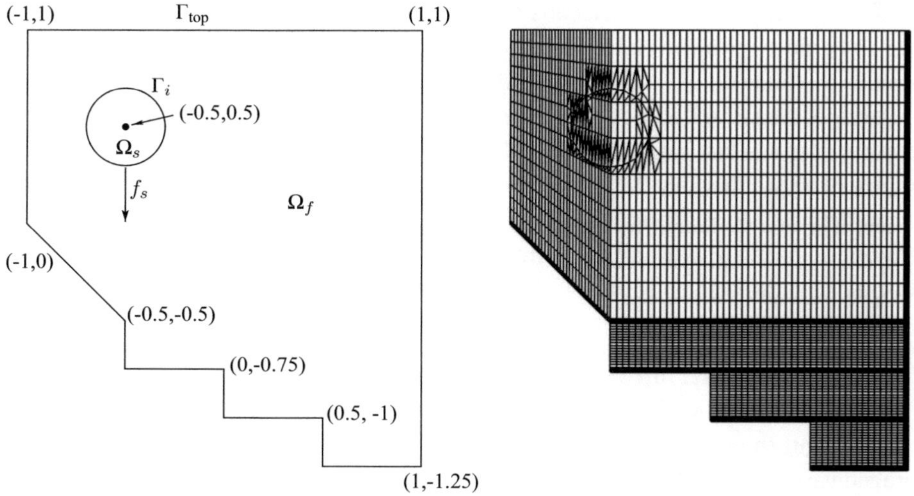

Fig. 3.24: Sketch of the configuration of the second example and the initial mesh. To simulate the contact dynamics accurately, fine mesh cells are used in the contact regions.

Numerical parameters

In this example, structural stabilization is necessary. Furthermore, compared to the previous example we need a considerably larger pressure stabilization in order to obtain stable results. For both pressure and structural stabilization, we use exactly the same parameters as for the FSI3 benchmark problem in Section 3.5.4, i.e., $\gamma = 5 \cdot 10^{-3}$, $\gamma_p = 10^{-3}$, $\alpha = 10$ and $s = 1$.

In order to capture the contact dynamics accurately, we use a mesh with 7905 nodes and very fine mesh cells around all walls the ball can possibly touch, see the right part of Figure 3.24. While the coarsest patches in the upper right part have a size of approximately 0.06×0.18, the size of the finest patches at the lower right corner is approximately $3 \cdot 10^{-3} \times 3 \cdot 10^{-3}$. For time discretization, we use again the modified $dG(0)$ scheme with small time steps $k = 5 \cdot 10^{-4}$.

Results

We show snapshots of the horizontal velocity at 12 different times in Figure 3.25 for $\rho_f = 300$. The ball drops onto the inclined plane and bounces to the right. The next contact is on the right part of the first stair. Afterwards, we observe three bounces on the second stair and two on the third one, before the ball comes to rest.

In Figure 3.26, we show contours of the ball for calculations with $\rho_f = 100, 150, 300$ and 1000. For the two larger density values, we use a contact force with parameters $\gamma_c = 5 \cdot 10^3$ and $\text{dist}_0 = 10^{-2}$. For $\rho_f \leq 150$ this force was not large enough to prevent the contact (see the contact parameter studies below). Here, we use $\gamma_c = 10^4$.

The rebounds are higher, the smaller the fluid density is. For $\rho_f = 100$ the rebound at the first stair is so high that the ball jumps over the second stair and has its next contact on the third one. On the third stair we obtain six small bounces before the ball comes to rest. Because of the viscous fluid between the ball and the wall, the ball is never in contact with the right wall and the distance to the wall is never below the reference distance $\text{dist}_0 = 10^{-2}$, where the contact force would become active.

For $\rho_f = 150$, the ball bounces exactly once on the first and second stair. Before dropping onto the last stair, the ball gets quite close to the right wall with a minimal distance of approximately $2 \cdot 10^{-2}$. However, it is slowed down by fluid forces before the contact force corresponding to the right wall would become active. For $\rho_f = 300$, the rebounds are already significantly smaller and for $\rho_f = 1000$, the ball bounces once on each stair and continues rolling to the right.

Influence of the contact force

In Figure 3.27 (left sketch), we plot the trajectories of the lower bottom of the ball for $\rho_f = 150$ and three different contact force parameters γ_c. On the right, we plot an averaged vertical velocity $\overline{v}_y$ over time.

The contact force with parameter $\gamma_c = 5 \cdot 10^3$ prevents the contact on the inclined plane and on the first stair, but it fails on the second one. Furthermore, we observe that for larger contact parameters the velocity of the ball is slightly higher, and the ball bounces earlier on each stair. As a consequence, the ball almost touches the right wall for $\gamma_c = 10^4$ when it falls down towards the third stair, while it remains at a significantly larger distance of around $6 \cdot 10^{-2}$ for $\gamma_c = 2 \cdot 10^4$. The averaged velocities show good agreement until the bounce on the second stair and differ slightly afterwards due to the different trajectories.

We conclude that in this final example the influence of the contact force was relatively small. However, some quantities of interest, e.g., the contact distance or the distance at rest, cannot be determined with this approach and require more sophisticated contact algorithms.

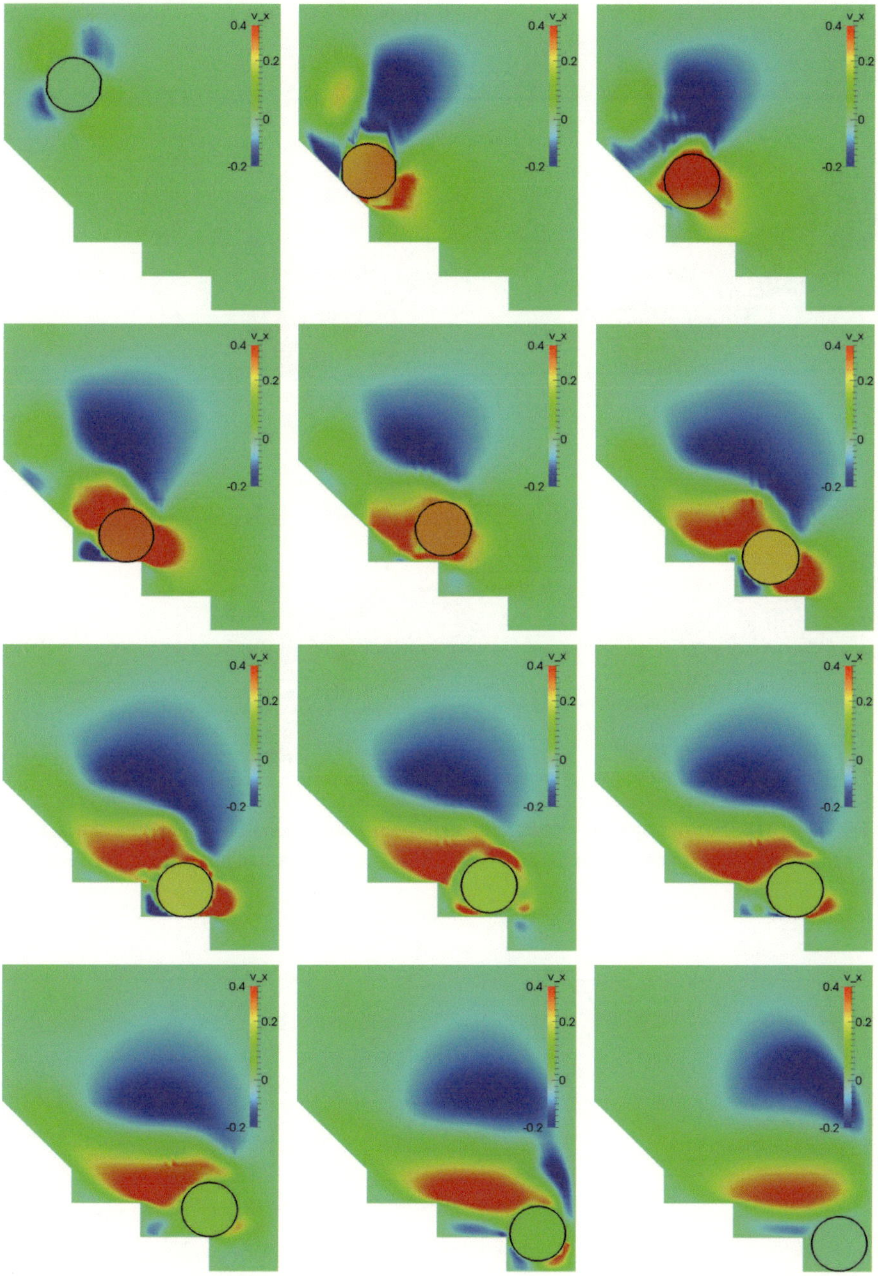

Fig. 3.25: Ball bouncing down three stairs for $\rho_f = 300$ at 12 different times. The color illustrates the horizontal velocity v_x, the black contour line is the discrete interface. First row: Free fall, contact with the inclined plane and rebound. Second row: Contact with the first stair and rebound. Third row: First contact with the second stair, small bounce and second contact. Last row: Third contact with the second stair, fall and position at rest.

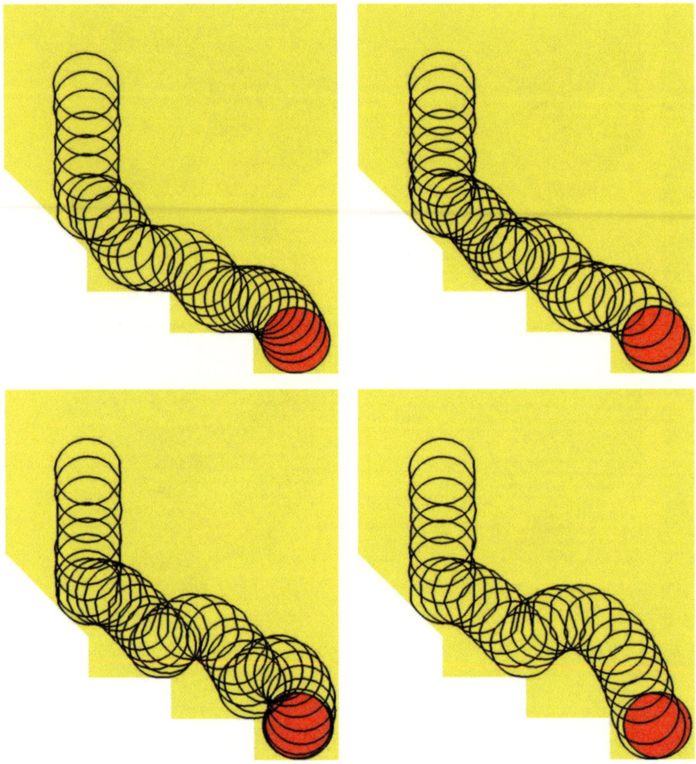

Fig. 3.26: Contour plots of the interface at several times. Top left: $\rho_f = 1000$, top right: $\rho_f = 300$, bottom left: $\rho_f = 150$, bottom right: $\rho_f = 100$. While for $\rho_f = 1000$ the ball rolls over the stairs, the ball bounces exactly once on each stair for $\rho_f = 150$. For $\rho_f = 100$, the ball jumps over the second stair.

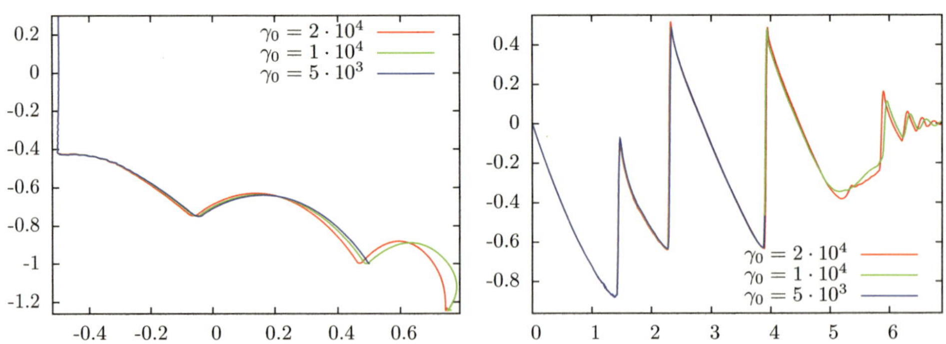

Fig. 3.27: Left: Position of the bottom of the ball. Right: Averaged velocity over time for $\rho_f = 150$ and different values of the contact force. For $\gamma_c = 5 \cdot 10^3$ the contact could not be prevented at the contact time with the second stair.

3.7 Conclusion

The aim of this paper was to present the recent advances made for the Fully Eulerian approach for fluid-structure interactions. For the first time, quantitative results could be given for a Eulerian simulation of the well-known benchmark problem by Hron and Turek. To achieve this result, special care had to be given to the various steps of temporal and spatial discretization. As an interface-problem with moving interfaces, the accurate resolution of the coupling surface has highest priority.

Further, we have demonstrated the ability of the Fully Eulerian method to treat elastic contact problems. Further research is necessary to fully understand the effects, that take place at contact or near contact, when a thin fluid layer remains between the domain's boundary and the elastic obstacle. In future work, we plan to apply more sophisticated contact algorithms that alter the physical contact dynamics as little as possible.

Bibliography

[1] F. P. Baaijens. A fictitious domain/mortar element method for fluid-structure interaction. *International Journal for Numerical Methods in Fluids*, 35(7):743–761, 2001.

[2] I. Babuška. The finite element method for elliptic equations with discontinuous coefficients. *Computing*, 5:207–213, 1970.

[3] I. Babuška, U. Banarjee, and J. E. Osborn. Generalized finite element methods: Main ideas, results, and perspective. *International Journal of Computational Methods*, 1:67–103, 2004.

[4] I. Babuška and A. Miller. The post-processing approach in the finite element method. Part 1. Calculation of displacements, stresses, and other higher derivatives of the displacements. Technical report, DTIC Document, 1982.

[5] J. Baiges and R. Codina. The fixed-mesh ALE approach applied to solid mechanics and fluid–structure interaction problems. *International Journal for Numerical Methods in Engineering*, 81(12):1529–1557, 2010.

[6] S. Basting, S. Canic, R. Glowinski, and A. Quaini. On the implementation and benchmarking of an extended ale method for fsi problems. In S. Frei, B. Holm, T. Richter, T. Wick, and H. Yang, editors, *Fluid-Structure Interaction: Modeling, Adaptive Discretization and Solvers*, Radon Series on Computational and Applied Mathematics. Walter de Gruyter, Berlin, 2017.

[7] Y. Bazilevs, K. Takizawa, and T. E. Tezduyar. *Computational fluid-structure interaction: methods and applications*. John Wiley & Sons, 2013.

[8] R. Becker and M. Braack. A finite element pressure gradient stabilization for the Stokes equations based on local projections. *Calcolo*, 38(4):173–199, 2001.

[9] R. Becker, M. Braack, D. Meidner, T. Richter, and B. Vexler. The finite element toolkit Gascoigne3d. http://www.gascoigne.uni-hd.de.

[10] T. Belytschko, J. M. Kennedy, and D. F. Schoeberle. Quasi-Eulerian finite element formulation for fluid-structure interaction. *Journal of Pressure Vessel Technology*, 102(1):62–69, 1980.

[11] D. Boffi and L. Gastaldi. A finite element approach for the immersed boundary method. *Computers & Structures*, 81(8):491–501, 2003.

[12] D. Boffi, L. Gastaldi, and L. Heltai. A finite element approach to the immersed boundary method. In *Progress in Engineering Computational Technology, B.H.V. Topping and C.A. Mota Soares Eds.*, pages 271–298. Saxe-Coburg Publications, Stirling, Scotland, 2004.

[13] M. Braack and T. Richter. Local projection stabilization for the Stokes system on anisotropic quadrilateral meshes. In B. de Castro et al., editor, *Enumath*, pages 770–778. Springer, 2005.

[14] M. Braack and T. Richter. Solutions of 3D Navier-Stokes benchmark problems with adaptive finite elements. *Computers and Fluids*, 35(4):372–392, 2006.

[15] E. Brummelen, K. Zee, and R. Borst. Space/time multigrid for a fluid-structure-interaction problem. *Applied Numerical Mathematics*, 58(12):1951–1971, 2008.

[16] E. Burman and M. A. Fernández. An unfitted Nitsche method for incompressible fluid–structure interaction using overlapping meshes. *Computer Methods in Applied Mechanics and Engineering*, 279:497–514, 2014.

[17] E. Burman, M. A. Fernández, and P. Hansbo. Edge stabilization for the incompressible Navier-Stokes equations: a continuous interior penalty method. Technical report, 2004.

[18] E. Burman and P. Hansbo. Edge stabilization for Galerkin approximations of convection–diffusion–reaction problems. *Computer Methods in Applied Mechanics and Engineering*, 193(15):1437–1453, 2004.

[19] E. Burman and P. Hansbo. Edge stabilization for the generalized Stokes problem: a continuous interior penalty method. *Computer Methods in Applied Mechanics and Engineering*, 195(19):2393–2410, 2006.

[20] P. Causin, J.-F. Gerbeau, and F. Nobile. Added-mass effect in the design of partitioned algorithms for fluid-structure problems. *Computer Methods in Applied Mechanics and Engineering*, 194:4506–4527, 2005.

[21] J. Chessa, P. Smolinski, and T. Belytschko. The extended finite element method (xfem) for solidication problems. *Int. J. Numer. Meth. Engrg.*, 53:1959–1977, 2002.

[22] R. Codina, G. Houzeaux, H. Coppola-Owen, and J. Baiges. The fixed-mesh ALE approach for the numerical approximation of flows in moving domains. *Journal of Computational Physics*, 228(5):1591–1611, 2009.

[23] C.-M. Colciago, S. Deparis, and D. Forti. Fluid-structure interaction for vascular flows: from supercomputers to laptops. In S. Frei, B. Holm, T. Richter, T. Wick, and H. Yang, editors, *Fluid-Structure Interaction: Modeling, Adaptive Discretization and Solvers*, Radon Series on Computational and Applied Mathematics. Walter de Gruyter, Berlin, 2017.

[24] G.-H. Cottet, E. Maitre, and T. Milcent. An Eulerian method for fluid-structure coupling with biophysical applications. In *ECCOMAS CFD 2006: Proceedings of the European Conference on Computational Fluid Dynamics, Egmond aan Zee, The Netherlands, September 5-8, 2006*, 2006.

[25] G.-H. Cottet, E. Maitre, and T. Milcent. Eulerian formulation and level set models for incompressible fluid-structure interaction. *ESAIM: Mathematical Modelling and Numerical Analysis*, 42(3):471–492, 2008.

[26] J. Donea, P. Fasoli-Stella, and S. Giuliani. Lagrangian and Eulerian finite element techniques for transient fluid-structure interaction problems. In *Structural mechanics in reactor technology*. 1977.

[27] J. Donea, A. Huerta, J.-P. Ponthot, and A. Rodríguez-Ferran. *Arbitrary Lagrangian-Eulerian Methods*. John Wiley & Sons, Ltd, 2004.

[28] N. D. dos Santos, J.-F. Gerbeau, and J.-F. Bourgat. A partitioned fluid–structure algorithm for elastic thin valves with contact. *Computer Methods in Applied Mechanics and Engineering*, 197(19):1750–1761, 2008.

[29] T. Dunne. An Eulerian approach to fluid-structure interaction and goal-oriented mesh refinement. *Int. J. Numer. Math. Fluids.*, 51:1017–1039, 2006.

[30] T. Dunne. *Adaptive Finite Element Approximation of Fluid-Structure Interaction Based on Eulerian and Arbitrary Lagrangian-Eulerian Variational Formulations*. PhD thesis, University of Heidelberg, 2007. urn:nbn:de:bsz:16-opus-79448.

[31] T. Dunne and R. Rannacher. Adaptive finite element approximation of fluid-structure interaction based on an Eulerian variational formulation. In H.-J. Bungartz and M. Schäfer, editors, *Fluid-Structure Interaction: Modeling, Simulation, Optimization*, Lecture Notes in Computational Science and Engineering, pages 110–145. Springer, 2006.

[32] T. Dunne, R. Rannacher, and T. Richter. Numerical simulation of fluid-structure interaction based on monolithic variational formulations. In G. Galdi and R. Rannacher, editors, *Comtemporary Challenges in Mathematical Fluid Mechanics*. World Scientific, Singapore, 2010.

[33] R. Dziri and J.-P. Zolésio. Eulerian derivative for non-cylindrical functionals. In *Shape optimization and optimal design*, volume 216, pages 87–108. J. Cagnol, M.P. Polis, J.-P. Zolésio, 2001.

[34] E. Feireisl. On the motion of rigid bodies in a viscous incompressible fluid. In *Nonlinear Evolution Equations and Related Topics*, pages 419–441. Springer, 2003.

[35] C. A. Felippa, K.-C. Park, and M. R. Ross. A classification of interface treatments for fsi. In *Fluid Structure Interaction II*, pages 27–51. Springer, 2011.

[36] R. M. Franck and R. B. Lazarus. Mixed Eulerian-Lagrangian method. In *Methods in Computational Physics, vol.3: Fundamental methods in Hydrodynamics*, pages 47–67. Academic Press: New York, 1964.

[37] S. Frei. *Eulerian finite element methods for interface problems and fluid-structure interactions*. PhD thesis, Heidelberg University, September 2016. urn:nbn:de:bsz:16-heidok-215905.

[38] S. Frei and T. Richter. A locally modified parametric finite element method for interface problems. *SIAM Journal on Numerical Analysis*, 52(5):2315–2334, 2014.

[39] S. Frei and T. Richter. A second-order time-stepping scheme for parabolic interface problems with moving interfaces. *ESAIM: Mathematical Modelling and Numerical Analysis*, accepted for publication (2016).

[40] S. Frei, T. Richter, and T. Wick. Eulerian techniques for fluid-structure interactions: Part i - modeling and simulation. In *Numerical Mathematics and Advanced Applications-ENUMATH 2013*, pages 745–753. Springer, 2013.

[41] S. Frei, T. Richter, and T. Wick. Eulerian techniques for fluid-structure interactions: Part ii – applications. In *Numerical Mathematics and Advanced Applications-ENUMATH 2013*, pages 755–762. Springer, 2013.

[42] S. Frei, T. Richter, and T. Wick. Long-term simulation of large deformation, mechano-chemical fluid-structure interactions in ALE and fully eulerian coordinates. *Journal of Computational Physics*, 321:874–891, 2016.

[43] T.-P. Fries and T. Belytschko. The extended/generalized finite element method: An overview of the method and its applications. *International Journal for Numerical Methods in Engineering*, 84(3):253–304, 2010.

[44] T.-P. Fries and A. Zilian. On time integration in the XFEM. *International Journal for Numerical Methods in Engineering*, 79(1):69–93, 2009.

[45] P. Gangl and U. Langer. Topology Optimization of Electric Machines based on Topological Sensitivity Analysis. *ArXiv e-prints*, 2013.

[46] M. W. Gee, U. Küttler, and W. A. Wall. Truly monolithic algebraic multigrid for fluid-structure interaction. *International Journal for Numerical Methods in Engineering*, 85(8):987–1016, 2011.

[47] D. Gérard-Varet, M. Hillairet, and C. Wang. The influence of boundary conditions on the contact problem in a 3d Navier–Stokes flow. *Journal de Mathématiques Pures et Appliquées*, 103(1):1–38, 2015.

[48] A. Gerstenberger and W. A. Wall. An extended finite element method/Lagrange multiplier based approach for fluid–structure interaction. *Computer Methods in Applied Mechanics and Engineering*, 197(19):1699–1714, 2008.

[49] A. J. Gil, A. Arranz Carreño, J. Bonet, and O. Hassan. An enhanced immersed structural potential method for fluid–structure interaction. *Journal of Computational Physics*, 250:178–205, 2013.

[50] R. Glowinski, T.-W. Pan, and J. Periaux. A fictitious domain method for Dirichlet problem and applications. *Computer Methods in Applied Mechanics and Engineering*, 111(3):283–303, 1994.

[51] C. Großmann, H.-G. Roos, and M. Stynes. *Numerical treatment of partial differential equations*. Springer, 2007.

[52] A. Hansbo and P. Hansbo. An unfitted finite element method, based on Nitsche's method, for elliptic interface problems. *Comput. Methods Appl. Mech. Engrg.*, 191(47-48):5537–5552, 2002.

[53] A. Hansbo and P. Hansbo. A finite element method for the simulation of strong and weak discontinuities in solid mechanics. *Comput. Methods Appl. Mech. Engrg.*, 193:3523–3540, 2004.

[54] P. Hansbo and J. Hermansson. Nitsche's method for coupling non-matching meshes in fluid-structure vibration problems. *Computational Mechanics*, 32(1-2):134–139, 2003.

[55] P. He and R. Qiao. A full-eulerian solid level set method for simulation of fluid-structure interactions. *Microfluidics and Nanofluidics*, 11:557–567, 2011.

[56] M. Heil, A. L. Hazel, and J. Boyle. Solvers for large-displacement fluid-structure interaction problems: Segregated vs. monolithic approaches. *Computational Mechanics*, 43, 2008.

[57] M. Hillairet. Lack of collision between solid bodies in a 2d incompressible viscous flow. *Communications in Partial Differential Equations*, 32:1345–1371, 2007.

[58] J. Hoffman, B. Holm, and T. Richter. The locally adapted parametric finite element method for interface problems on triangular meshes. In S. Frei, B. Holm, T. Richter, T. Wick, and H. Yang, editors, *Fluid-Structure Interaction: Modeling, Adaptive Discretization and Solvers*, Radon Series on Computational and Applied Mathematics. Walter de Gruyter, Berlin, 2017.

[59] G. A. Holzapfel. *Nonlinear Solid Mechanics: A Continuum Approach for Engineering*. Wiley-Blackwell, 2000.

[60] J. Hron and S. Turek. A monolithic FEM/multigrid solver for an ALE formulation of fluid-structure interaction with applications in biomechanics. In H.-J. Bungartz and M. Schäfer, editors, *Fluid-Structure Interaction: Modeling, Simulation, Optimization*, Lecture Notes in Computational Science and Engineering, pages 146–170. Springer, 2006.

[61] J. Hron and S. Turek. Proposal for numerical benchmarking of fluid-structure interaction between an elastic object and laminar incompressible flow. In H.-J. Bungartz and M. Schäfer, editors, *Fluid-Structure Interaction: Modeling, Simulation, Optimization*, Lecture Notes in Computational Science and Engineering, pages 371–385. Springer, 2006.

[62] T. J. Hughes, W. K. Liu, and T. K. Zimmermann. Lagrangian-Eulerian finite element formulation for incompressible viscous flows. *Computer Methods in Applied Mechanics and Engineering*, 29(3):329–349, 1981.

[63] S. Ii, K. Sugiyama, S. Takagi, and Y. Matsumoto. A computational blood flow analysis in a capillary vessel including multiple red blood cells and platelets. *Journal of Biomechanical Science and Engineering*, 7(1):72–83, 2012.

[64] J. Jim Douglas and T. F. Russell. Numerical methods for convection-dominated diffusion problems based on combining the method of characteristics with finite element or finite difference procedures. *SIAM Journal on Numerical Analysis*, 19(5):871–885, 1982.

[65] D. Jodlbauer and T. Wick. A monolithic fsi solver applied to the fsi 1,2,3 benchmarks. In S. Frei, B. Holm, T. Richter, T. Wick, and H. Yang, editors, *Fluid-Structure Interaction: Modeling, Adaptive Discretization and Solvers*, Radon Series on Computational and Applied Mathematics. Walter de Gruyter, Berlin, 2017.

[66] A. Laadhari, R. Ruiz-Baier, and A. Quarteroni. Fully Eulerian finite element approximation of a fluid-structure interaction problem in cardiac cells. *International Journal for Numerical Methods in Engineering*, 96(11):712–738, 2013.
[67] U. Langer and H. Yang. Recent development of robust monolithic fluid-structure interaction solvers. In S. Frei, B. Holm, T. Richter, T. Wick, and H. Yang, editors, *Fluid-Structure Interaction: Modeling, Adaptive Discretization and Solvers*, Radon Series on Computational and Applied Mathematics. Walter de Gruyter, Berlin, 2017.
[68] A. Legay, J. Chessa, and T. Belytschko. An Eulerian-Lagrangian method for fluid-structure interaction based on level sets. *Comput. Methods Appl. Mech. Engrg.*, 195:2070–2087, 2006.
[69] C. Lehrenfeld and A. Reusken. Analysis of a Nitsche XFEM-DG discretization for a class of two-phase mass transport problems. *SIAM Journal on Numerical Analysis*, 51(2):958–983, 2013.
[70] X. S. Li and J. W. Demmel. SuperLU_DIST: A scalable distributed-memory sparse direct solver for unsymmetric linear systems. *ACM Transactions on Mathematical Software*, 29(2):110–140, 2003.
[71] A. Massing, M. G. Larson, A. Logg, and M. E. Rognes. A Nitsche-based cut finite element method for a fluid–structure interaction problem. *ArXiv e-prints*, 2015.
[72] U. M. Mayer, A. Popp, A. Gerstenberger, and W. A. Wall. 3d fluid–structure-contact interaction based on a combined XFEM FSI and dual mortar contact approach. *Computational Mechanics*, 46(1):53–67, 2010.
[73] N. Moës, J. Dolbow, and T. Belytschko. A finite element method for crack growth without remeshing. *Int. J. Numer. Meth. Engrg.*, 46:131–150, 1999.
[74] J. A. Nitsche. Über ein Variationsprinzip zur Lösung von Dirichlet-Problemen bei Verwendung von Teilräumen, die keinen Randbedingungen unterworfen sind. *Abhandlungen aus dem Mathematischen Seminar der Universität Hamburg*, 36:9–15, 1970.
[75] W. F. Noh. Cel: A time-dependent, two-space-dimensional, coupled Eulerian-lagrange code. In *Methods in Computational Physics, vol.3: Fundamental methods in Hydrodynamics*, pages 117–179. Academic Press: New York, 1964.
[76] K.-C. Park, C. A. Felippa, and R. Ohayon. Partitioned formulation of internal fluid–structure interaction problems by localized Lagrange multipliers. *Computer Methods in Applied Mechanics and Engineering*, 190(24):2989–3007, 2001.
[77] S. V. Patankar. A numerical method for conduction in composite materials, flow in irregular geometries and conjugate heat transfer. In *Proc. 6th Int. Heat Transfer Conf.*, volume 3, pages 297–302, Toronto, Canada, 1978.
[78] C. S. Peskin. Flow patterns around heart valves: a numerical method. *Journal of Computational Physics*, 10(2):252–271, 1972.
[79] O. Pironneau. On the transport-diffusion algorithm and its applications to the navierstokes equations. *Numerische Mathematik*, 38:309–332, 1982.
[80] O. Pironneau. An energy preserving monolithic eulerian fluid-structure numerical scheme. *submitted to Chinese Annals of Mathematics*, 2016. arXiv:1607.08083.
[81] R. Rannacher and T. Richter. An adaptive finite element method for fluid-structure interaction problems based on a fully eulerian formulation. In H. Bungartz, M. Mehl, and M. Schäfer, editors, *Fluid-Structure Interaction II, Modelling, Simulation, Optimization*, number 73 in Lecture notes in computational science and engineering, pages 159–192. Springer, 2010.
[82] T. Richter. A fully eulerian formulation for fluid-structure interactions. *J. Comp. Phys.*, 223:227–240, 2013.
[83] T. Richter. A monolithic geometric multigrid solver for fluid-structure interactions in ALE formulation. *International Journal for Numerical Methods in Engineering*, 104(5):372–390, 2015.

[84] T. Richter. *Finite Elements for Fluid-Structure Interactions*. Springer, 2016. in preparation.

[85] T. Richter and T. Wick. Finite elements fo fluid-structure interaction in ale and fully eulerian coordinates. *Computer Methods in Applied Mechanics and Engineering*, 2010. doi: 10.1016/j.cma.2010.04.016).

[86] M. Sala, K. S. Stanley, and M. A. Heroux. On the design of interfaces to sparse direct solvers. *ACM Transactions on Mathematical Software (TOMS)*, 34(2):9, 2008.

[87] S. Sathe and T. E. Tezduyar. Modeling of fluid–structure interactions with the space–time finite elements: contact problems. *Computational Mechanics*, 43(1):51–60, 2008.

[88] J. A. Sethian. *Level Set Methods and Fast Marching Methods Evolving Interfaces in Computational Geometry*. Fluid mechanics, Computer Vision and Material Science. Cambridge University Press, 1999.

[89] K. Stein, T. E. Tezduyar, and R. Benney. Mesh moving techniques for fluid-structure interactions with large displacements. *Journal of Applied Mechanics*, 70(1):58–63, 2003.

[90] K. Sugiyama, S. Ii, S. Takeuchi, S. Takagi, and Y. Matsumoto. Full Eulerian simulations of biconcave neo-Hookean particles in a Poiseuille flow. *Computational Mechanics*, 46:147–157, 2010.

[91] K. Sugiyama, S. Ii, S. Takeuchi, S. Takagi, and Y. Matsumoto. A full Eulerian finite difference approach for solving fluid-structure coupling problems. *Journal of Computational Physics*, 230:596–627, 2011.

[92] T. E. Tezduyar, M. Behr, and J. Liou. A new strategy for finite element computations involving moving boundaries and interfaces - the deforming-spatial-domain/space-time procedure: I the concept and the preliminary numerical tests. *Comput. Methods Appl. Mech. Engrg.*, 94:339–351, 1992.

[93] S. Turek, J. Hron, M. Razzaq, H. Wobker, and M. Schäfer. *Numerical benchmarking of Fluid-Structure Interaction: A comparison of different discretization and solution approaches*. Springer, 2010.

[94] R. Van Loon, P. Anderson, F. Van de Vosse, and S. Sherwin. Comparison of various fluid–structure interaction methods for deformable bodies. *Computers & structures*, 85(11):833–843, 2007.

[95] X. Wang and W. K. Liu. Extended immersed boundary method using FEM and RKPM. *Computer Methods in Applied Mechanics and Engineering*, 193(12):1305–1321, 2004.

[96] T. Wick. *Adaptive Finite Element Simulation of Fluid-Structure Interaction with Application to Heart-Valve Dynamics*. PhD thesis, University of Heidelberg, 2011.

[97] T. Wick. Fluid-structure interactions using different mesh motion techniques. *Computers and Structures*, 89:1456–1467, 2011.

[98] T. Wick. Fully Eulerian fluid-structure interaction for time-dependent problems. *Comp. Methods Appl. Mech. Engrg.*, 255(0):14–26, 2012.

[99] T. Wick. Flapping and contact fsi computations with the fluid–solid interface-tracking/interface-capturing technique and mesh adaptivity. *Computational Mechanics*, 53(1):29–43, 2014.

[100] T. Wick. Coupling fluid-structure interaction with phase-field fracture: algorithmic details. In S. Frei, B. Holm, T. Richter, T. Wick, and H. Yang, editors, *Fluid-Structure Interaction: Modeling, Adaptive Discretization and Solvers*, Radon Series on Computational and Applied Mathematics. Walter de Gruyter, Berlin, 2017.

[101] L. Zhang, A. Gerstenberger, X. Wang, and W. K. Liu. Immersed finite element method. *Computer Methods in Applied Mechanics and Engineering*, 193(21):2051–2067, 2004.

[102] P. Zunino. Analysis of backward Euler/extended finite element discretization of parabolic problems with moving interfaces. *Computer Methods in Applied Mechanics and Engineering*, 258:152–165, 2013.

Part II: **Solvers**

Philipp Birken and Azahar Monge

4 Numerical methods for unsteady thermal fluid structure interaction

Abstract: We discuss thermal fluid–structure interaction processes, where a simulation of the time-dependent temperature field is of interest. Thereby, we consider partitioned coupling schemes with a Dirichlet–Neumann method. We present an analysis of the method on a model problem of discretized coupled linear heat equations. This shows that for large quotients in the heat conductivities, the convergence rate will be very small. The time dependency makes the use of time-adaptive implicit methods imperative. This gives rise to the question as to how accurately the appearing nonlinear systems should be solved, which is discussed in detail for both the nonlinear and linear case. The efficiency of the resulting method is demonstrated using realistic test cases.

Keywords: Thermal fluid–structure interaction, Conjugate heat transfer, Dirichlet–Neumann method, Time adaptivity, Termination criteria

Classification: 65F10, 65L04, 65M22, 74F04, 74F10

4.1 Introduction

Fluid–structure interaction occurs when a deformable or moving structure interacts with a surrounding or internal fluid flow. Our specific field of interest is thermal interaction between fluids and structures, also called conjugate heat transfer. Examples of thermal fluid–structure interaction are cooling of gas-turbine blades, thermal anti-icing systems of airplanes [12], supersonic re-entry of vehicles from space [25, 31], gas quenching, which is an industrial heat treatment of metal workpieces [23, 40], or the cooling of rocket nozzles [26, 27]. These problems are usually too complex to solve analytically, and therefore, numerical simulations of conjugate heat transfer are essential in many applications.

The efficient numerical simulation of fluid–structure interaction (FSI) models is one of the important current challenges in scientific computing as stated in [11]:

> "The issue of coupling models of different events at different scales and governed by different physical laws is largely wide open and represents an enormously challenging area for future research."

Philipp Birken, Lund University, Center for the Mathematical Sciences, Numerical Analysis, Box 118, 22100 Lund, Sweden, philipp.birken@na.lu.se
Azahar Monge, Lund University, Center for the Mathematical Sciences, Numerical Analysis, Box 118, 22100 Lund, Sweden, azahar.monge@na.lu.se

https://doi.org/10.1515/9783110494259-004

(a) Inductive heating. (b) Thermo-mechanical forming. (c) Local air-cooling.

Fig. 4.1: Gas quenching. Left and center picture: Institute of Mechanics and Dynamics, University of Kassel. Right picture: Institute of Metal Forming Technology, University of Kassel.

In this article, we focus on the coupling between air and an alloy. When being cooled or heated, the alloy experiences thermomechanical effects that change its internal structure. A first example are steel-forging processes. One possibility is to use cold high-pressured air as a cooling medium (Figure 4.1). Knowledge about the time-dependent temperature field is imperative to predict where martensite will be in a finished steel part. This allows one to predict material properties generated by the forging process.

Another example is the cooling of rocket thrust chambers. The temperature achieved in the turbines has increased over the years due to progress in building materials and therefore, advanced cooling methods are needed. This is both to avoid critical damage to the rocket nozzle when in use and to develop reusable rocket stages. An important case is the Ariane 5, see Figure 4.2a, which is used to deliver payloads into geostationary transfer orbit (GTO) or low Earth orbit (LEO). The first stage rocket engine for the Ariane 5 is the Vulcain 2, see Figure 4.2b.

The first-stage rocket engine is only used during the launch of the rocket. Therefore, recovering and reusing this first stage would reduce the cost of space access and the environmental impact. Related to this, the company SpaceX is developing a set of new technologies for an orbital launch system that may be reused many times in a manner similar to the reusability of aircraft. The first controlled vertical splashdown of an orbital rocket stage on the ocean surface was achieved in April 2014. The next two flights in January and April 2015 attempted to land the returning first stage on a floating platform. Both of them were guided accurately to the target, but they did not succeed in landing vertically on the floating platform and were destroyed [3]. Finally, the first vertical landing was achieved on December 21, 2015, when the first stage of Falcon 9 Flight 20 successfully landed on solid ground [1]. On April 8, 2016, Flight 23 achieved the first soft landing on a drone ship in the Atlantic Ocean [2], see Figure 4.3.

In view of these achievements, it is of interest to simulate several cycles of the combustion-cooling process of the thrust chamber. This will tell one how many times

(a) Ariane 5 on the launch pad. (b) The Vulcain engine in a museum. (c) Sketch of the rocket thrust chamber.

Fig. 4.2: Cooling of rocket thrust chambers. Left picture: DLR German Aerospace Center, CC BY 2.0. Center picture: Pline, CC BY-SA 3.0.

(a) Unsuccessful vertical landing attempt. (b) First stage landing vertically on solid ground in December 2015. (c) First stage landed on autonomous droneship in April 2016.

Fig. 4.3: SpaceX reusable launch system development program. SpaceX Photos, CC0 1.0.

the engine can be used without damaging the structure. More details about this can be found in [26, 27].

Figure 4.2c shows a combustion chamber with the nozzle, through which hot gas can escape. The nozzle is delimited by a structure that can be damaged due to the

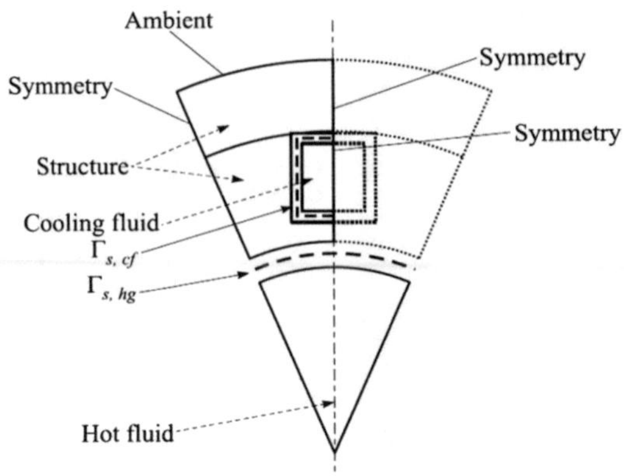

Fig. 4.4: Sketch of the coupling surfaces. Figure taken from [27].

high temperature of the gas flowing inside. In order to avoid this, a cooling fluid flows through small channels contained inside the structure. This results in a system with two thermal interactions between fluids and structures. On one hand, between the hot gas coming out from the combustion chamber and the structure covering the nozzle. On the other hand, between the cooling fluid and the structure. A sketch of the coupling surfaces can be seen in Figure 4.4 where $\Gamma_{s,\mathrm{cf}}$ corresponds to the interface between the structure and the cooling fluid and $\Gamma_{s,\mathrm{hg}}$ to the interface between the structure and the hot gas.

With regards to space discretization, the use of the finite element method (FEM) is ubiquitous for the structure. For the fluid, typically finite volume methods (FVM) are used. However, there are some approaches that use finite elements for both problems, in particular when using a monolithic method [4].

4.1.1 Partitioned coupling methods

Two main methods exist to simulate FSI problems: monolithic and partitioned methods. In the monolithic method, a new code is tailored for the coupled equations, whereas the partitioned approach allows one to reuse existing software for each subproblem. The coupling is done by a master program which calls interface functions of the other codes [16]. If the data transfer between the subsolvers is done only once per time step, we are using a loosely coupled scheme [17]. However, for stability reasons, often a strongly coupled scheme needs to be used [28]. In this case, the data exchange at every time step is repeated until a convergence criterion is satisfied. Here, we focus on partitioned methods.

At the boundary, one imposes that the temperature and the heat flux have to be continuous across the interface. To obtain such a solution, a Dirichlet–Neumann iteration can be employed. This consists of solving the fluid problem with Dirichlet boundary conditions at the interface and then the structure problem with Neumann

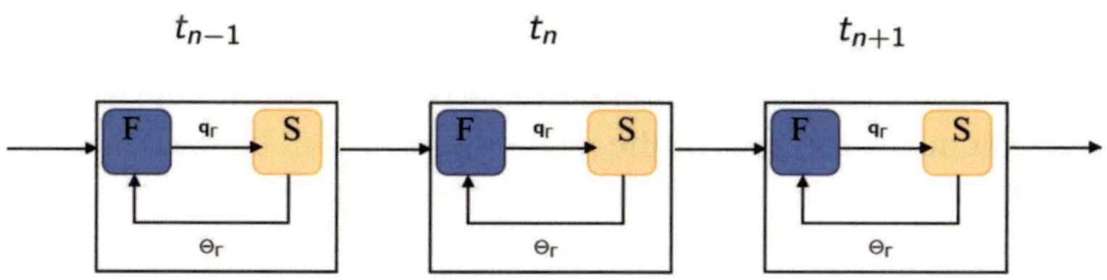

Fig. 4.5: Illustration of the FSI solver.

boundary conditions at the interface, resulting in a fixed-point iteration. Figure 4.5 illustrates this. This iteration is a basic method in both domain decomposition and fluid–structure interaction.

In the domain decomposition context, there are two main problems, namely slow convergence and the need for an implementation using a red-black coloring. The slow convergence can be improved using a relaxation procedure [34]. In fluid–structure interaction, there are only two domains, coupled along an interface, making the application straightforward. However, the convergence rate is not great for the coupling between a compressible fluid and a structure [14], which is why a lot of effort goes into convergence acceleration. Conversely, the Dirichlet–Neumann iteration was reported to be a very fast solver for thermal fluid–structure interactions [8].

As mentioned before, we are studying transient processes and therefore a numerical method needs to be chosen for the time discretization. In [29], the implicit midpoint rule is used in a monolithic scheme to analyze energy conservation of an aeroelasticity problem. Already in [5, 13], it is suggested to use an explicit high-order Runge–Kutta scheme for both subproblems with data exchange at each stage. However, the resulting scheme has limited time steps due to the explicit nature of the method. The order of coupling schemes on moving meshes is analyzed in [22], but only first-order convergence is proved for p-th order schemes. Moreover, higher order implicit Runge–Kutta schemes on moving meshes are analyzed in [43] (in 1D) and in [44] (in 3D). There, so-called explicit first-stage, singly diagonally implicit Runge–Kutta schemes (ESDIRK) are employed and higher order in time is proved by numerical results. The master program of the FSI procedure can be extended to SDIRK methods and furthermore, time adaptivity can be added into this framework as explained in [9, 10]. Another interesting alternative is subcycling, where in the subsolvers, different time-step sizes are chosen [21, 37]. So far, this has not been brought together with time-adaptive methods.

In summary, a partitioned method for unsteady thermal FSI involves two subsolvers (fluid and structure) and two embedded iterations. An outer loop exists corresponding to the time integration. Then, at each time step, a subiteration (the Dirichlet–Neumann method) couples the subsolvers, which have another nonlinear iteration so provide solutions to the subproblems.

An outline of this chapter follows. In Section 2, we describe the model and discretization, as well as the coupling conditions and the partitioned solution method, the Dirichlet–Neumann iteration. A model problem, namely two coupled discretized heat equations, is presented in Section 3. Our analysis to determine the convergence rate of the Dirichlet–Neumann method for the model problem can be found in Section 4. Numerical results are included to illustrate the theoretical analysis. In Section 5, we describe how a fast solver is obtained using time adaptivity, followed by an in-depth analysis of how tolerances in the various nested iterations should be chosen to avoid oversolving while giving the desired accuracy in Section 6. Numerical results to demonstrate the efficiency of the obtained method are presented in Section 7.

4.2 Treating thermal FSI problems

The basic setting we are in is that on a domain $\Omega_1 \subset \mathbb{R}^d$ where d corresponds to the spatial dimension, the physics is described by a fluid model, whereas on a domain $\Omega_2 \subset \mathbb{R}^d$, a different model describing the structure is used. The two domains are almost disjoint in that they are connected via an interface. The part of the interface where the fluid and the structure are supposed to interact is called the coupling interface $\Gamma \subset \partial\Omega_1 \cup \partial\Omega_2$. Note that Γ might be a true subset of the intersection, because the structure could be insulated. At the interface Γ, coupling conditions are prescribed that model the interaction between fluid and structure. For the thermal coupling problem, these conditions are that temperature and the normal component of the heat flux are continuous across the interface.

4.2.1 Fluid model

We model the fluid using the time-dependent Reynolds Averaged Navier–Stokes equations (URANS), which are a second-order system of conservation laws (mass, momentum, energy) modeling turbulent compressible flow. We consider the two-dimensional case, written in conservative variables density ρ, momentum $\mathbf{m} = \rho\mathbf{v}$, and energy per unit volume ρE, where a $\tilde{}$ denotes the Favre average and the overbar the ensemble average [6]:

$$\begin{aligned}
&\partial_t\overline{\rho} + \nabla\cdot\overline{\rho}\tilde{\mathbf{v}} = 0,\\
&\partial_t\overline{\rho}\tilde{\mathbf{v}} + \sum_{j=1}^{2}\partial_{x_j}(\overline{\rho}\tilde{v}_i\tilde{v}_j) = -\partial_{x_j}\overline{p}\delta_{ij} + \frac{1}{Re}\sum_{j=1}^{2}\partial_{x_j}\left(\tilde{S}_{ij} + S_{ij}^{\mathrm{R}}\right), \quad i = 1,2 \qquad (4.1)\\
&\partial_t\overline{\rho}\tilde{E} + \nabla\cdot(\overline{\rho}\tilde{H}\tilde{v}_j) = \sum_{j=1}^{2}\partial_{x_j}\left(\overline{(\frac{1}{Re}S_{ij} - S_{ij}^{\mathrm{R}})v_i} - \overline{\rho}\widetilde{v_j''} + \widetilde{S_{ij}v_i''} - \overline{\rho}\widetilde{v_j''k} + \frac{\overline{W}_j}{RePr}\right).
\end{aligned}$$

The Reynolds stresses

$$S_{ij}^{\mathrm{R}} = -\overline{\rho}\widetilde{v_i''v_j''}$$

and the turbulent energy

$$k = \frac{1}{2}\sum_{j=1}^{d} \overline{v_j' v_j'}$$

are modeled using the Spallart–Allmaras model [39]. Furthermore, $\mathbf{q}_f = (q_1, q_2)^T$ represents the heat flux and $\mathbf{S} = (S_{ij})_{i,j=1,2}$ the viscous shear stress tensor. As the equations are dimensionless, the Reynolds number Re and the Prandtl number Pr appear. The system is closed by the equation of state for the pressure $p = (\gamma - 1)\rho e$, the Sutherland law representing the correlation between temperature and viscosity, as well as the Stokes hypothesis. Additionally, we prescribe appropriate boundary conditions at the boundary of Ω_1 except for Γ, where we have the coupling conditions. In the Dirichlet–Neumann coupling, a temperature value is enforced at Γ.

4.2.2 Structure model

Regarding the structure model, we will consider heat conduction only. Thus, we have the nonlinear heat equation for the structure temperature Θ

$$\rho(\mathbf{x})c_p(\Theta)\frac{d}{dt}\Theta(\mathbf{x}, t) = -\nabla \cdot \mathbf{q}(\mathbf{x}, t)\,, \tag{4.2}$$

where

$$\mathbf{q}_s(\mathbf{x}, t) = -\lambda(\Theta)\nabla\Theta(\mathbf{x}, t)$$

denotes the heat flux vector. For alloys, the specific heat capacity c_p and heat conductivity λ are temperature-dependent and highly nonlinear. How to model thermomechanical effects is the subject of a lot of current research with an important question being how to relate the microstructure to macroscopical models.

As an example, an empirical model for the steel 51CrV4 was suggested in [35]. This was obtained simply by doing measurements and then a least squares fit to a chosen curve. The coefficient functions are then

$$\lambda(\Theta) = 40.1 + 0.05\Theta - 0.0001\Theta^2 + 4.9 \cdot 10^{-8}\Theta^3 \tag{4.3}$$

and

$$c_p(\Theta) = -10\ln\left(\frac{e^{-c_{p1}(\Theta)/10} + e^{-c_{p2}(\Theta)/10}}{2}\right) \tag{4.4}$$

with

$$c_{p1}(\Theta) = 34.2e^{0.0026\Theta} + 421.15 \tag{4.5}$$

and

$$c_{p2}(\Theta) = 956.5e^{-0.012(\Theta-900)} + 0.45\Theta\,. \tag{4.6}$$

For the mass density one has $\rho = 7836\ \text{kg/m}^3$.

Finally, on the boundary, we have Neumann conditions $\mathbf{q}_s(\mathbf{x}, t) \cdot \mathbf{n}(\mathbf{x}) = q_b(\mathbf{x}, t)$.

4.2.3 Coupling conditions

As mentioned before at the beginning of this section, the coupling conditions are that temperature and the normal component of the heat flux are continuous across the interface, i.e.,

$$T(\mathbf{x}, t) = \Theta(\mathbf{x}, t), \ \ \mathbf{x} \in \Gamma \, , \tag{4.7}$$

where T is the fluid temperature and Θ the structure temperature and

$$\mathbf{q}_f(\mathbf{x}, t) \cdot \mathbf{n}(x) = \mathbf{q}_s(\mathbf{x}, t) \cdot \mathbf{n}(x), \ \ \mathbf{x} \in \Gamma \, . \tag{4.8}$$

4.2.4 Discretization in space

Following the partitioned coupling approach, we discretize the two models separately in space. For the fluid, we use a finite volume method, leading to

$$\frac{d}{dt}\mathbf{u} + \mathbf{h}(\mathbf{u}, \boldsymbol{\Theta}_\Gamma) = \mathbf{0} \, , \tag{4.9}$$

where $\mathbf{h}(\mathbf{u}, \boldsymbol{\Theta}_\Gamma)$ represents the spatial discretization and its dependence on the temperatures on the discrete interface to the structure, here denoted by $\boldsymbol{\Theta}_\Gamma$.

Regarding structural mechanics, the use of finite element methods is ubiquitous. Therefore, we will also follow that approach here, leading to the nonlinear equation for all unknowns on Ω_2:

$$\mathbf{M}(\Theta)\frac{d}{dt}\Theta + \mathbf{A}(\Theta)\Theta = \mathbf{q}_b^f + \mathbf{q}_b^\Gamma(\mathbf{u}) \, . \tag{4.10}$$

Here, $\mathbf{M}$ is the mass matrix, also called the heat capacity matrix for this problem and $\mathbf{A}$ is the heat conductivity matrix. The vector Θ consists of all discrete temperature unknowns and $\mathbf{q}_b^\Gamma(\mathbf{u})$ is the discrete heat flux vector on the coupling interface to the fluid, whereas $\mathbf{q}_b^f$ corresponds to boundary heat fluxes independent of the fluid, for example at insulated boundaries.

4.2.5 Time discretization

For the sake of completeness, we now write down a discretization using the implicit Euler method. More advanced time integration schemes are discussed in Section 4.5. For the coupled system (4.9)–(4.10) we obtain

$$\mathbf{u}^{n+1} - \mathbf{u}^n + \Delta t_n \mathbf{h}(\mathbf{u}^{n+1}, \Theta_\Gamma^{n+1}) = \mathbf{0} \, , \tag{4.11}$$

$$\mathbf{M}(\Theta^{n+1})(\Theta^{n+1} - \Theta^n) + \Delta t_n \mathbf{A}(\Theta^{n+1})\Theta^{n+1} = \Delta t_n(\mathbf{q}_b^f + \mathbf{q}_b^\Gamma(\mathbf{u}^{k+1})) \, . \tag{4.12}$$

4.2.6 The Dirichlet–Neumann method

The Dirichlet–Neumann method is a basic iterative substructuring method in domain decomposition and it is a common choice for treating FSI problems. Therefore, we now employ it to solve the system (4.11)–(4.12). This corresponds to alternately solving equation (4.11) on Ω_1 with Dirichlet data on Γ and (4.12) on Ω_2 with Neumann data on Γ.

Thus, one gets for the k-th iteration the two decoupled equation systems

$$\mathbf{u}^{k+1} - \mathbf{u}^n + \Delta t_n \mathbf{h}(\mathbf{u}^{k+1}, \Theta_\Gamma^k) = \mathbf{0}\,, \tag{4.13}$$

$$\mathbf{M}(\Theta^{k+1})(\Theta^{k+1} - \Theta^n) + \Delta t_n \mathbf{A}(\Theta^{k+1})\Theta^{k+1} = \Delta t_n(\mathbf{q}_b^f + \mathbf{q}_b^\Gamma(\mathbf{u}^{k+1}))\,, \tag{4.14}$$

with some initial condition Θ_Γ^0. The iteration is terminated according to the standard criterion

$$\|\Theta_\Gamma^{k+1} - \Theta_\Gamma^k\| \leq \tau \tag{4.15}$$

where τ is a user-defined tolerance.

The convergence rate of the Dirichlet–Neumann iteration is not great for the coupling between a compressible fluid and a structure [14], which is why a lot of effort goes into convergence acceleration. Conversely, the Dirichlet–Neumann iteration was reported to be a very fast solver for thermal fluid–structure interaction. More specifically, in [8] the iteration is extremely efficient and achieves a very accurate solution with at most two iterations per time step. To analyze the convergence behavior of the Dirichlet–Neumann iteration will explain why this coupling method is very efficient in some FSI models and very inefficient in others.

In principle, the convergence rate of the Dirichlet–Neumann method is analyzed in any standard book on domain decomposition methods, e.g., [34, 41]. There, the iteration matrix of the discretized equations is derived with respect to the interface unknowns and the convergence rate is the spectral radius of that. However, because of the nonlinearity of the thermal fluid–structure interaction model explained previously, it is not possible to compute the spectral radius of the iteration matrix in this case. For this reason, we perform in the next section a convergence analysis of the coupling of two linear heat equations. We chose this model because it is a basic building block in fluid–structure interaction.

4.3 A model problem: Coupled heat equations

In FSI models, the solid is typically discretized using finite elements. Conversely, although the FEM is applicable to computational fluid dynamics, the FVM is generally a better choice for the discretization of the fluid. This method guarantees the conservation of fluxes through a particular control volume. Therefore, we present here a convergence analysis of the unsteady transmission problem with mixed discretizations (FVM-FEM).

For this model problem, Henshaw and Chand provided in [24] a method to analyze stability and convergence speed of the Dirichlet–Neumann iteration in 2D based on applying the continuous Fourier transform to the semidiscretized equations. Their result depends on ratios of thermal conductivities and diffusivities of the materials. However, in the fully discrete case we have observed that the iteration behaves differently in some cases. Therefore, we propose a complementary analysis for the fully discrete case here.

4.3.1 Model problem

The unsteady transmission problem is as follows, where we consider a domain $\Omega \subset \mathbb{R}^d$ which is cut into two subdomains $\Omega_1 \cup \Omega_2 = \Omega$ with transmission conditions at the interface $\Gamma = \Omega_1 \cap \Omega_2$:

$$
\begin{aligned}
\alpha_m \frac{\partial u_m(\mathbf{x}, t)}{\partial t} - \nabla \cdot (\lambda_m \nabla u_m(\mathbf{x}, t)) &= 0, \quad t \in [t_0, t_f], \quad \mathbf{x} \in \Omega_m \subset \mathbb{R}^2, \; m = 1, 2, \\
u_m(\mathbf{x}, t) &= 0, \quad t \in [t_0, t_f], \quad \mathbf{x} \in \partial\Omega_m \backslash \Gamma, \\
u_1(\mathbf{x}, t) &= u_2(\mathbf{x}, t), \quad \mathbf{x} \in \Gamma, \\
\lambda_2 \frac{\partial u_2(\mathbf{x}, t)}{\partial \mathbf{n}_2} &= -\lambda_1 \frac{\partial u_1(\mathbf{x}, t)}{\partial \mathbf{n}_1}, \quad \mathbf{x} \in \Gamma, \\
u_m(\mathbf{x}, 0) &= u_m^0(\mathbf{x}), \quad \mathbf{x} \in \Omega_m,
\end{aligned} \tag{4.16}
$$

where $\mathbf{n}_m$ is the outward normal to Ω_m for $m = 1, 2$.

The constants λ_1 and λ_2 describe the thermal conductivities of the materials on Ω_1 and Ω_2, respectively. D_1 and D_2 represent the thermal diffusivities of the materials and they are defined by

$$
D_m = \frac{\lambda_m}{\alpha_m}, \quad \text{with } \alpha_m = \rho_m c_{p_m} \tag{4.17}
$$

where ρ_m represents the density and c_{p_m} the specific heat capacity of the material placed in Ω_m, $m = 1, 2$.

We discretize this problem with a constant mesh width with respect to both spatial components ($\Delta y := \Delta x = 1/(N+1)$) resulting in N^2 interior space discretization points in both Ω_1 and Ω_2. We use the implicit Euler method for the time discretization.

4.3.2 Semidiscrete analysis

Before we present in the next section an analysis for the fully discrete equations, we want to describe previous results that analyze the behavior of the Dirichlet–Neumann iteration for the transmission problem in the semidiscrete case.

On one hand, a one-dimensional stability analysis was presented by Giles [20]. There, while using the implicit Euler method in the subsolvers, an explicit time integration method was chosen with respect to the interface unknowns.

On the other hand, Henshaw and Chand provided in [24] a method to analyze convergence speed of the Dirichlet–Neumann iteration. There, they applied the implicit Euler method for the time discretization on both equations in (4.16) while keeping the space continuous. Then, they applied the Fourier transform in space in order to transform the second-order derivatives into algebraic expressions. This converts the partial differential equations into a system of purely algebraic equations. Once we have a coupled system of algebraic equations, we can insert one into the other and obtain the convergence rate, called amplification factor β in [24]. They then derive the formula

$$\beta = \left| -\frac{\lambda_1}{\lambda_2}\sqrt{\frac{D_2}{D_1}}\frac{\tanh\left(-\frac{1}{\sqrt{D_2\Delta t}}\right)}{\tanh\left(\frac{1}{\sqrt{D_1\Delta t}}\right)}\right| . \tag{4.18}$$

For Δt big enough, we have $\tanh\left(-1/\sqrt{D_2\Delta t}\right) \approx -1/\sqrt{D_2\Delta t}$ and $\tanh\left(1/\sqrt{D_1\Delta t}\right) \approx 1/\sqrt{D_1\Delta t}$ and therefore:

$$\beta \approx \frac{\lambda_1}{\lambda_2}\sqrt{\frac{D_2}{D_1}}\frac{\sqrt{D_1\Delta t}}{\sqrt{D_2\Delta t}} = \frac{\lambda_1}{\lambda_2} . \tag{4.19}$$

One observes in (4.19) that the rates of the iteration behave as the quotient of thermal conductivities when $\Delta t \to 0$. This suggests that strong jumps in the thermal conductivities of the materials cause fast convergence.

4.3.3 Space discretization

We now describe a rather general space discretization of the model problem. The core property that we need is that the meshes of Ω_1 and Ω_2 are compatible on Γ (they share the same nodes on Γ) as shown in Figure 4.6. Furthermore, we need that there is a specific set of unknowns associated with the interface nodes.

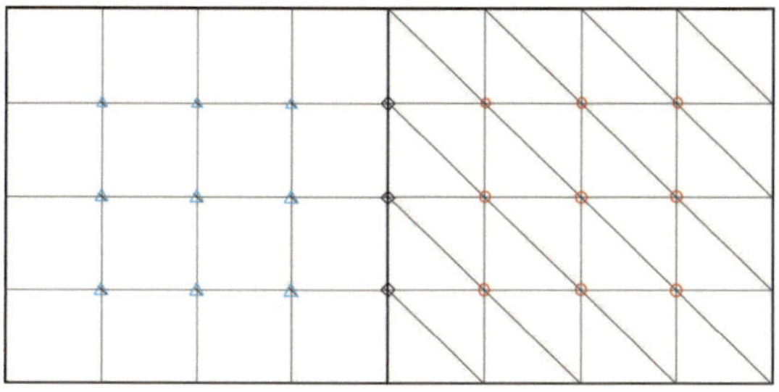

Fig. 4.6: Splitting of Ω between finite volumes and finite elements.

Then, letting $\mathbf{u}_I^{(1)}$ correspond to the unknowns on Ω_1 and $\mathbf{u}_\Gamma$ to the unknowns at the interface Γ, we can write a general discretization of the first equation in (4.16) in a compact form as:

$$\mathbf{M}_1\dot{\mathbf{u}}_I^{(1)} + \mathbf{M}_{I\Gamma}^{(1)}\dot{\mathbf{u}}_\Gamma + \mathbf{A}_1\mathbf{u}_I^{(1)} + \mathbf{A}_{I\Gamma}^{(1)}\mathbf{u}_\Gamma = \mathbf{0}\,. \tag{4.20}$$

Conversely, a general discretization of the first equation in (4.16) on Ω_2 can be written as:

$$\mathbf{M}_2\dot{\mathbf{u}}_I^{(2)} + \mathbf{M}_{I\Gamma}^{(2)}\dot{\mathbf{u}}_\Gamma + \mathbf{A}_2\mathbf{u}_I^{(2)} + \mathbf{A}_{I\Gamma}^{(2)}\mathbf{u}_\Gamma = \mathbf{0}\,, \tag{4.21}$$

where $\mathbf{u}_I^{(2)}$ correspond to the unknowns on Ω_2.

However, the system (4.20)–(4.21) is not enough to describe (4.16). Additionally, we need an approximation of the normal derivatives at Γ. If a FVM is used over Ω_1, we approximate the normal derivative with respect to u_1 using second-order one-sided finite differences:

$$-\lambda_1\frac{\partial u_1}{\partial \mathbf{n}_1} \approx \frac{\lambda_1}{2\Delta x}(4u_{1,N}(t) - u_{1,N-1}(t) - 3u_\Gamma)\,. \tag{4.22}$$

Conversely, if an FEM is used over Ω_1 and ϕ_j is a nodal basis function for a node on Γ we observe that the normal derivative with respect to u_2 can be written as linear functionals using Green's formula [41, pp. 3]. Thus, the approximation of the normal derivative is given by

$$\begin{aligned}\lambda_2\int_\Gamma \frac{\partial u_2}{\partial \mathbf{n}_2}\phi_j\mathrm{d}S &= \lambda_2\int_{\Omega_2}(\Delta u_2\phi_j + \nabla u_2\nabla\phi_j)\mathrm{d}\mathbf{x}\\ &= \alpha_2\int_{\Omega_2}\frac{\mathrm{d}}{\mathrm{d}t}u_2\phi_j + \lambda_2\int_{\Omega_2}\nabla u_2\nabla\phi_j\mathrm{d}\mathbf{x}\,.\end{aligned} \tag{4.23}$$

Consequently, the equation

$$\mathbf{M}_{\Gamma\Gamma}^{(2)}\dot{\mathbf{u}}_\Gamma + \mathbf{M}_{\Gamma I}^{(2)}\dot{\mathbf{u}}_I^{(2)} + \mathbf{A}_{\Gamma\Gamma}^{(2)}\mathbf{u}_\Gamma + \mathbf{A}_{\Gamma I}^{(2)}\mathbf{u}_I^{(2)} = -\mathbf{M}_{\Gamma\Gamma}^{(1)}\dot{\mathbf{u}}_\Gamma - \mathbf{M}_{\Gamma I}^{(1)}\dot{\mathbf{u}}_I^{(1)} - \mathbf{A}_{\Gamma\Gamma}^{(1)}\mathbf{u}_\Gamma - \mathbf{A}_{\Gamma I}^{(1)}\mathbf{u}_I^{(1)} \tag{4.24}$$

is a discrete version of the fourth equation in (4.16) and completes the system (4.20)–(4.21). We now reformulate the coupled equations (4.20), (4.21), and (4.24) into an ODE (ordinary differential equation) for the vector of unknowns $\mathbf{u} = \left(\mathbf{u}_I^{(1)}, \mathbf{u}_I^{(2)}, \mathbf{u}_\Gamma\right)^T$

$$\tilde{\mathbf{M}}\dot{\mathbf{u}} + \tilde{\mathbf{A}}\mathbf{u} = \mathbf{0}\,, \tag{4.25}$$

where

$$\tilde{\mathbf{M}} = \begin{pmatrix}\mathbf{M}_1 & \mathbf{0} & \mathbf{M}_{I\Gamma}^{(1)}\\ \mathbf{0} & \mathbf{M}_2 & \mathbf{M}_{I\Gamma}^{(2)}\\ \mathbf{M}_{\Gamma I}^{(1)} & \mathbf{M}_{\Gamma I}^{(2)} & \mathbf{M}_{\Gamma\Gamma}^{(1)} + \mathbf{M}_{\Gamma\Gamma}^{(2)}\end{pmatrix},\quad \tilde{\mathbf{A}} = \begin{pmatrix}\mathbf{A}_1 & \mathbf{0} & \mathbf{A}_{I\Gamma}^{(1)}\\ \mathbf{0} & \mathbf{A}_2 & \mathbf{A}_{I\Gamma}^{(2)}\\ \mathbf{A}_{\Gamma I}^{(1)} & \mathbf{A}_{\Gamma I}^{(2)} & \mathbf{A}_{\Gamma\Gamma}^{(1)} + \mathbf{A}_{\Gamma\Gamma}^{(2)}\end{pmatrix}.$$

4.3.4 Time discretization

Applying the implicit Euler method with time-step Δt to the system (4.24), we get for the vector of unknowns $\mathbf{u}^{n+1} = (\mathbf{u}_I^{(1),n+1}, \mathbf{u}_I^{(2),n+1}, \mathbf{u}_\Gamma^{n+1})^T$

$$\mathbf{A}\mathbf{u}^{n+1} = \tilde{\mathbf{M}}\mathbf{u}^n , \tag{4.26}$$

where

$$\mathbf{A} = \tilde{\mathbf{M}} + \Delta t\tilde{\mathbf{A}} = \begin{pmatrix} \mathbf{M}_1 + \Delta t\mathbf{A}_1 & \mathbf{0} & \mathbf{M}_{I\Gamma}^{(1)} + \Delta t\mathbf{A}_{I\Gamma}^{(1)} \\ \mathbf{0} & \mathbf{M}_2 + \Delta t\mathbf{A}_2 & \mathbf{M}_{I\Gamma}^{(2)} + \Delta t\mathbf{A}_{I\Gamma}^{(2)} \\ \mathbf{M}_{\Gamma I}^{(1)} + \Delta t\mathbf{A}_{\Gamma I}^{(1)} & \mathbf{M}_{\Gamma I}^{(2)} + \Delta t\mathbf{A}_{\Gamma I}^{(2)} & \mathbf{M}_{\Gamma\Gamma} + \Delta t\mathbf{A}_{\Gamma\Gamma} \end{pmatrix},$$

with $\mathbf{M}_{\Gamma\Gamma} = \mathbf{M}_{\Gamma\Gamma}^{(1)} + \mathbf{M}_{\Gamma\Gamma}^{(2)}$ and $\mathbf{A}_{\Gamma\Gamma} = \mathbf{A}_{\Gamma\Gamma}^{(1)} + \mathbf{A}_{\Gamma\Gamma}^{(2)}$.

4.3.5 Fixed-point iteration

We now employ a Dirichlet–Neumann iteration to solve the discrete system (4.26). This corresponds to alternately solving the discretized equations of the transmission problem (4.16) on Ω_1 with Dirichlet data on Γ and the discretization of (4.16) on Ω_2 with Neumann data on Γ.

Therefore, from (4.26) one gets for the k-th iteration the two equation systems

$$(\mathbf{M}_1 + \Delta t\mathbf{A}_1)\mathbf{u}_I^{(1),n+1,k+1} = -(\mathbf{M}_{I\Gamma}^{(1)} + \Delta t\mathbf{A}_{I\Gamma}^{(1)})\mathbf{u}_\Gamma^{n+1,k} + \mathbf{M}_1\mathbf{u}_I^{(1),n} + \mathbf{M}_{I\Gamma}^{(1)}\mathbf{u}_\Gamma^n , \tag{4.27}$$

$$\hat{\mathbf{A}}\hat{\mathbf{u}}^{k+1} = \hat{\mathbf{M}}\mathbf{u}^n - \mathbf{b}^k , \tag{4.28}$$

to be solved in succession. Here,

$$\hat{\mathbf{A}} = \begin{pmatrix} \mathbf{M}_2 + \Delta t\mathbf{A}_2 & \mathbf{M}_{I\Gamma}^{(2)} + \Delta t\mathbf{A}_{I\Gamma}^{(2)} \\ \mathbf{M}_{\Gamma I}^{(2)} + \Delta t\mathbf{A}_{\Gamma I}^{(2)} & \mathbf{M}_{\Gamma\Gamma}^{(2)} + \Delta t\mathbf{A}_{\Gamma\Gamma}^{(2)} \end{pmatrix}, \quad \hat{\mathbf{M}} = \begin{pmatrix} \mathbf{0} & \mathbf{M}_2 & \mathbf{M}_{I\Gamma}^{(2)} \\ \mathbf{M}_{\Gamma I}^{(1)} & \mathbf{M}_{\Gamma I}^{(2)} & \mathbf{M}_{\Gamma\Gamma} \end{pmatrix},$$

and

$$\mathbf{b}^k = \begin{pmatrix} \mathbf{0} \\ (\mathbf{M}_{\Gamma I}^{(1)} + \Delta t\mathbf{A}_{\Gamma I}^{(1)})\mathbf{u}_I^{(1),n+1,k+1} + (\mathbf{M}_{\Gamma\Gamma}^{(1)} + \Delta t\mathbf{A}_{\Gamma\Gamma}^{(1)})\mathbf{u}_\Gamma^{n+1,k} \end{pmatrix}, \tag{4.29}$$

$$\hat{\mathbf{u}}^{k+1} = \begin{pmatrix} \mathbf{u}_I^{(2),n+1,k+1} \\ \mathbf{u}_\Gamma^{n+1,k+1} \end{pmatrix},$$

with some initial condition, here $\mathbf{u}_\Gamma^{n+1,0} = \mathbf{u}_\Gamma^n$. The iteration is terminated according to the standard criterion $\|\mathbf{u}_\Gamma^{k+1} - \mathbf{u}_\Gamma^k\| \leq \tau$ where τ is a user-defined tolerance [7].

One way to analyze this method is to write it as a splitting method for (4.26) and try to estimate the spectral radius of that iteration. However, the results obtained in this way are much too inaccurate. For that reason, we now rewrite (4.27)–(4.28) as an iteration for $\mathbf{u}_\Gamma^{n+1}$ to restrict the size of the space to the dimension of $\mathbf{u}_\Gamma$. To this end, we isolate the term $\mathbf{u}_I^{(1),n+1,k+1}$ in (4.27) and $\mathbf{u}_I^{(2),n+1,k+1}$ in the first equation in (4.28) and we insert the resulting expressions into the second equation in (4.28). Consequently, the iteration $\mathbf{u}_\Gamma^{n+1,k+1} = \Sigma \mathbf{u}_\Gamma^{n+1,k} + \psi^n$ is obtained with iteration matrix

$$\Sigma = -\mathbf{S}^{(2)^{-1}}\mathbf{S}^{(1)}\,, \tag{4.30}$$

where

$$\mathbf{S}^{(m)} = (\mathbf{M}_{\Gamma\Gamma}^{(m)} + \Delta t \mathbf{A}_{\Gamma\Gamma}^{(m)}) - (\mathbf{M}_{\Gamma I}^{(m)} + \Delta t \mathbf{A}_{\Gamma I}^{(m)})(\mathbf{M}_m + \Delta t \mathbf{A}_m)^{-1}(\mathbf{M}_{I\Gamma}^{(m)} + \Delta t \mathbf{A}_{I\Gamma}^{(m)})\,, \tag{4.31}$$

for $m = 1, 2$ and ψ^n contains terms that depend only on the solutions at the previous time step. Notice that Σ is a discrete version of the Steklov–Poincaré operator.

Thus, the Dirichlet–Neumann iteration is a linear iteration and the rate of convergence is described by the spectral radius of the iteration matrix Σ.

4.4 Convergence analysis

The derivation so far was for a rather general discretization. It is now possible to look at specific discretizations. In this section we study the iteration matrix Σ for a specific FVM-FEM discretization.

The subdomains are here $\Omega_1 = [-1, 0] \times [0, 1]$, $\Omega_2 = [0, 1] \times [0, 1]$. An equidistant grid is chosen i.e., $\Delta x = \Delta y = 1/(N+1)$. At each discrete point of the finite volume discretization $x_{i,j}$, we integrate over the cell $I_{i,j} = [x_{i-1/2,j}, x_{i+1/2,j}] \times [x_{i,j-1/2}, x_{i,j+1/2}]$ and use the flux function

$$F(u_L, u_R) = -\frac{\lambda_1}{\Delta x}(u_R - u_L) \tag{4.32}$$

to approximate the flux, which results in a second-order scheme. For the FEM discretization, we use triangular elements distributed as sketched in Figure 4.7 and the following pyramidal test functions

$$\phi_k(x, y) = \begin{cases} \frac{x+y}{\Delta x} - 1, & \text{if} \quad \mathbf{x} = (x, y) \in \text{Region } 1\,, \\ \frac{y}{\Delta x}, & \text{if} \quad \mathbf{x} \in \text{Region } 2\,, \\ \frac{\Delta x - x}{\Delta x}, & \text{if} \quad \mathbf{x} \in \text{Region } 3\,, \\ 1 - \frac{x+y}{\Delta x}, & \text{if} \quad \mathbf{x} \in \text{Region } 4\,, \\ \frac{\Delta x - y}{\Delta x}, & \text{if} \quad \mathbf{x} \in \text{Region } 5\,, \\ \frac{x}{\Delta x}, & \text{if} \quad \mathbf{x} \in \text{Region } 6\,, \\ 0, & \text{otherwise}\,. \end{cases} \tag{4.33}$$

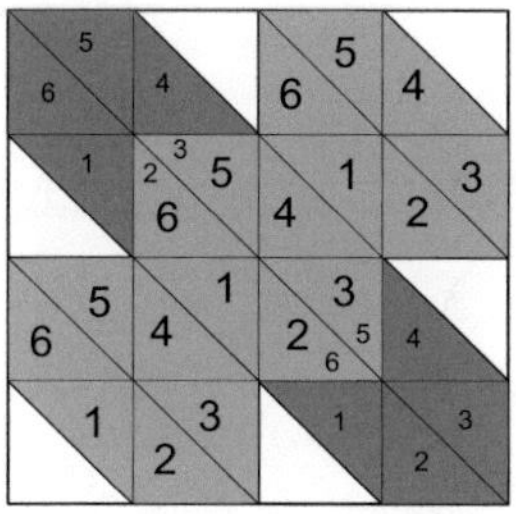

Fig. 4.7: Sketch of the regions for the pyramidal test functions defined in (4.33).

The discretization matrices are given in this case by

$$\mathbf{A}_1 = \frac{\lambda_1}{\Delta x^2}\begin{pmatrix} -\mathbf{B} & \mathbf{I} & & \mathbf{0} \\ \mathbf{I} & -\mathbf{B} & \ddots & \\ & \ddots & \ddots & \mathbf{I} \\ \mathbf{0} & & \mathbf{I} & -\mathbf{B} \end{pmatrix}, \quad \mathbf{A}_2 = \frac{\lambda_2}{\Delta x^2}\begin{pmatrix} \mathbf{B} & -\mathbf{I} & & \mathbf{0} \\ -\mathbf{I} & \mathbf{B} & \ddots & \\ & \ddots & \ddots & -\mathbf{I} \\ \mathbf{0} & & -\mathbf{I} & \mathbf{B} \end{pmatrix},$$

where

$$\mathbf{B} = \begin{pmatrix} 4 & -1 & & 0 \\ -1 & 4 & \ddots & \\ & \ddots & \ddots & -1 \\ 0 & & -1 & 4 \end{pmatrix},$$

and $\mathbf{I} \in \mathbb{R}^{N\times N}$ is an identity matrix. Note that each block of the matrices $\mathbf{A}_m \in \mathbb{R}^{N^2\times N^2}$ has size $N \times N$.

$$\mathbf{M}_2 = \alpha_2 \begin{pmatrix} \mathbf{N} & \tilde{\mathbf{N}} & & \mathbf{0} \\ \tilde{\mathbf{N}}^T & \mathbf{N} & \ddots & \\ & \ddots & \ddots & \tilde{\mathbf{N}} \\ \mathbf{0} & & \tilde{\mathbf{N}}^T & \mathbf{N} \end{pmatrix},$$

where

$$\mathbf{N} = \begin{pmatrix} 5/6 & -1/12 & & 0 \\ -1/12 & 5/6 & \ddots & \\ & \ddots & \ddots & -1/12 \\ 0 & & -1/12 & 5/6 \end{pmatrix}, \quad \tilde{\mathbf{N}} = \begin{pmatrix} -1/12 & 1/4 & & 0 \\ 0 & -1/12 & \ddots & \\ & \ddots & \ddots & 1/4 \\ 0 & & 0 & -1/12 \end{pmatrix}.$$

Each block of the matrix $\mathbf{M}_2 \in \mathbb{R}^{N^2\times N^2}$ has size $N \times N$ as well. We consider here $\mathbf{E}_j = \begin{pmatrix} \mathbf{0} & \cdots & \mathbf{0} & \mathbf{I} & \mathbf{0} & \cdots & \mathbf{0} \end{pmatrix}^T \in \mathbb{R}^{N^2\times N}$ where the only nonzero block is the j-th block of size $N \times N$. Thus,

$$\mathbf{A}_{I\Gamma}^{(1)} = \frac{\lambda_1}{\Delta x^2}\mathbf{E}_N,\ \mathbf{A}_{I\Gamma}^{(2)} = -\frac{\lambda_2}{\Delta x^2}\mathbf{E}_1,\ \mathbf{M}_{I\Gamma}^{(2)} = \alpha_2 \mathbf{E}_1 \tilde{\mathbf{N}},$$

$$\mathbf{M}_{\Gamma\Gamma}^{(2)} = \alpha_2 \begin{pmatrix} 5/12 & -1/24 & & 0 \\ -1/24 & 5/12 & \ddots & \\ & \ddots & \ddots & -1/24 \\ 0 & & -1/24 & 5/12 \end{pmatrix},$$

$$\mathbf{A}_{\Gamma\Gamma}^{(1)} = \frac{3\lambda_1}{2\Delta x^2}\mathbf{I},\ \mathbf{A}_{\Gamma\Gamma}^{(2)} = \frac{\lambda_2}{\Delta x^2} \begin{pmatrix} 2 & -1/2 & & 0 \\ -1/2 & 2 & \ddots & \\ & \ddots & \ddots & -1/2 \\ 0 & & -1/2 & 2 \end{pmatrix},$$

where $\mathbf{M}_{\Gamma\Gamma}^{(2)}$ and $\mathbf{A}_{\Gamma\Gamma}^{(m)} \in \mathbb{R}^{N\times N}$ for $m = 1, 2$.

$$\mathbf{A}_{\Gamma I}^{(1)} = \frac{\lambda_1}{2\Delta x^2}(4\mathbf{E}_N^T - \mathbf{E}_{N-1}^T),\ \mathbf{A}_{\Gamma I}^{(2)} = -\frac{\lambda_2}{\Delta x^2}\mathbf{E}_1^T,\ \mathbf{M}_{\Gamma I}^{(2)} = \alpha_2 \mathbf{E}_1^T \tilde{\mathbf{N}}.$$

In this specific case, $\mathbf{M}_1 = \alpha_1 \mathbf{I}$, $\mathbf{M}_{I\Gamma}^{(1)} = \mathbf{M}_{\Gamma\Gamma}^{(1)} = \mathbf{M}_{\Gamma I}^{(1)} = \mathbf{0}$. In particular,

$$\mathbf{S}^{(1)} = \Delta t \mathbf{A}_{\Gamma\Gamma}^{(1)} - \Delta t^2 \mathbf{A}_{\Gamma I}^{(1)} (\alpha_1 \mathbf{I} - \Delta t \mathbf{A}_1)^{-1} \mathbf{A}_{I\Gamma}^{(1)}\ , \tag{4.34}$$

$$\mathbf{S}^{(2)} = (\mathbf{M}_{\Gamma\Gamma}^{(2)} + \Delta t \mathbf{A}_{\Gamma\Gamma}^{(2)}) - (\mathbf{M}_{\Gamma I}^{(2)} + \Delta t \mathbf{A}_{\Gamma I}^{(2)})(\mathbf{M}_2 + \Delta t \mathbf{A}_2)^{-1}(\mathbf{M}_{I\Gamma}^{(2)} + \Delta t \mathbf{A}_{I\Gamma}^{(2)})\ . \tag{4.35}$$

One computes $\mathbf{S}^{(1)}$ and $\mathbf{S}^{(2)}$ by inserting the corresponding matrices specified above in (4.34) and (4.35) obtaining

$$\mathbf{S}^{(1)} = \frac{3\lambda_1 \Delta t}{2\Delta x^2}\mathbf{I} - \frac{\lambda_1^2 \Delta t^2}{2\Delta x^4}(4\mathbf{E}_N^T - \mathbf{E}_{N-1}^T)(\alpha_1 \mathbf{I} - \Delta t \mathbf{A}_1)^{-1}\mathbf{E}_N\ , \tag{4.36}$$

$$\begin{aligned} \mathbf{S}^{(2)} &= \left(\alpha_2 \text{tridiag}\left(-\frac{1}{24}, \frac{5}{12}, -\frac{1}{24}\right) + \frac{\lambda_2 \Delta t}{\Delta x^2}\text{tridiag}\left(-\frac{1}{2}, 2, -\frac{1}{2}\right)\right) \\ &- \left(\alpha_2 \mathbf{N}_2 - \frac{\lambda_2 \Delta t}{\Delta x^2}\mathbf{I}\right) \mathbf{E}_1^T (\mathbf{M}_2 + \Delta t \mathbf{A}_2)^{-1} \mathbf{E}_1 \left(\alpha_2 \mathbf{N}_2 - \frac{\lambda_2 \Delta t}{\Delta x^2}\mathbf{I}\right)\ . \end{aligned} \tag{4.37}$$

In the two-dimensional case, the iteration matrix Σ is a matrix of size $N \times N$. This makes the iteration matrix Σ difficult to compute for several reasons. First of all, the matrices $\alpha_1 \mathbf{I} - \Delta t \mathbf{A}_1$ and $\mathbf{M}_2 + \Delta t \mathbf{A}_2$ are sparse block tridiagonal matrices, and consequently, their inverses are not straight forward to compute. A block-by-block algorithm for inverting a block tridiagonal matrix is explained in [36]. However, the algorithm is based on the iterative application of the Schur complement [46], and it results in a sequence of block matrices and inverses of block matrices that we did not find possible to compute exactly. Moreover, the diagonal blocks of $\alpha_1 \mathbf{I} - \Delta t \mathbf{A}_1$ and $\mathbf{M}_2 + \Delta t \mathbf{A}_2$ are tridiagonal but their inverses are full matrices [18].

Because of these difficulties, we propose here to approximate Σ. One can observe that $\alpha_1\mathbf{I}-\Delta t\mathbf{A}_1$ and $\mathbf{M}_2+\Delta t\mathbf{A}_2$ are strictly diagonally dominant matrices, and therefore, we propose to approximate them by their block diagonal. Thus,

$$\mathbf{S}^{(1)} \approx \frac{3\lambda_1\Delta t}{2\Delta x^2}\mathbf{I} - \frac{2\lambda_1^2\Delta t^2}{\Delta x^4}\left(\text{tridiag}\left(-\frac{\lambda_1\Delta t}{\Delta x^2}, \frac{\alpha_1\Delta x^2 + 4\lambda_1\Delta t}{\Delta x^2}, -\frac{\lambda_1\Delta t}{\Delta x^2}\right)^{-1}\right), \tag{4.38}$$

$$\begin{aligned}\mathbf{S}^{(2)} \approx &\left(\alpha_2\text{tridiag}\left(-\frac{1}{24}, \frac{5}{12}, -\frac{1}{24}\right) + \frac{\lambda_2\Delta t}{\Delta x^2}\text{tridiag}\left(-\frac{1}{2}, 2, -\frac{1}{2}\right)\right)\\ &-\left(\frac{\alpha_2\Delta x^2 + 12\lambda_2\Delta t}{12\Delta x^2}\right)^2\left(\text{tridiag}\,(b, a, b)^{-1}\right),\end{aligned} \tag{4.39}$$

with

$$a = \frac{5\alpha_2\Delta x^2 + 24\lambda_2\Delta t}{6\Delta x^2}, \quad b = -\left(\frac{\alpha_2\Delta x^2 + 12\lambda_2\Delta t}{12\Delta x^2}\right).$$

Now, we compute the eigenvalues of the proposed approximations of $\mathbf{S}^{(1)}$ and $\mathbf{S}^{(2)}$. The eigenvalues of a tridiagonal Toeplitz matrix are known and given e.g., in [32, pp. 514–516]:

$$\mu_j^1 = \frac{3\lambda_1\Delta t(\alpha_1\Delta x^2 + 2\lambda_1\Delta t(2-\cos(j\pi\Delta x))) - 4\lambda_1^2\Delta t^2}{2\Delta x^2(\alpha_1\Delta x^2 + 2\lambda_1\Delta t(2-\cos(j\pi\Delta x)))}, \tag{4.40}$$

$$\mu_j^2 = \frac{2(\alpha_2\Delta x^2(5-\cos(j\pi\Delta x)) + 12\lambda_2\Delta t(2-\cos(j\pi\Delta x)))^2 - (\alpha_2\Delta x^2 + 12\lambda_2\Delta t)^2}{24\Delta x^2(\alpha_2\Delta x^2(5-\cos(j\pi\Delta x)) + 12\lambda_2\Delta t(2-\cos(j\pi\Delta x)))}, \tag{4.41}$$

for $j = 1, \ldots, N$. Here μ_j^1 are the eigenvalues of the approximation of $\mathbf{S}^{(1)}$ and μ_j^2 the eigenvalues of the approximation of $\mathbf{S}^{(2)}$. Note that the eigenvectors are common for symmetric tridiagonal Toeplitz matrices, i.e., the eigenvectors do not depend on the specific entries.

Thus, we obtain an estimate of the spectral radius of the iteration matrix Σ:

$$\begin{aligned}\rho(\Sigma) &= \rho\left(\mathbf{S}^{(2)^{-1}}\mathbf{S}^{(1)}\right) = \mu_{\max}\left(\mathbf{S}^{(2)^{-1}}\mathbf{S}^{(1)}\right) \approx \frac{\mu_1^1}{\mu_1^2}\\ &= \left(\frac{12(\alpha_2\Delta x^2(5-\cos(\pi\Delta x)) + 12\lambda_2\Delta t(2-\cos(\pi\Delta x)))}{\alpha_1\Delta x^2 + 2\lambda_1\Delta t(2-\cos(\pi\Delta x))}\right)\\ &\cdot\left(\frac{3\lambda_1\Delta t(\alpha_1\Delta x^2 + 2\lambda_1\Delta t(2-\cos(\pi\Delta x))) - 4\lambda_1^2\Delta t^2}{2(\alpha_2\Delta x^2(5-\cos(\pi\Delta x)) + 12\lambda_2\Delta t(2-\cos(\pi\Delta x)))^2 - (\alpha_2\Delta x^2 + 12\lambda_2\Delta t)^2}\right) := \sigma.\end{aligned} \tag{4.42}$$

Furthermore, computing the limits of (4.42) when $\Delta t \to 0$ and $\Delta x \to 0$ we get

$$\lim_{\Delta t\to 0}\rho(\Sigma) \approx \lim_{\Delta t\to 0}\sigma = \frac{(12\alpha_2\Delta x^2(5-\cos(\pi\Delta x)))\cdot 0}{\alpha_1\Delta x^2(2\alpha_2^2\Delta x^4(5-\cos(\pi\Delta x))^2 - \alpha_2^2\Delta x^4)} = 0, \tag{4.43}$$

$$\lim_{\Delta x\to 0}\rho(\Sigma) \approx \lim_{\Delta x\to 0}\sigma = \frac{12^2\lambda_2\Delta t(6\lambda_1^2\Delta t^2 - 4\lambda_1^2\Delta t^2)}{2\lambda_1\Delta t(2\cdot 12^2\lambda_2^2\Delta t^2 - 12^2\lambda_2^2\Delta t^2)} = \frac{\lambda_1}{\lambda_2} =: \delta. \tag{4.44}$$

Therefore, from the result obtained in (4.44) we expect that strong jumps in the physical properties of the materials placed in Ω_1 and Ω_2 will imply fast convergence. This is the case when modeling thermal fluid–structure interaction, where often a fluid with low thermal conductivity and density is coupled with a structure having higher thermal conductivity and density.

When $c \to \infty$, $\Delta t \gg \Delta x^2$ and (4.19) matches the asymptotic computed in (4.44). However, when $c < 1$, the semicontinuous analysis fails and the discrete analysis just presented fills the gap.

4.4.1 Numerical results

In this section we present a set of numerical experiments designed to show the validity of the approximation of $\rho(\Sigma)$ as an estimator for the rates of the coupled problem formulated above. We also show that the theoretical asymptotics deduced in (4.43) and (4.44) match with the numerical experiments.

We first compare the semidiscrete estimator β with the discrete estimator σ and the experimental convergence rates. The experimental convergence rates C_R are computed with respect to a reference solution u_{ref} over the whole domain Ω using the formula

$$C_R = \frac{\|u^3 - u_{\text{ref}}\|_2}{\|u^2 - u_{\text{ref}}\|_2},$$

where u^2 and u^3 are the second and third iterates of the Dirichlet–Neumann iteration.

Figure 4.8 shows a comparison between β and σ. On the left we plot β, σ, and the experimental convergence rates with $c \ll 1$ and on the right we plot the same but

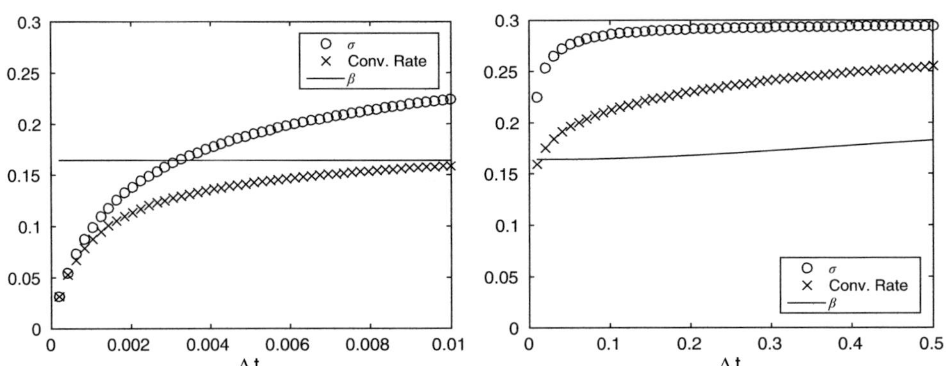

(a) $c \ll 1$ (b) $c \gg 1$

Fig. 4.8: Here, $D_1 = 1$, $D_2 = 0.5$, $\lambda_1 = 0.3$, and $\lambda_2 = 1$. The circles correspond to σ, the crosses to the experimental convergence rates and the continuous line to β. $\Delta x = 1/20$ and on the left the curves are restricted to the discrete values $\Delta t = 1e-2/50, 2 \cdot 1e-2/50, \ldots, 50 \cdot 1e-2/50$ and on the right to the values $\Delta t = 1e-2, 2 \cdot 1e-2, \ldots, 50 \cdot 1e-2$.

Tab. 4.1: Physical properties of the materials. λ is the thermal conductivity, ρ the density, C the specific heat capacity, and $\alpha = \rho C$.

Material	λ (W/mK)	ρ (kg/m^3)	C (J/kg K)	α (J/K m^3)
Air	0.0243	1.293	1005	1299.5
Water	0.58	999.7	4192.1	41908·10^6
Steel	48.9	7836	443	3471348

Tab. 4.2: The limits of the convergence rates when $\Delta t \to 0$ and $\Delta x \to 0$.

Case	$\Delta t \to 0$	$\Delta x \to 0$
Air–Steel	0	$4.9693 \cdot 10^{-4}$
Water–Steel	0	$1.19 \cdot 10^{-2}$

with $c \gg 1$. We can conclude that the estimator for the convergence rates presented in the previous section is minimally better than the one proposed by the semidiscrete analysis in [24] when $c \gg 1$. Moreover, when $c \ll 1$ our estimator also predicts the rates accurately and the semidiscrete estimator deviates.

We now want to illustrate how the formula (4.42) predicts the convergence rates and tends to the limits computed previously. To this end, we present two real data examples. We consider here the thermal interaction of air at 273 K with steel at 900 K and water at 283 K with steel at 900 K. Physical properties of the materials and resulting asymptotics for these two cases are shown in Table 4.1 and 4.2, respectively.

Figures 4.9 and 4.10 show the convergence rates for the interactions between air–steel and water–steel, respectively. On the left we always plot the rates with respect to

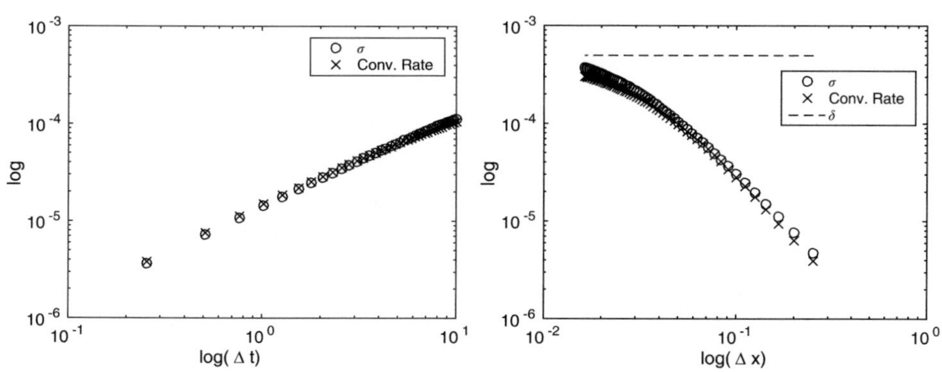

(a) The curves are restricted to the discrete values $\Delta t = 10/40, 2 \cdot 10/40, ..., 40 \cdot 10/40$ and $\Delta x = 1/20$.

(b) The curves are restricted to the discrete values $\Delta x = 1/3, 1/4, ..., 1/60$ and $\Delta t = 10$.

Fig. 4.9: Air–steel thermal interaction with respect to Δt on the left and to Δx on the right.

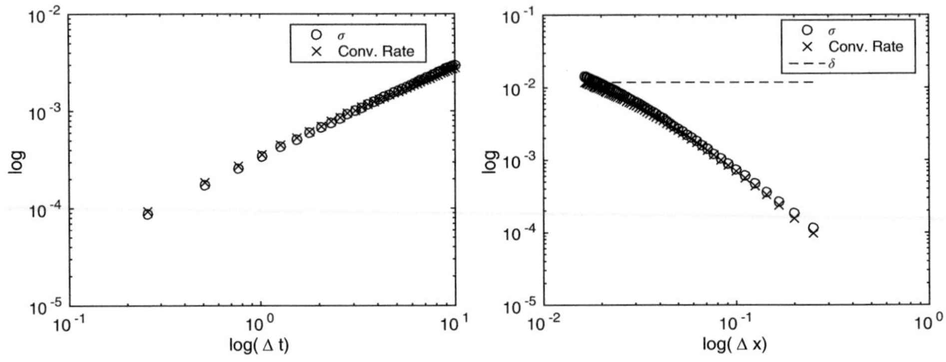

(a) The curves are restricted to the discrete values $\Delta t = 10/40, 2 \cdot 10/40, \ldots, 40 \cdot 10/40$ and $\Delta x = 1/20$.

(b) The curves are restricted to the discrete values $\Delta x = 1/3, 1/4, \ldots, 1/60$ and $\Delta t = 10$.

Fig. 4.10: Water–steel thermal interaction with respect to Δt on the left and to Δx on the right.

the variation of Δt and for a fixed Δx. On the right we plot the behavior of the rates for a fixed Δt and varying Δx.

We observe from figures 4.9 and 4.10 that the approximation σ predicts the convergence rates quite well because the difference with respect to the experimental rates is really small. Moreover, the rates in 4.9a and 4.10a tend to 0 as predicted in (4.43) and the rates in 4.9b and 4.10b tend to δ as predicted in (4.44).

From Figure 4.9 we can observe that the convergence rates are really fast (factor of $\sim 1e-4$) when strong jumps in the coefficient of the materials exist. For instance, when performing the thermal coupling between air and steel the Dirichlet–Neumann iteration only needs two iterations to achieve a tolerance of $1e-10$.

4.5 Time-adaptive methods

The standard in FSI is to use fixed time-step sizes. We find that it is much more efficient to use time-adaptive methods. Thus, the question arises how such a method can be implemented in a partitioned way. The method described here has been suggested in [9, 10]. Basis is a master program that does the global time stepping and calls functions available in the subsolvers.

Here, an A-stable singly diagonally implicit Runge–Kutta method (SDIRK) is used. Consider an autonomous initial value problem

$$\dot{\mathbf{u}}(\mathbf{t}) = \mathbf{f}(\mathbf{u}(t)), \quad \mathbf{u}(0) = \mathbf{u}_0 .$$

An SDIRK method is then defined as

$$\mathbf{U}^i = \mathbf{u}^n + \Delta t_n \sum_{j=1}^{i} a_{ij}\mathbf{f}(\mathbf{U}^j)\,, \quad i = 1, \ldots, s \tag{4.45}$$

$$\mathbf{u}^{n+1} = \mathbf{u}^n + \Delta t_n \sum_{i=1}^{s} b_i \mathbf{f}(\mathbf{U}^k)$$

with given coefficients a_{ij}, b_j, c_j. Here, we use the two-stage method SDIRK2, which is defined by the coefficients in the Butcher array in Table 4.3. For this method, $a_{si} = b_i$, which implies that the last line is superfluous (first-is-last-property). Besides saving computational effort, this guarantees L-stability. The vectors

$$\mathbf{k}_i = \mathbf{f}(\mathbf{U}^i)$$

are called stage derivates and s is the number of stages. Since the starting vector

$$\mathbf{s}_i = \mathbf{u}^n + \Delta t_n \sum_{j=1}^{i-1} a_{ij}\mathbf{k}_j\,, \quad i = 1, \ldots, s-1$$

is known, (4.45) is just a sequence of implicit Euler steps.

Tab. 4.3: Butcher array for SDIRK2.

α	α	0
1	$1-\alpha$	α
b_i	$1-\alpha$	α
$\hat{b}_i$	$1-\hat{\alpha}$	$\hat{\alpha}$
$b_i - \hat{b}_i$	$\hat{\alpha}-\alpha$	$\alpha-\hat{\alpha}$

$\alpha = 1 - \sqrt{2}/2$
$\hat{\alpha} = 2 - \frac{5}{4}\sqrt{2}$
$\alpha - \hat{\alpha} = -1 + \frac{3}{4}\sqrt{2}$

In the following it is assumed that at time t_n, the step size Δt_n is prescribed globally. Applying a DIRK method to equations (4.9)–(4.10) results in the coupled system of equations to be solved at Runge–Kutta stage i:

$$\mathbf{u}_i - \mathbf{s}_i^{\mathbf{u}} - \Delta t_n a_{ii}\mathbf{h}(\mathbf{u}_i, \boldsymbol{\Theta}_i^{\Gamma}) = \mathbf{0}\,, \tag{4.46}$$

$$[\mathbf{M} + \Delta t_n a_{ii}\mathbf{A}(\boldsymbol{\Theta})]\boldsymbol{\Theta}_i - \mathbf{M}\mathbf{s}_i^{\Theta} - \mathbf{q}_b^f - \mathbf{q}_b^{\Gamma}(\mathbf{u}_i) = \mathbf{0}\,. \tag{4.47}$$

Here, $\mathbf{s}_i^{\mathbf{u}}$ and $\mathbf{s}_i^{\Theta}$ are the given starting vectors in the subproblems.

The coupled equations (4.46)–(4.47) are then solved using a Dirichlet–Neumann coupling, as would be done for the implicit Euler method. Following the analysis just presented, temperature is prescribed for the equation with smaller heat conductivity, here the fluid, and heat flux is given on Γ for the structure. Choosing these conditions the other way round leads to an unstable scheme. The canonical starting guess for the Dirichlet–Neumann iteration at stage i is the starting vector $\mathbf{s}_i$.

Regarding implementation, it is a safe assumption that both the fluid and the solid solver are able to carry out time steps of implicit Euler type. Then, the master program of the FSI procedure can be extended to SDIRK methods very easily. The master program just has to call at stage i in iteration k the backward Euler routines with time-step size $a_{ii}\Delta t_n$ and starting vectors $\mathbf{s}_i$ and with the boundary data appropriate for iteration k.

To obtain time adaptivity, the technique of embedded Runge–Kutta methods is used. Thereby, the local error $\hat{\mathbf{l}}$ is estimated by the solvers separately:

$$\hat{\mathbf{l}}^{(i)} \approx \mathbf{l}^{(i)} = \sum_{j=1}^{2} (b_j - \hat{b}_j)\mathbf{k}_j^{(i)} , \quad i = 1, 2 , \tag{4.48}$$

where $\mathbf{k}_j^{(i)}$ are the stage derivatives on the subdomains for each Runge–Kutta stage. The computation of (4.48) will typically not be implemented in the subsolvers and has to be added to both codes, as well as that all stage derivatives have to be stored by the subsolvers. These error estimates are then reported back to the master program and aggregated in the form of a weighted scaled norm [9]:

$$\|\mathbf{l}\|_{\mathrm{WSN}} = \sqrt{\frac{1}{n}\sum_{j=1}^{n} \frac{l_i}{TOLu_j + TOL}} = \sqrt{\frac{1}{n}(n_1\|\mathbf{l}^{(1)}\|_{\mathrm{WSN}}^2 + n_2\|\mathbf{l}^{(2)}\|_{\mathrm{WSN}}^2)} . \tag{4.49}$$

Thus, the master solver either needs to know the number of unknowns n_1 and n_2 in the subsolvers, or instead the quantities $n_i\|\mathbf{l}^{(i)}\|_{\mathrm{WSN}}^2$ have to be provided to it. On this basis, the new time step is chosen to comply with a user-defined error tolerance TOL for the time integration:

$$\Delta t_{n+1} = \Delta t_n \cdot \|\mathbf{l}\|_{\mathrm{WSN}}^{-1/k} . \tag{4.50}$$

Finally, if the possibility of rejected time steps is taken into account, the current solution pair $(\mathbf{u}, \Theta)$ has to be stored as well.

4.5.1 Extrapolation from time integration

To find better starting values for iterative processes in implicit time integration schemes, it is common to use extrapolation based on knowledge about the trajectory of the solution of the initial value problem [15, 33]. In the spirit of partitioned solvers, the use of extrapolation of interface temperatures alone was suggested in [8].

Moreover, this strategy could be used as well within the subsolvers; we will not consider that here and leave it to the discretion of the developers of those codes. We now present extrapolation methods for SDIRK2 from [8].

At the first stage, we have the old time-step size Δt_{n-1} with value Θ_{n-1} and the current time-step size Δt_n with value Θ_n. We are looking for the value Θ_n^1 at the next stage time $t_n + c_1\Delta t_n$. Linear extrapolation results in

$$\Theta_n^1 \approx \Theta_n + c_1\Delta t_n(\Theta_n - \Theta_{n-1})/\Delta t_{n-1} = \left(1 + \frac{c_1\Delta t_n}{\Delta t_{n-1}}\right)\Theta_n - \frac{c_1\Delta t_n}{\Delta t_{n-1}}\Theta_{n-1} \, . \tag{4.51}$$

At the second stage, we linearly extrapolate Θ_n at t_n and Θ_n^1 at $t_n + c_1\Delta t$ to obtain

$$\Theta_{n+1} \approx \Theta_n + \Delta t_n(\Theta_n^1 - \Theta_n)/(c_1\Delta t_n) = \left(1 - \frac{1}{c_1}\right)\Theta_n + \frac{1}{c_1}\Theta_n^1 \, . \tag{4.52}$$

4.6 Choosing tolerances

In the time-adaptive setting, a core input parameter is a tolerance *TOL*. This, based on the error estimate, steers the time-step size according to (4.49)–(4.50). For this to work, it is imperative that the iteration error is smaller than the time integration error, so as not to destroy the error estimate of the latter. The Dirichlet–Neumann iteration is formulated as a fixed-point iteration in the interface unknowns with termination criterion (4.15). Thus, the tolerance there has to be chosen so that the iteration error not only on the interface, but on the whole domain is smaller than the time integration error. At the same time, oversolving should be avoided to prevent unnecessary computational cost. A good rule of thumb is to divide the tolerance for the time integration procedure by five [38], ending up with

$$\tau = TOL/5 \, . \tag{4.53}$$

On the assumption of exact solves in the subsolvers, errors in the interface will nevertheless be translated into errors in the subproblems. The boundary condition from the interface enters the right-hand side (rhs) of the nonlinear equations on the subproblem. Thus, by way of the inverse operator, errors in the interface values are translated into errors in the subdomains. For parabolic problems as in heat transfer, these are bounded operators and thus the error transfer can be controlled. Right now, we assume that it is of order one and do not adjust (4.53) further. However, a more careful analysis of this point to get better estimates is required, for example using discrete trace inequalities.

Actually, it is common to use iterative solvers for the subproblems as well, typically Newton or Multigrid methods (compare [6] for an overview on the state of the art). So the question needs to be raised again, what type of termination criteria should be chosen and what kind of tolerance. For a nonlinear subproblem of the form

$$\mathbf{F}(\mathbf{u}) = \mathbf{0} \, ,$$

the standard would be to use a relative termination criterion

$$\|\mathbf{F}(\mathbf{u}^k)\| \le \tau_r \|\mathbf{F}(\mathbf{u}^0)\|$$

with relative tolerance τ_r. The errors introduced from these systems into the fixed-point iteration will be on the order of the residual. Thus, it makes sense to use (4.53) for the tolerances in the subsolvers as well.

In the linear case, the same reasoning applies when we use the relative termination criterion

$$\|\mathbf{A}(\mathbf{x}^k)\mathbf{x}^{k+1} - \mathbf{b}\| \le \tau_r \|\mathbf{b}\| \,. \tag{4.54}$$

However, it turns out that this has some pitfalls and that in fact the use of the non-standard relative criterion

$$\|\mathbf{A}(\mathbf{x}^k)\mathbf{x}^{k+1} - \mathbf{b}\| \le \tau_r \|\mathbf{A}(\mathbf{x}^k)\mathbf{x}^k - \mathbf{b}\| \tag{4.55}$$

can speed up the iteration significantly. In the following analysis, taken from [7], an absolute criterion

$$\|\mathbf{A}(\mathbf{x}^k)\mathbf{x}^{k+1} - \mathbf{b}\| \le \tau_a \,, \tag{4.56}$$

with an absolute tolerance τ_a is possible as well. The first criterion, in case of convergence of the iteration, leads to an automatic tightening of the tolerance over the course of the iteration.

To determine good tolerances, we will now analyze the errors arising from these more carefully.

4.6.1 Perturbed nested fixed-point iteration

Consider two functions $\mathbf{F}: \Omega_1 \to \Omega_2$ and $\mathbf{S}: \Omega_2 \to \Omega_1$ with $\Omega_1, \Omega_2 \subset \mathbb{R}^n$ closed and the fixed-point equation

$$\mathbf{x} = \mathbf{S}(\mathbf{F}(\mathbf{x})) \tag{4.57}$$

which we assume to have a unique solution $\mathbf{x}^*$. Let then both the evaluation of $\mathbf{F}$ and of $\mathbf{S}$ be perturbed, namely $\mathbf{S}$ by δ_k and $\mathbf{F}$ by ϵ_k, and consider the iteration:

$$\mathbf{x}^{k+1} = \mathbf{S}(\mathbf{F}(\mathbf{x}^k) + \epsilon_k) + \delta_k \,. \tag{4.58}$$

We assume that this iteration is well defined and that this sequence has the limit $\mathbf{x}_\epsilon$. Then, we have the following theorem [7]:

Theorem 4.1. *Let* $\mathbf{F}$ *and* $\mathbf{S}$ *be Lipschitz continuous with Lipschitz constants* L_F *and* L_S, *respectively. Assume that* $L_F L_S < 1$. *Then we have, if* $\epsilon_k = \delta_k = \epsilon$ *for all k, that*

$$\|\mathbf{x}_\epsilon - \mathbf{x}^*\| \le \epsilon \frac{1 + L_S}{1 - L_S L_F} \,. \tag{4.59}$$

In the case $\epsilon_k = \epsilon$ and $\delta_k = \delta$, we obtain

$$\|\mathbf{x}_\epsilon - \mathbf{x}^*\| \leq \frac{\epsilon L_S + \delta}{1 - L_S L_F} . \tag{4.60}$$

Finally, $\mathbf{x}_\epsilon = \mathbf{x}^*$ if and only if both δ_k and ϵ_k converge to zero.

Proof. We have due to the Lipschitz continuity

$$\begin{aligned}
\|\mathbf{x}^{k+1} - \mathbf{x}^*\| &= \|\mathbf{S}(\mathbf{F}(\mathbf{x}^k) + \epsilon_k) + \delta_k - \mathbf{x}^*\| = \|\mathbf{S}(\mathbf{F}(\mathbf{x}^k) + \epsilon_k) + \delta_k - \mathbf{S}(\mathbf{F}(\mathbf{x}^*))\| \\
&\leq L_S\|\mathbf{F}(\mathbf{x}^k) - \mathbf{F}(\mathbf{x}^*) + \epsilon_k\| + \delta_k \leq L_S L_F\|\mathbf{x}^k - \mathbf{x}^*\| + L_S\epsilon_k + \delta_k \\
&\leq (L_S L_F)^2\|\mathbf{x}^{k-1} - \mathbf{x}^*\| + L_S^2 \mathrm{L_F}\epsilon_{k-1} + L_S L_F \delta_{k-1} + L_S\epsilon_k + \delta_k \\
&\leq (L_S L_F)^{k+1}\|\mathbf{x}^0 - \mathbf{x}^*\| + \left(\sum_{j=0}^{k} L_S^{j+1} L_F^j \epsilon_{k-j}\right) + \left(\sum_{j=0}^{k} L_S^j L_F^j \delta_{k-j}\right)
\end{aligned}$$

and thus in the limit $\mathbf{x}^{k+1} \to \mathbf{x}_\epsilon$,

$$\|\mathbf{x}_\epsilon - \mathbf{x}^*\| \leq L_S \lim_{k\to\infty} \sum_{j=0}^{k} (L_S L_F)^j \epsilon_{k-j} + \lim_{k\to\infty} \sum_{j=0}^{k} (L_S L_F)^j \delta_{k-j} . \tag{4.61}$$

For a constant perturbation overall, e.g., $\epsilon_k = \delta_k = \epsilon$ for all k, we obtain in the limit

$$\|\mathbf{x}_\epsilon - \mathbf{x}^*\| \leq \epsilon(1 + L_S) \lim_{k\to\infty} \sum_{j=0}^{k} (L_S L_F)^j = \epsilon \frac{1 + L_S}{1 - L_S L_F} ,$$

which proves the inequality (4.59). If we have constant but separate perturbations ϵ and δ of $\mathbf{S}$ and $\mathbf{F}$, we obtain (4.60) from

$$\|\mathbf{x}_\epsilon - \mathbf{x}^*\| \leq \epsilon L_S \lim_{k\to\infty} \sum_{j=0}^{k} (L_S L_F)^j + \delta \lim_{k\to\infty} \sum_{j=0}^{k} (L_S L_F)^j = \frac{\epsilon L_S + \delta}{1 - L_S L_F} .$$

In the general case, due to positivity, the right-hand side of (4.61) is zero if and only if both ϵ_k and δ_k are such that for $\phi_k = \epsilon_k$ or $\phi_k = \delta_k$,

$$\lim_{k\to\infty} \sum_{j=0}^{k} (L_S L_F)^j \phi_{k-j} = 0 .$$

This is the case if and only if both ϵ_k and δ_k converge to zero. □

4.6.2 Application: Dirichlet–Neumann coupling for a transmission problem

As an application of the theory just presented, we consider the steady transmission problem, where the Laplace equation with right-hand side $f(x, y)$ on a domain Ω is cut into two domains $\Omega = \Omega_1 \cup \Omega_2$ using transmission conditions at the interface $\Gamma = \Omega_1 \cap \Omega_2$:

$$\begin{aligned} \Delta u_m(x,y) &= f(x,y)\,, \quad (x,y) \in \Omega_m \subset \mathbb{R}^2,\ m = 1,2, \\ u_m(x,y) &= 0\,, \quad (x,y) \in \partial\Omega_m \Gamma\,, \\ u_1(x,y) &= u_2(x,y)\,, \quad (x,y) \in \Gamma, \\ \frac{\partial u_2(x,y)}{\mathbf{n}_2} &= -\frac{\partial u_1(x,y)}{\mathbf{n}_1}\,, \quad (x,y) \in \Gamma. \end{aligned} \tag{4.62}$$

We now employ a standard Dirichlet–Neumann iteration to solve it. This corresponds to alternately solving the problems

$$\mathbf{A}_1 \mathbf{u}_1^{k+1} = \mathbf{b}_1(\mathbf{u}_2^k) \tag{4.63}$$

and

$$\mathbf{A}_2 \mathbf{u}_2^{k+1} = \mathbf{b}_2(\mathbf{u}_1^{k+1})\,, \tag{4.64}$$

where problem (4.63) originates from a linear discretization of the transmission problem (4.62) on Ω_1 only with Dirichlet data on Γ given by $\mathbf{u}_2^k$ on the coupling interface and problem (4.64) corresponds to a linear discretization of (4.62) on Ω_2 only with Neumann data on Γ given by the discrete normal derivative of $\mathbf{u}_1$ on Γ.

By considering (4.63)–(4.64) as a nested iteration, we obtain a fixed-point formulation

$$\mathbf{u}_\Gamma = \mathbf{S}(\mathbf{F}(\mathbf{u}_\Gamma))$$

where $\mathbf{u}_\Gamma = \mathbf{u}_2|_\Gamma$, $\mathbf{F} = \mathbf{D}_{\mathbf{n}_\Gamma}\mathbf{A}_1^{-1}\mathbf{b}_1(\mathbf{u}_\Gamma)$ and $\mathbf{S} = \mathbf{P}_\Gamma \mathbf{A}_2^{-1}\mathbf{b}_2(\mathbf{u}_1)$. Hereby $\mathbf{D}_{\mathbf{n}_\Gamma}$ is the matrix that computes the discrete normal derivatives in Ω_1 on Γ and $\mathbf{P}_\Gamma$ is the discrete trace operator with respect to Γ.

In practice, the linear equation systems are solved iteratively, typically using the conjugate gradient method (CG) up to a relative tolerance of τ. Thus, we obtain a perturbed nested fixed-point iteration of the form (4.58) and the question is now again if we can quantify this perturbation. We have

$$\mathbf{u}_{1_\epsilon}^{k+1} = \mathbf{A}_1^{-1}\mathbf{b}_1(\mathbf{u}_\Gamma^k) + \epsilon_k \tag{4.65}$$

and

$$\mathbf{u}_{2_\epsilon}^{k+1} = \mathbf{A}_2^{-1}\mathbf{b}_2(\mathbf{u}_{1_\epsilon}^{k+1}) + \delta_k\,. \tag{4.66}$$

For the iteration (4.65) we obtain

$$\|\epsilon_k\| = \|\mathbf{u}_{1_\epsilon}^{k+1} - \mathbf{A}_1^{-1}\mathbf{b}_1(\mathbf{u}_\Gamma^k)\| \le \|\mathbf{A}_1^{-1}\|\|\mathbf{A}_1\mathbf{u}_{1_\epsilon}^{k+1} - \mathbf{b}_1(\mathbf{u}_\Gamma^k)\|\,.$$

Again, the second factor is what is tested in the termination criterion. In the case of the relative criterion (4.55), this results in

$$\|\mathbf{A}_1\mathbf{u}_{1_\epsilon}^{k+1} - \mathbf{b}_1(\mathbf{u}_\Gamma^k)\| \le \tau_r\|\mathbf{A}_1\mathbf{u}_{1_\epsilon}^{k} - \mathbf{b}_1(\mathbf{u}_\Gamma^k)\|\,.$$

We thus have managed to shift the index in $\mathbf{A}_1\mathbf{u}_{1_\epsilon}^k$, but not in $\mathbf{b}_1(\mathbf{u}_\Gamma^k)$. For this we add zero and use the triangle inequality:

$$\tau_\text{r}\|\mathbf{A}_1\mathbf{u}_{1_\epsilon}^k - \mathbf{b}_1(\mathbf{u}_\Gamma^k)\| \leq \tau_\text{r}(\|\mathbf{A}_1\mathbf{u}_{1_\epsilon}^k - \mathbf{b}_1(\mathbf{u}_\Gamma^{k-1})\| + \|\mathbf{b}_1(\mathbf{u}_\Gamma^{k-1}) - \mathbf{b}_1(\mathbf{u}_\Gamma^k)\|) .$$

We can now repeat this argument on $\|\mathbf{A}_1\mathbf{u}_{1_\epsilon}^k - \mathbf{b}_1(\mathbf{u}_\Gamma^k)\|$, thereby multiplying again with τ_r and adding a difference of right-hand sides:

$$\|\epsilon_k\| \leq \|\mathbf{A}_1^{-1}\| \left(\tau_\text{r}^k\|\mathbf{A}_1\mathbf{u}_{1_\epsilon}^1 - \mathbf{b}_1(\mathbf{u}_\Gamma^0)\| + \sum_{j=1}^{k} \tau_\text{r}^j\|\mathbf{b}_1(\mathbf{u}_\Gamma^{k-j}) - \mathbf{b}_1(\mathbf{u}_\Gamma^{k-j+1})\| \right) .$$

The right-hand sides are equal except at the boundary and we have

$$\|\mathbf{b}_1(\mathbf{u}_\Gamma^{k-j}) - \mathbf{b}_1(\mathbf{u}_\Gamma^{k-j+1})\| = 1/\Delta x^2\|\mathbf{u}_\Gamma^{k-j} - \mathbf{u}_\Gamma^{k-j+1}\| .$$

Furthermore, the last system is tested against the unperturbed initial data leading to

$$\|\epsilon_k\| \leq \|\mathbf{A}_1^{-1}\| \left(\tau_\text{r}^{k+1}\|\mathbf{A}_1\mathbf{u}_1^0 - \mathbf{b}_1(\mathbf{u}_\Gamma^0)\| + \sum_{j=1}^{k} \tau_\text{r}^j/\Delta x^2\|\mathbf{u}_\Gamma^{k-j} - \mathbf{u}_\Gamma^{k-j+1}\| \right) .$$

For k to infinity, the first term on the right converges to zero if and only if $\tau_\text{r} < 1$ and from Lemma 1 we know that the second term on the right converges to zero if and only if $\|\mathbf{u}_\Gamma^{k-j} - \mathbf{u}_\Gamma^{k-j+1}\|$ converges to zero, which is the case if $\tau_\text{r} < 1$ and if $\mathbf{u}_\Gamma^k$ is a convergent sequence. Note that these are the perturbed iterates of the twin iteration. Thus, we have to assume that the iteration converges.

If we choose the absolute termination criterion (4.56) or the relative one based on the right-hand side (4.54), we obtain a bound of the form

$$\|\epsilon_k\| \leq \|\mathbf{A}_1^{-1}\|\tau_\text{r}\|\mathbf{b}(\mathbf{u}_\Gamma^k)\| ,$$

respectively

$$\|\epsilon_k\| \leq \|\mathbf{A}_1^{-1}\|\tau_\text{a} .$$

Here, we cannot make a statement on the limit of ϵ_k.

In the second case, meaning the iteration with Neumann data (4.66), we obtain

$$\|\delta_k\| = \|\mathbf{u}_{2_\epsilon}^{k+1} - \mathbf{A}_2^{-1}\mathbf{b}_2(\mathbf{u}_{1_\epsilon}^{k+1})\| \leq \|\mathbf{A}_2^{-1}\|\|\mathbf{A}_2\mathbf{u}_{2_\epsilon}^{k+1} - \mathbf{b}_2(\mathbf{u}_{1_\epsilon}^{k+1})\|$$

and analogous arguments produce the same results for δ_k, meaning that we obtain convergence to zero under the assumptions that $\tau_\text{r} < 1$ and that the perturbations ϵ_k are convergent. By theorem 4.1, we then have that when using the relative criterion we obtain convergence to the exact solution for any $\tau_\text{r} < 1$.

4.6.3 Test case: Transmission problem

We now consider the transmission problem (4.62). Specifically, we use $\Omega_1 = [0,1] \times [0,1]$, $\Omega_2 = [1,2] \times [0,1]$ and

$$f(x,y) = \sin \pi y^2 (\pi \cos \frac{\pi}{2} x^2 - \pi^2 x^2 \sin \frac{\pi}{2} x^2) + \sin \frac{\pi}{2} x^2 (2\pi \cos \pi y^2 - 4\pi^2 y^2 \sin \pi y^2). \tag{4.67}$$

This was chosen such that the solution is

$$u(x,y) = \sin \pi y^2 \sin \frac{\pi}{2} x^2 , \tag{4.68}$$

which satisfies the boundary conditions.

We discretize this problem using central differences with a constant mesh width of $\Delta x = \Delta y$. As an initial guess for the Dirichlet–Neumann procedure, we employ a vector of all zeros. All linear systems are solved using a multigrid method. Specifically, we use a V cycle with exact solves for systems of dimension smaller than 16. Restriction and prolongation are standard full weighting, respectively bilinear interpolation [42], appropriately extended for the Neumann problem. The smoother is a damped Jacobi method with $\omega = 2/3$, which is applied thrice both as a pre- and a postsmoother. The exact solution and the discrete solution with $\Delta x = 1/64$ can be seen in Figure 4.11.

We now look at the convergence properties of the fixed-point schemes for different termination criteria. Thus, we choose $TOL = 1e-10$ and $\Delta x = 1/64$ and then run the iteration for different tolerances in multigrid. In the case of the termination criterion (4.55), the iteration recovers the reference solution for any tolerance τ_r, as predicted by the theory. For the relative termination criterion (4.54) and the absolute termination criterion (4.56) the situation is more complicated. We observe that for a given tolerance, $\|\mathbf{u}_\Gamma^{k+1} - \mathbf{u}_\Gamma^k\|$ does not converge to zero, but instead oscillates around a value that is proportional to the tolerance in multigrid. The error $\|\mathbf{u}_\Gamma - \mathbf{u}_\Gamma^*\|_2$ is then about half that value. This means that in practice, these schemes behave as if there were a constant perturbation and that the error can be controlled by choosing an appropriately small tolerance in the linear solver. Note that (4.54) is implemented in the MATLAB version of CG (conjugate gradient) and that thus, a native implementation of the Dirichlet–Neumann iteration with CG in MATLAB will produce questionable results if not used with care.

We now compare the different termination criteria with regards to efficiency for values of TOL more relevant in practice. Hereby, we assume that the user wants to have a solution that is TOL close to the exact one. On the basis of the theory discussed here, there are three choices: Using (4.55) with $\tau_r = 1e-1$ independent of TOL, using (4.56) with $\tau_a = TOL$ or (4.54) with $\tau_r = TOL$, meaning that for the latter ones, we have to solve the inner iteration more accurately the more accurate we want the outer one. Hereby, we choose $\Delta x = 1/128$ and $\Delta x = 1/256$. As it turns out, the error is

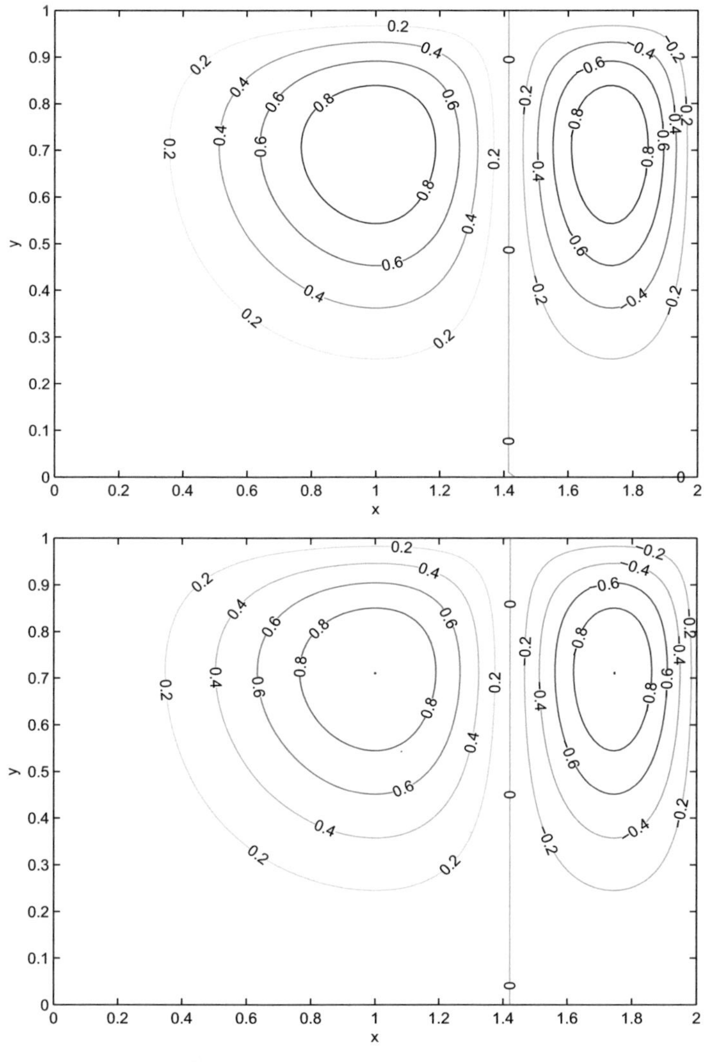

Fig. 4.11: Exact and discrete solution with $\Delta x = 1/64$.

in all cases about $0.4 \cdot TOL$. Since cost in the fixed-point iterations not arising from solving the linear systems is negligible and the cost of one multigrid iteration is fixed, we compare the total number of multigrid iterations needed until termination. The results are depicted in Figure 4.12. As we can see, the scheme corresponding to (4.54) is the slowest. This is more pronounced on the finer grid, where it is about a factor of two slower than the other schemes. The only scheme where the computational effort has a linear behavior with regards to tolerance is that corresponding to (4.55). Thus, while slower for large tolerances, it beats the scheme with (4.56) at some point.

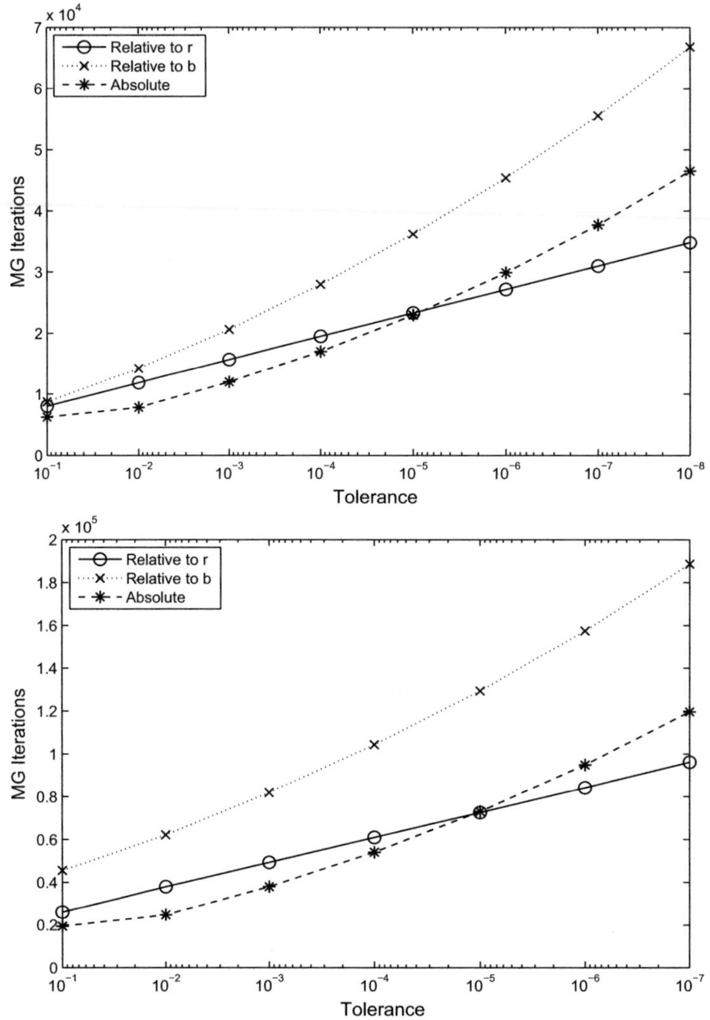

Fig. 4.12: Total multigrid iterations over tolerance for different termination criteria for the transmission problem with rhs (4.67) and $\Delta x = 1/128$ (top) and $\Delta x = 1/256$ (bottom).

An absolute termination criterion can lead to oversolving. This is illustrated in the next test, where we just multiply the right-hand side (4.67) with 100. The results are shown in Figure 4.13. Again, criterion (4.54) leads to the slowest scheme. However, the scheme corresponding to (4.55) is now faster than that with criterion (4.56) for $TOL < 1e-1$.

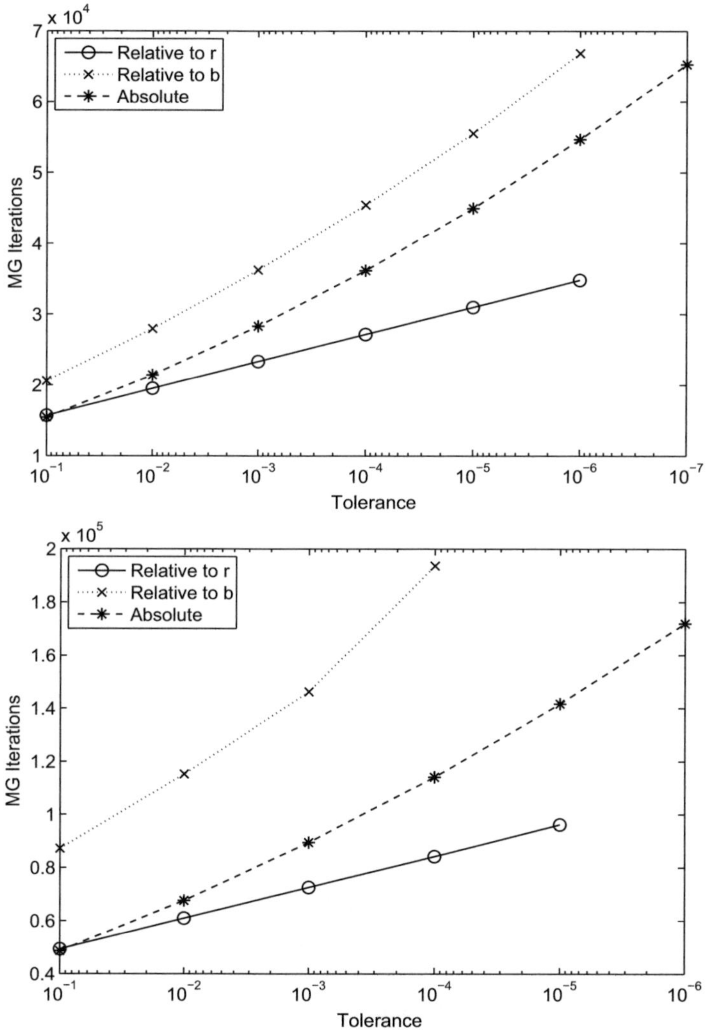

Fig. 4.13: Total multigrid iterations over tolerance for different termination criteria for the transmission problem with rhs (4.67) multiplied by 100 and $\Delta x = 1/128$ (top) and $\Delta x = 1/256$ (bottom).

4.7 Numerical results

The following results illustrate the performance of the complete solver described so far for realistic test cases of a steel–air coupling. Thus, the models from Section 4.2 are employed and discretized using FEM in the structure and FVM in the fluid. In particular, the DLR (German Aerospace Center, german: Deutsches Zentrum für Luft und Raumfahrt) TAU-Code is employed [19], which is a cell-vertex-type finite volume method with AUSMDV (AUSM with FDS (flux difference splitting) and FVS (flux vector splitting)) as flux function and a linear reconstruction to increase the order of accuracy. The finite element code uses quadratic finite elements [47] and is the inhouse code "Native" of the Institute for Static and Dynamic at the University of Kassel. The coupling between the solvers is done using the Component Template Library (CTL) of the University of Braunschweig [30]. The following numerical results are taken from [8].

4.7.1 Flow over a plate

As a first test case, the cooling of a flat plate resembling a simple work piece is considered. The work piece is initially at a much higher temperature than the fluid and then cooled by a constant air stream, see Figure 4.14.

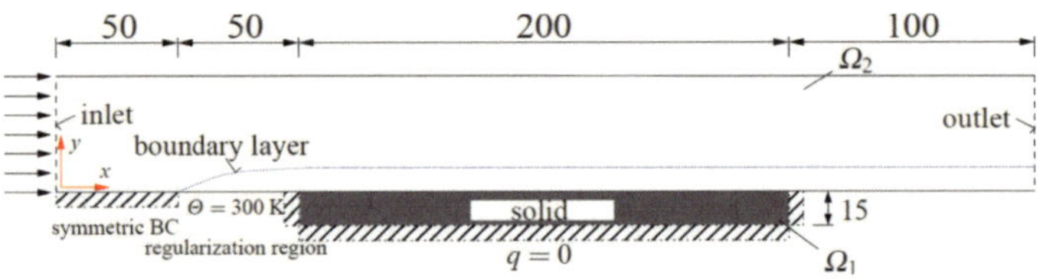

Fig. 4.14: Test case for the coupling method.

The inlet is given on the left, where air enters the domain with an initial velocity of $\mathrm{Ma}_\infty = 0.8$ in the horizontal direction and a temperature of 273 K. Then, there are two succeeding regularization regions of 50 mm to obtain an unperturbed boundary layer. In the first region, $0 \leq x \leq 50$, symmetry boundary conditions, $v_y = 0, q = 0$, are applied. In the second region, $50 \leq x \leq 100$, a constant wall temperature of 300 K is specified. Within this region the velocity boundary layer develops fully. The third part is the solid (work piece) of length 200 mm, which exchanges heat with the fluid, but is assumed insulated otherwise, thus $q_b = 0$. Therefore, Neumann boundary conditions are applied throughout. Finally, the fluid domain is closed by a second regularization region of 100 mm with symmetry boundary conditions and the outlet.

Regarding the initial conditions in the structure, a constant temperature of 900 K at $t = 0$ s is chosen throughout. To specify reasonable initial conditions within the fluid, a steady state solution of the fluid with a constant wall temperature $\Theta = 900$ K is computed.

The grid is chosen cartesian and equidistant in the structural part, where in the fluid region the thinnest cells are on the boundary and then become coarser in the y-direction (see Figure 4.15). To avoid additional difficulties from interpolation, the points of the primary fluid grid, where the heat flux is located in the fluid solver, and the nodes of the structural grid are chosen to match on the interface Γ.

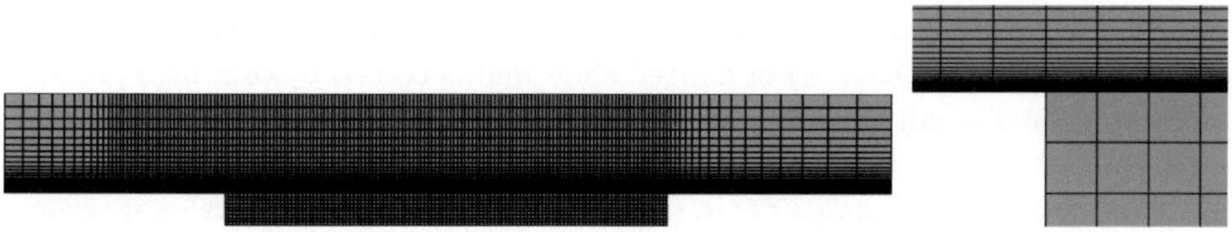

(a) Entire mesh (b) Mesh zoom

Fig. 4.15: Full grid (a) and zoom into coupling region (b)

We now compare the different schemes for a whole simulation of 100 seconds real time. If not mentioned otherwise, the initial time-step size is $\Delta t = 0.5\,s$. To first give an impression of the effect of the time-adaptive method, we look at fixed time step versus adaptive computations in Table 4.4. Thus, the different tolerances for the time-adaptive case lead to different time-step sizes and tolerances for the nonlinear system over the course of the algorithm, whereas in the fixed time-step size, they steer only how accurately the nonlinear systems are solved. For the fixed time-step case, we chose $\Delta t = 0.5\,s$ and $\Delta t = 5\,s$, which roughly corresponds to an error of 10^{-2} and 10^{-3}, respectively 10^{-4}. Thus, computations in one line of Table 4.4 correspond to similar errors. As can be seen, the time-adaptive method is in the worst case a factor of two faster and in the best case a factor of eight. Thus, the time-adaptive computation serves from now on as the base method for the construction of a fast solver.

Finally, we consider extrapolation based on the time integration scheme. As can be seen in Table 4.4, linear extrapolation speeds up the computations between 20% and 50%. Overall, we are thus able to simulate 100 seconds of real time for this problem for an engineering tolerance using only 19 calls to fluid and the structure solver each.

Tab. 4.4: Total number of iterations for 100 seconds of real time. Fixed time-step sizes versus adaptive steering. For time-adaptive calculations, the initial time step is $\Delta t_0 = 0.5\,s$.

TOL	Fixed time	step size	Time adapt.	Time adapt. with lin. extr.
10^{-2}	$\Delta t = 5\,s$	64	31	19
10^{-3}	$\Delta t = 5\,s$	82	39	31
10^{-4}	$\Delta t = 0.5\,s$	802	106	73
10^{-5}	$\Delta t = 0.5\,s$	1014	857	415

4.7.2 Cooling of a flanged shaft

To illustrate a realistic example of gas quenching, the cooling of a flanged shaft by cold high-pressure air is used as a second test case. The complete process consists of the inductive heating of a steel rod, the forming of the hot rod into a flanged shaft, transport to a cooling unit, and the cooling process. Here, we consider only the cooling, meaning that we have a hot flanged shaft that is cooled by cold high-pressure air coming out of small tubes [45]. We perform a two-dimensional cut through the domain and assume symmetry along the vertical axis, resulting in one half of the flanged shaft and two tubes blowing air at it, see Figure 4.16. Since the air nozzles are evenly distributed around the flanged shaft, we use an axisymmetric model in the structure. The heat flux from the two-dimensional simulation of the fluid at the boundary of the flanged shaft is impressed axially symmetrical on the structure.

We assume that the air leaves the tube in a straight and uniform way at a Mach number of 1.2. Furthermore, we assume a freestream in the x-direction of Mach 0.005.

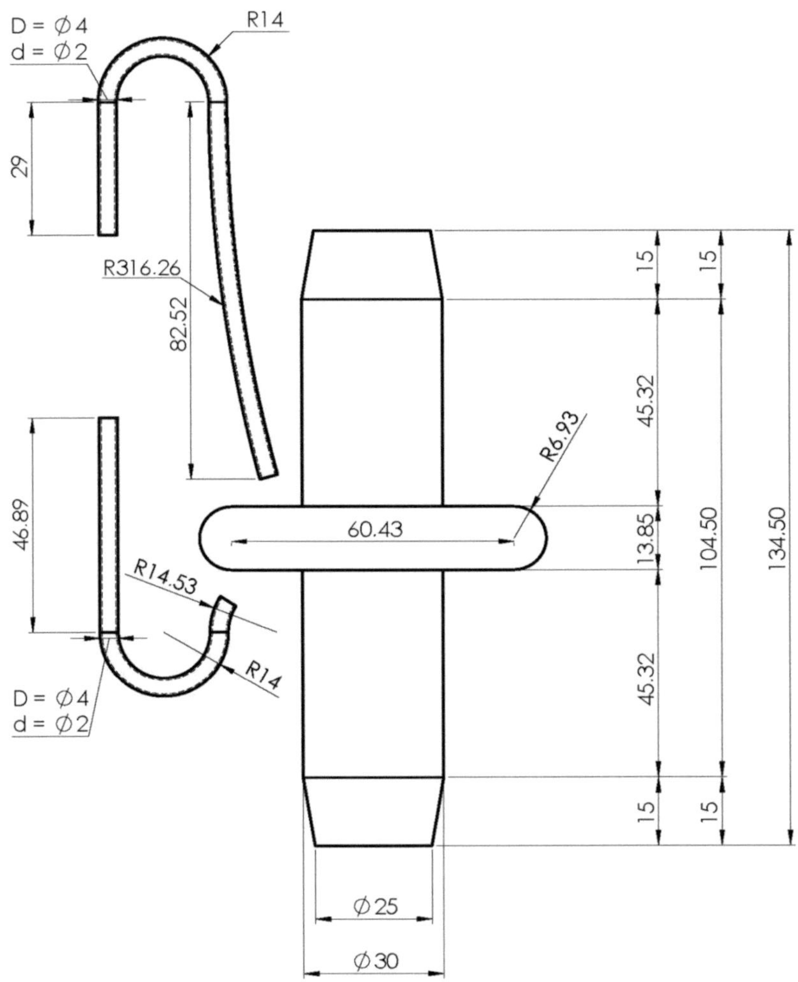

Fig. 4.16: Sketch of the flanged shaft.

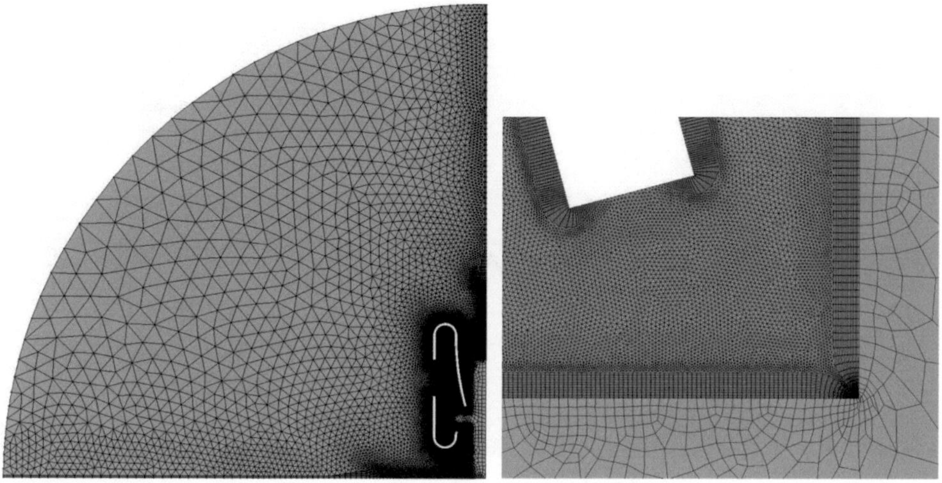

(a) Entire mesh (b) Mesh zoom

Fig. 4.17: Full grid (left) and zoom into shaft region (right).

This is mainly to avoid numerical difficulties at Mach 0, but could model a draft in the workshop. The Reynolds number is $Re = 2500$ and the Prandtl number $Pr = 0.72$.

The grid consists of 279,212 cells in the fluid, which is the dual grid of an unstructured grid of quadrilaterals in the boundary layer and triangles in the rest of the domain, and 1,997 quadrilateral elements in the structure. It is illustrated in Figure 4.17.

To obtain initial conditions for the subsequent tests, we use the following procedure: We define a first set of initial conditions by setting the flow velocity to zero throughout and choose the structure temperatures at the boundary points to be equal to temperatures that have been measured by a thermographic camera. Then, setting the y-axis on the axis of revolution of the flange, we set the temperature at each horizontal slice to the temperature at the corresponding boundary point. Finally, to determine the actual initial conditions, we compute 10^{-5} seconds of real time using the coupling solver with a fixed time-step size of $\Delta t = 10^{-6}$s. This means, that the high-pressure air is coming out of the tubes and the first front has already hit the flanged shaft. This solution is illustrated in Figure 4.18 (left).

Now, we compute 1 second of real time using the time-adaptive algorithm with different tolerances and an initial time-step size of $\Delta t = 10^{-6}$s. This small initial step size is necessary to prevent instabilities in the fluid solver. During the course of the computation, the time-step size is increased until it is of the order of $\Delta t = 0.1$ s, which demonstrates the advantages of the time-adaptive algorithm and reaffirms that it is this algorithm that we need to compare to. In total, the time-adaptive method needs 22, 41, 130, and 850 time steps to reach $t = 1$ s for the different tolerances, compared to the 10^6 steps the fixed time-step method would need. The solution at the final time

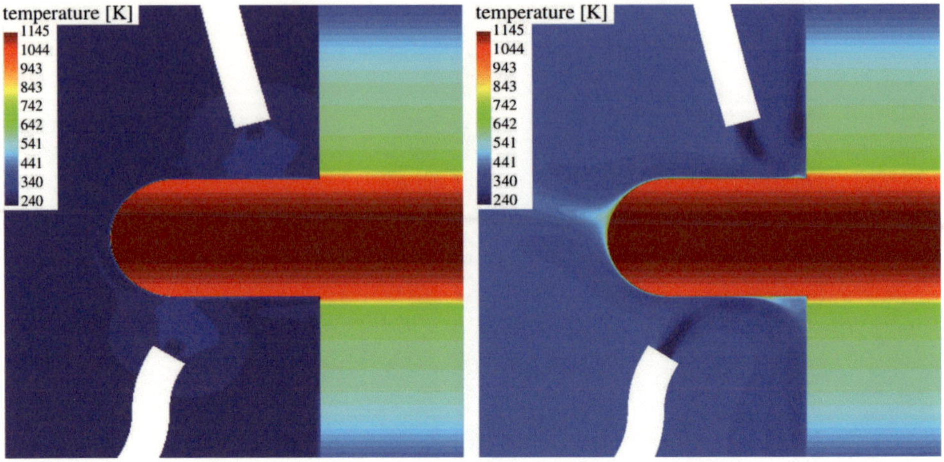

Fig. 4.18: Temperature distribution in fluid and structure at $t = 0$ s (left) and $t = 1$ s (right).

is depicted in Figure 4.18 (right). As can be seen, the stream of cold air is deflected by the shaft.

Finally, we consider extrapolation based on the time integration scheme. In Table 4.5, the total number of iterations for 1 second of real time is shown. As before, the extrapolation causes a noticeable decrease in the total number of fixed-point iterations. The speedup from linear extrapolation is between 18% and 34%, compared to the results obtained without extrapolation.

Tab. 4.5: Total number of iterations for 1 sec of real time for different extrapolation methods in time.

TOL	none	lin.
10^{-2}	51	42
10^{-3}	126	97
10^{-4}	414	309
10^{-5}	2768	1805

4.8 Summary and conclusions

We considered numerical methods for unsteady thermal fluid–structure interaction. As a basic paradigm, we employed a partitioned method with the use of the Dirichlet–Neumann iteration. In this framework, a time-adaptive SDIRK2 method with linear extrapolation of initial values is a fast solver. We cite three reasons for this: the time adaptivity, the very fast convergence rate of the Dirichlet–Neumann iteration for this problem, and a smart choice of tolerances in the various nested solvers.

The fast convergence rate can be explained by estimating the spectral radius of the iteration matrix based on a model problem. This shows that the quotients of material properties play a crucial role. An analysis of the tolerances to be chosen shows that in the linear case, a nonstandard relative termination criterion based on the current residual leads to better performance compared to a standard termination criterion and to avoidance of pitfalls with possible wrong limits.

Bibliography

[1] Background on Tonight's Launch. http://www.spacex.com/news/2015/12/21/background-tonights-launch. Accessed: 2016-08-23.

[2] CRS-8 Launch and Landing. http://www.spacex.com/news/2016/04/09/crs-8-launch-and-landing. Accessed: 2016-08-23.

[3] The Why and How of Landing Rockets. http://www.spacex.com/news/2015/06/24/why-and-how-landing-rockets. Accessed: 2016-08-23.

[4] Y. Bazilevs, K. Takizawa, and T. E. Tezduyar. *Computational Fluid-Structure Interaction: Methods and Applications*. Wiley, 2013.

[5] O. Bendiksen. A new approach to computational aeroelasticity. *AIAA Paper*, 91-09:1712–1727, 1991.

[6] P. Birken. *Numerical Methods for the Unsteady Compressible Navier-Stokes Equations*. Habilitation Thesis. University of Kassel, 2012.

[7] P. Birken. Termination criteria for inexact fixed point schemes. *Numer. Linear Algebra Appl.*, 2015.

[8] P. Birken, T. Gleim, D. Kuhl, and A. Meister. Fast Solvers for Unsteady Thermal Fluid Structure Interaction. *Int. J. Numer. Meth. Fluids*, 79(1):16–29, 2015.

[9] P. Birken, K. Quint, S. Hartmann, and A. Meister. Choosing norms in adaptive FSI calculations. *PAMM*, 10:555–556, 2010.

[10] P. Birken, K. Quint, S. Hartmann, and A. Meister. A time-adaptive fluid-structure interaction method for thermal coupling. *Comp. Vis. in Science*, 13(7):331–340, 2011.

[11] D. Brown et al. *Applied Mathematics at the U.S. Department of Energy: Past, Present and a View to the Future*. tech. rep., Office of Science, U.S. Department of Energy, 2008.

[12] J. Buchlin. Convective Heat Transfer and Infrared Thermography. *J. Appl. Fluid Mech.*, 3:55–62, 2010.

[13] G. Davis and O. Bendiksen. Transonic Panel Flutter. *AIAA Paper*, 93-1476, 1993.

[14] S. Deparis, M. Fernández, and L. Formaggia. Acceleration of a fixed point algorithm for fluid-structure interaction using transpiration conditions. *M2AN*, 37(4):601–616, 2003.

[15] P. Erbts and A. Düster. Accelerated staggered coupling schemes for problems of thermoelasticity at finite strains. *Comp. & Math. with Appl.*, 64(8):2408–2430, 2012.

[16] C. Farhat. CFD-based Nonlinear Computational Aeroelasticity. *in Encyclopedia of Computational Mechanics*, pages 459–480, ch. 13, 2004.

[17] C. Farhat, K. van der Zee, and P. Geuzaine. Provably second-order time-accurate loosely-coupled solution algorithms for transient nonlinear computational aeroelasticity. *Comput. Methods Appl. Mech. Engrg.*, 195:1973–2001, 2006.

[18] C. Fonseca and J. Petronilho. Explicit inverses of some tridiagonal matrices. *Linear Algebra Appl.*, 325(1-3):7–21, 2001.

[19] T. Gerhold, O. Friedrich, J. Evans, and M. Galle. Calculation of Complex Three-Dimensional Configurations Employing the DLR-TAU-Code. *AIAA Paper*, 97-0167, 1997.

[20] M. Giles. Stability Analysis of Numerical Interface Conditions in Fluid-Structure Thermal Analysis. *Int. J. Numer. Meth. Fluids*, 25:421–436, 1997.

[21] M. Guenther, A. Kvaerno, and P. Rentrop. Multirate partitioned Runge-Kutta methods. *BIT*, 41(3):504–514, 2001.

[22] H. Guillard and C. Farhat. On the significance of the geometric conservation law for flow computations on moving meshes. *Comput. Methods Appl. Mech. Engrg.*, 190:1467–1482, 2000.

[23] U. Heck, U. Fritsching, and B. K. Fluid flow and heat transfer in gas jet quenching of a cylinder. *Int. J. Numer. Methods Heat Fluid Flow*, 11:36–49, 2001.

[24] W. Henshaw and K. Chand. A composite grid solver for conjugate heat transfer in fluid-structure systems. *J. Comput. Phys.*, 228:2708–3741, 2009.

[25] M. Hinderks and R. Radespiel. Investigation of Hypersonic Gap Flow of a Reentry Nosecap with Consideration of Fluid Structure Interaction. *AIAA Paper*, 6:2708–3741, 2006.

[26] D. Kowollik, V. Tini, S. Reese, and M. Haupt. 3D fluid-structure interaction analysis of a typical liquid rocket engine cycle based on a novel viscoplastic damage model. *Int. J. Numer. Methods Engrg.*, 94:1165–1190, 2013.

[27] D. Kowollik, P. Horst, and M. Haupt. Fluid-structure interaction analysis applied to thermal barrier coated cooled rocket thrust chambers with subsequent local investigation of delamination phenomena. *Progress in Propulsion Physics*, 4:617–636, 2013.

[28] P. Le Tallec and J. Mouro. Fluid structure interaction with large structural displacements. *Comput. Methods Appl. Mech. Engrg.*, 190:3039–3067, 2001.

[29] R. Massjung. Discrete conservation and coupling strategies in nonlinear aeroelasticity. *Comput. Methods Appl. Mech. Engrg.*, 196:91–102, 2006.

[30] H. G. Matthies, R. Niekamp, and J. Steindorf. Algorithms for strong coupling procedures. *Comput. Methods Appl. Mech. Engrg.*, 195:2028–2049, 2006.

[31] R. Mehta. Numerical Computation of Heat Transfer on Reentry Capsules at Mach 5. *AIAA-Paper*, 178, 2005.

[32] C. Meyer. *Matrix Analysis and Applied Linear Algebra*. 2000.

[33] H. Olsson and G. Söderlind. Stage value predictors and efficient Newton iterations in implicit Runge-Kutta methods. *SIAM J. Sci. Comput.*, 20(1):185–202, 1998.

[34] A. Quarteroni and A. Valli. *Domain Decomposition Methods for Partial Differential Equations*. Oxford Science Publications, 1999.

[35] K. J. Quint, S. Hartmann, S. Rothe, N. Saba, and K. Steinhoff. Experimental validation of high-order time integration for non-linear heat transfer problems. *Comput. Mech.*, 48(1):81–96, 2011.

[36] M. Reuter and J. Hill. An efficient, block-by-block algorithm for inverting a block tridiagonal, nearly block toeplitz matrix. *Comp. Sci. Disc.*, 5, 2012.

[37] V. Savcenco, W. Hundsdorfer, and J. G. Verwer. A multirate time stepping strategy for stiff ordinary differential equations. *BIT Numerical Mathematics*, 47:137–155, 2007.

[38] G. Söderlind and L. Wang. Adaptive time-stepping and computational stability. *J. Comp. Appl. Math.*, 185:225–243, 2006.

[39] P. R. Spalart and S. R. Allmaras. A One-Equation Turbulence Model for Aerodynamic Flows. *AIAA 30th Aerospace Science Meeting*, 92–0439, 1992.

[40] P. Stratton, I. Shedletsky, and M. Lee. Gas Quenching with Helium. *Solid State Phenomena*, 118:221–226, 2006.

[41] A. Toselli and O. Widlund. *Domain Decomposition Methods - Algorithms and Theory*. Springer, 2004.

[42] U. Trottenberg, C. W. Oosterlee, and S. Schüller. *Multigrid*. Elsevier Academic Press, 2001.

[43] A. van Zuijlen and H. Bijl. Implicit and explicit higher order time integration schemes for structural dynamics and fluid-structure interaction computations. *Comp. & Struct.*, 83:93–105, 2005.

[44] A. van Zuijlen, A. de Boer, and H. Bijl. Higher-order time integration through smooth mesh deformation for 3d fluid-structure interaction simulations. *J. Comp. Phys.*, 224:414–430, 2007.

[45] U. Weidig, N. Saba, and K. Steinhoff. Massivumformprodukte mit funktional gradierten Eigenschaften durch eine differenzielle thermo-mechanische Prozessführung. *WT-Online*, pages 745–752, 2007.

[46] F. Zhang. *The Schur complement and its applications*. Springer, 2005.

[47] O. Zienkiewicz and R. Taylor. *The Finite Element Method*. Butterworth Heinemann, 2000.

Ulrich Langer and Huidong Yang

5 Recent development of robust monolithic fluid-structure interaction solvers

Abstract: In the last few years, from the modeling point of view, the monolithic approach for fluid-structure interaction problems in many different application fields has been adopted by more and more researchers. Meanwhile, the development of monolithic solvers in the solution procedure for solving such coupled fluid-structure interaction problems all at once is in general a very hard task and has received a lot of attention. Because of the coupling conditions on the interface, it is challenging to design efficient preconditioners for the linearized coupled system of equations, that are robust with respect to the mesh size, time-step size and material parameters. Further, it is nontrivial to realize scalable parallel implementations for solving such large-scale coupled systems, which requires special care for handling the interface conditions. In this survey, we present an overview of some recent results on robust monolithic fluid-structure interaction solvers, that are mainly based on the block factorization, geometric and algebraic multigrid, and domain decomposition methods.

Keywords: Fluid-structure interaction, monolithic solvers, block factorization, geometric multigrid methods, algebraic multigrid methods, domain decomposition methods, preconditioners

Classification: 74S05, 74F10, 65M55, 65M60

5.1 Introduction

Partitioned fluid-structure interaction (FSI) methods are traditionally adopted in engineering applications since it is easy to combine available solvers for fluid and solid mechanical problems; see, e.g., [4, 9, 10, 18–20, 23, 27, 31, 32, 37, 53, 59, 79] for recent development of partitioned methods. However, the separation of the solid from the fluid part in the solution process usually yields a loss in efficiency and robustness mainly due to the so-called added-mass effect; see, e.g., [23, 34, 60, 75, 89]. Another critical point is the control of the systematic errors caused by the accuracy of the fluid, solid (structure) and mesh subproblem solvers in each iteration step for solving the nonlinear "Schur complement system" (discrete Steklov–Poicaré problem). That is the main reason why the monolithic methods for solving the FSI problems have received more and more interest over the last few years; see, e.g., [13] for aspects of

---

Ulrich Langer, Huidong Yang, Johann Radon Institute for Computational and Applied Mathematics, Altenberger Str. 69, 4040 Linz

https://doi.org/10.1515/9783110494259-005

algorithms and computations in monolithic methods for FSI problems. The methods may be mainly classified into two groups (A and B),

A: the first one relies on the fluid, structure and mesh subproblem solvers, where efficient subproblem solvers on the underlying basis have to be called in some way, see, e.g., [5–7, 25, 26, 31, 36, 43, 44, 54, 55, 58, 61, 66];

B: the second one relies on solving many small local problems, which are characterized by the same form as the global problem, i.e., a restriction of the global problem onto the local subdomains or patches, and featured as black box-type solvers, see, e.g., [11, 12, 47, 64, 93].

The methods mentioned in these two groups have mainly been developed from the classical methods for solving individual field problems, e.g., block factorization [51, 68, 86], geometric multigrid [41], algebraic multigrid [67, 74] and domain decomposition [63, 73, 78] methods. Therefore, we will discuss the monolithic methods in the following sections:

1. Section 5.2, block factorization; see [5–7, 25, 26, 43, 44, 55, 61];
2. Section 5.3, geometric multigrid; see [47, 64, 66];
3. Section 5.4, algebraic multigrid; see [36, 54, 58];
4. Section 5.5, domain decomposition; see [11, 12, 93].

Note that some of the methods are a combination of two classical ones, e.g., [7, 25, 26] based on block factorization and domain decomposition methods, and [36, 54, 55] based on block factorization and algebraic multigrid methods. We characterize the methods by their main features. Recently, in addition to the above-mentioned work, which we will discuss in detail from the practical point of view, some results on the analysis of optimal FSI preconditioners have been published in [94]. Therein, optimal FSI preconditioners have been developed within the framework given in [57, 97], where the uniform well-posedness of the saddle point system and corresponding Riesz operator are exploited.

Concerning temporal, spatial discretization and linearization for the coupled model FSI problems, we refer the reader to [13] for details. In short, the computational FSI domain $\overline{\Omega^t} = \overline{\Omega_\mathrm{f}} \cup \overline{\Omega_\mathrm{s}}$ is decomposed into two deformable subdomains, the fluid Ω_f and structure Ω_s. The intersection of boundaries of two subdomains yields the moving interface $\Gamma_\mathrm{fs} = \partial\Omega_\mathrm{f} \cap \partial\Omega_\mathrm{s}$. By notations $\Gamma_\mathrm{f}^\mathrm{D}$, $\Gamma_\mathrm{f}^\mathrm{N}$, and $\Gamma_\mathrm{s}^\mathrm{D}$, $\Gamma_\mathrm{s}^\mathrm{N}$, we denote the Dirichlet and Neumann boundaries for the fluid and structure subproblem, respectively. We refer to Fig. 5.1 for a schematic illustration of the subdomains, interface and boundaries.

Depending on the particular applications, the incompressible Navier–Stokes equations under the arbitrary Lagrangian–Eulerian (ALE) framework [28, 33, 49, 52] are considered in the fluid subdomain Ω_f; the elasticity and hyperelasticity models [17, 45, 92] under the Lagrangian framework are usually used in the structure reference subdomain $\hat{\Omega}_\mathrm{s}$. On the interface Γ_fs, the no-slip condition and balance of

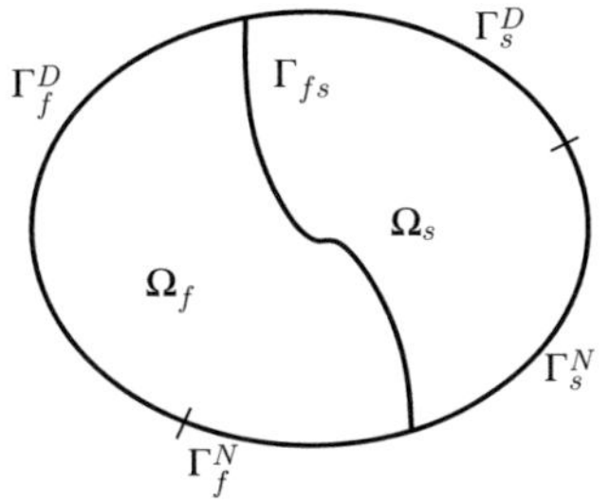

Fig. 5.1: A schematic illustration of the computational FSI domain: $\overline{\Omega^t} = \overline{\Omega_f} \cup \overline{\Omega_s}$.

surface tractions from the fluid and structure sides are fulfilled, i.e., the velocities match and the surface tractions balance. Further, a geometric adherence condition is imposed, i.e., the fluid domain displacement and structure displacement on the interface are equal to each other.

In this chapter, our focus is a short overview of recent developments of monolithic methods for solving the FSI linear system of algebraic equations:

$$Kx = b \, . \tag{5.1}$$

In order to ease the demonstration of each method, K denotes the Jacobian system matrix at the current state, which will be detailed for each case in the respective section, x the unknown corrections of possible combinations of fluid velocity, fluid pressure, structure displacement, structure velocity, structure pressure, mesh displacement, and b the corresponding residual. Further, we use the notation P with the corresponding subscription to indicate the respective preconditioner. In addition, Δt indicates the time-step size.

The remainder of the chapter is organized as follows: Section 5.2 deals with FSI preconditioners based on block factorizations. In Section 5.3, we discuss geometric multigrid methods for solving FSI problems. Section 5.4 concerns algebraic multigrid methods applied to FSI problems. Section 5.5 is devoted to FSI preconditioners based on domain decomposition techniques. Finally, some conclusions are drawn in Section 5.7.

5.2 Block factorization-based preconditioners

5.2.1 Direct *LU* decomposition

In [48], with a skyline profile technique, a direct LU decomposition of the system matrix is used for preconditioning in solving discretized FSI problems (at a small scale) using the space-time finite element methods [15, 42, 77]. The BiCGStab solver (see, e.g., [82]) combined with this preconditioner is quite efficient for smaller examples, i.e., it only needs a few iterations. However, this method is hardly applicable to large 3D examples.

5.2.2 Block-triangular approximations

In [43], three block-triangular approximations for the Jacobian matrix

$$K = \begin{bmatrix} S & C_{\text{su}} & C_{\text{sp}} \\ C_{\text{us}} & F & G \\ C_{\text{ps}} & D & \end{bmatrix} \tag{5.2}$$

are used as FSI (fluid in 2D coupled with a 1D thin-walled elastic beam) preconditioners by neglecting certain fluid-structure interaction blocks, i.e.,

$$\tilde{P}_{\text{sup}} = \begin{bmatrix} S & C_{\text{su}} & C_{\text{sp}} \\ & F & G \\ & D & \end{bmatrix}, \; \tilde{P}_{\text{sub}} = \begin{bmatrix} S & & \\ C_{\text{us}} & F & G \\ C_{\text{ps}} & D & \end{bmatrix}, \; \tilde{P}_{\text{diag}} = \begin{bmatrix} S & & \\ & F & G \\ & D & \end{bmatrix}. \tag{5.3}$$

Here, S represents the Jacobian matrix from the contributions of the tangent stiffness matrix of the wall equations and the fluid-structure interaction, F the fluid momentum equation, G the discrete gradient operator, D the discrete divergence operator, and the remaining nonzero off-diagonal blocks the coupling between different fields.

The most expensive fluid sub-block in the approximations (5.3) are replaced by a cheap global pressure Schur complement preconditioner [80], which leads to three FSI preconditioners

$$P_{\text{sup}} = \begin{bmatrix} S & C_{\text{su}} & C_{\text{sp}} \\ & F & G \\ & & X \end{bmatrix}, \; P_{\text{sub}} = \begin{bmatrix} S & & \\ C_{\text{us}} & F & G \\ C_{\text{ps}} & & X \end{bmatrix}, \; P_{\text{diag}} = \begin{bmatrix} S & & \\ & F & G \\ & & X \end{bmatrix}, \tag{5.4}$$

where $X = -(DG)(DFG)^{-1}(DG)$ is Elman's BFBt approximation [29] to the pressure Schur complement $-DF^{-1}G$. These preconditioners show relatively good performance with respect to the Reynolds number and the mesh size.

Further, in [44], two block-triangular FSI preconditioners P_1 and P_2 are proposed by neglecting the fluid-structure and structure-fluid interaction blocks in the reordered Jacobian matrix compared to (5.2)

$$K = \begin{bmatrix} F & G & C_{\text{us}} \\ D & & C_{\text{ps}} \\ C_{\text{su}} & C_{\text{sp}} & S \end{bmatrix}, \tag{5.5}$$

i.e.,

$$P_1 = \begin{bmatrix} F & G & \\ D & & \\ C_{\text{su}} & C_{\text{sp}} & S \end{bmatrix}, \; P_2 = \begin{bmatrix} F & G & C_{\text{us}} \\ D & & C_{\text{ps}} \\ & & S \end{bmatrix}. \tag{5.6}$$

Each preconditioning step requires solving the fluid and structure subproblems approximately. Compared to the earlier work [43], for the fluid subproblem, the least squares commutator (LSC) preconditioner [30] is employed with inclusion of a scaling factor $\hat{Q}$ in the pressure Schur complement, that is the diagonal of the velocity mass matrix. To be more precise, the LSC preconditioner for the fluid subproblem has the following form:

$$P_{\mathrm{LSC}} = \begin{bmatrix} F & G \\ & \tilde{X} \end{bmatrix}, \tag{5.7}$$

where $\tilde{X} = (D\hat{Q}G)^{-1}(D\hat{Q}^{-1}F\hat{Q}^{-1}G)(D\hat{Q}^{-1}G)^{-1}$ represents a special choice of proper approximations to the pressure Schur complement. The mesh-independent convergence rates of these two preconditioners are observed using uniform and adaptive mesh refinement.

5.2.3 Combination with parallel preconditioners

In [25, 26], the two serial block Gauss–Seidel preconditioners $P_{\mathrm{GS}}^{(1)}$ and $P_{\mathrm{GS}}^{(2)}$ are further developed in parallel context by performing block factorization of the augmented Jacobian matrix

$$K = \begin{bmatrix} C_{\mathrm{ff}} & C_{f\Gamma} & & & \\ C_{\Gamma f} & C_{\Gamma\Gamma} & & & I \\ & & N_{\mathrm{ss}} & N_{s\Gamma} & \\ & & N_{s\Gamma} & N_{\Gamma\Gamma} & -I \\ & I & & -I/\Delta t & \end{bmatrix} \tag{5.8}$$

using the geometry-convective explicit time discretization, neglecting the term $-I$,

$$\begin{aligned} P_{\mathrm{GS}}^{(1)} &= \begin{bmatrix} C_{\mathrm{ff}} & C_{f\Gamma} & & & \\ C_{\Gamma f} & C_{\Gamma\Gamma} & & & I \\ & & N_{\mathrm{ss}} & N_{s\Gamma} & \\ & & N_{s\Gamma} & N_{\Gamma\Gamma} & \\ & I & & -I/\Delta t & \end{bmatrix} \\ &= \begin{bmatrix} I & & & & \\ & I & & & \\ & & N_{\mathrm{ss}} & N_{s\Gamma} & \\ & & N_{s\Gamma} & N_{\Gamma\Gamma} & \\ & & & & I \end{bmatrix} \begin{bmatrix} C_{ff} & C_{f\Gamma} & & & \\ C_{\Gamma f} & C_{\Gamma\Gamma} & & & I \\ & & I & & \\ & & & I & \\ & I & & -I/\Delta t & \end{bmatrix} \\ &= P_{GS,1}^{(1)} P_{GS,2}^{(1)}, \end{aligned} \tag{5.9}$$

and $I/\Delta t$

$$
\begin{aligned}
P_{\mathrm{GS}}^{(2)} &= \begin{bmatrix} C_{\mathrm{ff}} & C_{f\Gamma} & & & \\ C_{\Gamma f} & C_{\Gamma\Gamma} & & & I \\ & & N_{\mathrm{ss}} & N_{s\Gamma} & \\ & & N_{s\Gamma} & N_{\Gamma\Gamma} & -I \\ & I & & & \end{bmatrix} \\
&= \begin{bmatrix} C_{\mathrm{ff}} & C_{f\Gamma} & & & \\ C_{\Gamma f} & C_{\Gamma\Gamma} & & & I \\ & & I & & \\ & & & I & \\ & I & & & \end{bmatrix} \begin{bmatrix} I & & & & \\ & I & & & \\ & & N_{\mathrm{ss}} & N_{s\Gamma} & \\ & & N_{s\Gamma} & N_{\Gamma\Gamma} & -I \\ & & & & I \end{bmatrix} \\
&= P_{GS,1}^{(2)} P_{GS,2}^{(2)} ,
\end{aligned}
\tag{5.10}
$$

on the off-diagonal part, respectively, and replacing the fluid and structure sub-blocks with proper parallel preconditioners. Here C with subscripts f and Γ denotes the fluid stiffness matrix associated with the degrees of freedom in the fluid domain and on the interface. The same applies to the structure stiffness matrix N. We mention that, in the augmented system, the two interface conditions are imposed explicitly. In [25], the same technique is used to construct the FSI preconditioners for the geometry-convective explicit time discretized FSI system.

Particularly, the three parallel preconditioners $P_{\mathrm{AS}}(K)$, $P_{\text{GS-AS}}^{(1)}$ and $P_{\text{GS-AS}}^{(2)}$ are then built by applying the one-level algebraic additive Schwarz preconditioner [63, 73, 78] to each factor of the full Jacobian matrix, the approximated one $P_{\mathrm{GS}}^{(1)}$ and $P_{\mathrm{GS}}^{(2)}$, respectively. The preconditioners $P_{\text{GS-AS}}^{(1)}$ and $P_{\text{GS-AS}}^{(2)}$ preserve modularity and have demonstrated similar or better performance than $P_{\mathrm{AS}}(K)$ with increasing number of processors.

A similar preconditioning technique is employed in [7] for the FSI simulation using anisotropic polyconvex hyperelastic and viscoelastic arterial models at finite strains. Such an approximated monolithic Dirichlet–Neumann preconditioner P_{DN} is obtained by performing a block factorization to the Jacobian matrix

$$
K = \begin{bmatrix} F & & C_1^T & D_{d_f} F \\ & D_{d_s} S & C_3^T & \\ C_1 & C_2 & & \\ & C_4 & & H \end{bmatrix}
\tag{5.11}
$$

neglecting the coupling block C_3^T, i.e.,

$$
\begin{aligned}
P_{\mathrm{DN}} &= \begin{bmatrix} F & & C_1^T & D_{d_\mathrm{f}}F \\ & D_{d_\mathrm{s}}S & & \\ C_1 & C_2 & & \\ & C_4 & & H \end{bmatrix} \\
&= \begin{bmatrix} I & & & \\ & D_{d_\mathrm{s}}S & & \\ & & I & \\ & & & I \end{bmatrix} \begin{bmatrix} I & & & \\ & I & & \\ & & I & \\ & C_4 & & H \end{bmatrix} \begin{bmatrix} F & & C_1^T & D_{d_\mathrm{f}}F \\ & I & & \\ C_1 & C_2 & & \\ & & & I \end{bmatrix},
\end{aligned} \tag{5.12}
$$

where F denotes the fluid stiffness matrix, S the structure stiffness matrix, H the geometry stiffness matrix, the remaining stemming from the continuity of two interface conditions and geometric adherence, $D_{d_\mathrm{f}}F$ the linearization of the fluid subproblem with respect to the fluid domain displacement, and $D_{d_\mathrm{s}}S$ the linearization of the structure subproblem with respect to the structure displacement. The one level algebraic additive Schwarz preconditioners from [73] are used to approximate the inverse of each factor in P_{DN}.

5.2.4 The augmentation preconditioner

In [61], for the linearized saddle-point system

$$
K_{\mathrm{PS}} = \begin{bmatrix} E & C_{\mathrm{xl}} \\ C_{\mathrm{lx}} & \end{bmatrix} \tag{5.13}
$$

arising from the pseudosolid mesh problem (large-displacement elasticity), the so-called augmentation preconditioner [38, 39, 65]

$$
P_{\mathrm{saug}} = \begin{bmatrix} E_{\mathrm{aug}} & \\ & W \end{bmatrix} = \begin{bmatrix} E + C_{\mathrm{xl}}W^{-1}C_{\mathrm{lx}} & \\ & W \end{bmatrix} \tag{5.14}
$$

is analyzed, which is used to update the moving fluid mesh. Here, E denotes the tangent stiffness matrix of the unconstrained pseudosolid problem, C_{xl} and C_{lx} results from the imposition of the displacement constraint by the Lagrange multipliers [14], W is a matrix such that P_{aug} is effective and E_{aug} sparse. In [61], a new optimal preconditioner is developed by choosing a special block diagonal form of W.

For the fully coupled FSI system, the argumentation preconditioner P_{fsiaug} is obtained by replacing the bottom-right 2×2 block (corresponding to the pseudosolid mesh problem) in the FSI Jacobian matrix

$$
K = \begin{bmatrix} F & & C_{\mathrm{fx}} & \\ C_{\mathrm{sf}} & S & C_{\mathrm{sx}} & \\ & & E & C_{\mathrm{xl}} \\ & C_{\mathrm{ls}} & C_{\mathrm{lx}} & \end{bmatrix} \tag{5.15}
$$

with the above pseudosolid augmentation preconditioner P_{saug}, i.e.,

$$P_{\mathrm{fsiaug}} = \begin{bmatrix} F & & C_{\mathrm{fx}} & \\ C_{\mathrm{sf}} & S & C_{\mathrm{sx}} & \\ & & E_{\mathrm{aug}} & \\ & C_{\mathrm{ls}} & & W \end{bmatrix} . \tag{5.16}$$

Here, F denotes the fluid Jacobian, S the tangent stiffness matrix of the structure, and the nonzero off-diagonal blocks the coupling between the various fields. Subject to the constraint that the tangent stiffness matrix of the unconstrained pseudosolid problem is symmetric positive definite, the nonunit eigenvalues of the preconditioned FSI system are bounded by the ones of the preconditioned pseudosolid problem. This efficient FSI preconditioner leads to a nearly optimal solver with respect to computational cost, i.e., the time for solving the linearized systems scales linearly with the number of unknowns.

5.2.5 Block *LU*-type factorization

In [6], an inexact block LU factorization is employed to construct an FSI preconditioner for the FSI system using 1-dimensional structures

$$K = \begin{bmatrix} C_{\mathrm{ff}} & G_{\mathrm{f}} & C_{\mathrm{f}\sigma} \\ D_{\mathrm{f}} & & D_{\sigma} \\ C_{\sigma\mathrm{f}} & G_{\sigma} & C_{\sigma\sigma} + N \end{bmatrix} , \tag{5.17}$$

where $C_{\alpha\beta}$, $\alpha, \beta \in \{\sigma, f\}$ represents the fluid matrix associated with the viscous, convective and $1/\Delta t$ scaled mass terms, with σ being nodes on the interface and f in the fluid domain, G_{α} the divergence term, $D_{\alpha} = G_{\alpha}^{T}$, and N the structure matrix corresponding to the no-slip interface condition. For the d-dimensional structures, the arising linear system has the same block structure as (5.17). An exact factorization leads to the following splitting:

$$K = \begin{bmatrix} C_{\mathrm{ff}} & & \\ D_{\mathrm{f}} & S_{\mathrm{pp}} & S_{\mathrm{p}\sigma} \\ C_{\sigma\mathrm{f}} & S_{\sigma\mathrm{p}} & S_{\sigma\sigma} \end{bmatrix} \begin{bmatrix} I & C_{\mathrm{ff}}^{-1} G_{\mathrm{f}} & C_{\mathrm{ff}}^{-1} C_{\mathrm{f}\sigma} \\ & I & \\ & & I \end{bmatrix} = LU , \tag{5.18}$$

where the Schur complement matrices $S_{\mathrm{pp}} = -D_{\mathrm{f}} C_{\mathrm{ff}}^{-1} G_{\mathrm{f}}$, $S_{\mathrm{p}\sigma} = D_{\sigma} - D_{\mathrm{f}} C_{\mathrm{ff}}^{-1} C_{\mathrm{f}\sigma}$, $S_{\sigma\mathrm{p}} = G_{\sigma} - C_{\sigma\mathrm{f}} C_{\mathrm{ff}}^{-1} G_{\mathrm{f}}$ and $S_{\sigma\sigma} = C_{\sigma\sigma} + N - C_{\sigma\mathrm{f}} C_{\mathrm{ff}}^{-1} C_{\mathrm{f}\sigma}$.

In the inexact factorization, the inverse of the fluid sub-block C_{f}^{-1} is replaced by the zero-order term of its Neumann expansion:

$$C_{\mathrm{ff}}^{-1} = \left(\frac{1}{\Delta t} M_{\mathrm{ff}} + K_{\mathrm{ff}} \right)^{-1} = \Delta t M_{\mathrm{ff}}^{-1} + \mathcal{O}\left(\Delta t^2\right) \simeq \Delta t M_{\mathrm{ff}}^{-1} , \tag{5.19}$$

where K_{ff} represents the fluid stiffness matrix associated with the viscous and convective terms, and M_{ff} the fluid velocity mass matrix. The approximations of the Schur complements follow $S_{\mathrm{pp}} \simeq T_{\mathrm{pp}} = -\Delta t D_{\mathrm{f}} M_{\mathrm{ff}}^{-1} G_{\mathrm{f}}$, $S_{\mathrm{p}\sigma} \simeq T_{\mathrm{p}\sigma} = D_{\sigma} - \Delta t D_{\mathrm{f}} M_{\mathrm{ff}}^{-1} C_{\mathrm{f}\sigma}$, $S_{\sigma\mathrm{p}} \simeq T_{\sigma\mathrm{p}} = G_{\sigma} - \Delta t C_{\sigma\mathrm{f}} M_{\mathrm{ff}}^{-1} G_{\mathrm{f}}$ and $S_{\sigma\sigma} \simeq T_{\sigma\sigma} = C_{\sigma\sigma} + N - \Delta t C_{\sigma\mathrm{f}} M_{\mathrm{ff}}^{-1} C_{\mathrm{f}\sigma}$. Now the approximated factors $\hat{L}$ and $\hat{U}$ are given by

$$\hat{L} = \begin{bmatrix} C_{\mathrm{ff}} & & \\ D_{\mathrm{f}} & T_{\mathrm{pp}} & T_{\mathrm{p}\sigma} \\ C_{\sigma\mathrm{f}} & T_{\sigma\mathrm{p}} & T_{\sigma\sigma} \end{bmatrix}, \quad \hat{U} = \begin{bmatrix} I & \Delta t M_{\mathrm{ff}}^{-1} G_{\mathrm{f}} & \Delta t M_{\mathrm{ff}}^{-1} C_{\mathrm{f}\sigma} \\ & I & \\ & & I \end{bmatrix}. \tag{5.20}$$

If the fluid sub-block approximation (5.19) is used in both the lower and upper block-triangular factors, a pressure-correction-like method (Chorin–Temam projection scheme) is obtained,

$$P_{\mathrm{PIC}} = \hat{L}\hat{U}\,, \tag{5.21}$$

that was originally designed for the numerical solution of the Navier–Stokes equations [24, 76], and was recently employed and analyzed in the explicit-implicit splitting algorithm for FSI problems [31]. If the approximation is only used in the lower block-triangular factor, a Yosida-like method is arrived at

$$P_{\mathrm{FSY}} = \hat{L}U\,, \tag{5.22}$$

that was originally proposed as an inexact factorization of the matrix arising from the numerical solution of the Navier–Stokes equations [62]. These two preconditioners are rather stable for the simulation of a pressure pulse propagation in a blood flow vessel.

In [5], an incomplete *LU* factorization P_{ILU}, the so-called *ILUT* preconditioner [68] is applied to the diagonally scaled system:

$$P_{\mathrm{ILU}}^{-1} D^{-1} K x = P_{\mathrm{ILU}}^{-1} D^{-1} b\,, \tag{5.23}$$

where D denotes the diagonal coefficients of K. In addition, an inexact block *LU* factorization is used to construct the FSI preconditioner and the same zeroth-order term as in (5.19) is utilized in both the factors. To reduce the computational cost, a lumped mass matrix in (5.19) is used. These preconditioners show good behavior in the large range of the added-mass effect [60].

5.2.6 A *LDU* block factorization

In [55], a new FSI preconditioner is constructed in the form of $\hat{L}\hat{D}\hat{U}$ with $\hat{L}$, $\hat{D}$ and $\hat{U}$ being proper approximations to the matrices L, D and U in the *LDU* block factorization of the fully coupled and reordered Jacobian matrix

$$K = \begin{bmatrix} A_{\mathrm{m}} & A_{\mathrm{ms}} & \\ & A_{\mathrm{s}} & A_{\mathrm{sf}} \\ A_{\mathrm{fm}} & A_{\mathrm{fs}} & A_{\mathrm{f}} \end{bmatrix}, \tag{5.24}$$

where the stiffness matrices A_{m}, A_{s} and A_{f} for the mesh movement, structure and fluid subproblem are respectively lying on the main diagonal, and the coupling matrices $A_{\beta\alpha}$, $\beta, \alpha \in \{f, m, s\}$ and $\beta \neq \alpha$ among different fields on the off-diagonal. An exact LDU factorization of K reads:

$$K = \begin{bmatrix} I & & \\ & I & \\ A_{\mathrm{fm}}A_{\mathrm{m}}^{-1} & \tilde{A}_{\mathrm{fs}}A_{\mathrm{s}}^{-1} & I \end{bmatrix} \begin{bmatrix} A_{\mathrm{m}} & & \\ & A_{\mathrm{s}} & \\ & & S \end{bmatrix} \begin{bmatrix} I & A_{\mathrm{m}}^{-1}A_{\mathrm{ms}} & \\ & I & A_{\mathrm{s}}^{-1}A_{\mathrm{sf}} \\ & & I \end{bmatrix}, \tag{5.25}$$

with the fluid Schur complement

$$S = A_{\mathrm{f}} - \tilde{A}_{\mathrm{fs}}A_{\mathrm{s}}^{-1}A_{\mathrm{sf}} = A_{\mathrm{f}} - (A_{\mathrm{fs}} - A_{\mathrm{fm}}A_{\mathrm{m}}^{-1}A_{\mathrm{ms}})A_{\mathrm{s}}^{-1}A_{\mathrm{sf}} \,.$$

The new FSI preconditioner $P_{\mathrm{LDU}} = \hat{L}\hat{D}\hat{U}$ is formed as

$$P_{\mathrm{LDU}} = \begin{bmatrix} I & & \\ & I & \\ A_{\mathrm{fm}}\hat{A}_{\mathrm{m}}^{-1} & \hat{\tilde{A}}_{\mathrm{fs}}\hat{A}_{\mathrm{s}}^{-1} & I \end{bmatrix} \begin{bmatrix} \hat{A}_{\mathrm{m}} & & \\ & \hat{A}_{S} & \\ & & \hat{S} \end{bmatrix} \begin{bmatrix} I & \hat{A}_{\mathrm{m}}A_{\mathrm{ms}} & \\ & I & \hat{A}_{\mathrm{s}}^{-1}A_{\mathrm{sf}} \\ & & I \end{bmatrix}, \tag{5.26}$$

where the Schur complement approximation $\hat{S}$ corresponds to the perturbed fluid subproblem, which is obtained by modifying corresponding entries in the original fluid matrix with an explicitly constructed approximation to the exact perturbation stemming from the sparse matrix-matrix multiplications:

$$\hat{S} = A_{\mathrm{f}} - \hat{\tilde{A}}_{\mathrm{fs}}\hat{\hat{A}}_{\mathrm{s}}^{-1}A_{\mathrm{sf}} = A_{\mathrm{f}} - (A_{\mathrm{fs}} - A_{\mathrm{fm}}\hat{\hat{A}}_{\mathrm{m}}^{-1}A_{\mathrm{ms}})\hat{\hat{A}}_{\mathrm{s}}^{-1}A_{\mathrm{sf}} \,, \tag{5.27}$$

where $\hat{\hat{A}}_{\mathrm{m}} = \mathrm{diag}[A_{\mathrm{m}}]$ and $\hat{\hat{A}}_{\mathrm{s}} = \mathrm{diag}[A_{\mathrm{s}}]$ denote the point-wise block diagonal of the corresponding mesh movement and structure matrices.

The inverse operations of $\hat{A}_{\mathrm{m}}^{-1}$, $\hat{A}_{\mathrm{s}}^{-1}$ and $\hat{A}_{\mathrm{f}}^{-1}$ in each factor of the Jacobian matrix are realized by applying a few W-cycles of a special class of algebraic multigrid methods [40, 50, 87, 88, 95] to the corresponding mesh, fluid and structure subproblems with 0 as the initial guess.

Further, dropping $\hat{L}$ leads to a reduced FSI preconditioner $\hat{P}_{\mathrm{LDU}}$ in the form of $\hat{P}_{\mathrm{LDU}} = \hat{D}\hat{U}$:

$$\hat{P}_{\mathrm{LDU}} = \begin{bmatrix} \hat{A}_{\mathrm{m}} & & \\ & \hat{A}_{S} & \\ & & \hat{S} \end{bmatrix} \begin{bmatrix} I & \hat{A}_{\mathrm{m}}A_{\mathrm{ms}} & \\ & I & \hat{A}_{\mathrm{s}}^{-1}A_{\mathrm{sf}} \\ & & I \end{bmatrix}, \tag{5.28}$$

which is based on an inexact Uzawa method (e.g., [16]) and originally analyzed in [72] for the fluid problem. The preconditioners demonstrate relatively good robustness with respect to temporal and spatial discretization parameters, and certain material parameters.

5.3 Geometric multigrid methods

5.3.1 A Vanka-like smoother

In [47, 64], a special monolithic ALE FSI system with respect to displacement, velocity and pressure unknowns on the whole computational FSI domain is formulated in a unified manner; see also [90]. For such a formulation, the conforming biquadratic (for displacement and velocity), discontinuous linear (for pressure) Q_2P_1 finite element pair is chosen. This leads to a natural fulfillment of the interface conditions, namely, the no-slip condition and the balance of the surface tractions. In addition, it allows the fluid and structure pressure jumps across the interface, which usually occurs in the FSI simulation. A geometrical multigrid method is then used to solve the linearized system of equations with the Jacobian matrix:

$$K = \begin{bmatrix} S_{\mathrm{uu}} & S_{\mathrm{uv}} & \\ S_{\mathrm{vu}} & S_{\mathrm{vv}} & B \\ B_{\mathrm{s}}^T & B_{\mathrm{f}}^T & \end{bmatrix}, \tag{5.29}$$

where S represents the discrete reactive, diffusive and convective terms, B the discrete gradient term, and B_{s}^T and B_{f}^T the discrete divergence term. In particular a Vanka-like smoother [80, 85] is performed on the mesh level l, i.e., on each patch (element) Ω_i, a local 39×39 system matrix K_{Ω_i} (in 2D) is inverted, with 3 pressure degrees of freedom and 36 associated displacement and velocity degrees of freedom:

$$x_l^{k+1} = x_l^k + \omega^k \sum_{\Omega_i} K_{\Omega_i}^{-1} r_{\Omega_i}^n, \quad k = 1, \ldots, \nu, \tag{5.30}$$

where ν denotes the smoothing steps, ω^k the relaxation parameter, $r_{\Omega_i}^n$ the defection on the patch Ω_i, and K_{Ω_i} has the following structure:

$$K_{\Omega_i} = \begin{bmatrix} S_{\mathrm{uu}|\Omega_i} & S_{\mathrm{uv}|\Omega_i} & \\ S_{\mathrm{vu}|\Omega_i} & S_{\mathrm{vv}|\Omega_i} & B_{|\Omega_i} \\ B_{\mathrm{s}|\Omega_i} & B_{\mathrm{f}|\Omega_i} & \end{bmatrix}, \tag{5.31}$$

which is a restriction of the global stiffness matrix K onto the patch Ω_i. This geometric multigrid method shows the accuracy and robustness with respect to the constitutive models.

5.3.2 A Dirichlet–Neumann smoother

In [66], a geometric multigrid preconditioner within the GMRES (Generalized Minimal Residual [69]) method is used in solving the linearized FSI system of algebraic equations. A smoothing step consists of approximations of the structure Neumann and fluid Dirichlet subproblems on the mesh level l:

$$x_l^{k+1} = \mathbf{S}_l(x^k, b_l), \quad k = 1, \ldots, \nu, \tag{5.32}$$

where $\mathbf{S}_l$ denotes the corresponding smoothing operator. In fact, the Neumann and Dirichlet subproblem approximations correspond to the dynamic (balance of surface traction) and kinematic (no-slip) interface conditions, respectively. For approximation of each subproblem in the smoothing step, several preconditioned Richardson iterations are applied. In particular, a stabilized incomplete lower-upper decomposition of the stiffness matrix for each subproblem is used as a corresponding preconditioner. The numerical results have demonstrated the robustness and efficiency with respect to the mesh size and problem parameters, e.g., fluid density, average inflow velocity, shear modulus and Poisson's ratio.

5.4 Algebraic multigrid methods

5.4.1 Smoothed aggregation multigrid methods

In [36], the block Gauss–Seidel preconditioner combined with approximate block inverses for respective fluid, mesh and structure subproblems by algebraic multigrid (AMG) methods are used to solve the FSI system with the Jacobian matrix:

$$K = \begin{bmatrix} S & & S^{\mathrm{SF}} \\ & G & G^{\mathrm{GF}} \\ F^{\mathrm{FS}} & F^{\mathrm{FG}} & F \end{bmatrix}, \tag{5.33}$$

where on the main diagonal, S, G and F represent the stiffness matrix from the structure, mesh and fluid subproblem, respectively, the off-diagonal blocks the coupling matrix between two different fields indicated with the superscriptions S-structure, F-fluid and G-mesh. In particular, the preconditioner P_{BGS} has the following form:

$$P_{\mathrm{BGS}} = \begin{bmatrix} S & & \\ & G & \\ F^{\mathrm{FS}} & F^{\mathrm{FG}} & F \end{bmatrix}. \tag{5.34}$$

As pointed out in [36], the preconditioner P_{BGS} only models a one-way interaction from the structure to the fluid without feedback from the fluid to the structure. To approximate the inverses of subproblems in the preconditioning steps, in particular, the so-called smoothed aggregation multigrid for symmetric positive definite elasticity equations [56, 83, 84] and the Petrov–Galerkin smoothed algebraic multigrid for nonsymmetric Navier–Stokes equations [70] are used.

Furthermore, a truly monolithic algebraic multigrid method was developed for solving the FSI system of equations in [36]. The combined level transfer operators P_l^{FSI} and R_l^{FSI} using the individual prolongation and restriction operators for each subproblem on a level l are constructed as

$$P_l^{FSI} = \begin{bmatrix} P_l^{\mathrm{S}} & & \\ & P_l^{\mathrm{G}} & \\ & & P_l^{\mathrm{F}} \end{bmatrix}, \quad R_l^{FSI} = \begin{bmatrix} R_l^{\mathrm{S}} & & \\ & R_l^{\mathrm{G}} & \\ & & R_l^{\mathrm{F}} \end{bmatrix}, \tag{5.35}$$

where P_l^{S}, P_l^{G} and P_l^{F} are prolongation operators on level l for the structure, mesh and fluid subproblem, respectively; R_l^{S}, R_l^{G} and R_l^{F} are the corresponding restriction operators. The matrix on the coarse level $l+1$ is then built as:

$$\begin{aligned} K_{l+1} &= \begin{bmatrix} R_l^{\mathrm{S}} & & \\ & R_l^{\mathrm{G}} & \\ & & R_l^{\mathrm{F}} \end{bmatrix} \begin{bmatrix} S & & S^{\mathrm{SF}} \\ & G & G^{\mathrm{GF}} \\ F^{\mathrm{FS}} & F^{\mathrm{FG}} & F \end{bmatrix}_l \begin{bmatrix} P_l^{\mathrm{S}} & & \\ & P_l^{\mathrm{G}} & \\ & & P_l^{\mathrm{F}} \end{bmatrix} \\ &= \begin{bmatrix} R_l^{\mathrm{S}} S_l P_l^{\mathrm{S}} & & R_l^{\mathrm{S}} S_l^{\mathrm{SF}} P_l^{\mathrm{F}} \\ & R_l^{\mathrm{G}} G_l P_l^{\mathrm{G}} & R_l^{\mathrm{G}} G_l^{\mathrm{GF}} P_l^{\mathrm{F}} \\ R_l^{\mathrm{F}} F_l^{\mathrm{FS}} P_l^{\mathrm{S}} & R_l^{\mathrm{F}} F_l^{\mathrm{FG}} P_l^{\mathrm{G}} & R_l^{\mathrm{F}} F_l P_l^{\mathrm{F}} \end{bmatrix} . \end{aligned} \tag{5.36}$$

The damped Richardson iteration with block Gauss–Seidel preconditioner $\tilde{P}_{\mathrm{BGS}}$ is used as a smoother: For $k = 0$,

$$x^{k+1} = x^k + \omega_k \tilde{P}_{\mathrm{BGS}}^{-1}(b - Kx^k) \tag{5.37}$$

with the Richardson iteration index k and a level-dependent damping parameter ω_k. In the preconditioning step, the inverse approximations for each subproblem are realized by simple smoothing steps, e.g., *ILU* or Gauss–Seidel. The monolithic FSI solution method combined with the new monolithic algebraic multigrid shows the best performance.

In [58], a novel generalized monolithic solution procedure is proposed, which maintains consistency in the nonconforming temporal and spatial discretizations. The above algebraic multigrid methods are utilized in solving the condensed monolithic linear system by eliminating the unknown Lagrange multipliers λ discretized by dual mortar shape functions [91].

In [35], these block Gauss-Seidel and monolithic algebraic multigrid preconditioners are employed in the preconditioners for the coupled system of the lung model problem, which contains the standard FSI system and additional algebraic constraints [96]. Using the idea of semi-implicit methods for the pressure-linked equations (SIMPLE, e.g., [22]) of the coupled system, the predictor equation, arising there, actually corresponds to the FSI system and is approximated by these FSI preconditioners.

5.4.2 Matrix-graph-based algebraic multigrid methods

In [54], a similar technique has been employed in developing algebraic multigrid methods for solving the linearized FSI system using three hyperelastic structure models, namely neo-Hookean, Mooney–Rivlin and anisotropic two-layer thick-walled artery (see [1, 8, 46]). In particular, the W-cycles of the special matrix-graph-based algebraic multigrid methods for individual fields have been applied in each smoothing step [40, 50, 87, 88, 95]. Furthermore, algebraic multilevel methods have been developed for solving the coupled FSI system [2, 3, 51, 86]. These methods show robustness with respect to the mesh size and different structure models.

5.5 Domain decomposition preconditioners

5.5.1 A restricted additive Schwarz preconditioner

In [93], a fully implicit domain-decomposition-based Newton–Krylov–Schwarz method is proposed for solving the nonlinear coupled FSI system of algebraic equations in 3D. In solving the right-preconditioned Jacobian system with the GMRES method, a restricted additive Schwarz preconditioner [21] is used, which is an overlapping variant of the classical additive Schwarz preconditioner for general sparse linear systems:

$$P_{\mathrm{RAS}}^{-1} = \sum_{l=1}^{N} (R_l^0)^T B_l^{-1} R_l \,, \tag{5.38}$$

where R_l denotes the restriction operator (a mapping of global unknowns to the corresponding ones on the local overlapping subdomain), $B_l = R_l K R_l^T$ the local subdomain Jacobian, and R_l^0 the restrictions to the unknowns on the nonoverlapping subdomain. Such a preconditioner requires a particular partition of the finite element meshes into overlapping subdomains containing both the fluid and structure elements. Note that the overlapping regions are only used to construct the subdomain Jacobians and provide information to solve the local subproblems, but not considered in the prolongation procedure. Although increasing overlapping size will reduce iteration numbers of the GMRES method, the computational time may increase due to higher local subproblem solving cost. The preconditioner has shown parallel scalability up to 2048 processors and robustness with respect to material parameters, e.g., fluid density, structure density, Young's modulus and Poisson ratio.

The same technique has been used in earlier work [11] for solving the FSI problems in 2D, which shows the robustness with respect to some material parameters, and parallel scalability up to hundreds of processors.

5.5.2 A two-level Schwarz preconditioner

In [12], a two-level Newton–Krylov–Schwarz algorithm is used to solve the nonlinear FSI system at each time step, which is given in an abstract form:

$$G_h(x_h) = 0 \,. \tag{5.39}$$

The solution of the analogous coarse grid problem

$$G_H(x_H) = 0 \tag{5.40}$$

is taken as a good initial guess for solving the nonlinear problem (5.39) on the fine mesh. We mention that the nonlinear system (5.40) on the coarse mesh is solved by an inexact Newton method. To solve the linearized system on the coarse and fine meshes,

Tab. 5.1: Three finite element meshes.

Level	#Nod	#Tet	#Dof
L_1	1913	8400	14,073
L_2	13,166	67,200	93,114
L_3	97,292	537,600	670,944

the one-level restricted additive Schwarz preconditioner P_H as (5.38) and the two-level hybrid preconditioner P_h [73, 78] are employed, respectively. In particular, on the fine mesh level, a multiplicative preconditioner combining both the coarse-level and fine-level ones is proposed:

$$P_h^{-1} = I_H^h B_0^{-1} I_h^H + \sum_{l=1}^{N} (R_l^0)^T B_l^{-1} R_l \,, \tag{5.41}$$

where the second term is the restricted additive Schwarz preconditioner on the fine level, I_h^H denotes the restriction from the fine grid to the coarse one, $I_H^h = (I_h^H)^T$ the corresponding interpolation operator from the coarse grid to the fine one, and B_0^{-1} the solution procedure on the coarse grid. The preconditioner has demonstrated robustness with respect to the structure's Young's modulus and the time-step size, and the nearly perfect weak scaling up to some thousand processors.

5.6 Some numerical studies

In this section we test some of monolithic FSI solvers that were developed in our previous work [54, 55]. We solve a 3D benchmark problem using block Gauss–Seidel, complete LDU factorization and AMG preconditioned GMRES solvers. The computational FSI domain and the mesh, generated by Netgen [71], are illustrated in Fig. 5.2. Three mesh levels are used to test the robustness of the preconditioners; see mesh information in Table 5.1, where we provide the number of nodes (#Nod), tetrahedral elements (#Tet) and degrees of freedom (#Dof).

In this experiment, we extend the FSI2 test, which was originally proposed in [81] as a two-dimensional FSI test, to three dimensions. The geometry configuration is il-

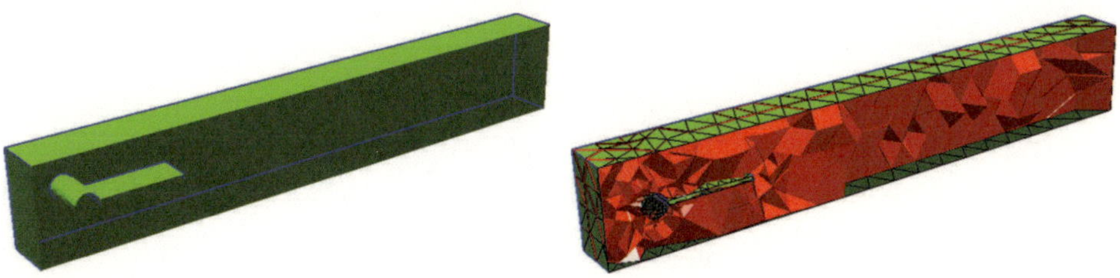

Fig. 5.2: Geometry (left) and mesh (right) for half of the computational FSI domain.

lustrated in Fig. 5.3. The structure domain is an elasticity plate with length 40 cm, 30 cm and 2 cm in the x-, y- and z-direction, respectively, attached to a cylinder. The fluid domain is a cuboid with length 250 cm, 41 cm and 41 cm in the three directions, respectively, cut by the cylinder along the y-direction. The cylinder has length 41 cm in the y-direction and radius 5 cm for the rest, which is fixed as in Turek's benchmark problem. We mention that the elasticity plate is a bit away from the center of the cuboid in either y- or z-direction.

For the structure, we use the nonlinear hyperelastic model of the St. Venant–Kirchhoff material with density $\rho_s = 1\text{g/cm}^3$ and Lamé constants $\mu_s = 1.15 \times 10^6$ dyn/cm^2 and $\lambda_s = 1.73 \times 10^6$ dyn/cm^2. The structure is fixed on the touching part with the cylinder, i.e., Γ_d in Fig. 5.3. The rest of the boundary is the interface Γ^0.

We use the incompressible Navier–Stokes equations to model the fluid. The fluid has density $\rho_f = 1.0\text{g/cm}^3$ and kinematic viscosity $\nu = 0.035\text{cm}^2/\text{s}$. A smooth parabolic velocity profile in time is prescribed at the inflow:

$$u(0, y, z, t) = \begin{cases} 1.5\bar{U}\frac{yz(H-y)(H-z)}{(\frac{H}{2})^2(\frac{H}{2})^2}\frac{1-\cos(\frac{\pi}{2}t)}{2}\ \text{cm/s} & \text{if } t < 2.0 \\ \\ 1.5\bar{U}\frac{yz(H-y)(H-z)}{(\frac{H}{2})^2(\frac{H}{2})^2}\ \text{cm/s} & \text{if } t \geq 2.0 \end{cases} \tag{5.42}$$

with $H = 41$ and $\bar{U} = 100$. At the outflow, we use the homogeneous Neumann boundary condition. The no-slip condition is prescribed on the rest of the fluid domain boundary.

The structure is initially at rest. The initial condition for the fluid is homogeneous zero.

For spatial and temporal discretization, and linearization of the nonlinear coupled FSI system, we refer to our work in [54, 55]. The time-step size is set to 0.625 ms.

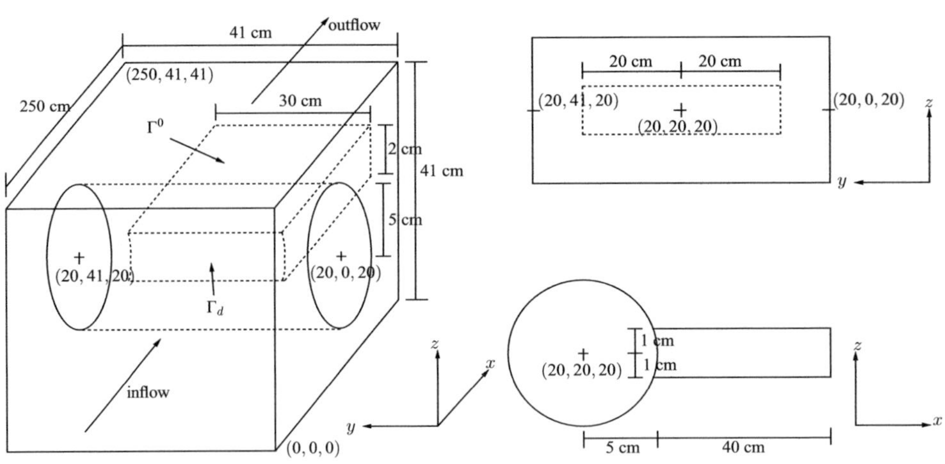

Fig. 5.3: A schematic illustration of the geometry and boundary configuration.

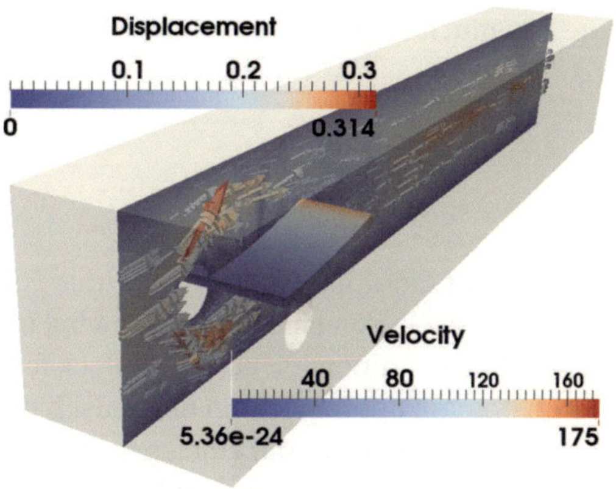

Fig. 5.4: Visualization of fluid velocity field and structure displacement.

The velocity fields and structure deformation (enlarged by a factor of 16) at time $t = 1.625$ s are plotted in Fig. 5.4.

For illustration purposes, we only show the iteration numbers and cost in second (s) for solving one linearized FSI system of equations. We stop solving the system when the relative residual error of the preconditioned system in the GMRES method is reduced by a factor of 10^7. For the performance of the complete simulation of some 3D benchmarks, we refer to the detailed numerical studies in [55]. In Table 5.2, we record the number of iterations and computational time in seconds (the value in brackets) using three preconditioned GMRES solvers on the three mesh levels.

It is easy to see from our numerical experiments that the LDU factorization preconditioned GMRES solver shows better performance than the other two. It is worth mentioning that the cost of both GS_GMRES and LDU_GMRES solvers scales well with respect to the number of degrees of freedom and iteration numbers. For the AMG preconditioned GMRES solver, we observe a better scaling factor than expected. That is because we have the same problem size on the coarsest AMG level for all three mesh levels, which is relatively large (around 2500×2500) and dominates computational cost on the level L_1. The problem size on the coarsest level is not easily reduced fur-

Tab. 5.2: Number of iterations and cost in second (s) of the solvers: GS_GMRES (GMRES solver preconditioned by block Gauss–Seidel), AMG_GMRES (GMRES solver preconditioned by AMG) and LDU_GMRES (GMRES solver preconditioned by complete LDU factorization.

	GS_GMRES	AMG_GMRES	LDU_GMRES
L_1	23 (26.1 s)	15 (262.1 s)	4 (7.5 s)
L_2	54 (493.6 s)	12 (461.6 s)	7 (183.3 s)
L_3	164 (12,133.3 s)	16 (3107.9 s)	9 (2426.3 s)

ther, since in the benchmark, the structure has much fewer degrees of freedom than the fluid one.

5.7 Conclusions

In this work, we present a short overview of recent developments of monolithic FSI solvers in the last decade, which are developed from classical methods for solving individual field problems. Several well-developed preconditioners, based on block factorization, geometric and algebraic multigrid, domain decomposition methods, have been discussed. These methods have shown their robustness and efficiency with respect to temporal and spatial discretization parameters as well as material parameters for a certain class of FSI problems. The topic is relatively new and is still under further investigation, although a lot of progress has been made by many research groups.

Acknowledgment: The second author would like to thank for the support of the working environment provided by Johann Radon Institue for Computational and Applied Mathematics (RICAM) as well as the Institut für Numerische Mathematik, Technische Universität Graz for preparing this book chapter.

Bibliography

[1] C. Augustin, G. Holzapfel, and O. Steinbach. Classical and all-floating FETI methods for the simulation of arterial tissues. *Int. J. Numer. Meth. Engng.*, 99(4):290–312, 2014.

[2] O. Axelsson and P. Vassilevski. Algebraic multilevel preconditioning methods. I. *Numer. Math.*, 56(2-3):157–177, 1989.

[3] O. Axelsson and P. Vassilevski. Algebraic multilevel preconditioning methods, II. *SIAM J. Numer. Anal.*, 27(6):1569–1590, 1990.

[4] S. Badia, F. Nobile, and C. Vergara. Robin-Robin preconditioned Krylov methods for fluid-structure interaction problems. *Comput. Methods. Appl. Mech. Engrg.*, 198:1768–2784, 2009.

[5] S. Badia, A. Quaini, and A. Quarteroni. Modular vs. non-modular preconditioners for fluid-structure systems with large added-mass effect. *Comput. Methods Appl. Mech. Engrg.*, 197(49–50):4216–4232, 2008.

[6] S. Badia, A. Quaini, and A. Quarteroni. Splitting methods based on algebraic factorization for fluid-structure interaction. *SIAM J. Sci. Comput.*, 30(4):1778–1805, 2008.

[7] D. Balzani, S. Deparis, S. Fausten, D. Forti, A. Heinlein, A. Klawonn, A. Quarteroni, O. Rheinbach, and J. Schröder. Numerical modeling of fluid-structure interaction in arteries with anisotropic polyconvex hyperelastic and anisotropic viscoelastic material models at finite strains. *Int. J. Numer. Methods Biomed. Engrg.*, 2015. DOI:10.1002/cnm.2756.

[8] D. Balzani, P. Neff, J. Schröder, and G. Holzapfel. A polyconvex framework for soft biological tissues. Adjustment to experimental data. *Int. J. Solids Struct.*, 43(20):6052–6070, 2006.

[9] J. Banks, W. Henshaw, and D. Schwendeman. An analysis of a new stable partitioned algorithm for fsi problems. part i: Incompressible flow and elastic solids. *J. Comput. Phys.*, 269:108–137, 2014.

[10] J. Banks, W. Henshaw, and D. Schwendeman. An analysis of a new stable partitioned algorithm for fsi problems. part ii: Incompressible flow and structural shells. *J. Comput. Phys.*, 268:399–416, 2014.

[11] A. Barker and X. Cai. Scalable parallel methods for monolithic coupling in fluid-structure interaction with application to blood flow modeling. *J. Comput. Phys.*, 229(3):642–659, 2010.

[12] A. Barker and X. Cai. Two-level newton and hybrid schwarz preconditioners for fluid-structure interaction. *SIAM J. Sci. Comput.*, 32(4):2395–2417, 2010.

[13] Y. Bazilevs, V. Calo, T. Hughes, and Y. Zhang. Isogeometric fluid-structure interaction: theory, algorithms, and computations. *Comput. Mech.*, 43(1):3–37, 2008.

[14] E. Béchet, N. Moës, and B. Wohlmuth. A stable Lagrange multiplier space for stiff interface conditions within the extended finite element method. *Int. J. Numer. Meth. Engng.*, 78(8):931–954, 2009.

[15] M. Behr and T. Tezduyar. Finite element solution strategies for large-scale flow simulations. *Comput. Methods Appl. Mech. Engrg.*, 112(1–4):3–24, 1994.

[16] M. Benzi, G. Golub, and J. Liesen. Numerical solution of saddle point problems. *Acta Numerica*, 14:1–137, 2005.

[17] J. Bonet and R. Wood. *Nonlinear Continuum Mechanics for Finite Element Analysis*. Cambridge University Press, New York, 2008.

[18] M. Bukač, S. Čanić, R. Glowinski, B. Muha, and A. Quaini. A modular, operator-splitting scheme for fluid-structure interaction problems with thick structures. *Int. J. Numer. Meth. Fluids*, 74(8):577–604, 2014.

[19] E. Burman and M. Fernández. Stabilization of explicit coupling in fluid-structure interaction involving fluid incompressibility. *Comput. Methods Appl. Mech. Engrg.*, 198(5–8):766–784, 2009.

[20] E. Burman and M. Fernández. Explicit strategies for incompressible fluid-structure interaction problems: Nitsche type mortaring versus robin-robin coupling. *Int. J. Numer. Meth. Engng.*, 97(10):739–758, 2014.

[21] X. Cai and M. Sarkis. A restricted additive schwarz preconditioner for general sparse linear systems. *SIAM J. Sci. Comput.*, 21(2):792–797, 1999.

[22] L. Caretto, A. Gosman, S. Patankar, and D. Spalding. *Two calculation procedures for steady, three-dimensional flows with recirculation*, pages 60–68. Springer Berlin Heidelberg, Berlin, Heidelberg, 1973.

[23] P. Causin, J. Gerbeau, and F. Nobile. Added-mass effect in the design of partitioned algorithms for fluid-structure problems. *Comput. Methods Appl. Mech. Engrg.*, 194(42–44):4506–4527, 2005.

[24] A. Chorin. Numerical solution of the Navier-Stokes equations. *Math. Comput.*, 22(104):745–762, 1968.

[25] P. Crosetto, S. Deparis, G. Fourestey, and A. Quarteroni. Parallel algorithms for fluid-structure interaction problems in haemodynamics. *SIAM J. Sci. Comput.*, 33(4):1598–1622, 2011.

[26] P. Crosetto, P. Reymond, S. Deparis, D. Kontaxakis, N. Stergiopulos, and A. Quarteroni. Fluid-structure interaction simulation of aortic blood flow. *Computers & Fluids*, 43(1):46–57, 2011.

[27] S. Deparis, M. Discacciati, G. Fourestey, and A. Quarteroni. Fluid-structure algorithms based on Steklov-Poincaré operators. *Comput. Methods Appl. Mech. Engrg.*, 195(41-43):5797–5812, 2006.

[28] J. Donea, A. Huerta, J. Ponthot, and A. Ferran. Arbitrary Lagrangian-Eulerian methods. In E. Stein, R. Borst, and T. Hughes, editors, *The Encyclopedia of Computational Mechanics*, volume 1, pages 413–437. Wiley& Sons, Ltd, 2004.

[29] H. Elman. Preconditioning for the steady-state navier-stokes equations with low viscosity. *SIAM J. Sci. Comput.*, 20(4):1299–1316, 1999.

[30] H. Elman, D. Silvester, and A. Wathen. *Finite Elements and Fast Iterative Solvers with Applications in Incompressible Fluid Dynamics.* Oxford University Press, New York, 2006.

[31] M. Fernández, J. Gerbeau, and C. Grandmont. A projection semi-implicit scheme for the coupling of an elastic structure with an incompressible fluid. *Int. J. Numer. Meth. Engng.*, 69(4):794–821, 2007.

[32] M. Fernández and M. Moubachir. A Newton method using exact jacobians for solving fluid-structure coupling. *Computers & Structures*, 83(2-3):127–142, 2005.

[33] L. Formaggia and F. Nobile. A stability analysis for the arbitrary Lagrangian Eulerian formulation with finite elements. *East-West J. Numer. Math.*, 7:105–132, 1999.

[34] C. Förster, W. Wall, and E. Ramm. Artificial added mass instabilities in sequential staggered coupling of nonlinear structures and incompressible viscous flows. *Comput. Methods Appl. Mech. Engrg.*, 196(7):1278–1293, 2007.

[35] V. Francesc, C. Roth, L. Yoshihara, and W. Wall. Efficient solvers for coupled models in respiratory mechanics. *Int. J. Numer. Meth. Biomed. Engng.*, 2016. DOI: 10.1002/cnm.2795.

[36] M. Gee, U. Küttler, and W. Wall. Truly monolithic algebraic multigrid for fluid-structure interaction. *Int. J. Numer. Meth. Engng.*, 85(8):987–1016, 2011.

[37] L. Gerardo-Giorda, F. Nobile, and C. Vergara. Analysis and optimization of robin-robin partitioned procedures in fluid-structure interaction problems. *SIAM J. Numer. Anal.*, 48(6):2091–2116, 2010.

[38] C. Greif and D. Schötzau. Preconditioners for saddel point linear systems with highly singular (1,1) blocks. *Elec. Trans. Numer. Ana.*, 22:114–121, 2006.

[39] C. Greif and D. Schötzau. Preconditioners for the discretized time-harmonic Maxwell equations in mixed form. *Numer. Lin. Alg. Appl.*, 14(4):281–297, 2007.

[40] G. Haase and U. Langer. *Modern Methods in Scientific Computing and Applications*, volume 75 of *NATO Science Series II. Mathematics, Physics and Chemistry*, chapter Multigrid Methods: From Geometrical to Algebraic Versions, pages 103–154. Kluwer Academic Press, Dordrecht, 2002.

[41] W. Hackbusch. *Multi-Grid Methods and Applications*. Springer, Berlin, 2003.

[42] P. Hansbo. The characteristic streamline diffusion method for the time-dependent incompressible navier-stokes equations. *Comput. Methods Appl. Mech. Engrg.*, 99(2):171–186, 1992.

[43] M. Heil. An efficient solver for the fully coupled solution of large-displacement fluid-structure interaction problems. *Comput. Methods Appl. Mech. Engrg.*, 193(1–2):1–23, 2004.

[44] M. Heil, A. Hazel, and J. Boyle. Solvers for large-displacement fluid-structure interaction problems: segregated versus monolithic approaches. *Comput. Mech.*, 43(1):91–101, 2008.

[45] G. Holzapfel. *Nonlinear Solid Mechanics: A Continuum Approach for Engineering*. John Wiley & Sons, Chichester, 2000.

[46] G. Holzapfel, T. Gasser, and R. Ogden. A new constitutive framework for arterial wall mechanics and a comparative study of material models. *J. Elasticity*, 61(1-3):1–48, 2000.

[47] J. Hron and S. Turek. A monolithic FEM/multigrid solver for an ALE formulation of fluid-structure interaction with applications in biomechanics. In *Fluid-Structure Interaction: Modelling, Simulation, Optimisation (ed. by H. Bungartz and M. Schäfer), Lecture Notes in Computational Science and Engineering, vol. 53*, pages 146–170, Berlin, Heidelberg, 2006. Springer.

[48] B. Hübner, E. Walhorn, and D. Dinkler. A monolithic approach to fluid-structure interaction using space-time finite elements. *Comput. Methods Appl. Mech. Engrg.*, 193(23–26):2087–2104, 2004.

[49] T. Hughes, W. Liu, and T. Zimmermann. Lagrangian-Eulerian finite element formulation for incompressible viscous flows. *Comput. Methods Appl. Mech. Engrg.*, 29(3):329–349, 1981.

[50] F. Kickinger. Algebraic multigrid for discrete elliptic second-order problems. In *Multigrid Methods V. Proceedings of the 5th European Multigrid conference (ed. by W. Hackbush), Lecture*

Notes in Computational Sciences and Engineering, vol. 3, pages 157–172, Berlin, Heidelberg, 1998. Springer.

[51] J. Kraus and S. Margenov. *Robust Algebraic Multilevel Methods and Algorithms*, volume 5 of *Radon Series on Computational and Applied Mathematics*. Walter de Gruyter, Berlin, New York, 2009.

[52] E. Kuhl, S. Hulshoff, and R. de Borst. An arbitrary Lagrangian Eulerian finite-element approach for fluid-structure interaction phenomena. *Int. J. Numer. Meth. Engng.*, 57:117–142, 2003.

[53] U. Küttler and W. Wall. Fixed-point fluid-structure interaction solvers with dynamic relaxation. *Comput. Mech.*, 43:61–72, 2008.

[54] U. Langer and H. Yang. Numerical simulation of fluid-structure interaction problems with hyperelastic models: A monolithic approach. *Math. Comput. Sim.*, 2016. DOI: 10.1016/j.matcom.2016.07.008.

[55] U. Langer and H. Yang. Robust and efficient monolithic fluid-structure-interaction solvers. *Int. J. Numer. Meth. Engng.*, 108(4):303–325, 2016.

[56] J. Mandel, M. Brezina, and P. Vaněk. Energy optimization of algebraic multigrid bases. *Computing*, 62(3):205–228, 1999.

[57] K. Mardal and R. Winther. Preconditioning discretizations of systems of partial differential equations. *Numer. Linear Algebra Appl.*, 18(1):1–40, 2011.

[58] M. Mayr, T. Klöppel, W. Wall, and M. Gee. A temporal consistent monolithic approach to fluid-structure interaction enabling single field predictors. *SIAM J. Sci. Comput.*, 37(1):B30–B59, 2015.

[59] M. Mehl, B. Uekermann, H. Bijl, D. Blom, B. Gatzhammer, and A. van Zuijlen. Parallel coupling numerics for partitioned fluid-structure interaction simulations. *Comput. Math. Appl.*, 71(4):869–891, 2016.

[60] H. Moran and R. Ohayon. *Fluid-Structure Interaction: Applied Numerical Methods*. John Wiley & Sons, 1995.

[61] R. Muddle, M. Mihajlović, and M. Heil. An efficient preconditioner for monolithically-coupled large-displacement fluid-structure interaction problems with pseudo-solid mesh updates. *J. Comput. Phys.*, 231(21):7315–7334, 2012.

[62] A. Quarteroni, F. Saleri, and A. Veneziani. Factorization methods for the numerical approximation of navier-stokes equations. *Comput. Methods Appl. Mech. Engrg.*, 188(1–3):505–526, 2000.

[63] A. Quarteroni and A. Valli. *Domain Decomposition Methods for Partial Differential Equations*. Oxford Science Publications, UK, 1999.

[64] M. Razzaq, H. Damanik, J. Hron, A. Ouazzi, and S. Turek. FEM multigrid techniques for fluid-structure interaction with application to hemodynamics. *Appl. Numer. Math.*, 62(9):1156–1170, 2012.

[65] T. Rees and C. Greif. A preconditioner for linear systems arising from interior point optimization methods. *SIAM J. Sci. Comput.*, 29(5):1992–2007, 2007.

[66] T. Richter. A monolithic geometric multigrid solver for fluid-structure interactions in ale formulation. *Int. J. Numer. Meth. Engng.*, 104(5):372–390, 2015.

[67] J. Ruge and K. Stüben. Algebraic multigrid. In S. McCormick, editor, *Multigrid Methods*, volume 3 of *Frontiers in Applied Mathematics*, pages 73–130. SIAM, Philadelphia, PA, 1987.

[68] Y. Saad. *Iterative Methods for Sparse Linear Systems*. PWS Publishing, Boston MA, 1996.

[69] Y. Saad and M. Schultz. Gmres: A generalized minimal residual algorithm for solving nonsymmetric linear systems. *SIAM J. Sci. Stat. Comput.*, 7(3):856–869, 1986.

[70] M. Sala and R. Tuminaro. A new petrov-galerkin smoothed aggregation preconditioner for nonsymmetric linear systems. *SIAM J. Sci. Comput.*, 31(1):143–166, 2008.

[71] J. Schöberl. NETGEN-An advancing front 2D/3D-mesh generator based on abstract rules. *Comput. Visual. Sci.*, 1:41–52, 1997.

[72] D. Silvester, H. Elman, D. Kay, and A. Wathen. Efficient preconditioning of the linearized navier-stokes equations for incompressible flow. *J. Comput. Appl. Math.*, 128(1–2):261–279, 2001.

[73] B. Smith, P. Bjorstad, and W. Gropp. *Domain Decomposition: Parallel Multilevel Methods for Elliptic Partial Differential Equations*. Cambridge University Press, UK, 1996.

[74] K. Stüben. A review of algebraic multigrid. *J. Comput. Appl. Math.*, 128(1–2):281–309, 2001.

[75] P. L. Tallec and J. Mouro. Fluid structure interaction with large structural displacements. *Comput. Methods Appl. Mech. Engrg.*, 190(24–25):3039–3067, 2001.

[76] R. Teman. Sur l'approximation de la solution des équations de Navier-Stokes par la méthode des pas fractionaires (I). *Arch. Ration. Mech. Anal.*, 32:135–153, 1969.

[77] T. Tezduyar, M. Behr, and J. Liou. A new strategy for finite element computations involving moving boundaries and interfaces–the deforming-spatial-domain/space-time procedure: I. the concept and the preliminary numerical tests. *Comput. Methods Appl. Mech. Engrg.*, 94(3):339–351, 1992.

[78] A. Toselli and O. Widlund. *Domain Decomposition Methods–Algorithms and Theroy*. Springer-Verlag, Berlin, 2005.

[79] P. Tricerri, L. Dedé, S. Deparis, A. Quarteroni, A. Robertson, and A. Sequeira. Fluid-structure interaction simulations of cerebral arteries modeled by isotropic and anisotropic constitutive laws. *Comput. Mech.*, 55(3):479–498, 2015.

[80] S. Turek. *Efficient Solvers for Incompressible Flow Problems*. Springer, Heidelberg, 1999.

[81] S. Turek and J. Hron. *Proposal for Numerical Benchmarking of Fluid-Structure Interaction between an Elastic Object and Laminar Incompressible Flow*, pages 371–385. Springer Berlin Heidelberg, Berlin, Heidelberg, 2006.

[82] H. van der Vorst. Bi-cgstab: A fast and smoothly converging variant of bi-cg for the solution of nonsymmetric linear systems. *SIAM J. Sci. Stat. Comput.*, 13(2):631–644, 1992.

[83] P. Vaněk, M. Brezina, and J. Mandel. Convergence of algebraic multigrid based on smoothed aggregation. *Numer. Math.*, 88(3):559–579, 2001.

[84] P. Vaněk, J. Mandel, and M. Brezina. Algebraic multigrid by smoothed aggregation for second and fourth order elliptic problems. *Computing*, 56(3):179–196, 1996.

[85] S. Vanka. Block-implicit multigrid solution of navier-stokes equations in primitive variables. *J. Comput. Phys.*, 65(1):138–158, 1986.

[86] P. Vassilevski. *Multilevel Block Factorization Preconditioners*. Springer, Heidelberg, 2008.

[87] M. Wabro. Coupled algebraic multigrid methods for the Oseen problem. *Comput. Visual. Sci.*, 7(3-4):141–151, 2004.

[88] M. Wabro. AMGe–coarsening strategies and application to the Oseen equations. *SIAM J. Sci. Comput.*, 27(6):2077–2097, 2006.

[89] W. Wall, D. Mok, and E. Ramm. Partitioned analysis approach of the transient coupled response of viscous fluids and flexible structure. In *Solids, Structures and Coupled Problems in Engineering (ed. by W. Wunderlich), Proceedings of the European Conference on Computational Mechanics ECCM '99, Munich*, 1999.

[90] T. Wick. Fluid-structure interactions using different mesh motion techniques. *Comput. Structures*, 89(13-14):1456–1467, 2011.

[91] B. Wohlmuth. A mortar finite element method using dual spaces for the Lagrange multiplier. *SIAM J. Numer. Anal.*, 38(3):989–1012, 2000.

[92] P. Wriggers. *Nonlinear Finite Element Methods*. Springer, Heidelberg, 2008.

[93] Y. Wu and X. Cai. A fully implicit domain decomposition based ale framework for three-dimensional fluid-structure interaction with application in blood flow computation. *J. Comput. Physics*, 258:524–537, 2014.

[94] J. Xu and K. Yang. Well-posedness and robust preconditioners for discretized fluid-structure interaction systems. *Comput. Methods Appl. Mech. Engrg.*, 292:69–91, 2015.

[95] H. Yang and W. Zulehner. Numerical simulation of fluid-structure interaction problems on hybrid meshes with algebraic multigrid methods. *J. Comput. Appl. Math.*, 235(18):5367–5379, 2011.

[96] L. Yoshihara, C. Roth, and W. Wall. Fluid-structure interaction including volumetric coupling with homogenised subdomains for modeling respiratory mechanics. *Int. J. Numer. Meth. Biomed. Engng.*, 2016. DOI: 10.1002/cnm.2812.

[97] W. Zulehner. Nonstandard norms and robust estimates for saddle point problems. *SIAM J. Matrix Anal. Appl.*, 32(2):536–560, 2011.

Daniel Jodlbauer and Thomas Wick

6 A monolithic FSI solver applied to the FSI 1,2,3 benchmarks

Abstract: Fluid-structure interaction problems have a wide range of applications, but their efficient solution – in particular of the linear systems – remains challenging. In this work, we provide all details necessary for a monolithic arbitrary Lagrangian–Eulerian (ALE) implementation using the finite element library deal.II. Moreover, we show different ways of incorporating the continuity conditions on the interface on the discrete level. To actually solve the arising linear systems, we develop a preconditioner based on an approximate block-wise LU-factorization, splitting the coupled system of equations into its natural fluid, solid and mesh subproblems. Alongside these developments, we briefly review various linear solver techniques. Numerical results illustrate the robust convergence with respect to different material parameters and mesh-size h, but with an acceptable dependence on the time-step size Δt. Furthermore, we observe that this iterative approach outperforms direct solvers even for a low number of degrees of freedoms and without parallelization.

Keywords: fluid-structure interaction, arbitrary Lagrangian–Eulerian technique, phase-field fracture, Newton's method, adaptive time-step control, numerical algorithms

Classification: 74R10; 74F10; 65M60; 49M15; 35Q74. See www.ams.org/msc

6.1 Introduction

In this work, we will present and analyze robust preconditioners for nonlinear transient fluid-structure interaction (FSI) problems. As the name suggests, these kind of problems consist of two parts, fluid and solid, that influence each other on a common interface.

A typical example of an FSI problem is the air flow around the wing of an aeroplane. This flow is certainly influenced by the shape of the wing, which in turn creates forces acting on it. These further result in a deformation of the wing which again influences the surrounding flow. This example already indicates one of the difficulties of FSI problems: every change to either the fluid or the solid part automatically influences

Note: This work was supported by the Austrian Academy of Sciences and Ecole Polytechnique in France

Daniel Jodlbauer, Institute of Computational Mathematics, JKU Linz, Altenberger Str. 69, 4040 Linz, Austria, daniel.jodlbauer@dk-compmath.jku.at
Thomas Wick, Centre de Mathématiques Appliquées, Ecole Polytechnique, Université Paris-Saclay, 91128 Palaiseau, France, thomas.wick@polytechnique.edu

https://doi.org/10.1515/9783110494259-006

the other part. FSI problems have also been used for computations in hemodynamics, like blood flow in vessels or the simulation of the human heart [7, 16, 17].

There are basically two different methods of solving such coupled problems:

A very natural attempt follows the description of the wing-flow above: first solve, e.g., the fluid subproblem, independent of the solid part, then compute the reaction of the solid subproblem and apply these changes again to the fluid part. If we continue to do so, we may converge to the solution of the whole FSI problem. This technique is called the “partitioned” or “iterative” approach and is very efficient if we can guarantee a fast convergence. However, in some applications, it may require a lot of iterations in order to converge or even fail completely to do so. In particular, this is the case if the fluid and solid densities are similar, like for many examples in hemodynamics. More details about this so-called added-mass effect and the partitioned method in general can be found in e.g., [10, 11, 25]).

A different class of approaches which overcome the added-mass effect are strongly coupled methods, and here in particular “monolithic” techniques. Here, one attempts to solve the whole coupled FSI problem at once. While this method tends to be more robust, it comes at the price of solving large and ill-conditioned linear systems. For a small number of degrees of freedom (dofs), a direct solver can be used to obtain a solution. However, as the size of the problem increases, direct solvers will require huge amounts of memory and additionally suffer from numerical instabilities. Therefore other solution methods must be considered.

Since direct solvers are going to fail at some point (because of too high memory consumption), we have to consider iterative solvers like the Generalized Minimal Residual Method (GMRES) [33] to solve the occurring linear systems. This class of solvers require less memory storage and are less susceptible to numerical breakdown. The disadvantage of iterative solvers is that they require a good preconditioner to converge in a reasonable number of iterations. While “black-box” preconditioners often work well for simple PDEs (partial differential equations) like the Laplace equation, no such general preconditioner works well enough for the fully coupled fluid-structure interaction problem. Therefore, we are going to have a look at more sophisticated attempts to precondition the occurring linear systems.

Similar developments in FSI solver techniques have been undertaken by various groups in the last 15 years. The most important are [21] who probably developed the first efficient monolithic FSI solver. Shortly after [24] presented several FSI benchmark problems in [9, 10, 37] using also a monolithic solver based on Vanka smoothing. Other solvers have been presented by [3, 38] and [7, 18]. More recently, papers by [26, 27, 31] and also the chapter [28] of this book, which we actually build upon, have been published. The technical differences of our FSI solver in comparison to [27] are rather small:

- A different interface-coupling strategy;
- Splitting the second-order-in-time solid equation into a first-order system;
- Presenting results for the most difficult, large deformation, FSI 2 benchmark.

The biggest difference has been the implementation itself:
- Using the finite element library deal.II [4].

The last point is somewhat important since a large community works in deal.II. Therefore, our developed solver could be added in the near future to deal.II in order to allow others to work with it or to develop further extensions. This is important because engineering problems are becoming more and more complex (as shown in various applications provided in this book) and the reuse of software is indispensable. In fact, in deal.II, many applications in fluid mechanics (step-22, step-35), solid mechanics (step-8, step-18), stationary fluid-structure interaction (step-46, [43]) and so forth have been demonstrated, and thus an FSI solver is of current need. As one example, our solver may be expanded to solving problems in which several multiphysics problems are coupled. Current examples are provided in the two last chapters of this book in which FSI is formulated within a binary-fluid-solid problem and coupling of FSI with phase-field fracture propagation. A final big advantage is the straightforward deal.II methodology to extend to 3D settings such that relevant practical problems can be tackled.

The outline of this chapter is as follows: in Section 6.2, we will state the FSI equations and have a closer look at discretization. The main emphasis of this work focuses on Sections 6.3 and 6.4, where we will describe methods for the solution of the arising linear systems. Finally, we will present the numerical results in Section 6.5.

6.2 FSI equations

First of all, we will briefly describe the FSI equations. These combine the Navier–Stokes equations in arbitrary Lagrangian–Eulerian coordinates (ALE), the solid equations using the STVK (Saint Venant-Kirchhoff) material model and the additional equations for the mesh motion. The ALE transformation introduces the frequently appearing terms $\hat{F} = \hat{I} + \hat{\nabla}\hat{u}$ and $\hat{J} = \det \hat{F}$. Boundary and initial conditions are given as stated in Section 6.5.1.

6.2.1 Domains and boundaries

As usual, we denote by $L^2(\Omega)$ the set of square-integrable functions with scalar-product $(f, g)_\Omega := \int_\Omega f(x)g(x)dx$. By $\langle x, y\rangle_\Gamma$, we indicate the L^2-scalar product over some part of the boundary. For matrix-valued functions, the corresponding integrals are interpreted as $\int_\Omega A : B dx$ with $A : B := \sum_{i,j} A_{ij}B_{ij}$. Further, $H^1(\Omega)$ denotes the usual Sobolev space consisting of L^2-functions with derivative in L^2 as well [1]. The corresponding vector-valued function spaces are denoted by $L^2(\Omega)^d = L^2(\Omega) \times \cdots \times L^2(\Omega)$, and analogously for all other spaces. A list of frequently used abbreviations is given in Table 6.1.

Tab. 6.1: List of abbreviations.

Notation	Description
$.\hat{.}.$	expression given in Lagrangian/ALE coordinates
$\hat{v}_f, \hat{p}_f$	fluid velocity and pressure
$\hat{u}_f$	fluid (mesh) displacement
$\hat{u}_s, \hat{v}_s$	solid displacement and velocity
$\hat{\varphi}_f^v, \hat{\varphi}_f^p, \hat{\varphi}_f^u, \hat{\varphi}_s^u, \hat{\varphi}_s^v$	test functions
$\hat{\mathcal{A}}$	ALE mapping $\hat{\mathcal{A}}(t,\hat{x}) := \hat{x} + \hat{u}$
$\hat{F}$	ALE gradient $\hat{F} := \hat{\nabla}\hat{\mathcal{A}} = \hat{I} + \hat{\nabla}\hat{u}$
$\hat{J}$	ALE determinant $J := \det \hat{F}$
$\hat{\rho}_f, \hat{\rho}_s$	fluid and solid density
ν_f	kinematic viscosity
ν_s	Poisson ratio
λ, μ	Lamé coefficients
$\hat{\Omega}_s, \hat{\Omega}_f$	solid and fluid domains (reference configuration)
$\hat{\Gamma}$	boundary (parts) of a domain
$\hat{\Gamma}_I$	fluid-solid interface
dim	spatial dimension
Δt	time-step size
h	mesh-size $h := \max_K \operatorname{diam}(\mathcal{T}_K)$
$\int dx$	domain integral
$\int ds$	boundary integral
$\partial_x p$	partial derivatives
dt	total time derivative
∇p	gradient of a (scalar-valued) function $(\partial_x p, \partial_y p, \dots)$
∇v	gradient of a (vector-valued) function $(\partial_j v_i)_{i,j=1}^{\dim}$
$\operatorname{div} v$	divergence of a vector-valued function $\sum_{i=1}^{\dim} \partial_i v_i = \operatorname{tr}(\nabla v)$
$\operatorname{div} \sigma$	divergence of a matrix-valued function (component-wise)
Δp	Laplacian of a scalar-valued function $\sum_{i=1}^{\dim} \partial_{i^2} p$
Δv	Laplacian of a vector-valued function $(\Delta v_1, \Delta v_2, \dots)$
$a(u, v)$	bilinear form, linear wrt. both arguments
$A(u)(v, \varphi)$	bilinear wrt. v and φ, nonlinear wrt. to u

For the discretization, we will use continuous, piecewise polynomial functions, precisely

$$Q(k) := \left\{ \sum_{\substack{\alpha \\ \max(\alpha_i) \le k}} c_{\alpha_i} x_i^{\alpha_i}, \; c_{\alpha_i} \in \mathbb{R}, \; \alpha \in \mathbb{N}_0^{\dim} \right\} \tag{6.1}$$

on each element. By DGQ (discontinuous Galerkin Q(k) elements) we denote the respective discontinuous versions, i.e., there are no common dofs between neighboring cells (Figure 6.1).

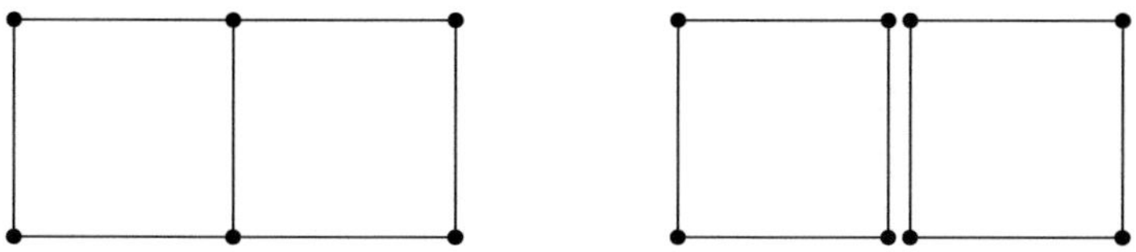

Fig. 6.1: Degrees of freedom for continuous Q(1) and discontinuous DGQ(1) elements.

6.2.2 Equations

Strong formulation

In strong formulation, the equations for $(\hat{v}_f, \hat{v}_s)$, $(\hat{u}_f, \hat{u}_s)$ and $\hat{p}_f$ are given as follows:

$$\begin{aligned}
\hat{J}\hat{\rho}_f\partial_t\hat{v}_f + \hat{J}\hat{\rho}_f\hat{\nabla}\hat{v}_f\hat{F}^{-1}\cdot(\hat{v}_f - \partial_t\hat{\mathcal{A}}) - \operatorname{div}_R(\hat{J}\hat{\sigma}_f\hat{F}^{-T}) &= 0 \quad \text{in } \hat{\Omega}_f\\
\hat{J}\hat{\rho}\ \operatorname{tr}(\hat{\nabla}\hat{v}_f\hat{F}^{-1}) &= 0 \quad \text{in } \hat{\Omega}_f\\
\hat{\rho}_s\partial_t\hat{v}_s - \operatorname{div}_R(\hat{F}\hat{\Sigma}) &= 0 \quad \text{in } \hat{\Omega}_s\\
\hat{\rho}_s\left(\partial_t\hat{u}_s - \hat{v}_f\right) &= 0 \quad \text{in } \hat{\Omega}_s\\
-\alpha_u \operatorname{div}_R(\hat{\sigma}_{MM}) &= 0 \quad \text{in } \hat{\Omega}_f
\end{aligned} \tag{6.2}$$

with the coupling conditions

$$\begin{aligned}
\hat{J}\hat{\sigma}_f\hat{F}^{-T}\hat{n}_f + \hat{F}\hat{\Sigma}_s &= 0 \ \text{ on } \hat{\Gamma}_I\\
\hat{v}_f - \hat{v}_s &= 0 \ \text{ on } \hat{\Gamma}_I\\
\hat{u}_f - \hat{u}_s &= 0 \ \text{ on } \hat{\Gamma}_I
\end{aligned} \tag{6.3}$$

and boundary conditions

$$\begin{aligned}
\hat{v}_f &= 0 \ \text{ on } \hat{\Gamma}_{top} \cup \hat{\Gamma}_{bottom} \cup \hat{\Gamma}_c\\
\hat{v}_f &= g \ \text{ on } \hat{\Gamma}_{in}\\
\hat{u}_f &= 0 \ \text{ on } \partial\hat{\Omega}_f\\
\hat{u}_s &= 0 \ \text{ on } \hat{\Gamma}_{cf}\\
\hat{v}_s &= 0 \ \text{ on } \hat{\Gamma}_{cf}\,.
\end{aligned} \tag{6.4}$$

On the outflow boundary, we employ the do-nothing condition

$$\hat{J}(-\hat{p}_f I + \hat{\rho}_f\hat{v}_f\hat{\nabla}\hat{v}_f\hat{F}^{-1})\hat{F}^{-T} = 0 \ \text{ on } \hat{\Gamma}_{out}\,. \tag{6.5}$$

The solid stress-tensor is given by $\hat{\Sigma}(\hat{E}) = 2\mu\hat{E} + \lambda\ \operatorname{tr}(\hat{E})\hat{I}$, with the nonlinear strain-tensor $\hat{E} = \frac{1}{2}(\hat{\nabla}\hat{u}_s + \hat{\nabla}\hat{u}_s^T + \hat{\nabla}\hat{u}_s\hat{\nabla}\hat{u}_s^T)$ and constant Lamé parameters λ and μ. For the fluid problem we have $\hat{\sigma}_f = -\hat{p}_f I + \hat{\rho}_f\hat{v}_f(\hat{\nabla}\hat{v}_f\hat{F}^{-1} + \hat{F}^{-T}\hat{\nabla}\hat{v}_f^T)$. The mesh motion is described using the nonlinear harmonic extension with $\hat{\sigma}_{MM} = \frac{1}{\hat{J}}\hat{\nabla}\hat{u}_f$, see [34, 42].

The interface conditions (6.3) describe the continuity of stresses, which propagates forces from one domain into the other, the no-slip condition and the continuity of the ALE mapping along the interface.

Weak formulation

For the numerical treatment of partial differential equations, we require a variational formulation of these equations. As usual, we first multiply each equation with a corresponding test function, integrate over the respective reference domain and perform integration by parts. The following paragraphs show the results of this procedure for the fluid, solid and mesh-motion equations.

Special care has to be taken on the interface when defining the test and ansatz functions. For all time-steps n, we have

$$\begin{aligned}(\hat{u}_s, \hat{u}_f) \in V_u := \{\hat{u}_s \in V_u^s(\hat{\Omega}_s)^{\dim}, \hat{u}_f \in H^1(\hat{\Omega}_f)^{\dim} :\ & \hat{u}_s = 0 \text{ on } \partial\hat{\Omega}_s \backslash \hat{\Gamma}_I ,\\ & \hat{u}_f = 0 \text{ on } \partial\hat{\Omega}_f \backslash \hat{\Gamma}_I ,\ \hat{u}_s = \hat{u}_f \text{ on } \hat{\Gamma}_I\} ,\end{aligned}$$

$$\begin{aligned}(\hat{v}_s, \hat{v}_f) \in V_v := \{\hat{v}_s \in V_v^s(\hat{\Omega}_s)^{\dim}, \hat{v}_f \in H^1(\hat{\Omega}_f)^{\dim} :\ & \hat{v}_s = 0 \text{ on } \partial\hat{\Omega}_s \backslash \hat{\Gamma}_I ,\\ & \hat{v}_f = 0 \text{ on } \partial\hat{\Omega}_f \backslash (\hat{\Gamma}_I \cup \hat{\Gamma}_{in}) ,\ \hat{v}_f = g(t) \text{ on } \hat{\Gamma}_{in} ,\ \hat{v}_s = \hat{v}_f \text{ on } \hat{\Gamma}_I\}\end{aligned}$$

and

$$\hat{p}_f \in V_p := L^2(\hat{\Omega}_f) ,$$

with suitable spaces V_u^s and V_v^s for the solid variables [47, 48].

The test spaces for $(\hat{\varphi}_s^u, \hat{\varphi}_f^u)$ and $\hat{\varphi}_f^p$ are the same, only for $(\hat{\varphi}_s^v, \hat{\varphi}_f^v)$ we get the homogeneous version $(\hat{\varphi}_s^v, \hat{\varphi}_f^v) \in V_v^0$, with the velocity test space

$$V_v^0 := V_v \quad \text{with } g(t) \equiv 0 . \tag{6.6}$$

Finally, this leads to the variational formulation of our FSI problem:

Formulation 6.1 (Weak formulation of FSI-ALE$_{fx}$). Find

$$((\hat{u}_s, \hat{u}_f), (\hat{v}_s, \hat{v}_f), \hat{p}_f) \text{ in } V_u \times V_v \times V_p$$

such that for all test functions $((\hat{\varphi}_s^u, \hat{\varphi}_f^u), (\hat{\varphi}_s^v, \hat{\varphi}_f^v), \hat{\varphi}_f^p)$ in $V_u \times V_v^0 \times V_p$ the following equations are satisfied for almost all times t:

$$\begin{aligned}(\hat{J}\hat{\rho}_f\hat{\partial}_t\hat{v}_f , \hat{\varphi}_f^v)_{\hat{\Omega}_f} + (\hat{J}\hat{\rho}_f\hat{\nabla}\hat{v}_f\hat{F}^{-1}\cdot(\hat{v}_f - \partial_t\hat{\mathcal{A}}) , \hat{\varphi}_f^v)_{\hat{\Omega}_f} &\\ + (\hat{J}\hat{F}^{-T}\hat{\sigma}_f , \hat{\nabla}\hat{\varphi}_f^v)_{\hat{\Omega}_f} - \langle \hat{\rho}_f\hat{\nu}_f\hat{J}\hat{F}^{-T}\hat{\nabla}\hat{v}_f^T\hat{F}^{-T}\cdot\hat{n}_f, \hat{\varphi}_f^v\rangle_{\hat{\Gamma}_{out}} &= 0\\ (\hat{J}\ \mathrm{tr}(\hat{\nabla}\hat{v}_f\hat{F}^{-1}) , \hat{\varphi}_f^p)_{\hat{\Omega}_f} &= 0\end{aligned}$$

$$\begin{aligned}(\hat{\rho}_s\hat{\partial}_t\hat{v}_s , \hat{\varphi}_s^v)_{\hat{\Omega}_s} + (\hat{F}\hat{\Sigma}_s , \hat{\nabla}\hat{\varphi}_s^v)_{\hat{\Omega}_s} &= 0\\ (\hat{\partial}_t\hat{u}_s - \hat{v}_s , \hat{\varphi}_s^u)_{\hat{\Omega}_s} &= 0\\ \alpha_u\left(\frac{1}{\hat{J}}\hat{\nabla}\hat{u}_f , \hat{\nabla}\hat{\varphi}_f^u\right)_{\hat{\Omega}_f} &= 0\end{aligned} \tag{6.7}$$

Since the test functions for the velocities are continuous along the interface, the continuity of stresses is incorporated by dropping the boundary integrals.

Remark 6.2. Instead of converting the strong fluid equations from the Eulerian system into ALE coordinates and then deriving the weak formulation, we could have also gone the other way: first compute the weak Eulerian formulation and then apply the ALE transformations. If we use this approach, we must not forget to transform the test functions into ALE coordinates as well.

Remark 6.3. Fluid-structure interaction in ALE coordinates can be realized in two ways:
- ALE_{dm} (explicit mesh moving): the fluid equations are computed on the deformed configuration Ω and the mesh is moved explicitly.
- ALE_{fx} (implicit mesh moving): all fluid equations are transformed onto the fixed reference configuration $\hat{\Omega}$ and the mesh movement is 'hidden' in the transformations $\hat{F}$ and $\hat{J}$.

In our work, we prefer the second choice originally proposed in [15, 24].

6.2.3 Discretization

In this section we discuss all aspects required for the discretization of the FSI equations. In addition to the spatial discretization, we have to deal with the time-dependence and occurring nonlinearities.

Time stepping

As a first step, we discretize the time-derivatives within the FSI equations using the one-step-theta scheme.

Definition 6.4 (One-Step-Theta Scheme). Given a differential equation $\partial_t u + A(u) = 0$, the one-step-theta scheme reads

$$u(t^n) - u(t^{n-1}) - \Delta t\,\theta\,A(u^n) - \Delta t\,(1-\theta)\,A(u^{n-1}) = 0$$

with $\theta \in [0, 1]$ and $\Delta t := t^n - t^{n-1}$.

Different values of θ result in time-stepping schemes with different properties. Popular choices are (see e.g., [44]):
- $\theta = 0$, which corresponds to the first-order explicit Euler scheme. This scheme requires extremely small time-step sizes for convergence. Thus, it is not feasible for FSI.

- $\theta = 0.5$ leads to the second-order Crank-Nicolson scheme (CN). This scheme is A-stable, however, for long-run simulations of the FSI 2 benchmark described in Section 6.5.1, it breaks down after some time.
- $\theta = 0.5 + \Delta t$ attempts to repair the CN scheme by shifting it to the implicit side. It is second order as well but now additionally strictly A-stable. No breakdown occurs for longer simulations (at least for FSI 2).
- $\theta = 1$ equals the first-order Implicit (Backward) Euler (BE) scheme, which is strongly A-stable. However, this time-stepping scheme introduces a lot of damping, therefore it is not applicable for the FSI 2 benchmark, as no oscillations would occur.

Applying this scheme to the fluid-structure interaction equations (6.2) yields (modified equations only)

$$
\begin{aligned}
\big(\hat{J}^\theta \hat{\rho}_{\mathrm{f}}(\hat{v}_{\mathrm{f}}^n - \hat{v}_{\mathrm{f}}^{n-1}), \hat{\varphi}_{\mathrm{f}}^v\big)_{\hat{\Omega}_f} - \big(\hat{J}\hat{\rho}_{\mathrm{f}}\hat{\nabla}\hat{v}_{\mathrm{f}}\hat{F}^{-1} \cdot (\hat{u}_{\mathrm{f}}^n - \hat{u}_{\mathrm{f}}^{n-1}), \hat{\varphi}_{\mathrm{f}}^v\big)_{\hat{\Omega}_f} &+ \\
\Delta t \left[\big(\hat{J}\hat{\rho}_{\mathrm{f}}\hat{\nabla}\hat{v}_{\mathrm{f}}\hat{F}^{-1} \cdot \hat{v}_{\mathrm{f}}, \hat{\varphi}_{\mathrm{f}}^v\big)_{\hat{\Omega}_f} + \big(\hat{J}\hat{\sigma}_{\mathrm{f}}^v \hat{F}^{-T}, \hat{\nabla}\hat{\varphi}_{\mathrm{f}}^v\big)_{\hat{\Omega}_f}\right]^\theta & \\
-\Delta t \left[\big\langle \hat{\rho}_{\mathrm{f}}\hat{\nu}_{\mathrm{f}}\hat{J}\hat{F}^{-T}\hat{\nabla}\hat{v}_{\mathrm{f}}^T\hat{F}^{-T} \cdot \hat{n}_{\mathrm{f}}, \hat{\varphi}_{\mathrm{f}}^v\big\rangle_{\hat{\Gamma}_{\mathrm{out}}}\right]^\theta &+ \\
\big(\hat{J}(-\hat{p}_{\mathrm{f}})\hat{F}^{-T}, \hat{\nabla}\hat{\varphi}_{\mathrm{f}}^v\big)_{\hat{\Omega}_f} &= 0
\end{aligned}
$$

$$
\begin{aligned}
\big(\hat{\rho}_{\mathrm{s}}(\hat{v}_{\mathrm{s}}^n - \hat{v}_{\mathrm{s}}^{n-1}), \hat{\varphi}_{\mathrm{s}}^v\big)_{\hat{\Omega}_s} + \Delta t \,\big(\hat{F}\hat{\Sigma}_{\mathrm{s}}, \hat{\nabla}\hat{\varphi}_{\mathrm{s}}^v\big)_{\hat{\Omega}_s}^\theta &= 0 \\
\big((\hat{u}_{\mathrm{s}}^n - \hat{u}_{\mathrm{s}}^{n-1}) - \Delta t\, \hat{v}_{\mathrm{s}}^\theta, \hat{\varphi}_{\mathrm{s}}^u\big)_{\hat{\Omega}_s} &= 0
\end{aligned}
\tag{6.7}
$$

with the modification

$$\hat{\sigma}_{\mathrm{f}}^v := \hat{\rho}_{\mathrm{f}}\hat{\nu}_{\mathrm{f}}(\hat{\nabla}\hat{v}_{\mathrm{f}}\hat{F}^{-1} + \hat{F}^{-T}\hat{\nabla}\hat{v}_{\mathrm{f}}^T)$$

as explained below. Further, we use the abbreviation $A^\theta := \theta A^n + (1-\theta)A^{n-1}$. The superscripts n and $n-1$ denote the evaluation at times t^n and t^{n-1}, respectively. For simplicity, we will skip the n superscripts from now on.

Remark 6.5 (Implicit Pressure). The pressure term contained in $\hat{\sigma}_{\mathrm{f}}$ is always treated fully implicitly, i.e., any appearing pressure variable within the fluid stress tensor is treated as if $\theta = 1$ holds. This is motivated by the theory of differential algebraic equations (DAE), see for example [40].

Linearization

In order to obtain the bilinear form required for the usual FE (finite element) approach, we have to get grid of the nonlinearities within the FSI equations (6.7, 6.7). A common way to do so is to linearize the equations using Newton's method (e.g., [46]).

Remark 6.6 (Convergence Properties of Newton's Method). Under several conditions on the differentiability and boundedness of the semilinear form A, Newton's method converges quadratically if provided with a good starting value. Theoretical results are given in e.g., the Newton–Kantorovich theorem in [30], however the requirements for this theorem are generally not fulfilled for FSI.

The term $A'(u)(\varphi, \delta u)$ denotes the directional derivative of A into the direction of the function δu (similar to the classical directional derivative in $\mathbb{R}^n$). This Jacobian is a bilinear form wrt. $(\varphi, \delta u)$, which is exactly what we need for discretization. For infinite dimensional function spaces, this derivative is typically called the Gâteaux differential.

Definition 6.7 (Gâteaux Derivative). The directional derivative of a semilinear form A is defined as

$$A'(u)(\delta u, \varphi) := \lim_{\epsilon \to 0} \frac{1}{\epsilon} \left(A(u + \epsilon \delta u)(\varphi) - A(u)(\varphi)\right) = \frac{d}{d\epsilon} A(u + \epsilon \delta u)(\varphi)\big|_{\epsilon=0}$$

Algorithm 6.1 Newton Linearization

Let $A(u)(\varphi)$ be a semilinear form (linear with respect to the second argument), $F(\varphi)$ a linear form. Then the solution of $A(u)(\varphi) = F(\varphi)$ can be obtained by the following iteration:

Initial guess u_0
for $k = 0, 1, \ldots$ until convergence **do**
Solve $A'(u_k)(\varphi, \delta u_k) = F(\varphi) - A(u_k)(\varphi)$ for δu_k
Update $u_{k+1} := u_k + \delta u_k$
end for

For the differentiation of the whole FSI operator (6.9), we need the derivatives of quantities like $\hat{J}$, $\hat{F}$, $\hat{F}^{-1}$ and others. Since we will use the notation A' for the full Jacobian, we denote the following derivatives by e.g., $\partial_u \hat{F}(u)$ or simply $\partial \hat{F}$, if it is clear by which variable we are differentiating.

Theorem 6.8 (FSI-Related Derivatives).

$$\begin{aligned} \partial \hat{F} &= \hat{\nabla} \delta \hat{u} \\ \partial \hat{J} &= \hat{J}\, \mathrm{tr}(\hat{F}^{-1} \hat{\nabla} \delta \hat{u}) \\ \partial \hat{F}^{-1} &= -\hat{F}^{-1} \hat{\nabla} \delta \hat{u} \hat{F}^{-1} \\ \partial \hat{F}^{-T} &= (\partial \hat{F}^{-1})^T \\ \partial\, \mathrm{tr}(E) &= \mathrm{tr}(\partial E) \end{aligned} \tag{6.8}$$

Proof. Unless trivial, see for example [2, 23]. □

Remark 6.9. All precise calculations of these derivatives can be found for example in [41].

Now we have the necessary ingredients to tackle the FSI problem. Its semilinear form $A(U)(\Phi)$ and linear form $F(\Phi)$ are given as the sum of all equations in (6.7). Note that the derivative only applies to functions evaluated at time-step n. Functions from previous time steps are constant with respect to δU and therefore belong to the right-hand side F. With this in mind, we can define the FSI operator A and right-hand side F after time discretization as

$$\begin{aligned} A(U)(\Phi) = {} & \big(\hat\rho_{\mathrm f}\hat J^{\theta}(\hat v_{\mathrm f} - \hat v_{\mathrm f}^{n-1}) \,,\, \hat\varphi_{\mathrm f}^{v}\big)_{\hat\Omega_f} - \big(\hat J\hat\rho_{\mathrm f}\hat F^{-1}\hat\nabla\hat v_{\mathrm f}\cdot\hat u_{\mathrm f} \,,\, \hat\varphi_{\mathrm f}^{v}\big)_{\hat\Omega_f} \\ & + \Delta t\,\theta\Big[\big(\hat J\hat\rho_{\mathrm f}\hat F^{-1}\hat\nabla\hat v_{\mathrm f}\cdot\hat v_{\mathrm f} \,,\, \hat\varphi_{\mathrm f}^{v}\big)_{\hat\Omega_f} + \big(\hat J\hat\sigma_{\mathrm f}^{v}\hat F^{-T} \,,\, \hat\nabla\hat\varphi_{\mathrm f}^{v}\big)_{\hat\Omega_f}\Big] \\ & - \Delta t\,\theta\,\big\langle \hat\rho_{\mathrm f}\hat v_{\mathrm f}\hat J\hat F^{-T}\hat\nabla\hat v_{\mathrm f}^{T}\hat F^{-T}\cdot\hat n_{\mathrm f},\, \hat\varphi_{\mathrm f}^{v}\big\rangle_{\hat\Gamma_{\mathrm{out}}} \\ & + \big(\hat J(-\hat p_{\mathrm f})\hat F^{-T} \,,\, \hat\nabla\hat\varphi_{\mathrm f}^{v}\big)_{\hat\Omega_f} \\ & + \big(\hat\rho_{\mathrm s}\hat v_{\mathrm s}^{n} \,,\, \hat\varphi_{\mathrm s}^{v}\big)_{\hat\Omega_s} + \Delta t\,\theta\,\big(\hat F\hat\Sigma_{\mathrm s} \,,\, \hat\nabla\hat\varphi_{\mathrm s}^{v}\big)_{\hat\Omega_s} \\ & + \big(\hat u_{\mathrm s} \,,\, \hat\varphi_{\mathrm s}^{u}\big)_{\hat\Omega_s} - \Delta t\,\theta\,\big(\hat v_{\mathrm s} \,,\, \hat\varphi_{\mathrm s}^{u}\big)_{\hat\Omega_s} \\ & + \alpha_u\left(\frac{1}{\hat J}\hat\nabla\hat u_{\mathrm f} \,,\, \hat\nabla\hat\varphi_{\mathrm f}^{u}\right)_{\hat\Omega_f} \end{aligned} \tag{6.9}$$

and

$$\begin{aligned} F(\Phi) = {} & \big(\hat\rho_{\mathrm s}\hat v_{\mathrm s}^{n-1} \,,\, \hat\varphi_{\mathrm s}^{v}\big)_{\hat\Omega_s} - \Delta t\,(1-\theta)\big(\hat F^{n-1}\hat\Sigma_{\mathrm s}^{n-1} \,,\, \hat\nabla\hat\varphi_{\mathrm s}^{v}\big)_{\hat\Omega_s} \\ & + \big(\hat u_{\mathrm s}^{n-1} \,,\, \hat\varphi_{\mathrm s}^{u}\big)_{\hat\Omega_s} + \Delta t\,(1-\theta)\,\big(\hat v_{\mathrm s}^{n-1} \,,\, \hat\varphi_{\mathrm s}^{u}\big)_{\hat\Omega_s} \\ & - \big(\hat J\hat\rho_{\mathrm f}\hat F^{-1}\hat\nabla\hat v_{\mathrm f}\cdot\hat u_{\mathrm f}^{n-1} \,,\, \hat\varphi_{\mathrm f}^{v}\big)_{\hat\Omega_f} \\ & - \Delta t\,(1-\theta)\Big[\big(\hat J\hat\rho_{\mathrm f}\hat F^{-1}\hat\nabla\hat v_{\mathrm f}\cdot\hat v_{\mathrm f} \,,\, \hat\varphi_{\mathrm f}^{v}\big)_{\hat\Omega_f} + \big(\hat J\hat\sigma_{\mathrm f}^{v}\hat F^{-T} \,,\, \hat\nabla\hat\varphi_{\mathrm f}^{v}\big)_{\hat\Omega_f}\Big]^{n-1} \\ & + \Delta t\,(1-\theta)\Big[\big\langle \hat\rho_{\mathrm f}\hat v_{\mathrm f}\hat J\hat F^{-T}\hat\nabla\hat v_{\mathrm f}^{T}\hat F^{-T}\cdot\hat n_{\mathrm f},\, \hat\varphi_{\mathrm f}^{v}\big\rangle_{\hat\Gamma_{\mathrm{out}}}\Big]^{n-1} \end{aligned} \tag{6.10}$$

with test function

$$\Phi := (\hat\varphi_{\mathrm f}^{v}, \hat\varphi_{\mathrm f}^{u}, \hat\varphi_{\mathrm f}^{p}, \hat\varphi_{\mathrm s}^{v}, \hat\varphi_{\mathrm s}^{u})$$

and solution variable

$$U := (\hat v_{\mathrm f}, \hat u_{\mathrm f}, \hat p_{\mathrm f}, \hat v_{\mathrm s}, \hat u_{\mathrm s})\,.$$

Because of the nested nonlinearities, which all require multiple applications of the product rule, the computation of all single terms of the Jacobian is quite lengthy. However, when implementing the derivatives we do not actually need to expand all terms, therefore we do not write them down explicitly here.

Spatial discretization

Now we are ready to discretize the Jacobian and the Newton residual given in Algorithm 6.1. We are going to use a quadrilateral subdivision $\hat{\Omega} = \bigcup_i \mathcal{T}_i$ of the reference domain with $Q(2)$ shape functions for displacements and velocities and discontinuous $Q(1)$ elements for the pressure. The subdivision matches the interface, i.e., every $\mathcal{T}_i$ is either part of the fluid or the solid domain, but not both.

Note that since A' is bilinear wrt. δU and test functions Φ, we have to represent the Newton correction δU using discrete subspaces, not the solution U itself.

As indicated above we use the following discrete function spaces

$$\begin{aligned} V_h^s &:= \left\{v \in H^1(\hat{\Omega}_s) : v|_{\mathcal{T}_i} \in Q(2) \quad \forall \mathcal{T}_i \in \hat{\Omega}_s\right\} \\ V_h^f &:= \left\{v \in H^1(\hat{\Omega}_f) : v|_{\mathcal{T}_i} \in Q(2) \quad \forall \mathcal{T}_i \in \hat{\Omega}_f\right\} \\ L_h^f &:= \left\{v \in L^2(\hat{\Omega}_f) : v|_{\mathcal{T}_i} \in Q(1)\ \forall \mathcal{T}_i \in \hat{\Omega}_f\right\} \end{aligned} \tag{6.11}$$

with the nodal basis functions

$$\begin{aligned} V_h^s &= \operatorname{span}\{\varphi_{s,h}^v[j],\ j = 1, \ldots, N_{v_s}\} = \operatorname{span}\{\varphi_{s,h}^u[j],\ j = 1, \ldots, N_{u_s}\} \\ V_h^f &= \operatorname{span}\{\varphi_{f,h}^v[j],\ j = 1, \ldots, N_{v_f}\} = \operatorname{span}\{\varphi_{f,h}^u[j],\ j = 1, \ldots, N_{u_f}\} \\ L_h^f &= \operatorname{span}\{\varphi_{f,h}^p[j],\ j = 1, \ldots, N_{p_f}\}\,. \end{aligned} \tag{6.12}$$

This leads to the space $X := (V_h^f)^{\dim} \times (V_h^f)^{\dim} \times L_h^f \times (V_h^s)^{\dim} \times (V_h^s)^{\dim}$ for the discrete Newton correction δU. The basis functions for the whole space X are the combinations of the basis functions of the components in the following way:

$$\begin{aligned} \Phi_j &= \left(\varphi_{f,h}^v[j], 0, 0, \ldots, 0\right),\ j = 1, \ldots N_{v_f} \\ \Phi_{j+N_{v_f}} &= \left(0, \varphi_{f,h}^v[j], 0, 0, \ldots, 0\right),\ j = 1, \ldots N_{v_f} \\ \Phi_{j+2N_{v_f}} &= \left(0, 0, \varphi_{f,h}^u[j], 0, \ldots, 0\right),\ j = 1, \ldots N_{u_f} \\ &\ldots \end{aligned} \tag{6.13}$$

Using the ansatz

$$\delta U := \sum_{j=1}^{N_{\mathrm{FSI}}} \delta U_j \Phi_j$$

for the equation at each Newton step k leads to

$$A'(U_k)\Big(\sum_{i=1}^{N_{\mathrm{FSI}}} \Phi_i, \sum_{j=1}^{N_{\mathrm{FSI}}} \delta U_j \Phi_j\Big) = F\Big(\sum_{i=1}^{N_{\mathrm{FSI}}} \Phi_i\Big) - A(U_k)\Big(\sum_{i=1}^{N_{\mathrm{FSI}}} \Phi_i\Big)$$

$$\begin{aligned} \Leftrightarrow \sum_{i,j=1}^{N_{\mathrm{FSI}}} A'(U_k)(\Phi_i, \Phi_j)\delta U_j &= \sum_{i=1}^{N_{\mathrm{FSI}}} \left(F(\Phi_i) - A(U_k)(\Phi_i)\right) \\ \Leftrightarrow A_h \delta U_h &= F_h \end{aligned} \tag{6.14}$$

with

$$\begin{aligned} \delta U_h &:= (\delta U_1, \ldots, \delta U_{N_{\mathrm{FSI}}}) \in \mathbb{R}^{N_{\mathrm{FSI}}}\,, \\ A_h &= \left(A'(U_k)(\Phi_i, \Phi_j)\right)_{i,j=1}^{N_{\mathrm{FSI}}} \in \mathbb{R}^{N_{\mathrm{FSI}} \times N_{\mathrm{FSI}}} \quad \text{and} \\ F_h &= (A(U_k)(\Phi_i) - F(\Phi_i))_{j=1}^{N_{\mathrm{FSI}}} \in \mathbb{R}^{N_{\mathrm{FSI}}}\,. \end{aligned} \tag{6.15}$$

Without any interface coupling, A_h omits the following structure:

$$\begin{bmatrix} \mathcal{M} & 0 & 0 \\ 0 & \mathcal{S} & 0 \\ \mathcal{B} & 0 & \mathcal{F} \end{bmatrix}, \tag{6.16}$$

where $\mathcal{M}$, $\mathcal{S}$ and $\mathcal{F}$ denote the discrete versions of the mesh motion, solid and fluid equations, respectively. The coupling term $\mathcal{B}$ arises because of the ALE transformation of the fluid equations.

Up to now, we have not included any boundary or interface conditions in our discrete formulation. This issue will be tackled in the next parts. Therein, we differentiate between dofs located on the interface (denoted by I) and dofs in the interior (Ω). Furthermore, we split the fluid and solid variables into $\hat{v}_{\mathrm{f}}$, $\hat{p}_{\mathrm{f}}$ and $\hat{u}_{\mathrm{s}}$, $\hat{v}_{\mathrm{s}}$. Note that we do not have pressure dofs on the interface because we are using discontinuous elements. This leads to the more detailed description of the Jacobian

$$\begin{bmatrix} \mathcal{M}^{\Omega\Omega} & \mathcal{M}^{\Omega I} & & & & & & & \\ \mathcal{M}^{I\Omega} & \mathcal{M}^{II} & & & & & & & \\ & & \mathcal{S}_{\mathrm{uu}}^{\Omega\Omega} & \mathcal{S}_{\mathrm{uu}}^{\Omega I} & \mathcal{S}_{\mathrm{uv}}^{\Omega\Omega} & \mathcal{S}_{\mathrm{uv}}^{\Omega I} & & & \\ & & \mathcal{S}_{\mathrm{uu}}^{I\Omega} & \mathcal{S}_{\mathrm{uu}}^{II} & \mathcal{S}_{\mathrm{uv}}^{I\Omega} & \mathcal{S}_{\mathrm{uv}}^{II} & & & \\ & & \mathcal{S}_{\mathrm{vu}}^{\Omega\Omega} & \mathcal{S}_{\mathrm{vu}}^{\Omega I} & \mathcal{S}_{\mathrm{vv}}^{\Omega\Omega} & \mathcal{S}_{\mathrm{vv}}^{\Omega I} & & & \\ & & \mathcal{S}_{\mathrm{vu}}^{I\Omega} & \mathcal{S}_{\mathrm{vu}}^{II} & \mathcal{S}_{\mathrm{vv}}^{I\Omega} & \mathcal{S}_{\mathrm{vv}}^{II} & & & \\ \mathcal{B}_{\mathrm{fm}}^{I\Omega} & \mathcal{B}_{\mathrm{fm}}^{II} & & & & & \mathcal{F}_{\mathrm{vv}}^{II} & \mathcal{F}_{\mathrm{vv}}^{I\Omega} & \mathcal{F}_{\mathrm{vp}}^{I} \\ \mathcal{B}_{\mathrm{fm}}^{\Omega\Omega} & \mathcal{B}_{\mathrm{fm}}^{\Omega I} & & & & & \mathcal{F}_{\mathrm{vv}}^{\Omega I} & \mathcal{F}_{\mathrm{vv}}^{\Omega\Omega} & \mathcal{F}_{\mathrm{vp}}^{\Omega} \\ \mathcal{B}_{\mathrm{pm}}^{\Omega} & \mathcal{B}_{\mathrm{pm}}^{I} & & & & & \mathcal{F}_{\mathrm{pv}}^{I} & \mathcal{F}_{\mathrm{pv}}^{\Omega} & 0 \end{bmatrix}. \tag{6.17}$$

Discrete boundary conditions

Basically, we have to differentiate between boundary conditions subject to the solution variable U_h and those that apply to the Newton corrections δU_h.

It is important to keep in mind that any constraints incorporated into the matrix A_h and right-hand side F_h only directly affect the Newton update and not the actual solution.

The first type of boundary conditions are the initial conditions applied to the solution itself. Typical examples include a given inflow velocity at some part of the boundary and zero values for the mesh displacement (e.g., fixed channel). These conditions are directly incorporated into the solution vector U_h (more specifically into the initial guess U_0 of the Newton algorithm at each time step):

$$U_h(i) := g(i)\,, \quad \text{for all degrees of freedom } i \text{ subject to given initial conditions } g\,.$$

Boundary conditions applied to the Newton correction are nothing else than homogeneous versions of the previous conditions. After some Newton iterations, we have for the solution vector

$$U_h = U_0 + \delta U_0 + \delta U_1 + \dots\,.$$

To retain the conditions imposed on U_0, we have to make sure that these are not disturbed by the following corrections. Therefore, we have to apply homogeneous boundary conditions to δU_k in every Newton step for all those dofs, that are set in U_0. Incorporating such homogeneous Dirichlet conditions can be done in various ways.

One of those is by the following modification of the matrix A_h and right-hand-side F_h:

$$\begin{aligned}
&\forall i \in BC := \{d\colon \text{dof d subject to boundary conditions}\}, j = 1, \dots, N_{\text{FSI}}: \text{set}\\
&\quad A_h(i,i) := 1, A_h(i,j) := 0, A_h(j,i) := 0\,,\\
&\quad F_h(i) := 0\,.
\end{aligned} \tag{6.18}$$

Trivially, the solution of $A_h x = F_h$ will have zero values for all constrained dofs. The diagonal element $A_h(i,i)$ may in fact be chosen arbitrarily nonzero. The disadvantage of this method is that we have to modify each row and column subject to constraints, which might be costly, depending on the representation of the sparse matrix.

A simpler variant of the above method would be

$$\forall i \in BC\colon \text{set } A_h(i,i) := \beta\,, \quad \text{with } \beta \text{ sufficiently large}\,.$$

The penalty parameter β has to be chosen such that it dominates the corresponding row and column of the original matrix. Then, the solution will be approximately the same as if we had eliminated these entries.

Interface conditions

Without interface conditions, the fluid and solid equations would be completely independent of each other and no interaction takes place. To resolve this, we will now have a look at the most important part of fluid-structure interaction.

As given in 6.3 we have three interface conditions given on the continuous level:
1. continuity of stresses;
2. continuity of velocities;
3. continuity of displacements.

For the continuous problem, the continuity of stresses (1) is implicitly fulfilled, if the test functions for the velocities match on the interface. Therefore, we only have to care about conditions (2) and (3), as long as the discrete velocity test functions have the same property.

Incorporating the interface conditions can be done in various ways, like assembling the continuity equations in a DG-like (discontinuous-Galerkin-like) manner (see e.g., [39]). Unfortunately it turned out that the preconditioner shown in Section 6.4 has a bad dependence on the used penalty parameter. The simplest choice would be to use global elements for the velocity and displacement, which guarantees all coupling conditions automatically, see [15]. A corresponding implementation based on deal.II has been provided in [43]. However, this restricts us heavily to choosing the same degrees for the fluid and solid elements. A more general approach has recently been provided by the FE-library deal.II in step-46 introducing a so-called `FE_Nothing element`.

deal.II ConstraintMatrix To realize boundary and interface coupling conditions, we employ the deal.II ConstraintMatrix class. Since this is a sensitive part of the implementation, we describe the basic behavior in the following.

Definition 6.10 (Constraints). Linear constraints on a degree of freedom x_i can be represented as

$$x_i = \sum_{j \neq i} c_{ij} x_j$$

or equivalently $C \cdot X = 0$ with $X = (x_1, x_2, \ldots)$.

Incorporating the constraints C into the solution of $A \cdot X = b$ stemming from a continuous problem $a(\varphi_i, \varphi_j) = f(\varphi_j)$ leads to the following equations to be solved:

$$\begin{aligned} AX &= b \\ CX &= 0\,. \end{aligned} \tag{6.19}$$

However, it is not clear whether this system has a solution at all. The method implemented in deal.II [6] replaces the original, unconstrained shape functions by conforming modifications carrying the constraints. In particular, the following changes are incorporated:

$$\tilde{\varphi}_j = \varphi_j + \sum_{x_i \text{ constr}} c_{ij} \varphi_i \tag{6.20}$$

$$\tilde{A}_{ij} = \begin{cases} a(\tilde{\varphi}_i, \tilde{\varphi}_i), \text{ i, j unconstrained} \\ 1, \text{i} = \text{j constrained} \\ 0, \text{ i or j unconstrained} \end{cases} \qquad \tilde{F}_i = \begin{cases} (f, \tilde{\varphi}_i), \text{ i unconstrained} \\ 0, \text{ i constrained} \end{cases} \tag{6.21}$$

At first glance, this method seems to be very costly as it involves a lot of manipulations of matrix entries. However, one can overcome this issue by applying these modifications when distributing the local contributions from each cell into the global matrix and vector. The only remaining thing to take care of are the additional entries in the sparsity pattern.

In the simplest case of uniform refinement (no hanging nodes) and equal elements for the fluid and solid pairs of displacement and velocity, the interface coupling is a simple 1-to-1 coupling of dofs (see Figure 6.2). Hence, we only have constraints of the form $x_i^f = x_j^s$ for corresponding degrees of freedom on the fluid and solid side. For the general case, we have to represent one set of basis functions by linear combinations of the others.

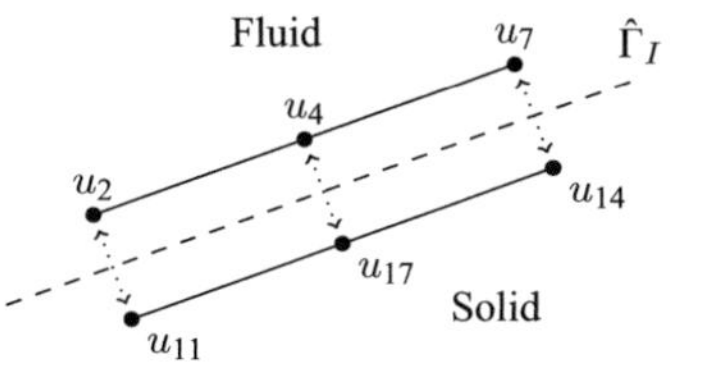

Fig. 6.2: 1-to-1 interface coupling for Q^2-elements.

Using this approach leads to the following modified Jacobian

$$\begin{bmatrix}
\mathcal{M}^{\Omega\Omega} & 0 & & \mathcal{M}^{\Omega I} & & & & & \\
0 & I & & & & & & & \\
 & & \mathcal{S}_{\mathrm{uu}}^{\Omega\Omega} & \mathcal{S}_{\mathrm{uu}}^{\Omega I} & \mathcal{S}_{\mathrm{uv}}^{\Omega\Omega} & \mathcal{S}_{\mathrm{uv}}^{\Omega I} & & & \\
\mathcal{M}^{I\Omega} & & \mathcal{S}_{\mathrm{uu}}^{I\Omega} & \mathcal{S}_{\mathrm{uu}}^{II} + \mathcal{M}^{II} & \mathcal{S}_{\mathrm{uv}}^{I\Omega} & \mathcal{S}_{\mathrm{uv}}^{II} & & & \\
 & & \mathcal{S}_{\mathrm{vu}}^{\Omega\Omega} & \mathcal{S}_{\mathrm{vu}}^{\Omega I} & \mathcal{S}_{\mathrm{vv}}^{\Omega\Omega} & \mathcal{S}_{\mathrm{vv}}^{\Omega I} & & & \\
\mathcal{B}_{\mathrm{fm}}^{I\Omega} & & \mathcal{S}_{\mathrm{vu}}^{I\Omega} & \mathcal{B}_{\mathrm{fm}}^{II} + \mathcal{S}_{\mathrm{vu}}^{II} & \mathcal{S}_{\mathrm{vv}}^{I\Omega} & \mathcal{S}_{\mathrm{vv}}^{II} + \mathcal{F}_{\mathrm{vv}}^{II} & & \mathcal{F}_{\mathrm{vv}}^{I\Omega} & \mathcal{F}_{\mathrm{vp}}^{I} \\
0 & 0 & & & & & I & 0 & 0 \\
\mathcal{B}_{\mathrm{fm}}^{\Omega\Omega} & 0 & & \mathcal{B}_{\mathrm{fm}}^{\Omega I} & & \mathcal{F}_{\mathrm{vv}}^{\Omega I} & 0 & \mathcal{F}_{\mathrm{vv}}^{\Omega\Omega} & \mathcal{F}_{\mathrm{vp}}^{\Omega} \\
\mathcal{B}_{\mathrm{pm}}^{\Omega} & 0 & & \mathcal{B}_{\mathrm{pm}}^{I} & & \mathcal{F}_{\mathrm{pv}}^{I} & 0 & \mathcal{F}_{\mathrm{pv}}^{\Omega} & 0
\end{bmatrix} \tag{6.22}$$

Note that the moved matrix blocks are only equal if the corresponding dofs have the very same ordering. Otherwise, the ordering of the rows and columns within these blocks is different.

Remark 6.11.

- Incorporating homogeneous Dirichlet constraints as presented in Equation (6.18) is a special case of this method. Therefore, the deal.II ConstraintMatrix may be used for interface coupling as well as boundary conditions.

6.2.4 Summary

We conclude this chapter by giving a rough overview of the necessary steps in a program solving the FSI equations in Algorithm 6.2.

Algorithm 6.2 FSI program.

Define problem parameters (Δt, θ, material parameters, ...)
Create mesh
Assign degrees of freedom
Set-up constraints (hanging nodes, homogeneous boundary conditions, interface coupling)
Initialize sparsity pattern, Jacobian A and vectors for the right-hand-side r, solution U, newton update δU and previous solution U_{-1}

$t := 0$
Initial guess for current and previous solution $U := 0$, $U_{-1} := 0$

while $t < T$ **do**
    Apply initial conditions of time t to U
    while not converged **do**
        Assemble r
        (if necessary): assemble Jacobian A
        Solve $A\delta U = r$
        Update $U := U + \delta U$
    end while
    Optional:
    – evaluate functionals (drag, lift, displacement, ...)
    – output results for current time-step
    $t := t + \Delta t$
    $U_{-1} := U$
end while

Remark 6.12 (Initial Guess). At the first time step, we chose a zero initial guess with appropriate initial conditions for the solution U. For larger time-step sizes this may lead to more Newton iterations at the beginning. For all subsequent time steps, we can reuse the solution from the previous time step with adjusted initial conditions as a starting value for the Newton iteration.

If we use an iterative solver like GMRES for the solution of the linear system, we also require an initial guess for the Newton update. In contrast to the previous situation, we have no good candidate from the time-step before, since the corrections are more or less “random”. Therefore, we use $\delta U_0 = 0$, which is a good choice for the last couple of Newton steps, as we would expect that $\delta U \to 0$.

Remark 6.13 (Quasi-Newton Method). Within the Newton iterations, the Jacobian is not necessarily assembled in every single step. If the iteration makes good progress in reducing the residual, we can assume that we are already very close to the eventual solution. Therefore, the Jacobian from the previous Newton step might be a good enough approximation to continue with (Quasi-Newton method), e.g., [13, 29]. This saves us some computational work for assembling the matrix and initializing the linear solver and preconditioner. In particular, if we use a direct solver, we can reuse the previous LU-factorization and get the solution of the current iteration almost for free.

6.3 Linear solvers

The most challenging remaining question is: how do we solve the linear system in every Newton step? While for a low number of dofs a direct solver is a suitable choice, it will eventually become too expensive once the size of the problem increases (Figures 6.6 and 6.7). This will become even worse for problems in three dimensions. As already mentioned, iterative solvers will come to the rescue.

6.3.1 Direct solvers

Although direct solvers are quite costly in terms of computational effort and memory usage, they also have some very big advantages (if applicable). First of all, the solution is more or less "exact" in a numerical sense. This may lead to fewer Newton steps per time step, depending on the convergence criterion of the iterative solver in use. More importantly, direct solvers are very robust: as long as the matrix is nonsingular, direct methods will (most likely) find the solution to the linear system independent of the choice of the parameters in the underlying PDEs.

Most available direct solvers perform a variant of the usual (sparse) LU-decomposition $A = LU$ with lower- and upper-triangular matrices L and U. Once we have such a factorization, we can easily solve the initial system $Ax = b$ by solving $Ly = b$ and $Ux = y$ by forward- and backward-substitution. The general algorithm of the LU-decomposition (without optimizations) is given in Algorithm 6.3.

Algorithm 6.3 LU-decomposition of $n \times n$ matrix A, zero-based indexing.

for $j = 0, \dots, n-1$ **do**
  $L_{jj} = 1$
  for $i = 0, \dots, j$ **do**
    $U_{ij} = A_{ij} - \sum_{k=0}^{i-1} L_{ik} U_{kj}$
  end for
  for $i = j, \dots, n-1$ **do**
    $L_{ij} = (A_{ij} - \sum_{k=0}^{j-1} L_{ik} U_{kj}) / U_{jj}$
  end for
end for

Unfortunately, the LU-decomposition of a sparse matrix is in general dense. This effect, also called fill-in, depends heavily on the ordering of the rows and columns of A (respectively the ordering of the underlying dofs). A somewhat extreme example is given in Figure 6.3. There, the LU-factorization of the original matrix A is dense, whereas for the inverse-ordered matrix $\tilde{A}$, the LU-factors have the same sparsity pattern as $\tilde{A}$.

$$A = \begin{bmatrix} * & * & * & * & * \\ * & * & & & \\ * & & * & & \\ * & & & * & \\ * & & & & * \end{bmatrix} \qquad \tilde{A} = \begin{bmatrix} * & & & & * \\ & * & & & * \\ & & * & & * \\ & & & * & * \\ * & * & * & * & * \end{bmatrix} \tag{6.23}$$

Fig. 6.3: Example of reordering. Original matrix (left) leads to dense LU-factors, whereas the inverse-ordered matrix (right) has no fill-in at all.

6.3.2 Reordering schemes

The previous section gives an immediate starting point for optimizations of direct solvers: ordering the dofs in such a way that the fill-in is minimized. Many reordering methods have been developed, but unfortunately their analysis is mostly heuristic. An introduction and comparison of these schemes as well as implementation issues for direct solvers can be found in [14, 20]. For uncoupled problems, such reordering methods can lead to significant improvements. However, one has to be careful: while the minimal-degree ordering is very efficient for the Laplace equation, it falls far behind the Cuthill–McKee algorithm for FSI problems; see Table 6.2 for a comparison. The resulting sparsity patterns are given in Figures 6.4 and 6.5. Further, we have to be aware that many available direct solvers perform their own reordering, therefore the effect of manually applying such schemes is not immediately obvious.

Remark 6.14 (Complexity). Following [20], the application of direct solvers to a $2-d$ problem requires about $\mathcal{O}(n \log n)$ memory and $\mathcal{O}(n^{3/2})$ operations. For $3-d$ problems, these complexity bounds are $\mathcal{O}(n^2)$ and $\mathcal{O}(n^{4/3})$. As we can see in Figures 6.6 and 6.7, the UMFPACK solver performs slightly worse for the FSI 2 test case.

6.3.3 ILU

A variant of the LU-decomposition that "avoids" the fill-in, is the so-called incomplete LU-factorization (ILU). This method follows the same algorithm as the LU-decomposition given in Algorithm 6.3. The only difference is, that the computation

Tab. 6.2: Effect of reordering on the LU-decomposition of the FSI 2 system-matrix (top) and the Laplace equation. Both tests were done with roughly 4000 degrees of freedom. The rows "nnz" and "nnz %" denote the number of nonzero entries in the L and U matrices and fill-in corresponds to the factor "*nnz* divided by *nnz of original matrix*".

Laplace	**Cuthill–McKee**	**Minimal**	**Random**
nnz	670,856	236,947	8,400,900
nnz %	3.7	1.3	47.0
fill-in	10.9	3.8	137.1

FSI	**Cuthill–McKee**	**Minimal**	**Random**
nnz	1,360,599	15,067,630	17,177,010
nnz %	7.9	87.5	99.8
fill-in	8.9	99.4	113.3

of the LU-factors is restricted to entries available in the sparsity pattern of A. Clearly, such an incomplete factorization will not lead to the exact solution in general, but it can be useful as a preconditioner for iterative solvers. As we saw in the previous

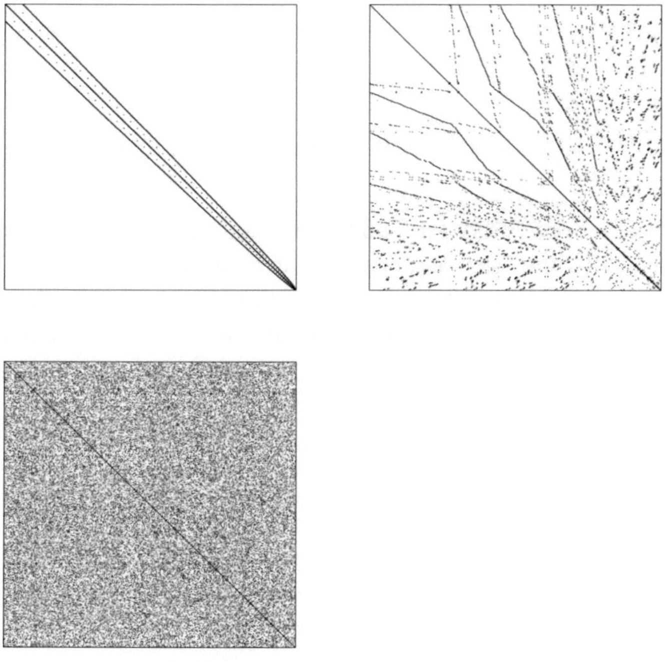

Fig. 6.4: Matrix patterns for the Laplace equation using Cuthill–McKee (top left), minimal-degree (top right), and random ordering of the degrees of freedom.

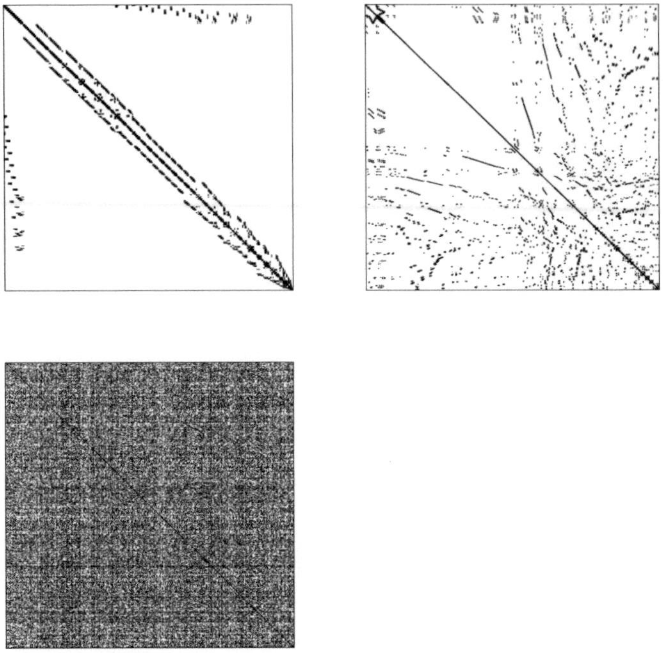

Fig. 6.5: Matrix patterns for the FSI equation using Cuthill–McKee (top left), minimal-degree (top right), and random ordering of the degrees of freedom.

sections, the efficiency of ILU can be improved for simple problems by choosing a suitable ordering.

Remark 6.15 (Variants). Another possibility for the improvement of ILU is the use of larger sparsity patterns. Typically, one uses a pattern generated by one of the matrices $A^2, A^3, \ldots$. These matrices are still sparse but have significantly more entries than A. These modifications are denoted by ILU(k) when using the pattern from A^{1+k}, i.e., ILU(0) denotes the “classical” ILU decomposition.

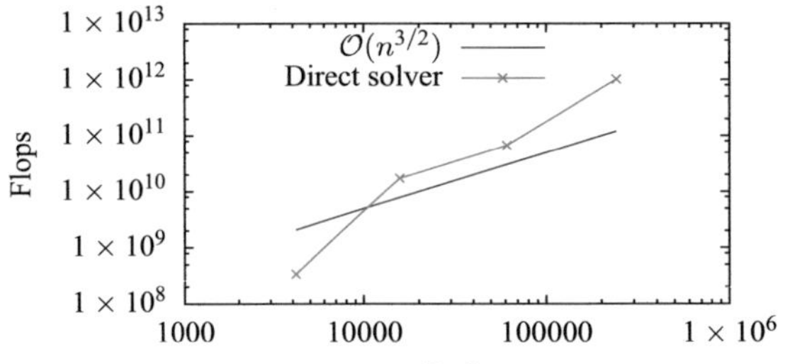

Fig. 6.6: Required flops for the sparse direct solver UMFPACK.

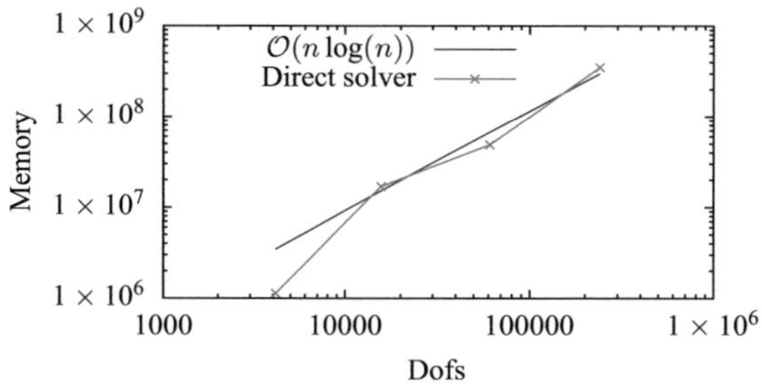

Fig. 6.7: Memory usage for the sparse direct solver UMFPACK.

6.3.4 Schur complement

Closely related to the LU-decomposition is the Schur complement. Considering a $(n + m) \times (n + m)$ block system

$$\begin{bmatrix} A & B \\ D & C \end{bmatrix} \begin{bmatrix} v \\ p \end{bmatrix} = \begin{bmatrix} f \\ g \end{bmatrix} \tag{6.24}$$

with a nonsingular matrix A, we obtain by Gaussian elimination (multiplying the first row by $-DA^{-1}$ and add to the second row)

$$\begin{bmatrix} A & B \\ 0 & C - DA^{-1}B \end{bmatrix} \begin{bmatrix} v \\ p \end{bmatrix} = \begin{bmatrix} f \\ g - DA^{-1}f \end{bmatrix}, \tag{6.25}$$

if $S := C - DA^{-1}B \in \mathbb{R}^{m\times m}$ is invertible. This matrix is usually called the Schur complement. The modified system can now be solved by computing $Sp = g - DA^{-1}f$ and $Av = f - Bp$. Therefore, the solution of the original $(n + m) \times (n + m)$ system reduces to the solution of a $m \times m$ and $n \times n$ system, which is generally cheaper. For the application of direct solvers or Algebraic Multigrid (AMG) methods, we need to compute S explicitly, hence we have to use some approximation of A^{-1}.

6.3.5 GMRES

In our application, we will use a preconditioned GMRES method [32] for solving the linear systems $A\delta U = r$, $A \in \mathbb{R}^{n\times n}$, $\delta U, r \in \mathbb{R}^n$. In the following section, we will have a short look at the algorithm and some theoretical properties of this method.

Similar to many powerful iterative methods (like the conjugate gradient method (CG)), the Generalized Minimal Residual Method (GMRES) is a Krylov subspace method. In every iteration the residual is minimized over an increasing subspace. More precisely, for the m-th step, the computed solution δU_m is a minimizer of $\|A\delta U - r\|_{l^2}$ over the space

$$\delta U_0 + \mathcal{K}_m(A, r_0) = \delta U_0 + \operatorname{span}\{r_0, Ar_0, A^2 r_0, \ldots, A^{m-1} r_0\},$$

Algorithm 6.4 GMRES without preconditioning for solving $A\delta U = b$

$\delta U_0 =$ initial guess, $r_0 = r - A\delta U_0$, $y_1 = \|r_0\|$, $v_1 = \frac{r_0}{y_1}$
for $j = 1, 2, \ldots$ **do** until convergence
  $h_{i,j} = (Av_j, v_i)$, for $i = 1, \ldots, j$
  $w_j = Av_j - \sum_{i=1}^{j} h_{i,j} v_i$
  $h_{j+1,j} = \|w_j\|$
  for $i = 1, \ldots, j-1$ **do**
    $\begin{bmatrix} h_{i,j} \\ h_{i+1,j} \end{bmatrix} = \begin{bmatrix} c_{i+1} & s_{i+1} \\ -s_{i+1} & c_{i+1} \end{bmatrix} \cdot \begin{bmatrix} h_{i,j} \\ h_{i+1,j} \end{bmatrix}$
  end for
  $\beta = \sqrt{h_{j,j}^2 + h_{j+1,j}^2}$
  $s_{j+1} = \frac{h_{j+1,j}}{\beta}$
  $c_{j+1} = \frac{h_{j,j}}{\beta}$
  $h_{j,j} = \beta$
  $y_{j+1} = -s_{j+1} y_j$
  $y_j = c_{j+1} y_j$
  if $|y_{j+1} > TOL|$ **then** $v_{j+1} = \frac{w_j}{h_{j+1,j}}$
  else
    for $i = j, \ldots, 1$ **do**
      $y_i = \frac{1}{h_{j,j}}(y_i - \sum_{k=i+1}^{j} h_{i,k} y_k)$
      $\delta U = \delta U_0 + \sum_{i=1}^{j} y_i v_i$
    end for
  end if
end for

leading to the exact solution after at most n steps when using exact arithmetic. The big advantage of GMRES compared to CG is that it can also cope with matrices that are not necessarily symmetric or positive definite.

As we can see from Algorithm 6.4, the required memory storage and computational effort increases in every iteration. One way to limit the amount of work is to restart the algorithm after a fixed number of iterations m (denoted by GMRES(m)). However, by doing so, one loses the theoretical minimization property and convergence can no longer be guaranteed. Therefore, any used preconditioner should keep the number of required iterations as low as possible such that a restart is not necessary.

Remark 6.16 (Flexible GMRES). The algorithm presented above assumes that a possible preconditioner is constant for all iterations. This prohibits the use of many iterative methods, like GMRES itself, within the preconditioner. The Flexible GMRES takes such nonlinear preconditioners into account, which comes at the cost of higher memory requirements.

In this work, we refer to the flexible variant when talking about GMRES.

Remark 6.17 (Convergence Criterion). For the GMRES solver we use a relative reduction of $\|r - A\delta U\|_{l^\infty} < 10^{-9}$ as the stopping criterion. This is good enough for the Newton solver to converge without too many additional iterations, compared to using a direct solver. Convergence of the Newton solver is achieved once we obtain a residual with $\|r\|_{l^\infty} < 10^{-8}$.

6.3.6 Multigrid methods

A different kind of solvers are the so called multigrid methods. The main idea among them is to solve the linear system $A\delta U = r$ rising from the FE discretization on a sequence of grids. Within those, the results from coarser grids are used to enhance the solution on finer grids. To construct such multigrid methods, we require a few things.

First of all, we need a hierarchy of grids. One way to obtain them is via the refinement of a given grid, which will then be used as the coarsest mesh. Since this method is based on available geometric information, it is called "Geometric Multigrid" (GMG). A different approach is to construct coarser "grids" using only information provided by the matrix A itself, which leads to the "Algebraic Multigrid" (AMG) methods. The construction of such algebraic coarsening schemes is a topic on its own, therefore we refer to [35] for an introduction to this issue.

The advantage of AMG over GMG is its flexibility. For GMG, one requires to create the necessary data structures to hold the information from all levels of refinement. Depending on the code, this may be difficult to incorporate later on. AMG methods do not require these structures, since these methods are able to compute the necessary information themselves. However, GMG methods are, if applicable, generally more efficient than AMG.

Once we have a hierarchy from either AMG or GMG, we need to be able to represent the residual of a linear system stemming from the finer grids in terms of coarser grids and vice-versa. The responsible operators for this procedure are the restriction and prolongation operators (matrices)

$$\mathcal{R}_l \colon \mathbb{R}^{N_l} \mapsto \mathbb{R}^{N_{l-1}} \quad \text{and} \quad \mathcal{P}_l \colon \mathbb{R}^{N_{l-1}} \mapsto \mathbb{R}^{N_l} \tag{6.26}$$

where N_l denotes the number of dofs on level l. For GMG methods, these operators can be naturally defined using standard FE-interpolation.

However, if we just project the residual onto a coarser mesh, we lose a lot of accuracy if the residual shows a highly oscillating behavior. The idea is to reduce these high-frequency parts first. If this is successful, we can assume that the vector can be represented on the coarser grid without too much loss of information. An illustrative example is given in Figure 6.8. Damping of the oscillations within the residual is achieved via so-called smoothers. Mostly, these are iterative solvers lacking a good overall convergence rate but with the property that they reduce the error in the high-frequency parts very quickly. Typical examples of such smoothers include the Jacobi method, (symmetric) Gauss–Seidel, polynomial smoothers like Chebyshev or MLS (Multilevel-Schwarz), but also methods like ILU can be used within multigrid methods. Once we are on the coarsest mesh, we solve the problem exactly using a (sparse) LU-decomposition described in Section 6.3.1.

A frequently used smoother is the Gauss–Seidel method, which we will also use later on. Its application to the linear system $A\delta U = r$ can be written as

$$\delta U_{k+1} = (D + L)^{-1}(r - U\delta U_k) , \tag{6.27}$$

with the decomposition $A = L + U + D$ into a lower-, upper-triangular and diagonal part and some initial guess δU_0.

Combining all the single components introduced above, we can now state a general recursive multigrid application in Algorithm 6.5. Here, $\mathcal{S}(A_l, \delta U_l, r_l)$ denotes the smoother depending on the level l matrix A_l, an initial guess and output δU_l and corresponding right-hand side r_l. Typical values for γ are 1 or 2. The resulting cycles are referred to as V-cycle and W-cycle due to their characteristic shape as seen in Figure 6.9.

Similar to the smoothers listed above, multigrid methods can solve linear systems on their own, yet they are mostly used as efficient preconditioners. In our application, we use the AMG-implementation provided by the Trilinos package ML [19, 22]. A

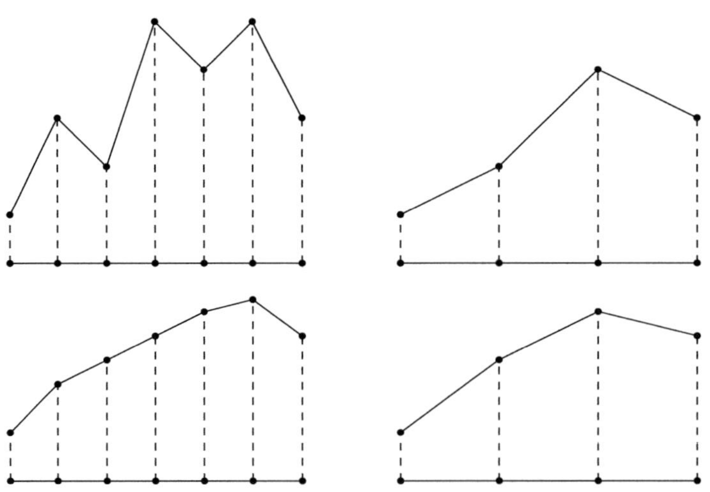

Fig. 6.8: Schematic 1-d multigrid illustration of a nonsmooth residual (top row) and smoothed version (bottom row) on the fine grid (left) and coarse grid (right). Most of the behavior of the nonsmoothed function cannot be represented on the coarser grid.

Algorithm 6.5 Multigrid method for solving $A\delta U = r$.

procedure $MG_l(\delta U_l, r_l)$
    if $l = 0$ **then**
        Solve coarsest level $A_l \delta U_l = r_l$
        return
    end if
    Pre-smoothing: $\delta U_l = \mathcal{S}(A_l, \delta U_l, r_l)$, $r_\mathrm{f} = r_l - A_l \delta U_l$
    Restriction: $rc = \mathcal{R}_l r_\mathrm{f}$
    Coarse-grid approximation: $w_{l-1} = 0$
    for $i = 1$ to γ **do**
        $MG_{l-1}(w_{l-1}, r_c)$
    end for
    Prolongation: $w_l = \mathcal{P}_l w_{l-1}$
    Correction: $\delta U_l = \delta U_l + w_l$
    Post-smoothing: $\delta U_l = \mathcal{S}(A_l, \delta U_l, r_l)$
end procedure

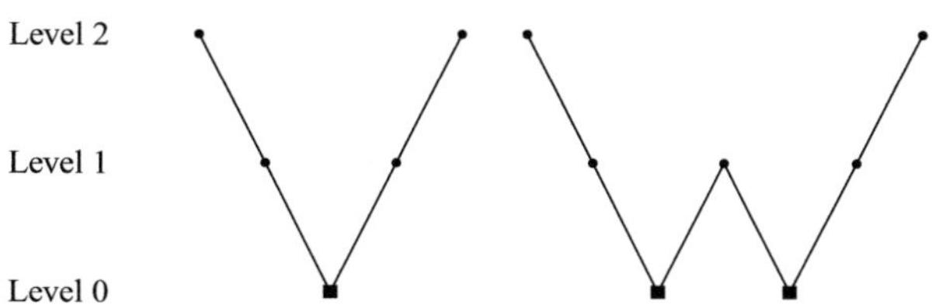

Fig. 6.9: V- and W-cycle. Circles denote an application of the smoother, squares correspond to the solution on the coarsest grid.

full geometric multigrid preconditioner for fluid-structure interaction can be found in e.g., [31].

6.4 Preconditioners

Iterative solvers as presented in the previous section like GMRES become most powerful if they are combined with a good preconditioner. The development of such a preconditioner depends very much on the structure of the underlying problem. Specifically, the crucial part in FSI is the treatment of the interface and the multidomain character since the equations only live in subdomains rather than in the entire domain. This is the characteristic feature in comparison to other multiphysics problems and makes the developments of preconditioners challenging. Algebraic "black-box" preconditioners like the AMG or ILU often work well for simple problems, but cease work for more complex PDEs like FSI, as illustrated in Figure 6.10. Therefore, we can not avoid spending some time on the development of preconditioners that specifically target this kind of problem.

6.4.1 Introduction

The principal idea of preconditioning is to rewrite a system of linear equations $A\delta U = r$ as $P^{-1}A\delta U = r$ (left preconditioning) or $AP^{-1}y$, $\delta U = P^{-1}y$ (right preconditioning), with an arbitrary but nonsingular operator P. Now we may choose P^{-1} such that the new systems $P^{-1}A$ or AP^{-1} are "easier" to solve, i.e., we would like to have something like $P^{-1}A \approx I$. Of course, this can be achieved by using $P^{-1} = A^{-1}$, which leads to a trivially solvable equation. However, the application of this preconditioner would correspond to solving the original equation, so we do not gain anything from that. The other extremal case would be to choose $P^{-1} = I$. This preconditioner is of course very cheap to apply, but it does not make the linear system easier to solve. While these two choices for P^{-1} have no practical use, they illustrate the problem of a good choice of the preconditioner:

On the one hand, its application should be as cheap as possible while on the other hand, it should "simplify" the equations a lot.

Remark 6.18. The meaning of "simplify" depends on the actual solver used. For example, a direct solver benefits from a block-diagonal structure and low bandwidth of $P^{-1}A$ (see Section 6.3.2), whereas iterative solvers like GMRES work best if the eigenvalues are clustered into just a few blocks.

6.4.2 First examples

First of all, we require a handleable block representation of the Jacobian, since the structure given in (6.22) is way to detailed to work with. This step already differs depending on how the resulting preconditioner is supposed to work. Some examples of

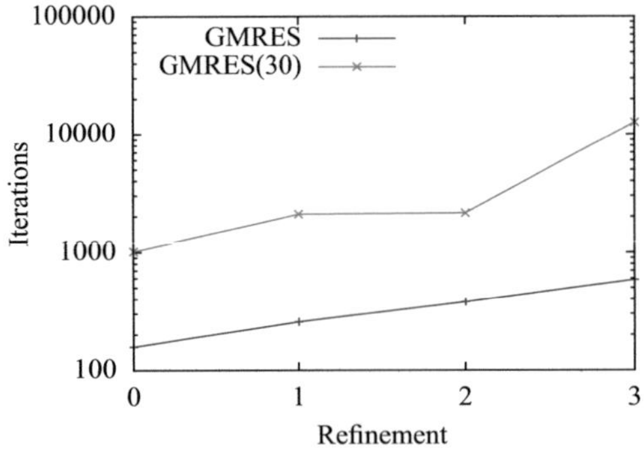

Fig. 6.10: Number of iterations using GMRES and restarted GMRES(30) with ILU(0)-preconditioning.

different block notations of the same matrix are given below (see [21, 27]):

$$A \equiv \begin{bmatrix} \mathcal{M} & \mathcal{C}_{\mathrm{ms}} & 0 \\ \mathcal{C}_{\mathrm{sm}} & \mathcal{S} & \mathcal{C}_{\mathrm{sf}} \\ \mathcal{C}_{\mathrm{fm}} & \mathcal{C}_{\mathrm{fs}} & \mathcal{F} \end{bmatrix} \tag{6.28}$$

or

$$A \equiv \begin{bmatrix} \mathcal{S} & \mathcal{C}_{\mathrm{sf}} & \mathcal{C}_{\mathrm{sp}} \\ \mathcal{C}_{\mathrm{fs}} & \mathcal{F} & \mathcal{C}_{\mathrm{fp}} \\ \mathcal{C}_{\mathrm{ps}} & \mathcal{C}_{\mathrm{pf}} & 0 \end{bmatrix} \tag{6.29}$$

Note that the first representation differentiates between mesh, solid and fluid (velocity and pressure) while the latter splits the degrees of freedom into solid, pressure and fluid (including mesh displacement and velocity).

A first idea shown in [21] considers replacing such a block matrix by an approximation. Several possibilities for such an approximation of (6.29) are given. All of these neglect some parts of the coupling terms to obtain the simpler systems (6.30).

$$P_{\mathrm{diag}} := \begin{bmatrix} \mathcal{S} & 0 & 0 \\ 0 & \mathcal{F} & \mathcal{C}_{\mathrm{fp}} \\ 0 & \mathcal{C}_{\mathrm{pf}} & 0 \end{bmatrix} P_{\mathrm{sup}} := \begin{bmatrix} \mathcal{S} & 0 & 0 \\ \mathcal{C}_{\mathrm{fs}} & \mathcal{F} & \mathcal{C}_{\mathrm{fp}} \\ \mathcal{C}_{\mathrm{ps}} & \mathcal{C}_{\mathrm{pf}} & 0 \end{bmatrix} P_{\mathrm{sub}} := \begin{bmatrix} \mathcal{S} & \mathcal{C}_{\mathrm{sf}} & \mathcal{C}_{\mathrm{sp}} \\ 0 & \mathcal{F} & \mathcal{C}_{\mathrm{fp}} \\ 0 & \mathcal{C}_{\mathrm{pf}} & 0 \end{bmatrix} \tag{6.30}$$

These simplified block systems can be solved a lot easier than the original one. For example, the application of P_{sup}^{-1} on a vector $r = (r_{\mathrm{s}}, r_{\mathrm{f}}, r_p)$ is given in Algorithm 6.6.

Algorithm 6.6 Evaluation of $x = P_{\mathrm{sup}}^{-1} r$

1: Solve $\begin{bmatrix} \mathcal{F} & \mathcal{C}_{\mathrm{fp}} \\ \mathcal{C}_{\mathrm{pf}} & 0 \end{bmatrix} \cdot \begin{bmatrix} x_{\mathrm{f}} \\ x_p \end{bmatrix} = \begin{bmatrix} r_{\mathrm{f}} \\ r_p \end{bmatrix}$

2: Update $r_{\mathrm{s}} = rs - \mathcal{C}_{\mathrm{sf}} x_{\mathrm{f}} - \mathcal{C}_{\mathrm{sp}} x_p$

3: Solve $\mathcal{S} x_{\mathrm{s}} = r_{\mathrm{s}}$

This approximate solver can then be used as a preconditioner for the original system (6.29) inside GMRES. However, it remains to find the solutions or some approximations of the occurring fluid and solid subsystems.

If we naively apply this technique using our ALE formulation, this preconditioner is less useful, as we have the variables for the mesh motion as a part of the fluid block. For problems like the FSI benchmark given in Section 6.5.1, the number of dofs is not well balanced, as we have a very small solid system $\mathcal{S}$ (only the thin flag) but a very big fluid system.

Remark 6.19. In [21], replacing the exact Jacobian A by one of its approximations in the manner of a quasi-Newton method is also considered. This requires a lot more Newton iterations and is hence not considered here.

6.4.3 LU-approach

A slight generalization of the above method is based on a block-wise approximate LU-decomposition presented in [27]. For a first starting point, we have a look at the LU-factorization of a 3x3 matrix with a zero entry in the top-right corner given in Equation (6.31). This corresponds to the same structure as our FSI matrix (6.28).

$$L = \begin{bmatrix} A_{11} & 0 & 0 \\ A_{21} & -\frac{A_{12}A_{21}}{A_{11}} + A_{22} & 0 \\ A_{31} & -\frac{A_{12}A_{31}}{A_{11}} + A_{32} & -\frac{A_{23}\left(-\frac{A_{12}A_{31}}{A_{11}} + A_{32}\right)}{-\frac{A_{12}A_{21}}{A_{11}} + A_{22}} + A_{33} \end{bmatrix}$$

$$U = \begin{bmatrix} 1 & \frac{A_{12}}{A_{11}} & 0 \\ 0 & 1 & \frac{A_{11}A_{23}}{-A_{12}A_{21}+A_{11}A_{22}} \\ 0 & 0 & 1 \end{bmatrix} \tag{6.31}$$

Remark 6.20 (Computation). The factorization was computed using a computer algebra system. This program assumes that the entries of the matrix are scalars, therefore we have to interpret fractions as multiplications with the inverse of the denominator. Further, we have to multiply the matrices in the correct order, which is not the case in Equation (6.31).

While this exact LU-decomposition is too complicated to apply to the FSI block matrix, we can simplify it in the same manner as before by dropping one of the off-diagonal coupling blocks.

In [27], the coupling block $\mathcal{C}_{\mathrm{sm}}$ (corresponding to A_{21}) does not appear due to the slightly different formulation, which leads to the factorization

$$\begin{aligned} A \approx P_{21} := & \begin{bmatrix} \mathcal{M} & \mathcal{C}_{\mathrm{ms}} & 0 \\ 0 & \mathcal{S} & \mathcal{C}_{\mathrm{sf}} \\ \mathcal{C}_{\mathrm{fm}} & \mathcal{C}_{\mathrm{fs}} & \mathcal{F} \end{bmatrix} \\ = & \begin{bmatrix} \mathcal{M} & 0 & 0 \\ 0 & \mathcal{S} & 0 \\ \mathcal{C}_{\mathrm{fm}} & \tilde{\mathcal{C}}_{\mathrm{fs}} & \mathcal{X} \end{bmatrix} \begin{bmatrix} I & \mathcal{M}^{-1}\mathcal{C}_{\mathrm{ms}} & 0 \\ 0 & I & \mathcal{S}^{-1}\mathcal{C}_{\mathrm{sf}} \\ 0 & 0 & I \end{bmatrix} =: LU \end{aligned} \tag{6.32}$$

with

$$\tilde{\mathcal{C}}_{\mathrm{fs}} = \mathcal{C}_{\mathrm{fs}} - \mathcal{C}_{\mathrm{fm}}\, \mathcal{M}^{-1}\, \mathcal{C}_{\mathrm{ms}} \tag{6.33}$$

and the fluid Schur complement

$$\mathcal{X} = \mathcal{F} - \tilde{\mathcal{C}}_{\mathrm{fs}}\, \mathcal{S}^{-1}\, \mathcal{C}_{\mathrm{sf}} = \mathcal{F} - (\mathcal{C}_{\mathrm{fs}} - \mathcal{C}_{\mathrm{fm}}\, \mathcal{M}^{-1}\, \mathcal{C}_{\mathrm{ms}})\, \mathcal{S}^{-1}\, \mathcal{C}_{\mathrm{sf}}\,. \tag{6.34}$$

Having such a decomposition, it is easy to compute the action of the inverse. From linear algebra we know that $P^{-1}r = U^{-1}L^{-1}r$. Consecutively solving with L and U finally leads to Algorithm 6.7.

Algorithm 6.7 Evaluation of $(LU)^{-1}r$.

1: Solve $x_m = \mathcal{M}^{-1}r_m$
2: Solve $x_s = \mathcal{S}^{-1}r_s$
3: Update $r_f = r_f - \mathcal{C}_{fm}x_m - \mathcal{C}_{fs}x_s$
4: Solve $x_f = \mathcal{X}^{-1}r_f$
5: Update $x_s = x_s - \mathcal{S}^{-1}\mathcal{C}_{sf}x_f$
6: Update $x_m = x_m - \mathcal{M}^{-1}\mathcal{C}_{ms}x_s$

Remark 6.21 (Connection to Partitioned Methods). This preconditioner based on an LU-factorization can be seen as an (approximate) iteration of a partitioned solver. The off-diagonal coupling blocks correspond to the incorporation of the results from one subproblem into the other. Depending on which block is neglected, one gets slightly different preconditioners. In this work, we will have a look at the preconditioners resulting from dropping $\mathcal{C}_{sm}(A_{21})$, $\mathcal{C}_{sf}(A_{23})$ and $\mathcal{C}_{ms}(A_{12})$. We will denote these preconditioners by the indices 21, 23 and 12, respectively. Most of the tests done in Section 6.5 provided similar results for these choices.

While this is basically the final preconditioner, we still have to clarify several things. First of all, we do not necessarily require the exact solution of the interior subproblems – a "good-enough" approximation may be enough. Nevertheless, we can use a direct solver for these subproblems in order to get a first idea of the capabilities of the preconditioner. It is reasonable to assume that the required GMRES iterations when using direct solvers are the lowest we can get from this particular preconditioner. Any approximate subsolution will most likely increase the number of iterations. However, as observed in [21], it is possible that the approximate version performs better than the exact one. (Un)fortunately, we were not able to observe such a behavior in any of our tests.

Although using direct solvers is not what we want at the very end due to their huge memory requirements, we still gain something. Since the computational effort of direct methods increases in a nonlinear fashion, solving three smaller systems is cheaper than computing the solution of the whole system. Hence, we might be able to solve larger problems than previously before we run out of memory. Further, this preconditioner with direct solvers for the subproblems is already significantly faster than a direct solver alone (Figure 6.11).

A small benefit when using direct solvers is that we only have to factorize the subproblems once every time the matrix changes. Thus, only the very first GMRES iter-

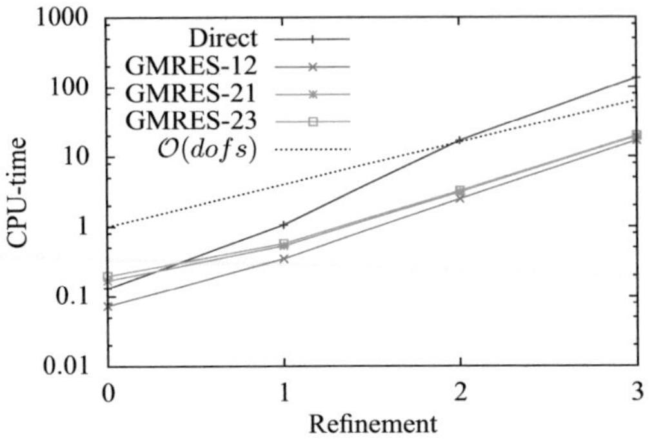

Fig. 6.11: Required time using a direct solver and GMRES with direct subsolvers.

ation is costly, while the effort for all subsequent iterations (and probably the next Newton steps as well) is relatively low (matrix-vector products and forward/backward substitution).

If we used iterative solvers for the approximation, every application of the preconditioner has approximately equal cost (there are still things that can be precomputed, like the coarsening strategy in AMG methods, but these do not contribute as much as a complete factorization). A closer look at the approximate solution of the subproblems follows in Section 6.4.4.

Another point worth a closer look is the computation of $\tilde{\mathcal{C}}_{fs}$ and the Schur complement $\mathcal{X}$. Both involve matrix-matrix multiplications and even worse, require the inverse of $\mathcal{M}$ and $\mathcal{S}$. If we use a direct solver or AMG method for the Schur complement, we have to compute these matrices explicitly since we need to access single entries. Thus, we would also need $\mathcal{M}^{-1}$ and $\mathcal{S}^{-1}$ in explicit form! Of course, this is not feasible to do, since the inverse is in general no longer sparse and the computational effort is way to much. The solution to this problem is the same as in the previous sections: we have to approximate these matrices respectively its inverses. One possible way of doing so is to choose a diagonal approximation like

$$\tilde{A}(i,i) := \sum_c |A(i,c)| \tag{6.35}$$

for $\mathcal{M}$ and $\mathcal{S}$, which can then be easily inverted. Another choice proposed in [27] is a block-diagonal version of the original matrix with low enough block size (e.g., 3), such that each diagonal block can still be inverted explicitly. The drawback of this approach is that the resulting approximation depends on the ordering of the dofs, which is not the case for the diagonal row-sum approximation. Also, the diagonal approach seems to perform better in most of our tests.

It turns out that we can approximate the Schur complement even more by neglecting the perturbation term $\tilde{\mathcal{C}}_{fs}\,\mathcal{S}^{-1}\,\mathcal{C}_{sf}$ completely. As shown in Figure 6.12, the average number of GMRES iterations is almost the same whether we compute the Schur complement or simply take the fluid matrix $\mathcal{F}$. The computations were done for the FSI 2 benchmark with time-step size $\Delta t = 0.01$ using direct solvers for the subproblems.

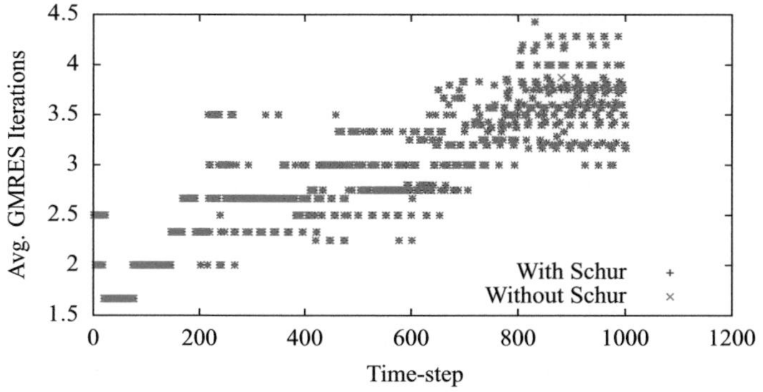

Fig. 6.12: Similar number of GMRES iterations with and without computing the Schur complement for the FSI 2 benchmark at each time step.

This approximation saves us the expensive matrix-matrix multiplication for the Schur complement. However, it is unclear whether this approximation will do well for all the different problems, geometries and elements used.

6.4.4 Solving the subproblems

In the following sections, we will have a look at the approximate solution of the arising subproblems $\mathcal{M}$, $\mathcal{S}$ and $\mathcal{X}$. Ideally, we want to obtain a good approximation with low computational effort that can eventually be run in parallel. Unfortunately, it is quite challenging to find solvers that fit all these requirements.

6.4.5 Mesh

For the solution of the mesh system, we apply a few AMG V-cycles with Gauss–Seidel smoothing. This is motivated by the fact that for small deformations, the Jacobian of the mesh equation reduces to $\left(\hat{\nabla}\hat{\varphi}_{\mathrm{f}}^{u}, \hat{\nabla}\hat{\varphi}_{\mathrm{f}}^{u}\right)_{\hat{\Omega}_f}$. This corresponds to the usual Laplace equation in weak form, which is known to be efficiently solved by this method. For large displacements, we may use the AMG cycles as a preconditioner for another GMRES solver to improve the convergence.

6.4.6 Fluid

The most challenging part is the robust solution of the fluid equations. Solving this linear system requires special care as this matrix is indefinite and includes a zero block.

Precisely, we have to solve the following equation

$$\mathcal{F} := \begin{bmatrix} \mathcal{F}^{\Omega\Omega} & \mathcal{F}^{\Omega}_{\text{vp}} \\ \mathcal{F}^{\Omega}_{\text{pv}} & 0 \end{bmatrix} \begin{bmatrix} x_v \\ x_p \end{bmatrix} = \begin{bmatrix} r_v \\ r_p \end{bmatrix} .$$

Several methods have been developed for the (linear) Stokes problem, like Vanka-type methods, Braess–Sarazin smoother and Uzawa-like methods (see e.g., [8, 36, 45]). All of these methods are closely related to the Schur complement, and differ mainly in the choice of arising approximations.

In our application, we use the Schur complement approach as described in Section 6.3.4 as a preconditioner inside GMRES, with the approximation $(\tilde{\mathcal{F}}^{\Omega\Omega})^{-1} := \text{diag}(\mathcal{F}^{\Omega\Omega})^{-1}$ for computing the Schur complement $\mathcal{X}_p$. The sub-subproblems $\mathcal{F}^{\Omega\Omega}$ and $\mathcal{X}_p$ are approximated using GMRES with AMG preconditioning.

6.4.7 Solid

Lacking a good alternative, we use the same technique for the solid system as in the fluid case, by splitting the dofs into velocity in displacement parts:

$$\begin{aligned} \mathcal{S} &= \begin{bmatrix} \mathcal{S}_{\text{uu}} & \mathcal{S}_{\text{uv}} \\ \mathcal{S}_{\text{vu}} & \mathcal{S}_{\text{vv}} \end{bmatrix} \begin{bmatrix} x_u \\ x_v \end{bmatrix} = \begin{bmatrix} r_u \\ r_v \end{bmatrix} \\ \Leftrightarrow \mathcal{P}_S &:= \begin{bmatrix} \mathcal{S}_{\text{uu}} & \mathcal{S}_{\text{uv}} \\ 0 & \mathcal{X}_S \end{bmatrix} \begin{bmatrix} x_u \\ x_v \end{bmatrix} = \begin{bmatrix} r_u \\ r_v - \mathcal{S}_{\text{vu}} r_u \end{bmatrix} , \end{aligned} \tag{6.36}$$

with the Schur complement $\mathcal{X}_S = \mathcal{S}_{\text{vv}} - \mathcal{S}_{\text{vu}}\, diag(\mathcal{S}_{\text{uu}})^{-1}\, \mathcal{S}_{\text{uv}}$ as described in Section 6.3.4.

6.5 Numerical tests – FSI benchmarks

Last but not least, we present some numerical results for the preconditioned solver. Here, we mainly focus on the well-known 2-d FSI benchmarks [9, 10, 37] and some variations. These test problems are presented in the next section. The implementation uses the deal.II FE-library [4, 5] and the Trilinos [19, 22] packages therein for the AMG method. For the direct solution of linear systems the package UMFPACK [12] was used. The basis of the FSI implementation of this chapter was provided by the FSI template [43] also based on deal.II. This code has been extended using a hp-element formulation (step-46 in deal.II) for more flexibility in the choice of finite element combinations for the subproblems and therefore this new programming code does allow for easier future extension to other multiphysics problems.

Tab. 6.3: FSI benchmark configurations.

Parameter	Symbol / Unit	FSI 1	FSI 2	FSI 3
Structure density	$\rho_s[\mathrm{kg/m^3}]$	10^3	10^4	10^3
Poisson ratio	ν_s	0.4	0.4	0.4
Lamé coefficient	$\mu_s[\mathrm{kg/m^2}]$	0.5×10^6	0.5×10^6	2.0×10^6
Fluid density	$\rho_f[\mathrm{kg/m^3}]$	10^3	10^3	10^3
Kinematic viscosity	$\nu_f[\mathrm{m^2/s}]$	10^{-3}	10^{-3}	10^{-3}
Inflow velocity	$v_{in}[\mathrm{m/s}]$	0.2	1.0	2.0
Channel length	$L := 2.5\,\mathrm{m}$			
height	$H := 0.41\,\mathrm{m}$			
Cylinder center	$M := (0.2\,\mathrm{m}, 0.2\,\mathrm{m})$			
radius	$r := 0.05\,\mathrm{m}$			
Flag width	$w := 0.35\,\mathrm{m}$			
height	$h := 0.02\,\mathrm{m}$			
Measurement point	$A := (0.6\,\mathrm{m}, 0.2\,\mathrm{m})$			

6.5.1 Setup

Frequently used test cases for FSI solvers are the benchmarks proposed by Turek and Hron in [37]. These problems consist of an incompressible laminar channel flow around a cylinder, with an elastic beam (flag) attached to it (Figure 6.13, Table 6.3).

Originally, three problems with different choices for the fluid and structure parameters were defined:

The point A denotes the right-most point of the flag at which the displacements are measured. The initial condition is given as the inflow profile at the left boundary

$$v(0, y) := 1.5\, v_{\text{in}}\, \frac{4y(H-y)}{H^2}\,,$$

with an additional smoothing of the form

$$v(t, 0, y) := \begin{cases} v(0, y)\, \dfrac{1}{2}\left(1 - \cos\left(\dfrac{\pi}{2}t\right)\right) & \text{for } t < 2 \\ v(0, y) & \text{for } t \geq 2 \end{cases}$$

for FSI 2 and 3. This improves the convergence of the Newton-solver for the first time steps. Furthermore, we have no-slip conditions $v_f = 0$ on the top and bottom boundary part of the channel, the circle boundary and on the fluid-structure interface.

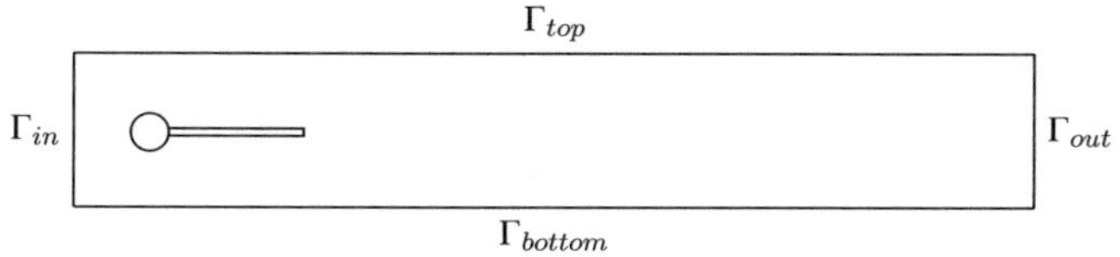

Fig. 6.13: Computational domain for the FSI benchmarks.

The flag is not exactly centered in the y-direction, which leads to differences in the forces acting on the flag from above and below. For FSI 3, the inflow velocity is large enough such that the fluid develops oscillations on its own. Within the FSI 2 benchmark, this behavior results from the interaction between fluid and solid, i.e., the small pressure differences caused by the fluid move the flag, which in turn increases these differences even more. Figure 6.14 shows this displacement of the flag as well as the velocity profile and Figures 6.15 and 6.16 give a more detailed view of the periodic behavior.

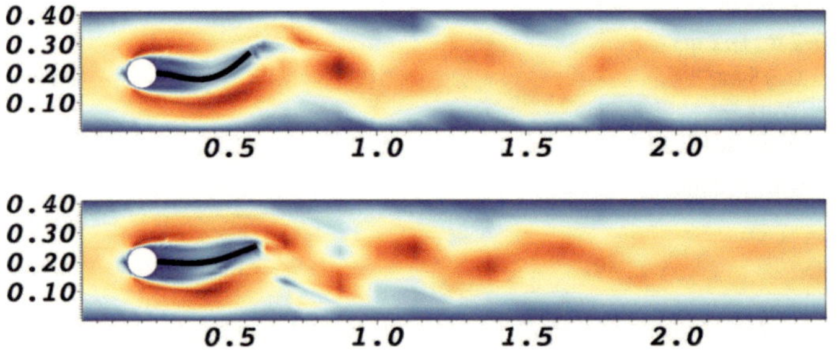

Fig. 6.14: Velocity profile of FSI 2 (top) at approx. $t = 9$ s and FSI 3 (bottom) at $t \approx 3.3$ s. One of the characteristics of the FSI 2 benchmark is the very large deformation requiring a careful choice of the ALE transformation.

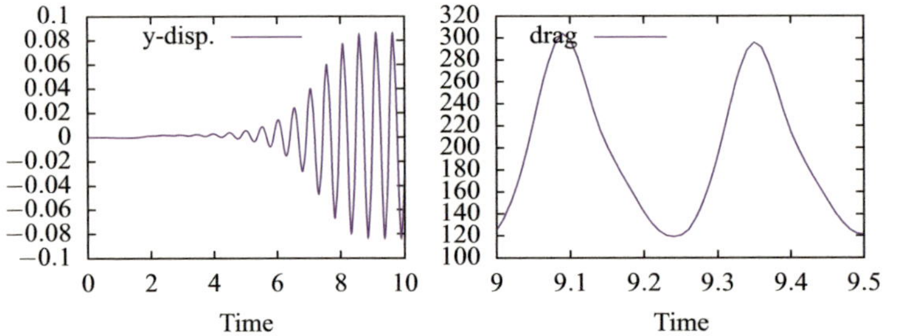

Fig. 6.15: y-displacement and drag for the FSI 2 benchmark.

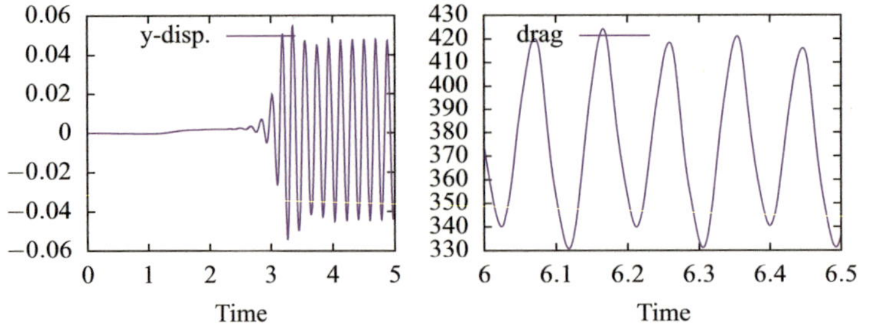

Fig. 6.16: y-displacement and drag for the FSI 3 benchmark.

For FSI 1, (almost) no such oscillations occur, which makes it a lot easier to solve. The solution of this benchmark converges to a stationary solution.

The most challenging benchmark is FSI 2, since the inflow velocity is quite high but the structure is still very flexible. This leads to large deformations, which can be problematic for many solution methods. For FSI 3, the Lamé coefficient μ is larger, leading to smaller deformation despite having the highest flow velocity.

Remark 6.22 (Time-Step Sizes). Because of its different characteristics, the various FSI benchmarks require according time-step sizes. For FSI 1, $\Delta t = 1.0$ is good enough, since this simulation is almost stationary. FSI 2 demands $\Delta t \approx 0.01$ in order to produce the correct results, whereas step sizes for the last benchmark should be around $\Delta t \approx 0.001$ to cope with the high velocity and fast oscillations. All computations were done using the Crank–Nicolson time-stepping scheme, but almost the same preconditioner behavior is achieved using different values for θ.

6.5.2 Test 1: Using direct solvers for the subproblems

In our first tests, we solve the subproblems with a direct solver to get an idea of the behavior of the preconditioners.

All tests were done on the same mesh, with the number of dofs depending on the level of refinement given as in Table 6.4.

Tab. 6.4: Number of dofs for different refinement levels.

Refinement	0	1	2	3	4	5
Dofs	4,148	15,684	60,932	240,132	953,348	3,799,044

Oscillations

As can be seen in Figure 6.17, the GMRES iterations tend to mimic the oscillatory behavior of the flag for the FSI 2 benchmark.

Dependence on refinement and time-step size

One important property of a preconditioner is its robustness with respect to the mesh-size h and time-step size Δt. As we can observe in Figure 6.18, the number of GMRES iterations remains almost constant after spatial refinement, but show a small dependence on Δt. All tests shown were done for the FSI 2 benchmark.

Dependence on material parameters

Furthermore, we need to be able to run tests with different kinds of material configurations. Because of its relation to partitioned methods, it is also interesting to investigate the added-mass effect on the preconditioner. In Figure 6.19, the number of iterations

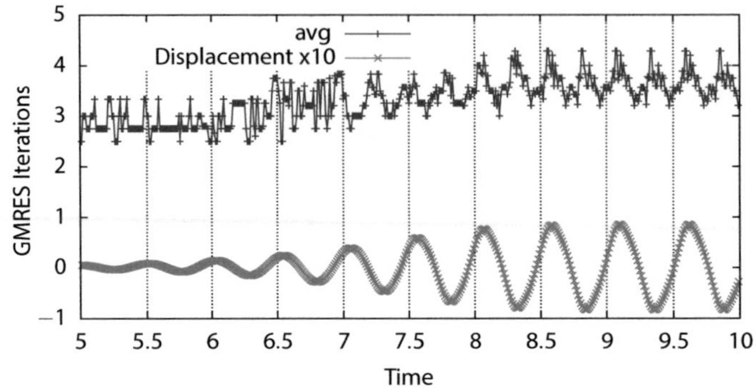

Fig. 6.17: GMRES iterations for the FSI 2 benchmark and the 21-preconditioner. The number of iterations behaves similar to the y-displacement of the flag. Displacement has been scaled by a factor of 10.

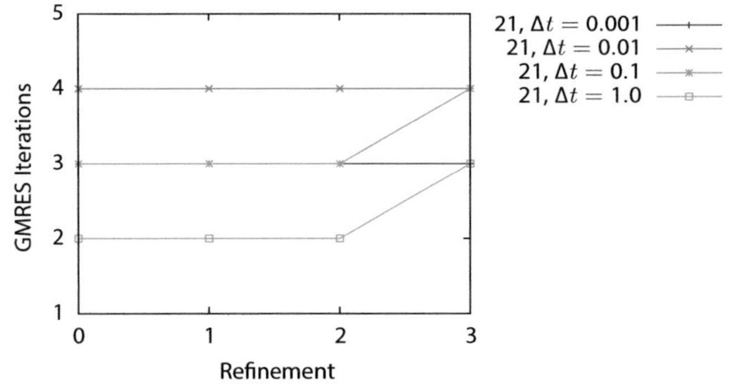

Fig. 6.18: Test 1: Number of GMRES iterations under h-refinement for different time-step values with direct solvers for subproblems.

required for the different FSI benchmarks is given. We clearly see that we are not completely independent of the choice of parameters, but the number of required iterations remains quite small. Various modifications of the FSI benchmarks yielded similar results.

6.5.3 Test 2: Iterative subsolvers

In this section we finally use a fully iterative solver, replacing the direct solvers for the subsystems in subsection 6.5.2 by iterative variants as described previously. We specifically focus on the FSI 2 benchmark since to date for monolithic schemes only [21] reported detailed results on the performance of the linear solver. Results of a similar benchmark with similar challenges can be found in [18] and the FSI 3 benchmark has been investigated in detail in [31].

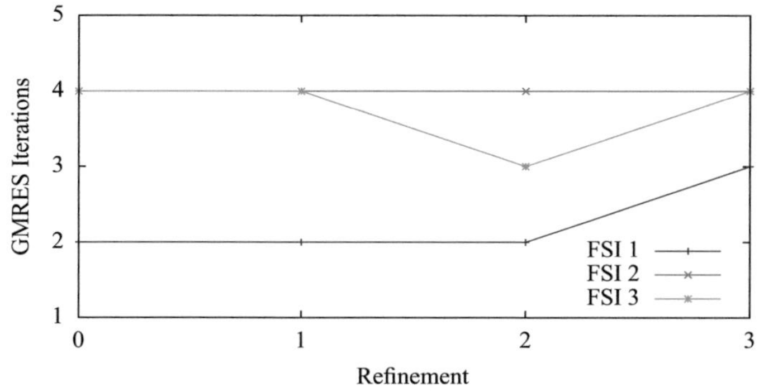

Fig. 6.19: Number of GMRES iterations for the different FSI benchmarks using Δt as proposed in Theorem 6.22 with direct subsolvers.

Discussion of our findings

Repeating all of our tests yields a similar behavior as before, but with more GMRES iterations due to the iterative approximations. As we can see in Figure 6.20, the time-step size Δt has a larger impact for the fully iterative solver. Nevertheless, the number of iterations is still within reasonable bounds. Interestingly, smaller values for Δt seem to decrease the convergence rate. For this purpose we conducted an additional test to study the behavior for very small time-step sizes. In these tests, an additional volume force was added acting onto the solid in order to speed up the displacement. Indeed, we observe in Figure 6.21 and 6.22 an improvement of the rates for step-sizes $\Delta t < 10^{-4}$. We note that we have no explanation for the slow convergence using $\Delta t = 10^{-4}$ when using the fully iterative solver.

In order to investigate the added-mass effect, we vary the structural density of the FSI 2 benchmark by using $\rho_s \in \{10^3, \ldots, 10^8\}$. Figure 6.23 shows that we require

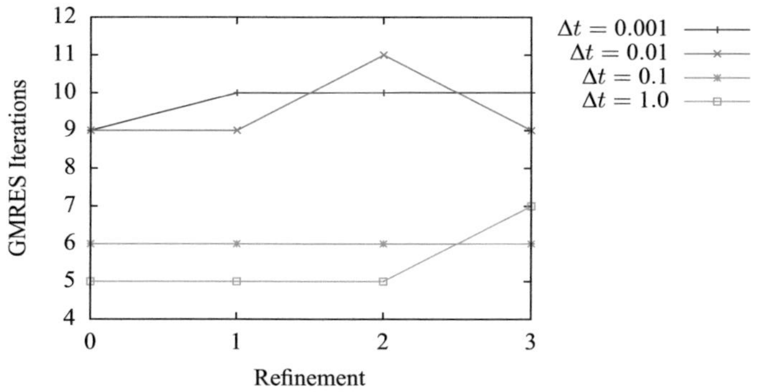

Fig. 6.20: Test 2: Number of GMRES iterations for the FSI 2 benchmark for different time-step values and refinements using the fully iterative variant.

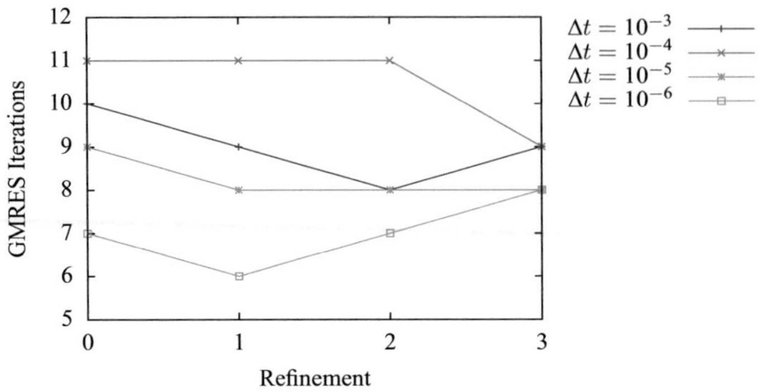

Fig. 6.21: Test 1: Number of GMRES iterations for small time-step values in a slightly modified FSI 2 setting.

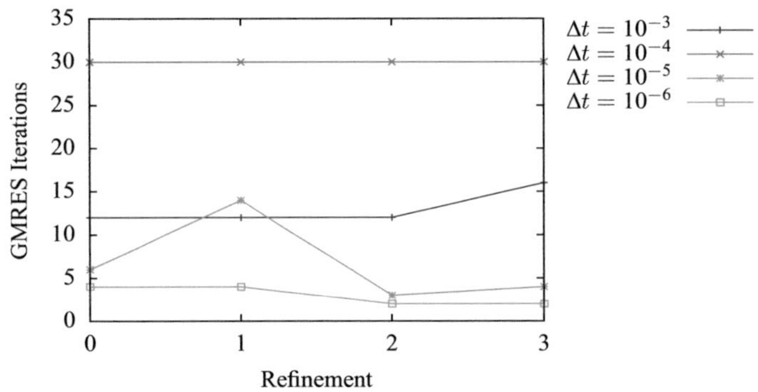

Fig. 6.22: Test 2: Number of GMRES iterations for small time-step values in a slightly modified FSI 2 setting.

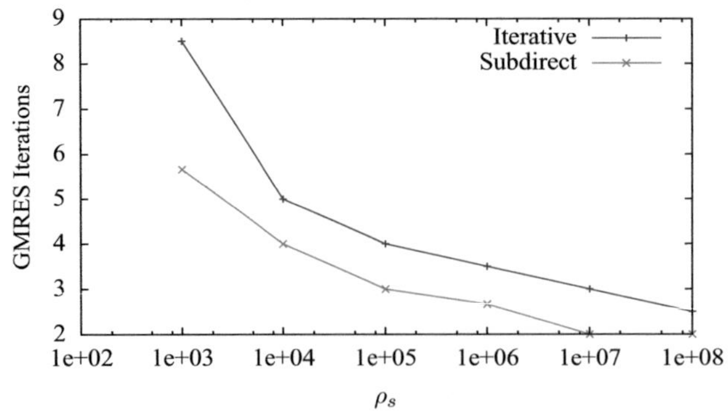

Fig. 6.23: Iterations depending on the solid density ρ_s with $\rho_f = 1000$ fixed with direct and iterative subsolvers. Other values were chosen as in FSI 2.

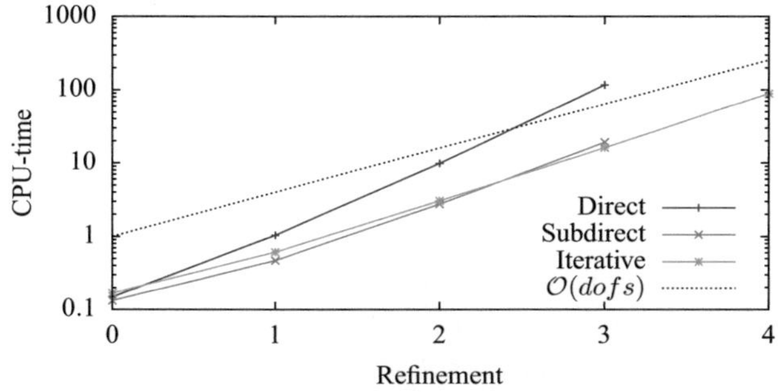

Fig. 6.24: Comparison of the computational time for various solution strategies. "Direct" refers to the use of a direct solver for the whole matrix, "subdirect" denotes the solution of the subproblems with direct methods, whereas "iterative" solves all these systems iteratively.

more iterations for $\rho_s \approx \rho_f$, indicating that the added-mass effect has some influence on the preconditioner. A comparison of the computational time for different solution strategies is provided in Figure 6.24

6.6 Conclusions and future work

6.6.1 Conclusions

In this work we adapted the idea presented in [27] to our FSI formulation and discretization. Contrary to [27], we split the second-order solid equations into a system of first-order equations employing the One-Step-θ scheme for temporal discretization instead of the Newmark scheme. Because of the different interface-coupling strategies used, we do not immediately arrive at a block-structure leading to a usable block-LU factorization. This issue is overcome by neglecting one of the coupling terms, similar to the idea presented in [21], yielding three slightly different preconditioners. From a variety of tests done, the 12-preconditioner seems to be the most robust one. This is particularly interesting, as this one is also the cheapest one to apply.

Furthermore, we neglect the computation of the global Schur complement completely. This splits the coupled system into its natural subproblems, which can then be solved by available techniques similar to the partitioned approach. However, it is unclear whether this approximation works for all kinds of problem configurations.

The solution of the fluid and solid subproblems remains challenging. In particular, the robustness of the preconditioner depends very much on the robustness of the solvers for the fluid and structural subproblems. This makes direct solvers for the (sub-)subproblems a viable option. Nevertheless, we were able to reduce the computational time significantly compared to a plain direct solver. The computations were done on an 8 GB laptop, which was just about enough for 3-times refinement ($\approx$ 250.000 dofs) when using a direct solver.

6.6.2 Future work

There are several aspects that can be improved. The most important one is to incorporate more efficient iterative solvers for the fluid and solid subproblems, which work robustly with respect to different material parameters, h and Δt. Several ideas are presented in [25, 36]. Furthermore, one could adapt the preconditioner for different mesh-motion techniques like a biharmonic version. This should be straightforward, but one will most likely have to replace the AMG mesh solver by a different method.

Currently, only global refinements are done in our application. For further reduction of computational cost, local mesh adaptivity can be employed. Here, we need to study the effects on the preconditioner.

Another point improvement consists in the adaptive control of the convergence criteria for the nested iterative subsolvers, which attempts to minimize the computational time. During testing, it was always some kind of guessing game to find a balance between speed and accuracy.

Furthermore, a very important issue is the parallelization on distributed memory machines. The limiting factor here is probably the scalability of the matrix-matrix multiplication required for the Schur complements. One may overcome this by using local approximations of the Schur complement or different solution techniques.

From the application point of view, we envisage tackling multiphysics problems beyond classical fluid-structure interaction. Therefore, this work – implemented in a modern software library – constitutes a crucial step.

Acknowledgment: We want to thank Professor Ulrich Langer for supporting this work at the Institute of Computational Mathematics at JKU Linz.

Bibliography

[1] Adams RA, Sobolev spaces, Pure and applied mathematics; 65, Acad. Press, New York, NY [u.a.], 1975.
[2] Antman SS, Nonlinear problems of elasticity, Applied mathematical sciences; 107, Springer, New York, NY [u.a.], 1995.
[3] Badia S, Quaini Q and Quarteroni A, Splitting methods based on algebraic factorization for fluid-structure interaction, SIAM J. Sci. Comput. 30 (2008), 1778–1805.
[4] Bangerth W, Davydov D, Heister T, Heltai L, Kanschat G, Kronbichler M, Maier M, Turcksin B and Wells D, The `deal.II` Library, Version 8.4, Journal of Numerical Mathematics 24 (2016).
[5] Bangerth W, Hartmann R and Kanschat G, deal.II – a General Purpose Object Oriented Finite Element Library, ACM Trans. Math. Softw. 33 (2007), 24/1–24/27.

[6] Bangerth W and Kayser-Herold O, Data Structures and Requirements for hp Finite Element Software, ACM Trans. Math. Softw. 36 (2009), 4/1–4/31.

[7] Bazilevs Y, Takizawa K and Tezduyar TE, Computational Fluid-Structure Interaction: Methods and Applications, Wiley, 2013.

[8] Braess D and Sarazin R, An Efficient Smoother for the Stokes Problem, Appl. Numer. Math. 23 (1997), 3–19.

[9] Bungartz HJ and Schäfer M, Fluid-Structure Interaction: Modelling, Simulation, Optimization, Lecture Notes in Computational Science and Engineering 53, Springer, 2006.

[10] Bungartz HJ, Miriam Mehl and Michael Schäfer, Fluid Structure Interaction II: Modelling, Simulation, Optimization, 73, Springer Science & Business Media, 2010.

[11] Causin P, Gerbeau JF and Nobile F, Added-mass effect in the design of partitioned algorithms for fluid-structure problems, Comput. Methods Appl. Mech. Engrg. 194 (2005), 4506–4527.

[12] Davis TA, Algorithm 832: UMFPACK V4.3 – an Unsymmetric-pattern Multifrontal Method, ACM Trans. Math. Softw. 30 (2004), 196–199.

[13] Deuflhard P, Newton Methods for Nonlinear Problems, Springer Series in Computational Mathematics 35, Springer Berlin Heidelberg, 2011.

[14] Duff IS, Erisman AM and John Ker Reid, Direct methods for sparse matrices, Clarendon press Oxford, 1986.

[15] Dunne T, Adaptive Finite Element Approximation of Fluid-Structure Interaction Based on Eulerian and Arbitrary Lagrangian-Eulerian Variational Formulations, Ph.D. thesis, University of Heidelberg, 2007.

[16] Formaggia L, Quarteroni A and Veneziani A, Cardiovascular Mathematics: Modeling and simulation of the circulatory system, Springer-Verlag, Italia, Milano, 2009.

[17] Galdi G and Rannacher R, Fundamental Trends in Fluid-Structure Interaction, World Scientific, 2010.

[18] Gee MW, Küttler U and Wall WA, Truly monolithic algebraic multigrid for fluid–structure interaction, International Journal for Numerical Methods in Engineering 85 (2011), 987–1016.

[19] Gee MW, Siefert CM, Hu JJ, Tuminaro RS and Sala MG, ML 5.0 Smoothed Aggregation User's Guide, Sandia National Laboratories, Report no. SAND2006-2649, 2006.

[20] George A and Liu J, Computer Solution of Sparse Linear Systems, 1994.

[21] Heil M, An efficient solver for the fully coupled solution of large-displacement fluid-structure interaction problems, Computer Methods in Applied Mechanics and Engineering 193 (2004), 1–23.

[22] Heroux M, Bartlett R, Howle V, Hoekstra R, Hu J, Kolda T, Lehoucq R, Long K, Pawlowski R, Phipps E, Salinger A, Thornquist H, Tuminaro R, Willenbring J and Williams A, An Overview of Trilinos, Sandia National Laboratories, Report no. SAND2003-2927, 2003.

[23] Holzapfel GA, Nonlinear solid mechanics; a continuum approach for engineering, reprint. ed, Wiley, Chichester, 2010.

[24] Hron J and Turek S, A monolithic FEM/Multigrid solver for ALE formulation of fluid structure with application in biomechanics, 53, pp. 146–170, Springer-Verlag, 2006.

[25] Langer U and Yang H, Partitioned solution algorithms for fluid-structure interaction problems with hyperelastic models, Journal of Computational and Applied Mathematics 276 (2015), 47–61.

[26] Langer U and Yang H, Numerical simulation of fluid–structure interaction problems with hyperelastic models: A monolithic approach, Mathematics and Computers in Simulation (2016).

[27] Langer U and Yang H, Robust and efficient monolithic fluid-structure-interaction solvers, International Journal for Numerical Methods in Engineering 108 (2016), 303–325.

[28] Langer U and Yang H, Recent development of robust monolithic fluid-structure interaction solvers, Fluid-Structure Interaction: Modeling, Adaptive Discretization and Solvers (Frei S,

Holm B, Richter T, Wick T and Yang H, eds.), Radon Series on Computational and Applied Mathematics, Walter de Gruyter, Berlin, 2017.
[29] Nocedal J and Wright SJ, Numerical optimization, Springer Ser. Oper. Res. Financial Engrg., 2006.
[30] Ortega JM, The Newton-Kantorovich Theorem, The American Mathematical Monthly 75 (1968), 658–660.
[31] Richter T, A monolithic geometric multigrid solver for fluid-structure interactions in ALE formulation, International Journal for Numerical Methods in Engineering 104 (2015), 372–390, nme.4943.
[32] Saad Y, Iterative Methods for Sparse Linear Systems, 2nd ed, Society for Industrial and Applied Mathematics, Philadelphia, PA, USA, 2003.
[33] Saad Y and Schultz MH, GMRES: A generalized minimal residual algorithm for solving nonsymmetric linear systems, SIAM J. SCI. STAT. COMPUT 7 (1986), 856–869.
[34] Stein K, Tezduyar T and Benney R, Mesh moving techniques for fluid-structure interactions with large displacements, J. Appl. Mech. 70 (2003), 58–63.
[35] Stüben K, Algebraic Multigrid (AMG): An Introduction with Applications, GMD-Report, GMD-Forschungszentrum Informationstechnik, 1999.
[36] Turek S, Efficient solvers for incompressible flow problems: An algorithmic approach, 1999.
[37] Turek S and Hron J, Proposal for Numerical Benchmarking of Fluid-Structure Interaction between an Elastic Object and Laminar Incompressible Flow, pp. 371–385, Springer Berlin Heidelberg, Berlin, Heidelberg, 2006.
[38] E. H. van Brummelen, K. G. van der Zee and R. de Borst, Space/Time Multigrid for a Fluid–structure-interaction Problem, Appl. Numer. Math. 58 (2008), 1951–1971.
[39] Wang H and Belytschko T, Fluid–structure interaction by the discontinuous-Galerkin method for large deformations, International Journal for Numerical Methods in Engineering 77 (2009), 30–49.
[40] Weickert J and Weickert J, Navier–Stokes equations as a differential-algebraic system, 1996.
[41] Wick T, Adaptive Finite Element Simulation of Fluid-Structure Interaction with Application to Heart-Valve Dynamics, Ph.D. thesis, University of Heidelberg, 2011.
[42] Wick T, Fluid-structure interactions using different mesh motion techniques, Computers & Structures 89 (2011), 1456–1467.
[43] Wick T, Solving Monolithic Fluid-Structure Interaction Problems in Arbitrary Lagrangian Eulerian Coordinates with the deal.II Library, Archive of Numerical Software 1 (2013), 1–19.
[44] Wood WL, Practical Time-stepping Schemes, Clarendon Press, 1990.
[45] Zulehner W, A Class of Smoothers for Saddle Point Problems, Computing 65 (2000), 227–246.
[46] Fernández MA and Moubachir M, A Newton method using exact jacobians for solving fluid–structure coupling, Computers & Structures 83 (2005), 127–142, Advances in Analysis of Fluid Structure Interaction.
[47] Ciarlet P, Mathematical elasticity, volume I: Three-dimensional elasticity, Acta Applicandae Mathematica 18 (1990), 190–195.
[48] Ball JM, Convexity conditions and existence theorems in nonlinear elasticity, Archive for Rational Mechanics and Analysis 63 (1976), 337–403.

Part III: **Applications**

Claudia M. Colciago, Simone Deparis, and Davide Forti

7 Fluid-structure interaction for vascular flows: From supercomputers to laptops

Abstract: Several models exist for the simulation of vascular flows; they span from simple circuit models to full three-dimensional ones that take into account detailed features of the blood and of the arterial wall. Each model comes with both benefits and drawbacks, the main denominator being a compromise between detailed resolution requirements versus computational time.

We first present a fluid-structure interaction computational model where both the fluid and the structure are three dimensional. In particular, the fluid includes modeling of large eddies by the variational multiscale method. After time and space discretizations carried out by finite differences and finite elements, respectively, we set up a parallel solver based on domain decomposition and a FaCSI preconditioner. These simulations allow one to capture details of the flow dynamics and of the structure deformation even in the transitional regime characterizing hemodynamics in the aorta. It takes roughly 10 hours to complete a simulation of one heartbeat with 35 million degrees of freedom on 2048 cores.

We then reduce both the model and its numerical complexity. The structural model is simplified to a two-dimensional membrane located at the fluid-structure interface and the fluid computational domain is fixed. For a fixed geometry and mesh, these assumptions allow one to apply proper orthogonal decomposition and generate a space discretization which has only a few dozen degrees of freedom. It is then possible to perform the simulation of one heartbeat on a laptop in less than one second.

The modeling and numerical reduction therefore allows a dramatic reduction of computational time. However, the price to pay comes, on the one hand, in terms of the preparation of a reduced basis specific to the patient and the geometry of the vessel and, on the other hand, with a detriment of certain quantities of interest. For example, when using a finite element discretization with 9 million degrees of freedom, the offline part takes about 12 hours on 720 cores for the example provided in this work; in this case, the flow profiles in the aorta are pretty close to the full three-dimensional

Note: We gratefully acknowledge the CSCS for providing us with CPU resources under project ID s635. This research has been supported in part by ERC Advanced grant N.227058 MATHCARD, SNF (no. 140184) and PASC.

Claudia M. Colciago, MATHICSE, Chair of Modeling and Scientific Computing, École Polytechnique Fédérale de Lausanne, Station 8, 1015 Lausanne, Switzerland, claudia.colciago@epfl.ch
Simone Deparis, MATHICSE, Chair of Modeling and Scientific Computing, École Polytechnique Fédérale de Lausanne, Station 8, 1015 Lausanne, Switzerland, simone.deparis@epfl.ch
Davide Forti, MATHICSE, Chair of Modeling and Scientific Computing, École Polytechnique Fédérale de Lausanne, Station 8, 1015 Lausanne, Switzerland, davide.forti@epfl.ch

https://doi.org/10.1515/9783110494259-007

model, but the wall shear stress is overestimated (although it follows the same temporal patterns).

Keywords: Fluid-structure interaction, reduced order modeling, hemodynamics, high-performance computing, algorithms for FSI, preconditioners for FSI

Classification: 65M60, 65M55, 76Z05, 76D05, 35Q30, 35Q35, 92C10

7.1 Introduction

The complexity of the cardiovascular system poses numerous challenges for its biological and mathematical modeling, numerical discretization and simulation. The study of the cardiovascular system is mainly motivated by the significant number of pathologies and deaths related to cardiovascular diseases. Heart attacks and strokes are responsible of half of human deaths in developed countries. Two of the main cardiovascular diseases are atherosclerosis and aneurysms. Atherosclerosis is an accumulation of fatty material, fibrous elements and calcium located in the intima which is the structural layer of the arterial wall that is in direct contact with the blood flow. Several studies have shown that the mechanical stimuli of the blood flow on the arterial wall are strictly related to the development of atherosclerosis [69, 80, 81, 108]. Aneurysms are another example of a widespread vascular disease. An aneurysm is a gradual stretching of the arterial wall which becomes thinner and weaker. Although aneurysms can develop in all the arteries of the cardiovascular system, they are more likely to appear in cerebral arteries and in the ascending and abdominal parts of the aorta. It has been shown that the pathogenesis of aortic aneurysms can be triggered by mechanical stimuli, such as low or oscillatory wall shear stress, or also by regions featuring flow recirculation or stagnation [19, 82, 97]. It is evident that the study of hemodynamics is of primary importance from the clinical standpoint, hence the interest in modeling these problems from the mathematical point of view and in numerical simulations.

Both blood flow and the arterial wall are modeled by continuum mechanics; their interaction is expressed by the action-reaction principle (equilibrium of stresses) and by fluid sticking to the wall (continuity of displacements and velocities). The modeling, the discretization and the design of numerical schemes to solve this coupled Fluid-Structure Interaction (FSI) problem efficiently are still challenging tasks and major topics of ongoing research. Despite the fact that cardiovascular FSI problems have been studied for almost two decades, their intrinsic complexity keeps the debate open about the set up of appropriate models and algorithms. Numerical simulations of such complex systems may provide detailed and quantitative information to improve our understanding of the underlying complex physical phenomena, thus helping and supporting the development of more effective clinical treatments without using invasive techniques.

Mathematical models of the constituents of the cardiovascular system such as the arterial wall and blood should be able to reproduce their most important physiological features and the biological interactions among them. The vascular system is composed of a network of vessels in which blood flows under the stimuli of a quasiperiodic pump, the heart. Blood is a heterogeneous medium, composed of a particulate phase and a liquid one. The fluid part is called plasma and it behaves like water. Despite its heterogeneity, as a common assumption, blood in large vessels is considered to be a continuous media since the dimension of the particles is of orders of magnitude smaller than the diameter of the vessel. Concerning the viscous properties of blood, at low shear rates it shows a non-Newtonian behavior: in small vessels like capillaries, blood particles tend to form aggregates which affect the fluid viscosity. Nevertheless, in large arteries with diameters greater than 0.3 cm, it is widely accepted to assume a Newtonian behavior. Since our goal is to focus on large vessels such as the aorta (about 3 cm diameter) and its main bifurcations, we also chose a constant value for blood viscosity [92].

The resulting mathematical model is expressed by the Navier–Stokes equations [44, 92]. The other component of our physical system is the arterial wall. Vessels are compliant pipes that convey the blood to all the organs of the body, they are complex living tissues that react and interact with the blood flow depending on several factors. Because of the difficulty of performing *in vivo* experiments on vessel tissues, the rheological behavior of the arterial wall is still an ongoing topic of research [112]. Several models have been proposed featuring increasing levels of complexity. In the reference literature both linear and nonlinear models have been used to perform FSI simulations of the cardiovascular system, featuring both isotropic and anisotropic properties (see e.g., [62, 96, 118, 119]). Some models are enhanced by multilayered structure, by the presence of fibers or of an intraluminal thrombus (see e.g., [36, 48, 116]). Also damage and growth models were a focus of researcher interest, in particular for their relevance in pathological conditions (see e.g., [63, 66, 71]). Not only the choice of the constitutive model is challenging, but also the assessment and tuning of the mechanical parameters which show a significant variation among both intra- and interpatient conditions. As a mathematical model for the artery we chose a quasi-incompressible continuous elastic material governed by an hyperbolic partial differential equation.

In this chapter, we focus on comparisons between different numerical choices rather than comparing them with experimental results. Thus, for the sake of our presentation, we will consider a rather simplistic structural model featuring a linear elastic and isotropic constitutive law, being aware that our numerical output can be distant from *in vivo* biomechanical behavior.

The coupled fluid-structure interaction problem can not be solved analytically and so it has to be studied by means of experiments or by numerical simulations; we are interested in the latter.

On the modeling side, a distinction between the different methodologies to deal with fluid-structure interaction problems comes from the choice of the frame of reference in which the problem itself is formulated. We choose to describe the structure in a Lagrangian reference and the fluid equations using an arbitrary Lagrangian-

Eulerian (ALE) frame of reference [14, 15, 26, 45, 49, 64, 73, 93, 109, 122], which requires a new equation for the fluid domain motion. Another possibility would consist of using a space-time formulation of the FSI problem. This involves a discretization of the computational domain in time slabs, and each solution in a time slab is then computed sequentially (see [15, 111, 113, 114] for a complete overview on this formulation). It is also possible to formulate the FSI problem in a fully Eulerian frame of reference [25, 103, 124], i.e., on a fixed fluid domain; in this case, one needs to keep track of the position of the fluid-structure interface. In [38, 101, 102], a fully Eulerian formulation of FSI is used in two dimensions to avoid the degeneration of ALE mapping and to facilitate mesh adaptivity. Yet another approach is the immersed boundary method, where the fluid is written in Eulerian coordinates, while the structure is still in a Lagrangian frame of reference [18, 89, 98]. A further alternative approach to ALE methods is based on the use of the Extended Finite Element method (XFEM) [53, 87, 123].

The use of an ALE formulation for the fluid, together with a Lagrangian frame for the structure, yields an FSI problem that is composed of three subproblems, namely the fluid problem, which allows for the computation of the velocity and pressure inside the fluid domain, the solid problem, which describes the deformation of the vessel wall, and the so-called geometry problem, which accounts for the moving computational fluid domain. A peculiarity in biomechanics is the similar density of the fluid and the structure, which, combined with incompressible flows, leads to increased numerical effort due to an added-mass effect [6, 22, 120, 121]. Because of the added-mass effect, staggered methods, where the fluid and the solid are explicitly coupled, are numerically unstable. Extensive work has been carried out on the development of algorithms for the solution of time-dependent FSI problems in the framework of biomechanics. The approaches are typically categorized either as segregated or as monolithic schemes although the distinction is not always straightforward. Segregated schemes can range from simple, loosely coupled fixed-point iterations to schemes which still use segregated solvers but apply a much stronger coupling. Monolithic schemes may include block preconditioners constructed from segregated solvers as well as preconditioners for the fully coupled problem which are not constructed from separate solvers. In monolithic schemes all the unknowns of the FSI problem are solved simultaneously. Monolithic algorithms have been investigated in the literature, e.g., in [12, 14, 26, 49, 64, 73, 88, 127]. In [33], a comparison of the performance of a strongly coupled Steklov–Poincaré algorithm using the Dirichlet–Neumann, Neumann–Dirichlet, and Neumann–Neumann preconditioners with a monolithic approach preconditioned by a Dirichlet–Neumann preconditioner is reported.

A modular approach to solving the FSI problem would consist of dealing with the three problems separately, while treating their coupling implicitly. For example, one can consider the FSI problem using different types of interface conditions (Dirichlet–Neumann [75–77, 79, 86], Robin–Robin [5, 6, 95], Robin–Neumann [42], Neumann–Neumann [33], FETI [30], etc. [117]) to ensure the coupling. A comparison of different coupling schemes for FSI problems in hemodynamics is carried out in [73]. A further approach makes use of a Steklov–Poincaré formulation [29] to enforce the coupling on the fluid-structure interface.

The way the coupled set of equations describing the FSI phenomena are discretized in time induces further distinctions. One approach may consist of dealing with all the nonlinearities of the problem in an implicit mode, leading to the so-called Fully Implicit (FI) strategy [12, 14, 35, 41, 49, 59, 60, 73, 115, 127]. Although this approach is the most stable, it is also the most expensive choice. A large variety of alternative time discretizations can be devised. For instance, a Convective Explicit time discretization is used in [26], in which the nonlinear convective term in the unsteady Navier–Stokes equations is treated using a semi-implicit approach. Further, a Geometry-Convective Explicit discretization is proposed in [4, 6, 26] wherein the moving geometry is taken at the previous time step and the convective term is treated semi-implicitly. A possible approach to deal with the nonlinearity makes use of the Aitken accelerated fixed-point algorithm, see e.g., [6, 22, 29, 74, 75, 79, 86]. In this way each fixed-point iteration requires one residual evaluation. Otherwise, the time-discretized problem can be linearized via the Newton method, either considering the full Jacobian matrix, as in [41, 49, 115], or neglecting some of its contributions, as in [28, 50, 59]. Each nonlinear Newton iteration requires a solve of the linearized subproblems. Thus, the cost of each nonlinear iteration corresponds to the cost of one residual evaluation plus a variable number of solutions of the linearized subproblems.

After spatial and temporal discretizations, one of the key aspects which characterizes the different methodologies is the choice of the preconditioner. For instance, relying on monolithic schemes, one may use block preconditioners built upon physics-specific solvers as well as preconditioners for the fully coupled problem that are not constructed from separate solvers. Several strategies have been proposed in the literature, see e.g., [6, 8, 12, 26, 49, 59, 127]. Overlapping Schwarz methods within monolithic approaches were studied [26] in different regimes of severity of the added-mass effect confirming successful results for 2D already reported in [12]. A monolithic FSI approach coupling nonlinear hyperelastic solid models with Navier–Stokes equations is presented in [64], considering the incompressible case for the solid. A block preconditioner with Schur complements for the monolithic system is presented in [68]. A scalable monolithic solver for an FSI problem coupling blood flow with a conforming arterial wall in two dimensions is presented in [11, 12]. There, a Newton scheme with an explicitly computed Jacobian is applied; see also [12, 14, 41]. For the solution of FSI linearized systems, in [12] the authors use a hybrid multilevel Schwarz preconditioner that uses restricted additive Schwarz on the fine level and multiplicative Schwarz on the coarse level. The parallel Newton-Krylov-Schwarz approach for the monolithic system is extended to three dimensions in [127], and scalability is shown for up to three thousand processors. Parallel algebraic multigrid preconditioners have recently been applied to fully monolithic ALE formulations of FSI problems in the setting of biomechanics, see, e.g., [49, 125]. Specifically, in [49], the authors propose two preconditioners that apply algebraic multigrid techniques to the linearized monolithic FSI system obtained after spatial and time discretizations.

It is clear that the numerical solution of full three-dimensional coupled FSI problems is extremely challenging and, as we discussed above, requires *ad hoc* strategies to lower computational costs. Despite the fact that advanced parallel solvers and preconditioners successfully managed to decrease the costs of numerical simulations, we are still far from having a real-time resolution of FSI problems using a finite element method. The intrinsic complexity of the system and the fine unstructured meshes yield computational costs that are orders of magnitude bigger than the simulated physical time: for the solution of a few heartbeats (order of seconds), several hours of CPU time on supercomputers are required. Also in terms of memory the FSI problem is expensive: the dimensions of the algebraic system and the discretized solutions are of orders of millions of degrees of freedom and they need a suitable amount of memory to be stored and manipulated. It is therefore necessary to use supercomputers and be able to solve the discretized model in parallel.
A major issue arises when we consider the final goal of these numerical simulations.

The desired features of numerical tools which may be used in everyday clinical practice includes reliability, parameter tuning and real-time computations. Indeed, numerical tools should allow specialists to compute accurate approximations of target medical outputs together with the possibility of tuning patient-specific physical parameters. Moreover, these computations should be performed in a small amount of time and, ideally, using a limited amount of computational resources (e.g., using a personal computer). These challenges have driven a research stream of numerical and applied mathematics in recent years: a great effort has been devoted to the study of *numerical reduction* techniques. These methods aim at dramatically lowering the computational costs through a likewise dramatic decrement of the dimension of the approximation spaces: numerical reduction techniques aim at describing the approximated solution using hundreds of degrees of freedom versus the millions required when using a *high-fidelity* approximation like the finite element one.

Our choice to perform numerical reduction relies on Proper Orthogonal Decomposition (POD) for *parametrized* PDEs, i.e., partial differential equations or systems of equations that depend on preselected physical and/or geometrical parameters. In cardiovascular applications, for example, physical parameters may be the mechanical properties of the structure like the Young's modulus, or the coefficients that affect the boundary conditions [58] like the average inlet flow rate or outlet resistance and compliance coefficients [17]. In scientific research many efforts have also been devoted to the investigation of geometrical parameters, like the inflow angle or the morphology of the domain itself, e.g., in [34, 52, 67, 78, 85, 105].

The POD allows one to select a set of basis functions which allows one to approximate well the manifold generated by the parameter-dependent solutions. In practice, suitable operators are built to project the high-fidelity functions onto the low-dimensional discretization space. The resulting *reduced-order model* can thus be solved with low computational costs in a many query context, i.e., for different values of the model parameters. To better quantify the efficiency of a reduced-order model, it is in general necessary to split the computational costs into two stages: an *offline* stage, expensive but performed once for the construction of the projection operators, and an *online* stage that is very rapid, in which reduced approximated solutions are computed for several realizations of the parameters.

For a detailed presentation of the reduced-order model we address the interested reader to the recent books [61, 99]. A lot of interest has been devoted to discretization reduction techniques for parametrized PDE problems in realistic applications (e.g., [100]). Recently, increasing attention has been devoted in particular to hemodynamics and cardiovascular applications (see e.g., [9, 10, 17, 20, 21, 56, 57, 84, 106]).

7.1.1 Outline

The chapter is organized as follows. In Section 2, we introduce the three-dimensional fluid and structure models, their coupling, as well as the finite element discretization and the resolution algorithm; the resulting approximation is highly accurate and the computational effort very expensive, requiring parallel solvers. Section 3 presents a possible reduced model for the structure mechanics and the integration of the structure variable into the fluid ones. This allows for a numerical reduction using POD as shown in Section 4. The resulting system is very small and the computational effort can be split into two separate steps: an offline one, requiring the resolution of finite element problems, a little less expensive than the full three-dimension model, and an online one, which is very cheap. The proposed methods are tested on a geometry including the whole aorta and few additional ramifications, see Section 5; we show numerical results, including comparisons between the full three-dimensional model and the reduced one.

7.2 Modeling fluid-structure interaction in moving domains

We are interested in simulating the hemodynamics in a piece of artery. Our model needs to account for the moving walls and at the same time focus on the region of interest. A natural choice for the structure frame of reference is the Lagrangian one. It allows for describing exactly the piece of artery we are interested in. The choice of the fluid frame of reference needs a little more care. If on one hand it has to follow the wall movement, a Lagrangian frame of reference would fit well, while on the other the fluid leaves the piece of artery after a short time, therefore a Eulerian frame would be advisable. The Arbitrary Lagrangian-Eulerian (ALE) frame of reference allows one to account for both constraints: near the arterial wall, the fluid computational domain sticks to the wall, whilst at the inlet and outlet sides the domain remains fixed.

7.2.1 Fluid-structure interaction modeling

Let $\widehat{\Omega}_{\mathrm{f}}$ and $\widehat{\Omega}_{\mathrm{s}}$ be the fluid and solid domains, respectively, in their reference configuration. We denote by $\widehat{\Gamma} = \partial\widehat{\Omega}_{\mathrm{f}} \cap \partial\widehat{\Omega}_{\mathrm{s}}$ the fluid-structure interface. The initial time is set equal to zero and the final time is denoted by T. At any time $t \in [0, T]$, we presume that the current fluid domain configuration $\Omega^{f}(t)$ can be retrieved from $\widehat{\Omega}_{\mathrm{f}}$ by the ALE mapping

$$\begin{aligned} \mathcal{A}_t\colon\ \widehat{\Omega}_{\mathrm{f}} &\to \Omega_{\mathrm{f}}(t) \\ \widehat{\mathbf{x}} &\mapsto \mathcal{A}_t(\widehat{\mathbf{x}}) = \widehat{\mathbf{x}} + \widehat{\mathbf{d}}_{\mathrm{f}}(\widehat{\mathbf{x}})\,, \end{aligned} \tag{7.1}$$

where $\widehat{\mathbf{d}}_{\mathrm{f}}$ represents the displacement of the fluid domain. The ALE mapping does not represent any physical quantity as it is a mathematical construct. We will need to construct a discrete version which is needed to describe the displacement and velocity of each mesh point. In fact, the use of the ALE formulation allows an arbitrary reconstruction of the volumetric computational grid of $\Omega_{\mathrm{f}}(t)$ from the one of the domain boundary $\partial\Omega_{\mathrm{f}}(t)$. For the sake of computation, this reconstruction is directly operated on the reference configuration by means of a harmonic extension of the structure displacement $\widehat{\mathbf{d}}_{\mathrm{s}}$ at the fluid-structure interface $\widehat{\Gamma}$ to the interior of the reference fluid domain $\widehat{\Omega}_{\mathrm{f}}$, i.e.,

$$\begin{cases} -\Delta\widehat{\mathbf{d}}_{\mathrm{f}} = 0 & \text{in } \widehat{\Omega}_{\mathrm{f}}, \qquad (7.2a) \\ \widehat{\mathbf{d}}_{\mathrm{f}} = \widehat{\mathbf{d}}_{\mathrm{s}} & \text{on } \widehat{\Gamma}. \qquad (7.2b) \end{cases}$$

Other types of extensions are also possible; for example one may use the linear elasticity equations [15, 110]. In our experience, this is necessary in the presence of large (but smooth) boundary displacements. As the solid displacement $\widehat{\mathbf{d}}_{\mathrm{s}}$ changes in time, equation (7.2a) allows one to define the current fluid domain configuration $\Omega_{\mathrm{f}}(t) = \mathcal{A}_t(\widehat{\Omega}_{\mathrm{f}})$ thanks to the ALE map parametrization (7.1).

In our FSI model, the fluid dynamics is governed by the incompressible Navier–Stokes equations written in the ALE frame of reference [15, 37, 45, 65], for almost all $t \in (0, T)$,

$$\begin{cases} \rho_{\mathrm{f}} \left.\dfrac{\partial \mathbf{u}}{\partial t}\right|_{\mathcal{A}_t} + \rho_{\mathrm{f}}((\mathbf{u}-\mathbf{w})\cdot\nabla)\mathbf{u} - \nabla\cdot\boldsymbol{\sigma}_{\mathrm{f}} = 0 & \text{in } \Omega_{\mathrm{f}}(t), \\ \nabla\cdot\mathbf{u} = 0 & \text{in } \Omega_{\mathrm{f}}(t), \\ \mathbf{u} = \mathbf{h}_{\mathrm{f}} & \text{on } \Gamma_{\mathrm{D}}^{f}, \\ \boldsymbol{\sigma}_{\mathrm{f}}\mathbf{n}_{\mathrm{f}} = \mathbf{g}_{\mathrm{f}} & \text{on } \Gamma_{\mathrm{N}}^{f}, \\ \mathbf{u}\circ\mathcal{A}_t = \dfrac{\partial\widehat{\mathbf{d}}_{\mathrm{s}}}{\partial t} & \text{on } \widehat{\Gamma}, \qquad (7.3a) \end{cases}$$

where $\frac{\partial}{\partial t}\big|_{\mathcal{A}_t} = \frac{\partial}{\partial t} + \mathbf{w}\cdot\nabla$ is the ALE derivative, $\mathbf{w} = \frac{\partial \mathcal{A}_t}{\partial t} \circ \mathcal{A}_t^{-1}$ is the fluid domain velocity, $\mathbf{u}$ and p are the velocity and pressure of the fluid, respectively, and $\mathbf{n}_\mathrm{f}$ the outward unit normal vector to $\partial\Omega_\mathrm{f}(t)$. We denoted by ρ_f the density of the fluid and by $\boldsymbol{\sigma}_\mathrm{f}$ the Cauchy stress tensor

$$\boldsymbol{\sigma}_\mathrm{f} = \mu_\mathrm{f}(\nabla\mathbf{u} + (\nabla\mathbf{u})^T) - p\mathbf{I}\,,$$

with $\mathbf{I}$ being the identity tensor, and μ_f the dynamic viscosity of the fluid. The functions $\mathbf{h}_\mathrm{f}$ and $\mathbf{g}_\mathrm{f}$ indicate the Dirichlet and Neumann data applied at the Dirichlet and Neumann boundaries Γ_D^f and Γ_N^f, respectively, of $\Omega_\mathrm{f}(t)$; equation 7.3a represents the continuity of the velocities at the interface.

We formulate the full three-dimensional structure problem in a Lagrangian frame of reference. The conservation of momentum for the structure reads

$$\begin{cases} \hat{\rho}_\mathrm{s}\dfrac{\partial^2\hat{\mathbf{d}}_\mathrm{s}}{\partial t^2} - \nabla_{\hat{\mathbf{x}}}\cdot\widehat{\boldsymbol{\Pi}}(\hat{\mathbf{d}}_\mathrm{s}) = 0 & \text{in } \widehat{\Omega}_\mathrm{s}\times(0,T], \quad (7.4a)\\ \hat{\mathbf{d}}_\mathrm{s} = \mathbf{h}_\mathrm{s} & \text{on } \widehat{\Gamma}_\mathrm{D}^s\times(0,T],\\ \widehat{\boldsymbol{\Pi}}(\hat{\mathbf{d}}_\mathrm{s})\hat{\mathbf{n}}_\mathrm{s} = 0 & \text{on } \widehat{\Gamma}_\mathrm{N}^s\times(0,T],\\ (\det[\mathbf{F}])\widehat{\boldsymbol{\Pi}}(\hat{\mathbf{d}}_\mathrm{s})\mathbf{F}^T\hat{\mathbf{n}}_\mathrm{s} + \boldsymbol{\sigma}_\mathrm{f}\mathbf{n}_\mathrm{f} = 0 & \text{on } \Gamma\times(0,T]. \quad (7.4b) \end{cases}$$

Here $\hat{\mathbf{n}}_\mathrm{s}$ represents the outward unit normal vector to $\partial\widehat{\Omega}_\mathrm{s}$ and $\mathbf{F} = \mathbf{I} + \nabla_{\hat{\mathbf{x}}}\hat{\mathbf{d}}_\mathrm{s}$ is the deformation gradient tensor. The function $\mathbf{h}_\mathrm{s}$ indicates the Dirichlet data applied at the Dirichlet boundary Γ_D^s of $\widehat{\Omega}_\mathrm{s}$. The material is characterized by a density $\hat{\rho}_\mathrm{s}$, Young's modulus E_s and the Poisson's ratio ν_s, which define in turn the first Piola-Kirchhoff stress tensor

$$\widehat{\boldsymbol{\Pi}}(\hat{\mathbf{d}}_\mathrm{s}) = \lambda_\mathrm{s}\,\mathrm{tr}\left(\frac{\nabla_{\hat{\mathbf{x}}}\hat{\mathbf{d}}_\mathrm{s} + (\nabla_{\hat{\mathbf{x}}}\hat{\mathbf{d}}_\mathrm{s})^T}{2}\right)\mathbf{I} + \mu_\mathrm{s}(\nabla_{\hat{\mathbf{x}}}\hat{\mathbf{d}}_\mathrm{s} + (\nabla_{\hat{\mathbf{x}}}\hat{\mathbf{d}}_\mathrm{s})^T)\,, \tag{7.5}$$

where

$$\lambda_\mathrm{s} = \frac{E_\mathrm{s}\nu_\mathrm{s}}{(1-2\nu_\mathrm{s})(1+\nu_\mathrm{s})} \quad\text{and}\quad \mu_\mathrm{s} = \frac{E_\mathrm{s}}{2(1+\nu_\mathrm{s})}$$

are the first and second Lamé coefficients, respectively. Other constitutive laws are of course also possible, here we chose the simplest one.

The coupling between these three subproblems is ensured by imposing the geometry adherence, the continuity of the velocity and the continuity of the normal component of the stresses at the interface through Equations (7.2b), (7.3a) and (7.4b), respectively. Defining the physical subproblems with (7.3a) and (7.4b) is arbitrary, one could choose to swap them. We notice that the resulting coupled system is nonlinear due to the convective term in the fluid momentum equation, to the possible use of nonlinear material laws to model the structure deformation and to the moving computational fluid domain.

7.2.2 Weak formulation

We recall the weak form of the FSI problem in nonconservative form. The velocity coupling condition is imposed strongly, while the continuity of the normal component of the stresses is imposed in weak form. We presume that $\mathcal{A}_t$ is a diffeomorphism for any time $t \in [0, T]$ and introduce the following functional spaces:

$$
\begin{aligned}
U^f &= \{\mathbf{v} = \widehat{\mathbf{v}} \circ \mathcal{A}_t^{-1} \mid \widehat{\mathbf{v}} \in [H^1(\widehat{\Omega}_\mathrm{f})]^3\}\,,\\
U^f_\mathrm{D} &= \{\mathbf{v} = \in U^f, \mathbf{v} = 0 \;\text{ on }\; \Gamma^f_\mathrm{D}(t)\}\,,\\
Q^f &= \{q = \widehat{q} \circ \mathcal{A}_t^{-1} \mid \widehat{q} \in L^2(\widehat{\Omega}_\mathrm{f})\}\,,\\
U^s &= [H^1(\widehat{\Omega}_\mathrm{s})]^3,\; U^s_\mathrm{D} = \{\mathbf{v} \in U^s \mid \mathbf{v} = 0 \;\text{ on }\; \Gamma^s_\mathrm{D}\}\,,\\
U^g &= [H^1(\widehat{\Omega}_\mathrm{f})]^3,\; U^g_\mathrm{D} = \{\mathbf{v} \in U^g \mid \mathbf{v} = 0 \;\text{ on }\; \Gamma^f_\mathrm{fixed}\}\,,\\
U^\lambda &= [H^{1/2}(\widehat{\Gamma})]^3\,.
\end{aligned}
$$

Γ^f_fixed is the part of the fluid boundary $\widehat{\Omega}_\mathrm{f}$ where the domain motion is prescribed (e.g., when a portion of the fluid domain boundary is fixed). The weak form of the fluid and structure equations is standard [45]. We introduce an auxiliary variable $\boldsymbol{\lambda} \in U^\lambda$ representing the normal stresses exerted by the fluid on the interface Γ and which, thanks to (7.4b), can be written as:

$$
\boldsymbol{\lambda} \in U^\lambda:\; \int_{\widehat{\Gamma}} \boldsymbol{\lambda} \cdot \boldsymbol{\eta}\, dy = \int_{\Gamma(t)} \sigma_\mathrm{f}\mathbf{n}_\mathrm{f} \cdot \boldsymbol{\eta} \circ \mathcal{A}_t^{-1}\, dy = -\int_{\widehat{\Gamma}} \Pi(\widehat{\mathbf{d}}_\mathrm{s})\widehat{\mathbf{n}}_\mathrm{s} \cdot \boldsymbol{\eta}\, d\widehat{y} \quad \forall \boldsymbol{\eta} \in U^\lambda\,. \tag{7.6}
$$

We recall the notation for the Dirichlet boundary data for the fluid, structure, and geometry subproblem: $\mathbf{h}_\mathrm{f}\colon \Gamma^f_\mathrm{D} \to \mathbb{R}^2$, $\mathbf{h}_\mathrm{s}\colon \Gamma^s_\mathrm{D} \to \mathbb{R}^2$, $\mathbf{h}_g\colon \Gamma^f_\mathrm{fixed} \to \mathbb{R}^2$, respectively. The weak form of the FSI problem reads: for all $t \in (0, T]$, find $(\mathbf{u}, p, \widehat{\mathbf{d}}_\mathrm{s}, \widehat{\mathbf{d}}_\mathrm{f}, \boldsymbol{\lambda}) \in U^f \times Q^f \times U^s \times U^g \times U^\lambda$ such that $\mathbf{u} = \mathbf{h}_\mathrm{f}$ on Γ^f_D, $\widehat{\mathbf{d}}_\mathrm{s} = \mathbf{h}_\mathrm{s}$ on Γ^s_D, $\widehat{\mathbf{d}}_\mathrm{f} = \mathbf{h}_g$ on Γ^f_fixed, satisfying

$$
\int_{\Omega_\mathrm{f}(t)} \left(\rho_\mathrm{f} \left.\frac{\partial \mathbf{u}}{\partial t}\right|_{\mathcal{A}_t} \cdot \mathbf{v}_\mathrm{f} + \rho_\mathrm{f}((\mathbf{u} - \mathbf{w}) \cdot \nabla)\mathbf{u} \cdot \mathbf{v}_\mathrm{f} + \boldsymbol{\sigma}_\mathrm{f} : \nabla \mathbf{v}_\mathrm{f} \right) d\Omega \tag{7.7}
$$

$$
+ \int_{\widehat{\Gamma}} \boldsymbol{\lambda} \cdot (\mathbf{v}_\mathrm{f} \circ \mathcal{A}_t)\, d\widehat{y} = \int_{\Gamma^f_\mathrm{N}} \mathbf{g}_\mathrm{f} \cdot \mathbf{v}_\mathrm{f}\, dy
$$

$$
\int_{\Omega_\mathrm{f}(t)} q \nabla \cdot \mathbf{u}\, d\Omega = 0 \tag{7.8}
$$

$$
\int_{\widehat{\Omega}_\mathrm{s}} \left(\widehat{\rho}_\mathrm{s} \frac{\partial^2 \widehat{\mathbf{d}}_\mathrm{s}}{\partial t^2} \cdot \mathbf{v}_\mathrm{s} + \widehat{\Pi}(\widehat{\mathbf{d}}_\mathrm{s}) : \nabla_{\widehat{\mathbf{x}}} \mathbf{v}_\mathrm{s} \right) d\Omega - \int_{\widehat{\Gamma}} \boldsymbol{\lambda} \cdot \mathbf{v}_\mathrm{s}\, d\widehat{y} = \int_{\Gamma^s_\mathrm{N}} \mathbf{g}_\mathrm{s} \cdot \mathbf{v}_\mathrm{s}\, dy
$$

$$
\int_{\widehat{\Gamma}} (\mathbf{u}_\mathrm{f} \circ \mathcal{A}_t) \cdot \boldsymbol{\eta}\, dy - \int_{\widehat{\Gamma}} \frac{\partial \widehat{\mathbf{d}}_\mathrm{s}}{\partial t} \cdot \boldsymbol{\eta}\, dy = 0
$$

$$\int_{\widehat{\Gamma}} \nabla_{\hat{\mathbf{x}}}\widehat{\mathbf{d}}_{\mathrm{f}} : \nabla_{\hat{\mathbf{x}}}\mathbf{v}_g \, d\Omega = 0 \qquad \forall\, (\mathbf{v}_{\mathrm{f}}, q, \mathbf{v}_{\mathrm{s}}, \mathbf{v}_g, \boldsymbol{\eta}) \in U_{\mathrm{D}}^f \times Q^f \times U_{\mathrm{D}}^s \times U_{\mathrm{D}}^g \times U^\lambda$$

$$\widehat{\mathbf{d}}_{\mathrm{f}} = \widehat{\mathbf{d}}_{\mathrm{s}} \qquad \text{on } \widehat{\Gamma}.$$

Note that $\boldsymbol{\eta}$ represents the Lagrange multiplier which enforces the equilibrium of the stresses (7.4b) in weak form. From the variational viewpoint, the normal stresses are in $H^{-1/2}(\Gamma)$, it is therefore natural to impose their equilibrium only weakly. A strong treatment would not be consistent. From (7.6) and (7.7) we observe that

$$\int_{\Gamma(t)} \sigma_{\mathrm{f}}\mathbf{n}_{\mathrm{f}} \cdot \boldsymbol{\eta} \circ \mathcal{A}_t^{-1} \, dy + \int_{\widehat{\Gamma}} \Pi(\widehat{\mathbf{d}}_{\mathrm{s}})\widehat{\mathbf{n}}_{\mathrm{s}} \cdot \boldsymbol{\eta} \, d\hat{y} = 0 \quad \forall \boldsymbol{\eta} \in U^\lambda \tag{7.9}$$

7.2.3 Fully implicit monolithic scheme

A large variety of different time-discretization schemes for the coupled set of equations describing the FSI phenomena can be found in the literature. One approach may consist in dealing with all the nonlinearities of the problem in a fully implicit fashion, leading to the so-called fully implicit strategy [14, 41, 49, 59, 127]. Although this approach is the most stable, it is also the most expensive. Alternative time discretizations, depending on the targeted application, can be devised. For instance, in the field of hemodynamics, a Convective Explicit (CE) time discretization is used in [27], in which the nonlinear convective term in the unsteady Navier–Stokes equations is linearized by means of extrapolation. Further, a Geometry-Convective Explicit (GCE) discretization is proposed in [7] wherein the moving geometry is taken at the previous time step and the convective term is treated semi-implicitly, thus leading to a linear problem to be solved at each discrete time step (provided that a linear material law is used for the solid).

We consider a Fully Implicit (FI) scheme for which all the nonlinearities are treated implicitly. In space, we consider a Galerkin finite elements approximation. Stabilized P1-P1 elements are used to approximate the fluid velocity and pressure variables with VMS-SUPG (Variational Multiscale–Streamline-Upwind Petrov-Galerkin) stabilization [13, 46]. The solid displacement and ALE fields are discretized by P1 finite elements. When considering the reduced model presented in section 7.3, we use Taylor-Hood P2-P1 finite elements without stabilization for the fluid, while the structure is integrated into the fluid variable.

For the sake of brevity, here we do not report on the finite element formulation since it differs from (7.8) only on the spaces. When using the VMS-SUPG stabilization, the following terms are added to the Galerkin formulation. Inside each element K of the fluid tetrahedral mesh $\mathcal{T}_{\mathrm{f}}^h$ representing the current fluid domain $\Omega_{\mathrm{f}}(t)$, the finite element solution is a linear function and it is possible to compute the strong residual of the momentum equation as

$$\mathbf{r}_K(\mathbf{u},p) = \rho_\mathrm{f}\frac{\partial \mathbf{u}}{\partial t}\bigg|_{\mathcal{A}_t} + \rho_\mathrm{f}((\mathbf{u}-\mathbf{w})\cdot\nabla)\mathbf{u} + \nabla p - \mu\Delta\mathbf{u})\,.$$

The SUPG-VMS stabilizing terms are then computed as

$$\begin{aligned}&\sum_{K\in\mathcal{T}_\mathrm{f}^h}\int_K ((\mathbf{u}-\mathbf{w})\cdot\nabla\mathbf{v}_\mathrm{f} + q)\ \tau_\mathrm{m}\,\mathbf{r}_K(\mathbf{u},p)\,d\Omega\\ &+\sum_{K\in\mathcal{T}_\mathrm{f}^h}\int_K \nabla\cdot\mathbf{v}_\mathrm{f}\,\tau_\mathrm{c}\nabla\cdot\mathbf{u}\,d\Omega\\ &+\sum_{K\in\mathcal{T}_\mathrm{f}^h}\int_K (\mathbf{u}-\mathbf{w})\cdot(\nabla\mathbf{v}_\mathrm{f})^T\,\tau_\mathrm{m}\,\mathbf{r}_K(\mathbf{u},p)\,d\Omega\,,\end{aligned}$$

where the first two rows represent classical Streamline-Upwind Petrov-Galerkin (SUPG) stabilization terms and the third row stems from the Variational Multiscale (VMS) technique applied to the Navier–Stokes equations. In the stabilization terms we denoted by $K\in\mathcal{T}_\mathrm{f}^h$ one tetrahedron of the fluid mesh, while τ_m and τ_c are two stabilization parameters defined as:

$$\tau_\mathrm{m} = \frac{1}{\sqrt{\dfrac{4\rho_\mathrm{f}^2}{\Delta t^2} + \rho_\mathrm{f}^2(\mathbf{u}-\mathbf{w})\cdot\mathbf{G}(\mathbf{u}-\mathbf{w}) + C_r\mu_\mathrm{f}^2\mathbf{G}:\mathbf{G}}}\,, \tag{7.10}$$

$$\tau_\mathrm{c} = (\tau_\mathrm{m}\mathbf{g}\cdot\mathbf{g})^{-1}\,. \tag{7.11}$$

In equations (7.10) and (7.11), the covariant metric tensor $\mathbf{G}$ is defined component-wise as $G_{i,j} = \sum_{k=1}^{3}\frac{\partial\eta_k}{\partial x_i}\frac{\partial\eta_k}{\partial x_j}$ and is related to the geometrical mapping with $\boldsymbol{\eta} = (\eta_1,\eta_2,\eta_3)$, while the vector $\mathbf{g}$ has components $g_i = \sum_{k=1}^{3}\frac{\partial\eta_k}{\partial x_i}$. The constant $C_r > 0$ depends on the degree p of the basis functions and we set it as an element-wise inverse estimate $C_r = 60\cdot 2^{p-2}$ as suggested by [13].

We discretize the time derivative of the fluid problem by second-order backward differentiation formulas [46, 54]

$$\frac{\partial\mathbf{u}}{\partial t}(t^{n+1})\bigg|_{\mathcal{A}_t} \approx \frac{3\mathbf{u}^{n+1} - 4\mathbf{u}^n + \mathbf{u}^{n-1}}{2\Delta t}\,, \tag{7.12}$$

where Δt is the time-step size, $\{t^n\}_{n=1,2,\dots}$ are the times at which we solve the discrete space problem and $\mathbf{u}^{n+1}$ is the approximation of $\mathbf{u}$ at time t^{n+1}. The time discretization of the structural problem is carried out by the Newmark method [91]

$$\frac{\partial^2\hat{\mathbf{d}}_\mathrm{s}}{\partial t^2}(t^{n+1}) \approx \frac{1}{\beta\Delta t^2}\hat{\mathbf{d}}_\mathrm{s}^{n+1} - \frac{1}{\beta\Delta t^2}(\hat{\mathbf{d}}_\mathrm{s}^n + \Delta t\,\dot{\hat{\mathbf{d}}}_\mathrm{s}^n) - \frac{1-2\beta}{2\beta}\,\ddot{\hat{\mathbf{d}}}_\mathrm{s}^n\,, \tag{7.13}$$

where $\beta = 0.25$, $\gamma = 0.5$ and:

$$\ddot{\widehat{\mathbf{d}}}_{\mathrm{s}}^{n} = \frac{1}{\beta \Delta t^2}\widehat{\mathbf{d}}_{\mathrm{s}}^{n} - \frac{1}{\beta \Delta t^2}(\widehat{\mathbf{d}}_{\mathrm{s}}^{n-1} + \Delta t\, \dot{\widehat{\mathbf{d}}}_{\mathrm{s}}^{n-1}) - \frac{1-2\beta}{2\beta}\ddot{\widehat{\mathbf{d}}}_{\mathrm{s}}^{n-1}, \tag{7.14}$$

$$\dot{\widehat{\mathbf{d}}}_{\mathrm{s}}^{n} = \dot{\widehat{\mathbf{d}}}_{\mathrm{s}}^{n-1} + \Delta t\,(\gamma\, \ddot{\widehat{\mathbf{d}}}_{\mathrm{s}}^{n} + (1-\gamma)\, \ddot{\widehat{\mathbf{d}}}_{\mathrm{s}}^{n-1}) . \tag{7.15}$$

At each time step, the resulting nonlinear system to be solved may be rewritten as

$$\begin{pmatrix} S(\widehat{\mathbf{d}}_{\mathrm{s}}^{n+1}) & + & 0 & + & 0 & - I_{\Gamma^s}^T \boldsymbol{\lambda}^{n+1} \\ -I_{\Gamma^s}\widehat{\mathbf{d}}_{\mathrm{s}}^{n+1} & + & G(\widehat{\mathbf{d}}_{\mathrm{f}}^{n+1}) & + & 0 & + \; 0 \\ 0 & + & 0 & + & F(\mathbf{u}_{\mathrm{f}}^{n+1}, p_{\mathrm{f}}^{n+1}, \widehat{\mathbf{d}}_{\mathrm{f}}^{n+1}) & + I_{\Gamma^f}^T \boldsymbol{\lambda}^{n+1} \\ -\frac{\gamma}{\beta\Delta t} I_{\Gamma^s}\widehat{\mathbf{d}}_{\mathrm{s}}^{n+1} & + & 0 & + & I_{\Gamma^f}\mathbf{u}_{\mathrm{f}}^{n+1} & + \; 0 \end{pmatrix} = \begin{pmatrix} \mathbf{b}_{\mathrm{s}} \\ 0 \\ \mathbf{b}_{\mathrm{f}} \\ \mathbf{b}_{\mathrm{c}} \end{pmatrix} . \tag{7.16}$$

We denoted by $\widehat{\mathbf{d}}_{\mathrm{s}}^{n+1}$, $\mathbf{d}^{n+1}$, $\mathbf{u}^{n+1}$, p^{n+1}, and $\boldsymbol{\lambda}^{n+1}$, the unknown displacement of the structure, the displacement of the fluid domain, the velocity and pressure of the fluid, and the Lagrange multipliers, respectively. We make use of an augmented formulation wherein the Lagrange multipliers are used to impose the continuity of the velocity at the fluid-structure interface.

The diagonal blocks on the left-hand side of (7.16) account for the discretized solid, geometry and fluid problems. We remark that F is nonlinear due to the convective term and the fact that the fluid domain moves. The matrices I_{Γ^f} and I_{Γ^s} are the restriction of fluid velocity and structure displacement vectors to the interface and in (7.16) account for the continuity of velocities and the geometry adherence, which are imposed strongly. Their transposes account for the continuity of the normal component of the stresses, which is imposed weakly. The last row of (7.16) represents the discretized kinematic coupling condition at the fluid-structure interface:

$$I_{\Gamma^f}\mathbf{u}^{n+1} - \frac{\gamma}{\beta\Delta t} I_{\Gamma^s}\widehat{\mathbf{d}}_{\mathrm{s}}^{n+1} = \mathbf{b}_{\mathrm{c}} , \tag{7.17}$$

where

$$\mathbf{b}_{\mathrm{c}} = I_{\Gamma^s}\dot{\widehat{\mathbf{d}}}_{\mathrm{s}}^{n} - \frac{\gamma}{\beta\Delta t}(I_{\Gamma^s}\widehat{\mathbf{d}}_{\mathrm{s}}^{n} + \Delta t\, I_{\Gamma^s}\dot{\widehat{\mathbf{d}}}_{\mathrm{s}}^{n}) - \Delta t\,\gamma \frac{1-2\beta}{2\beta} I_{\Gamma^s}\ddot{\widehat{\mathbf{d}}}_{\mathrm{s}}^{n} + \Delta t\,(1-\gamma)\, I_{\Gamma^s}\ddot{\widehat{\mathbf{d}}}_{\mathrm{s}}^{n} . \tag{7.18}$$

We assumed that the fluid and structural meshes, as well as the polynomial degree are conforming at the interface. To account for nonconforming fluid-structure interface discretizations, please refer to [47, 87].

7.2.4 Algebraic formulation

We solve the nonlinear fully coupled FSI problem (7.16) using the Newton method as in, e.g., [14, 41, 59]. The solution of (7.16) at time $t^n = n\,\Delta t$ is denoted by $\mathbf{X}^n = (\widehat{\mathbf{d}}_{\mathrm{s}}^n, \widehat{\mathbf{d}}_{\mathrm{f}}^n, \mathbf{u}^n, p^n, \boldsymbol{\lambda}^n)^T$. At each time step, we first extrapolate an approximate solution $\mathbf{X}_0^{n+1}$ from the previous time levels by Newton-Gregory backward polynomials. In particular, since we use second-order time integrators, we set $\mathbf{X}_0^{n+1} = 2\mathbf{X}^n - \mathbf{X}^{n-1}$. Then, we compute a sequence of approximations $\mathbf{X}_1^{n+1}$, $\mathbf{X}_2^{n+1}$, etc., until the numerical solution converges up to a prescribed tolerance. The generic $k+1$ iteration of the Newton method applied to (7.16) is described as follows. Starting from an approximation of $\mathbf{X}_k^{n+1}$, we compute the residual $\mathbf{R}_k^{n+1} = (\mathbf{r}_{\mathbf{d}_s,k}^{n+1}, \mathbf{r}_{\mathbf{d}_f,k}^{n+1}, \mathbf{r}_{u,k}^{n+1}, r_{p,k}^{n+1}, \mathbf{r}_{\boldsymbol{\lambda},k}^{n+1})^T$:

$$\mathbf{R}_k^{n+1} = \begin{pmatrix} \mathbf{b}_s \\ 0 \\ \mathbf{b}_f \\ \mathbf{b}_c \end{pmatrix} - \begin{pmatrix} S(\widehat{\mathbf{d}}_{s,k}^{n+1}) & -I_{\Gamma^s}^T \boldsymbol{\lambda}_k^{n+1} \\ -I_{\Gamma^s}\widehat{\mathbf{d}}_{s,k}^{n+1} & +G(\widehat{\mathbf{d}}_{f,k}^{n+1}) \\ F(\mathbf{u}_k^{n+1}, p_k^{n+1}, \widehat{\mathbf{d}}_{f,k}^{n+1}) & +I_{\Gamma^f}^T \boldsymbol{\lambda}_k^{n+1} \\ -\frac{\gamma}{\beta\Delta t} I_{\Gamma^s}\widehat{\mathbf{d}}_{s,k}^{n+1} & +I_{\Gamma^f}\mathbf{u}_k^{n+1} \end{pmatrix} . \tag{7.19}$$

Then, we compute the Newton correction vector $\delta\mathbf{X}_k^{n+1} = (\delta\widehat{\mathbf{d}}_{s,k}^{n+1}, \delta\widehat{\mathbf{d}}_{f,k}^{n+1}, \delta\mathbf{u}_k^{n+1}, \delta p_k^{n+1}, \delta\boldsymbol{\lambda}_k^{n+1})^T$ by solving the Jacobian linear system

$$J_{\mathrm{FSI}}\, \delta\mathbf{X}_k^{n+1} = -\mathbf{R}_k^{n+1} , \tag{7.20}$$

being

$$J_{\mathrm{FSI}} = \begin{pmatrix} \mathcal{S} & 0 & 0 & -I_{\Gamma^s}^T \\ -I_{\Gamma^s} & \mathcal{G} & 0 & 0 \\ 0 & \mathcal{D} & \mathcal{F} & I_{\Gamma^f}^T \\ -\frac{\gamma}{\beta\Delta t} I_{\Gamma^s} & 0 & I_{\Gamma^f} & 0 \end{pmatrix} , \tag{7.21}$$

where $\mathcal{S}$, $\mathcal{G}$ and $\mathcal{F}$ represent the linearized structure, geometry and fluid problems, respectively; $\mathcal{D}$ are the shape derivatives, i.e., the derivatives of $F(\mathbf{u}^{n+1}, p^{n+1}, \widehat{\mathbf{d}}_f^{n+1})$ with respect to $\widehat{\mathbf{d}}_f^{n+1}$ (see [41] for their exact computation).

Finally, we update the solution: $\mathbf{X}_{k+1}^{n+1} = \mathbf{X}_k^{n+1} + \delta\mathbf{X}_k^{n+1}$. The Newton iterations are stopped when

$$\frac{\|\mathbf{R}_k^{n+1}\|_\infty}{\|\mathbf{R}_0^{n+1}\|_\infty} \le \epsilon , \tag{7.22}$$

where $\mathbf{R}_0^{n+1}$ is the residual at the first Newton iteration and ϵ is a given tolerance.

7.2.5 Preconditioning by FaCSI

At each iteration of the Newton method, linear system (7.20) is solved by the GMRES (Generalized Minimal Residual) method preconditioned by FaCSI [32]. FaCSI exploits the factorized form of the FSI Jacobian matrix, the use of static condensation to formally eliminate the interface degrees of freedom of the fluid equations, and the use of SIMPLE (Semi-Implicit Method For Pressure-Linked Equations) for the condensed fluid system. The FaCSI preconditioner reads:

$$P_{\mathrm{FaCSI}} = P_{\mathcal{S}}^{ap} \cdot P_{\mathcal{G}}^{ap} \cdot P_{\mathcal{F}}^{ap} , \tag{7.23}$$

where:

$$P_{\mathcal{S}}^{ap} = \begin{pmatrix} \mathcal{H}_{\mathcal{S}} & 0 & 0 & 0 \\ 0 & I & 0 & 0 \\ 0 & 0 & I & 0 \\ 0 & 0 & 0 & I \end{pmatrix}, \quad P_{\mathcal{G}}^{ap} = \begin{pmatrix} I & 0 & 0 & 0 \\ -I_{\Gamma^s} & \mathcal{H}_{\mathcal{G}} & 0 & 0 \\ 0 & 0 & I & 0 \\ 0 & 0 & 0 & I \end{pmatrix},$$

$$P_{\mathcal{F}}^{ap} = \begin{pmatrix} I & 0 & 0 & 0 \\ 0 & I & 0 & 0 \\ 0 & \mathcal{D} & & \\ -\frac{\gamma}{\beta \Delta t} I_{\Gamma^s} & 0 & \multicolumn{2}{c}{\mathcal{H}_{\mathcal{F}}} \end{pmatrix}$$

In the preconditioner we have dropped the off-diagonal block $-I_{\Gamma^s}^T$ of J_{FSI}. In the application of FaCSI, the inverses of the diagonal blocks associated with the structure and geometry problems are approximated by efficient preconditioners denoted by $\mathcal{H}_{\mathcal{S}}$ (structure problem) and $\mathcal{H}_{\mathcal{G}}$ (geometry problem). The preconditioner for the fluid and Lagrange multiplier parts, denoted by $\mathcal{H}_{\mathcal{F}}$, is built upon static condensation to formally eliminate the interface degrees of freedom of the fluid equations and the use of the SIMPLE preconditioner for the unsteady Navier–Stokes equations.

For the sake of clarity, below we report how the preconditioner of the fluid problem $P_{\mathcal{F}}^{ap}$ applies to a given linear residual $\mathbf{r} = (\mathbf{r}_{\mathbf{d}_s}, \mathbf{r}_{\mathbf{d}_f}, \mathbf{r}_{\mathbf{u}}, \mathbf{r}_p, \mathbf{r}_\lambda)^T$:

1. compute $\mathbf{z}_F = (\mathbf{r}_{\mathbf{u}}, \mathbf{r}_p)^T - \mathcal{D}\mathbf{r}_{\mathbf{d}_f}$ and $\mathbf{z}_\lambda = \mathbf{r}_\lambda + \frac{\gamma}{\beta \Delta t} I_{\Gamma^s} \mathbf{r}_{\mathbf{d}_s}$
2. after denoting by $\mathbf{w}_{\mathbf{u}}$ and $\mathbf{w}_p$, $\mathbf{z}_{\mathbf{u}}$ and $\mathbf{z}_p$ the velocity and pressure components of $\mathbf{w}_F$ and $\mathbf{z}_F$, respectively, we set $\mathbf{z}_{\mathbf{u},\Gamma} = \mathbf{z}_\lambda$
3. application of the SIMPLE preconditioner, which involves the following steps:
   a) $\mathbf{y}_{\mathbf{u},i} = \mathcal{H}_{\mathcal{K}_{ii}}^{-1}(\mathbf{z}_{\mathbf{u},i} - \mathcal{K}_{i\Gamma}\mathbf{z}_{\mathbf{u},\Gamma})$,
   b) $\mathbf{w}_p = \mathcal{H}_{\tilde{\mathcal{S}}}^{-1}(\mathcal{B}_i \mathbf{y}_{\mathbf{u},i} - \mathbf{z}_p + \mathcal{B}_\Gamma \mathbf{z}_{\mathbf{u},\Gamma})$,
   c) $\mathbf{w}_{\mathbf{u}} = (\mathbf{w}_{\mathbf{u},i}, \mathbf{w}_{\mathbf{u},\Gamma})^T = (\mathbf{y}_{\mathbf{u},i} - D^{-1}\mathcal{B}_i^T \mathbf{w}_p, \mathbf{z}_{\mathbf{u},\Gamma})^T$.

   Finally, we compute $\mathbf{w}_\lambda = I_{\Gamma^{u_f}}(\mathbf{z}_{\mathbf{u}} - \mathcal{K}\mathbf{w}_{\mathbf{u}} - \mathcal{B}^T \mathbf{w}_p)$.

In the application of the SIMPLE preconditioner we denoted by $\mathcal{K}$ and $\mathcal{B}$ the matrices representing the linearized advection-diffusion-reaction and gradient terms of the fluid problem split into their internal (index i) and interface (index Γ) components. The matrix D is the diagonal of $\mathcal{K}_{ii}$, $\mathcal{H}_{\mathcal{K}_{ii}}$ is the preconditioner associated with $\mathcal{K}_{ii}$ and $\mathcal{H}_{\tilde{\mathcal{S}}}$ is an approximation of the Schur complement $\mathcal{B}_i D^{-1} \mathcal{B}_i^T$ of the fluid problem.

In this work, we make use of the one-level Algebraic Additive Schwarz (AAS) method for $\mathcal{H}_{\mathcal{S}}$, $\mathcal{H}_{\mathcal{G}}$, while the three-level Algebraic Multigrid (AMG) method is used for $\mathcal{H}_{\mathcal{K}_{ii}}$ and $\mathcal{H}_{\tilde{\mathcal{S}}}$. We remark that the exact local subdomain solves for AAS as well as the exact coarse solve of the AMG preconditioner are carried out by LU factorization using the library MUMPS (MUltifrontal Massively Parallel Sparse direct Solver) [2, 3].

7.3 Modeling reduction for FSI in arteries

In Section 2 we focused on parallel algorithms to solve the three-dimensional FSI problems. Now we make a trend reversal and we proceed with a complexity *reduction* of the FSI system. In particular, our goal is to compute an approximation of specific target outputs using a small amount of computational resources. As we said in the introduction, this challenge can be faced exploiting reduced-order models that are meant for approximating the solution of parametric PDE systems in real time.

Applying numerical reduction to fully three-dimensional realistic FSI problems may be even more challenging than solving the FSI problems themselves. To begin with, we start by introducing further modeling assumptions which aims at reducing the model complexity of the original 3D nonlinear FSI system.

Once the model reduction has been completed, it is possible to proceed with its numerical reduction by Reduced-Order Modeling (ROM) based on Proper Orthogonal Decomposition (POD).

7.3.1 Transpiration conditions

First, we focus on the complexity due to the motion of the fluid domain. We make use of transpiration conditions [31, 40] with the specific goal of formulating the FSI problem on a fixed computational domain. Transpiration conditions are based on a Taylor expansion of the system variables. For instance, the Taylor expansion of the velocity $\mathbf{u}$ at a point of the current configuration $\mathbf{x} = \mathcal{A}_t(\widehat{\mathbf{x}})$ in the neighborhood of the point $\widehat{\mathbf{x}}$ (Figure 7.1) reads:

$$\mathbf{u}(\mathcal{A}_t(\widehat{\mathbf{x}})) = \mathbf{u}(\widehat{\mathbf{x}}) + \nabla_{\widehat{\mathbf{x}}}\mathbf{u}|_{\widehat{\mathbf{x}}}\widehat{\mathbf{d}_s} + O(\|\widehat{\mathbf{d}_s}|_{\widehat{\Gamma}}\|^2)\,, \tag{7.24}$$

where (7.24) is a first-order approximation. Further, if we truncate the expansion at the zeroth order, we obtain:

$$\mathbf{u}(\mathcal{A}_t(\widehat{\mathbf{x}})) = \mathbf{u}(\widehat{\mathbf{x}}) + O(\|\widehat{\mathbf{d}_s}|_{\widehat{\Gamma}}\|)\,. \tag{7.25}$$

In equations (7.24) and (7.25) we used the fact that on the interface $\widehat{\Gamma}$ the following identity holds:

$$\widehat{\mathbf{d}}_s = \mathcal{A}_t(\widehat{\mathbf{x}}) - \widehat{\mathbf{x}}\,.$$

The Taylor expansion allows us to write the velocity $\mathbf{u}$ as a function of $\widehat{\mathbf{x}}$ on the reference fixed configuration $\widehat{\Omega}_f$ which usually corresponds to the initial configura-

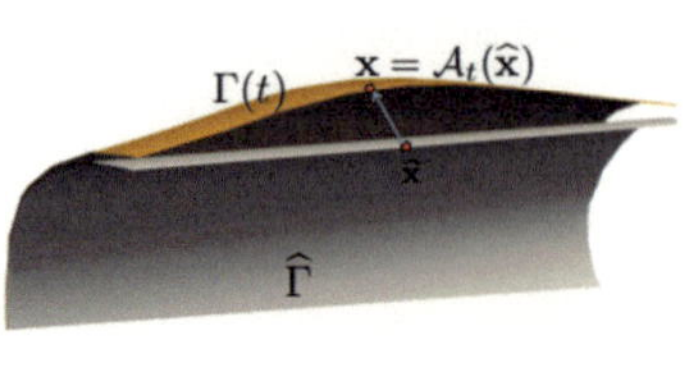

Fig. 7.1: Graphical representation of the transpiration conditions.

tion $\Omega_f(0)$. There is a crucial difference between choosing transpiration conditions of first or zeroth order. Indeed, if we choose first-order transpiration conditions, on the one hand, we would manage to fix the computational domain, while on the other, we would introduce several nonlinear terms $(\nabla_{\widehat{\mathbf{x}}}\mathbf{u}|_{\widehat{\mathbf{x}}}\widehat{\mathbf{d}}_s)$. We opt then for zeroth-order expansions, although this may affect the results with a modeling error of the same order as that of the interface displacement.

Since transpiration conditions allow us to approximate the current configuration with the reference one, there is no need to differentiate the notation. In what follows we simplify the notation removing the $\widehat{\cdot}$, e.g.:

$$\Omega_f(t) \approx \widehat{\Omega}_f =: \Omega_f \qquad \text{and} \qquad \Gamma(t) \approx \widehat{\Gamma} =: \Gamma$$

The strong formulation of the FSI problem wherein transpiration conditions are used can be derived from system (7.8) applying the following three modifications:

- the domain $\Omega_f(t)$ is substituted in the domain Ω_f;
- the ALE velocity $\mathbf{w}$ disappears since it is equal to zero;
- the kinematic coupling condition (7.3a) can be rewritten as follows:

$$\frac{\partial \mathbf{d}_s(\mathbf{x})}{\partial t} = \mathbf{u}(\mathbf{x}) \qquad \text{for} \quad \mathbf{x} \in \Gamma\,. \tag{7.26}$$

Finally, we remark that an equivalent Taylor expansion as the one written for the velocity $\mathbf{u}$ can be performed on the fluid Cauchy tensor σ_f, thus approximating $\sigma_f(\mathbf{x})$ with $\sigma_f(\widehat{\mathbf{x}})$. At this point, also the dynamic coupling condition (7.9) can be reformulated and reads as follows:

$$\int_\Gamma \sigma_f \mathbf{n}_f \cdot \boldsymbol{\eta}\, d\gamma + \int_\Gamma \Pi(\mathbf{d}_s)\mathbf{n}_s \cdot \boldsymbol{\eta}\, d\Gamma = 0 \quad \forall \boldsymbol{\eta} \in U^\lambda\,. \tag{7.27}$$

7.3.2 Thin-wall model

The second major source of complexity of FSI problems results from the presence of two coupled physics: the fluid and the solid. Because of the fact that our target output is mainly driven by the hemodynamics of the system, we concentrate the computational effort on the fluid domain.

In this regard, we aim at building a lower dimensional structural model, while keeping the effect that the compliant arterial wall produces on the blood flow. It is then possible to integrate the solid model as a boundary condition for the fluid one through specific assumptions on the biomechanics of the arterial wall, cf. [24, 43, 94].

We assume that the wall is a thin-walled structure, i.e., $h_s \ll D$, being h_s the wall thickness and D the lumen diameter. Using this approximation in system (7.8), volume integrals on Ω_s can be substituted with surface integrals on Γ multiplied by the thickness h_s. Furthermore, we assume that all the derivatives of structural quantities in the

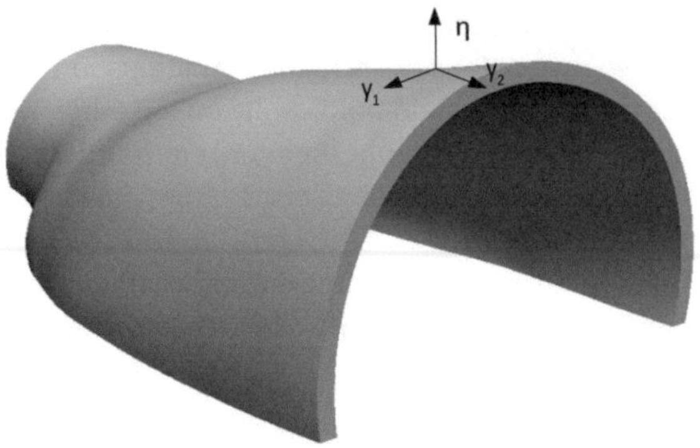

Fig. 7.2: Schematic representation of the structural domain

normal direction are negligible. In the following, the pedix y denotes the tangential derivatives involved in gradient operators. With respect to the notation of Figure 7.2, considering a generic scalar field ϕ, these assumptions yield the following mathematical formulas:

$$\phi(\eta, y_1, y_2) \approx \phi(y_1, y_2), \text{therefore} \int_{\Omega_s} \phi(\eta, y_1, y_2) d\Omega_s = \int_{\Gamma} h_s \phi(y_1, y_2) d\Gamma\,,$$

$$\nabla_{\mathbf{x}} \phi \approx \nabla_y \phi, \text{therefore } \nabla_{\mathbf{x}} \phi \cdot \mathbf{n} \approx 0\,,$$

with $\nabla_y \phi = \nabla_{\mathbf{x}} \phi - \nabla_{\mathbf{x}} \phi \cdot \mathbf{n}$ the tangential gradient of ϕ.

At this point, we can rewrite the structural equation of system (7.8) as follows:

$$\int_{\Gamma} h_s \rho_s \frac{\partial^2 \mathbf{d}_s}{\partial t^2} \cdot \mathbf{v}_s d\Gamma + \int_{\Gamma} h_s \Pi_y(\mathbf{d}_s) : \nabla_y \mathbf{v}_s d\Gamma = \int_{\Gamma} \Pi_y(\mathbf{d}_s) \mathbf{n}_s \cdot \mathbf{v}_s d\Gamma\,, \tag{7.28}$$

where $\Pi_y(\mathbf{d}_s)$ is the surface stress tensor that, with respect to (7.5), is modified as follows:

$$\Pi_y(\mathbf{d}_s) = \lambda_s \operatorname{tr}\left(\frac{\nabla_y \mathbf{d}_s + (\nabla_y \mathbf{d}_s)^T}{2}\right) + \mu_s(\nabla_y \mathbf{d}_s + (\nabla_y \mathbf{d}_s)^T)\,.$$

Using the coupling conditions (7.26) and (7.27), we replace the structural unknowns in the first and third term of equation (7.28) with fluid ones:

$$\int_{\Gamma} h_s \rho_s \frac{\partial \mathbf{u}}{\partial t} \cdot \mathbf{v}_f d\Gamma + \int_{\Gamma} h_s \Pi_y(\mathbf{d}_s) : \nabla_y \mathbf{v}_f d\Gamma = -\int_{\Gamma} \sigma_f(\mathbf{u}, p) \mathbf{n}_f \cdot \mathbf{v}_f d\Gamma\,. \tag{7.29}$$

We notice that the second term of equation (7.28) still depends on structural unknowns, i.e., on the displacement $\mathbf{d}_s$. It is possible to recover $\mathbf{d}_s$ from (7.26) and approximate, e.g., by exploiting a generic discretization of the first-order temporal derivatives of $\mathbf{d}_s$

$$\left.\frac{\partial \mathbf{d}_s}{\partial t}\right|_{t=t^{n+1}} \approx \frac{\alpha}{\Delta t} \mathbf{d}_s^{n+1} - \mathbf{f}_{\mathbf{d}_s}^n\,,$$

where $\mathbf{f}^n_{\mathbf{d}_s} = f(\mathbf{d}^n_s, \dot{\mathbf{d}}^n_s, \ldots)$ and similarly for $\mathbf{f}^n_{\mathbf{u}}$. Discretization of the kinematic condition (7.26) then reads: $\mathbf{u} = \frac{\partial \mathbf{d}_s}{\partial t}$ on Γ:

$$\mathbf{u}^{n+1} = \frac{\alpha}{\Delta t}\mathbf{d}^{n+1}_s - \mathbf{f}^n_{\mathbf{d}_s}$$

which then allow us to substitute $\mathbf{d}_s$ at the time t^{n+1} inside the stress tensor $\Pi_y(\mathbf{d}^{n+1}_s)$ as follows:

$$\Pi_y(\mathbf{d}^{n+1}_s) = \frac{\Delta t}{\alpha}\Pi_y(\mathbf{u}^{n+1} + \mathbf{f}^n_{\mathbf{d}_s}) .$$

Similarly, using

$$\left.\frac{\partial \mathbf{u}}{\partial t}\right|_{t=t^{n+1}} \approx \frac{\alpha}{\Delta t}\mathbf{u}^{n+1} - \mathbf{f}^n_{\mathbf{u}} ,$$

the temporal discretization of equation (7.29) at time t^{n+1} reads as follows

$$\int_\Gamma h_s\rho_s\Big(\frac{\alpha}{\Delta t}\mathbf{u}^{n+1} - \mathbf{f}^n_{\mathbf{u}}\Big)\cdot \mathbf{v}_f d\Gamma + \int_\Gamma h_s\frac{\Delta t}{\alpha}\Pi_y(\mathbf{u}^{n+1} + \mathbf{f}^n_{\mathbf{d}_s}) : \nabla_y \mathbf{v}_f d\Gamma$$
$$= -\int_\Gamma \sigma_f(\mathbf{u}, p)\mathbf{n}_f \cdot \mathbf{v}_f d\Gamma . \quad (7.30)$$

The above equation can be seen as a *generalized Robin* boundary condition on Γ for the fluid problem: in fact it states that the weak term referring to the fluid stress on Γ can be written as a function of a boundary mass and a boundary stiffness term. We remark that $\mathbf{d}^n_s$ is not a primal unknown since it is reconstructed as a weighted sum of the velocities at previous time levels.

7.3.3 Weak formulation of the reduced fluid-structure interaction problem

We call the simplification of the FSI formulation deriving from the transpiration conditions and the thin wall assumption the Reduced FSI (RFSI) problem. The RFSI system is composed of the standard Navier–Stokes equation, written on a fixed domain, enhanced by a generalized Robin boundary condition on the lateral surface of the lumen. The second-order differential operator on the boundary requires a higher regularity on the surface Γ (see [70] for more details on the analysis of partial differential equations with generalized Robin boundary conditions). Thus, to guarantee the well-posedness of the weak formulation, the velocity spaces (Section 7.2.2) needs to be modified as follows:

$$\widetilde{U} = \{\mathbf{v} \in H^1(\Omega) \mid \mathbf{v}|_\Gamma \in [H^1(\Gamma)]^3\},\ \widetilde{U}_D = \{\mathbf{v} \in \widetilde{U} \mid \mathbf{v} = 0 \text{ on } \Gamma_{\text{fixed}}\} .$$

We use P2-P1 Taylor-Hood finite element spaces, which fulfill the higher regularity requirement on the boundary Γ, i.e., $U_h \subset \widetilde{U}$, $U_{h,D} \subset \widetilde{U}_D$, and $Q_h \subset L^2(\Omega_f)$.

We suppose to divide the temporal interval $[0, T]$ in $N_T + 1$ points uniformly distributed with a time step $\Delta t = T/N_T$, such that $t_n = n\Delta t$. The fully discretized weak formulation of the RFSI problem then reads: for all $n = 0, \dots, N_T - 1$, find $(\mathbf{u}_h^{n+1}, p_h^{n+1}) \in U_h \times Q_h$ such that

$$\begin{cases} a(\mathbf{u}_h^{n+1}, \mathbf{v}_h) + c(\mathbf{u}_h^n, \mathbf{u}_h^{n+1}, \mathbf{v}_h) + b(\mathbf{v}_h, p_h^{n+1}) = f(\mathbf{v}_h; \mathbf{u}_h^n, \mathbf{d}_{s,h}^n) & \forall \mathbf{v}_h \in U_{h,D} \\ b(\mathbf{u}_h^{n+1}, q_h) = 0 & \forall q_h \in Q_h \,, \end{cases} \tag{7.31}$$

where

$$a(\mathbf{u}_h^{n+1}, \mathbf{v}_h) = \int_{\Omega_\mathrm{f}} \left(\rho_\mathrm{f} \alpha \frac{\mathbf{u}_h^{n+1}}{\Delta t} \cdot \mathbf{v}_h + \mu_\mathrm{f}(\nabla \mathbf{u}_h^{n+1} + (\nabla \mathbf{u}_h^{n+1})^T) : \nabla \mathbf{v}_h \right) d\Omega$$

$$+ \int_\Gamma h_\mathrm{s} \left(\rho_\mathrm{s} \frac{\alpha}{\Delta t} \mathbf{u}_h^{n+1} \cdot \mathbf{v}_h \frac{\Delta t}{\alpha} \Pi_y(\mathbf{u}_h^{n+1}) : \nabla_y \mathbf{v}_h \right) d\Gamma$$

$$c(\mathbf{u}_h^n, \mathbf{u}_h^{n+1}, \mathbf{v}_h) = \int_{\Omega_\mathrm{f}} \rho_\mathrm{f}((\mathbf{u}_h^n) \cdot \nabla) \mathbf{u}_h^{n+1} \cdot \mathbf{v}_h$$

$$b(\mathbf{v}_h, p_h^{n+1}) = \int_{\Omega_\mathrm{f}} p_h^{n+1} \nabla \cdot \mathbf{v}_h d\Omega$$

$$f(\mathbf{v}_h; \mathbf{u}_h^n, \mathbf{d}_{s,h}^n) = \int_{\Omega_\mathrm{f}} \rho_\mathrm{f} \mathbf{f}_{\mathbf{u}_h}^n \cdot \mathbf{v}_h d\Omega + \int_{\Gamma_N^f} \mathbf{g}_\mathrm{f} \cdot \mathbf{v}_h d\Gamma$$

$$+ \int_\Gamma h_\mathrm{s} \left(\rho_\mathrm{s} \mathbf{f}_{\mathbf{u}_h}^n \cdot \mathbf{v}_h + \frac{\Delta t}{\alpha} \Pi_y(\mathbf{f}_{\mathbf{d}_{s,h}}^n) : \nabla_y \mathbf{v}_h \right) d\Gamma \,.$$

Solutions $(\mathbf{u}_h^n, p_h^n)$ of problem (7.31) represent the so-called *high-fidelity* approximation which will be used as the reference solution in the numerical reduction phase described in the next section.

7.4 Reduced-order methods for reduced fluid-structure interaction models

7.4.1 Numerical reduction

Once the model equations have been simplified, we are ready to proceed with their numerical reduction. In this section we will recall only some basic concepts of reduced-order models, the full theoretical assessment of these methods can be found in [61, 99]. Numerical reduction techniques are particularly interesting when dealing with parametric PDEs. The parameters can be either geometrical and/or physical, for instance the diameter of the vessel, mean flowrate, frequency, mechanical properties.

Let us select a suitable set of parameters. We denote $\boldsymbol{\mu}$ the vector that collects all the chosen coefficients and $\mathcal{D}$ the set of possible parameter realizations. The high-fidelity approximations resulting from system (7.31) will thus depend not only on space and time but also on the specific values of the parameters: $\mathbf{u}_h^n(\mathbf{x};\boldsymbol{\mu})$ and $p_h^n(\mathbf{x};\boldsymbol{\mu})$ for $n = 1,\dots,N_T$.

The aim of reduction techniques is to find a suitable set of basis functions to approximate the high-fidelity solution by a Galerkin projection onto low-dimensional spaces U_N and Q_N. We therefore focus on the construction of discretization spaces $U_N \subset U_h^g$, $U_{N,D} \subset U_{h,D}^g$ and $Q_N \subset Q_h$, whose dimensions are orders of magnitude lower than the dimension of the finite element spaces. The dimension of the velocity and pressure spaces $U_N \times Q_N$ is noted by N and the one of $U_h^g \times Q_h$ by $\mathcal{N}_h$. The weak formulation of the RFSI problem in the reduced numerical framework reads: for all $n = 0,\dots,N_T - 1$, find $(\mathbf{u}_N^{n+1}, p_N^{n+1}) \in U_N \times Q_N$ such that

$$\begin{cases} a(\mathbf{u}_N^{n+1}, \mathbf{v}_N) + c(\mathbf{u}_N^n, \mathbf{u}_N^{n+1}, \mathbf{v}_N) + b(\mathbf{v}_N, p_N^{n+1}) = f(\mathbf{v}_N; \mathbf{u}_N^n, \mathbf{d}_{s,N}^n) & \forall \mathbf{v}_N \in U_{N,D} \\ b(\mathbf{u}_N^{n+1}, q_N) = 0 & \forall q_N \in Q_N\,, \end{cases} \tag{7.32}$$

We remark that to build a basis for the spaces U_N and Q_N, we need to explore the variability of the finite element solutions with respect to both time and parameters.

7.4.2 Proper orthogonal decomposition

To generate a set of basis functions for the spaces U_N and Q_N we rely on the Proper Orthogonal Decomposition (POD) based on snapshot temporal solutions. In the context of this work we only detail the specific numerical choices that we perform to reduce the RFSI problem; the interested reader can find more detail on the POD algorithm applied to fluid problems in e.g., [72, 104, 126]. The POD is performed assuming a fixed value for each parameter in $\boldsymbol{\mu}$, which will thus be removed from the notation.

Let us suppose that our solution is periodic and that the time indices $n = 0,\dots,N_T$ are related to temporal instants spanning one single period. We select as snapshots the high-fidelity solutions of problem (7.31) sampled at specific instants such that we cover the time lapse of one full cycle. To reduce the size of the eigenvalue problem involved in the POD reduction, the solutions are sampled at a lower frequency than the time discretization used in the simulation. Instead of using all of the $N_T + 1$ solutions, only a fraction $N_S = N_T/k$ are taken. The set of selected temporal indices is therefore $\mathcal{N}_S = \{0, k, 2k, 3k, \dots, N_T\}$. The solutions $\{(\mathbf{u}_h^s, p_h^s)\}_{s\in\mathcal{N}_S}$ will represent our snapshots, the starting point for the POD analysis. Since velocity and pressure have different dynamics, we split the POD into two separate eigenvalue decompositions as also done in [51]. We measure the energy associated with the snapshots using the following scalar products: for the velocity, we use

$$(\mathbf{u}_h, \mathbf{v}_h)_{U_h^g} := \int_{\Omega^f} \nabla\mathbf{u}_h : \nabla\mathbf{v}_h \, d\Omega + \int_{\Gamma} \nabla_y\mathbf{u}_h : \nabla_y\mathbf{v}_h \, d\Gamma$$

and for the pressure

$$(p_h, q_h)_Q := \int_\Omega p_h\, q_h\, d\Omega\,.$$

Then, we compute the two Gramian matrices

$$G^{\mathbf{u}}_{ij} = (\mathbf{u}^i_h, \mathbf{u}^j_h)_{\mathbf{V}} \quad \text{and} \quad G^p_{ij} = (p^i_h, p^j_h)_Q \qquad \forall j, i \in \mathcal{N}_S\,.$$

We remark that since we are dealing with RFSI problems, the internal product of the velocity space is enhanced by a boundary term. At this stage the indices i, j are not contiguous. Since $\mathcal{N}_S$ is ordered, it is easy to renumber the indices such that $G^{\mathbf{u}}$ and G^p are $N_S \times N_S$ matrices. We perform the eigenvalue decomposition of $G^{\mathbf{u}}$ and the one of G^p, obtaining the pairs $(\lambda^{\mathbf{u}}_k, \boldsymbol{\zeta}^{\mathbf{u}}_k)$ and $(\lambda^p_k, \boldsymbol{\zeta}^p_k)$ where $\lambda^{\mathbf{u}}_k, \lambda^p_k \in \mathbb{R}$ and $\boldsymbol{\zeta}^{\mathbf{u}}_k, \boldsymbol{\zeta}^p_k \in \mathbb{R}^{N_S}$ are the $k-th$ eigenvalues and eigenvectors of the velocity and pressure Gramian matrices, respectively, for $k = 1, \ldots, N_S$; we sort the eigenvalues from the largest to the smallest. Fixing the same tolerance $0 < \delta < 1$ for both velocity and pressure decompositions, we select the first $N^{\mathbf{u}}$ and N^p eigenpairs such that:

$$\frac{\sum_{j=1}^{N^{\mathbf{u}}} \lambda^{\mathbf{u}}_j}{\sum_{k=1}^{N_S} \lambda^{\mathbf{u}}_k} \geq 1 - \delta \quad \text{and} \quad \frac{\sum_{j=1}^{N^p} \lambda^p_j}{\sum_{k=1}^{N_S} \lambda^p_k} \geq 1 - \delta\,,$$

respectively. The j–th velocity eigenfunction $\boldsymbol{\phi}^{\mathbf{u}}_j \in U^g_h$ is reconstructed using the linear combination:

$$\boldsymbol{\phi}^{\mathbf{u}}_j = \frac{1}{\lambda_j} \sum_{s \in \mathcal{N}_S} [\boldsymbol{\zeta}^{\mathbf{u}}_j]_s \mathbf{u}^s_h\,, \quad \text{for } j = 1, \ldots, N^{\mathbf{u}}$$

and similarly for $\phi^p_j \in Q_h$, $j = 1, \ldots, N^p$. The velocity bases are linear combinations of solutions of problem (7.31), thus, they all verify: $\int_\Omega q_h \nabla \cdot \boldsymbol{\phi}^{\mathbf{u}}_j = 0$, $\forall q_h \in Q_h$ for $j = 1, \ldots, N^{\mathbf{u}}$. If we consider the functional spaces generated from the velocity functions $\boldsymbol{\phi}^{\mathbf{u}}_j$ and the pressure modes ϕ^p_j, linear system (7.32) would be singular. For Navier–Stokes systems, this issue can be solved by restricting the system and solving it only for the velocity unknown (see e.g., [20]). Unfortunately, this is not possible when considering problem (7.31). Indeed, the boundary condition applied on Γ derives from a structural model whose solution is driven by the pressure condition set on the external boundary of the structural model itself (see [24, 90]). If in the reduced system (7.32) we do not account for the pressure variable, we cannot recover the correct velocity on the boundary Γ and the related output functionals (e.g., wall shear stresses). In order to construct a nonsingular reduced system (7.32), for each selected pressure mode ϕ^p_j, we define the corresponding supremizer function $\boldsymbol{\sigma}_j \in U^g_{h,D}$ as the solution of the following problem (see [107]):

$$(\boldsymbol{\sigma}_j, \mathbf{v}_h) = \int_\Omega \phi^p_j \nabla \cdot \mathbf{v}_h d\Omega \qquad \forall \mathbf{v}_h \in U^g_h\,, \qquad \text{for } j = 1, \ldots, N^p\,,$$

and we add the supremizers to the POD velocity basis. We define U_N and Q_N the POD reduced spaces associated with the RFSI model that are generated by the basis functions $\{\boldsymbol{\phi}^{\mathbf{u}}_j\}_{j=1}^{N^{\mathbf{u}}} \oplus \{\boldsymbol{\phi}^{\boldsymbol{\sigma}}_j\}_{j=1}^{N^p}$ and $\{\phi^p_j\}_{j=1}^{N^p}$, respectively, where $\boldsymbol{\phi}^{\boldsymbol{\sigma}}_j$ for $j = 1, \ldots, N^p$ represent the orthonormalization of the supremizer functions $\boldsymbol{\sigma}_j$, obtained with a Gram-Schmidt algorithm with respect to the scalar product $(\cdot\,,\,\cdot)_{U_h^g}$. The other basis functions are already orthogonal thanks to the POD algorithm. The total dimension of the reduced system is $N = N^{\mathbf{u}} + 2 \cdot N^p$.

7.5 Fluid-structure interaction in a patient-specific aortofemoral artery

In this section we address the study of the blood flow dynamics in a patient-specific aortofemoral artery. Studying the hemodynamics of the vasculature distal to the abdominal aorta may be important in understanding common diseases in peripheral arteries downstream of the thoracic aorta. Diseases in the peripheral vasculature affect millions of people in the U.S and can have a profound effect on daily quality of life. Peripheral arterial disease is the build-up of fatty tissue, or atherosclerosis, in lower extremity arteries. The prevalence of peripheral arterial occlusive disease increases with age and can increase to up to 20% of the geriatric population. Atherosclerotic occlusive disease of the lower extremity arteries is a major cause of walking impairment, pain, ulcerations and gangrene.

The geometry of the aortofemoral artery used for the numerical simulations has been obtained from [1] and is shown in Figure 7.3, while the meshes for the fluid and structure physical domains were generated using a software library [39] based on VMTK. We consider the blood characterized by a density $\rho_f = 1.06 \text{g/cm}^3$ and by a dynamic viscosity $\mu_f = 0.04$ g/(cm s). The physical parameters for the structure are

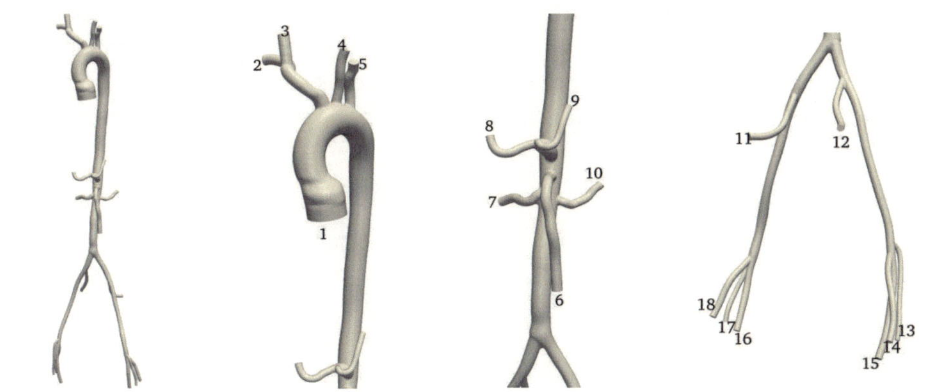

(a) Whole geometry. (b) Thoracic aorta. (c) Abdominal aorta. (d) Iliac and femoral arteries.

Fig. 7.3: Patient-specific aortofemoral geometry: from left to right we show the whole arterial geometry and a zoomed-in view of the thoracic and descending aorta, of the abdominal aorta and of the iliac arteries.

Tab. 7.1: Values of the resistances used at each outflow section of the fluid domain, values obtained from [1].

ID surface	Artery name	Resistance [dyne s cm^{-5}]
2	Right Subclavian	788.82
3	Right Common Carotid	6697.69
4	Left Common Carotid	6708.98
5	Left Subclavian	788.87
6	Superior Mesenteric	829.30
7	Right Renal	4141.67
8	Hepatic	1114.61
9	Splenic	1086.33
10	Left Renal	4102.96
11	Left Internal	3108.73
12	Right Internal	3109.10
13	Left Profunda	6620.49
14	Right External Circumflex	3299.25
15	Right Femoral	3312.96
16	Left Femoral	3322.05
17	Left Profunda	6620.49
18	Left External Circumflex	3316.24

set as follows: Young's modulus $E_s = 4 \cdot 10^6$ dyne/cm^2, Poisson's ratio $\nu_s = 0.45$ and density $\hat{\rho}_s = 1.2$g/cm^3. The boundaries of the computational fluid domain are reported in Figure 7.3a, 7.3b and 7.3c: see Table 7.1 for the identification of each artery. The inflow section of the fluid domain corresponds to surface ID 1 in Figure 7.3a and represents the aortic root.

In a preprocessing phase, starting from the geometry shown in Figure 7.3 we generated flow extensions at the inflow 1 to obtain an inflow section of circular shape, see Figure 7.4, where a patient-specific flowrate is prescribed using a Dirichlet boundary condition, i.e., mapping the flowrate through a parabolic velocity profile. The flowrate

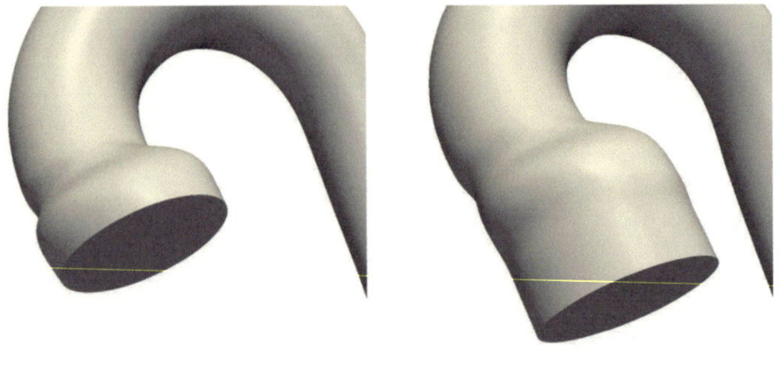

Fig. 7.4: Initial geometry (left) and flow extension generated at the inflow section of the fluid domain (right).

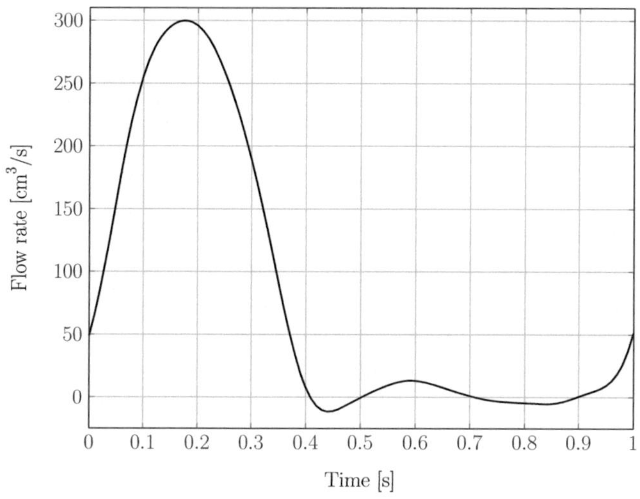

Fig. 7.5: Patient-specific inflow flowrate profile prescribed.

profile prescribed is shown in Figure 7.5. At each fluid outflow we impose the following RC-type boundary condition proposed in [55]:

$$\boldsymbol{\sigma}^{n+1}\mathbf{n}_\mathrm{f} = \frac{\Delta t}{\Delta t + RC}\left(RQ^n + P^n\frac{RC}{\Delta t}\right), \tag{7.33}$$

where $\Delta t = 10^{-3}$ s is time-step size used in our simulations, P^n and Q^n are the pressure and flowrate values, respectively, at time $t^n = n\,\Delta t$. In equation (7.33) R and C represent the resistance and capacity associated with a distal position of the vascular tree. In Table 7.1 we report the numerical values of the resistances used; for each outflow, the value of the capacity used is $C = 3\Delta t \backslash R$.

The solid domain is clamped at the inlet and outflow rings while at the external wall we impose a Robin-like boundary condition [83, 90] of the type:

$$p_0\,\widehat{\mathbf{n}}_\mathrm{s} + \widehat{\Pi}\,\widehat{\mathbf{n}}_\mathrm{s} + \kappa_\mathrm{s}\,\widehat{\mathbf{d}}_\mathrm{s} + c_\mathrm{s}\frac{\partial \widehat{\mathbf{d}}_\mathrm{s}}{\partial t} = 0 \quad \text{on } \widehat{\Gamma}^s_\mathrm{ext} \tag{7.34}$$

where $\kappa_\mathrm{s}\,\widehat{\mathbf{d}}_\mathrm{s}$ and $c_\mathrm{s}\frac{\partial \widehat{\mathbf{d}}_\mathrm{s}}{\partial t}$ represent the elastic and the viscoelastic components of the external tissue, respectively. Concerning the reduced structural model, equation (7.30) is modified as follows:

$$\begin{aligned}\int_\Gamma \left(\frac{h_\mathrm{s}\rho_\mathrm{s}\alpha}{\Delta t} + \kappa_\mathrm{s}\frac{\Delta t}{\alpha} + c_\mathrm{s}\right)\mathbf{u}^{n+1}\cdot \mathbf{v}_\mathrm{f}d\Gamma + \frac{\Delta t}{\alpha}\int_\Gamma h_\mathrm{s}\Pi_y(\mathbf{u}^{n+1} + \mathbf{f}^n_\mathbf{d})\colon \nabla_y \mathbf{v}_\mathrm{f}d\Gamma \\ = \int_\Gamma (-\sigma_\mathrm{f}(\mathbf{u}^{n+1}, p^{n+1})\mathbf{n} + h_\mathrm{s}\rho_\mathrm{s}\mathbf{f}^n_\mathbf{u})\cdot \mathbf{v}_\mathrm{f}d\Gamma - \int_\Gamma \kappa_\mathrm{s}\frac{\Delta t}{\alpha}\mathbf{f}^n_\mathbf{d}\cdot \mathbf{v}_\mathrm{f}d\Gamma\,.\end{aligned}$$

7.5.1 Numerical results obtained by the full three-dimensional FSI model

In this section we show the results obtained using the full 3D FSI model described in section 7.2. In the three-dimensional setting, we set the thickness of the arterial wall to 15% of the local radius of the lumen.

The details of the computational meshes realized for the fluid and solid domains are summarized in Table 7.2, while the number of degrees of freedom obtained after spatial discretization is reported in Table 7.3. The numerical simulations were carried out using PizDora, a Cray XC 40 supercomputer at the Swiss National Supercomputing Center, whose technical specifications are summarized in Table 7.4.

In Figure 7.6 and 7.7 we show a postprocessing of the solution obtained at several time instances during the third heart beat simulated. In particular in Figure 7.6 we show a volume rendering of the blood flow velocity, while in Figure 7.7 we report the displacement of the arterial wall.

The blood flow features transition to turbulent flow developing in the downstream part of the descending aorta. The typical peak Reynolds number $\mathrm{Re} \approx 2650$, based on an average velocity $U = 100\,\mathrm{cm/s}$ and length $D = 1\,\mathrm{cm}$. The very fine mesh and the SUPG-VMS model allow us to recover peculiarities of the transitional flow regime. We notice the presence of several recirculation zones, as shown in Figure 7.8, occurring between the end of the systolic phase and the diastolic one. To better visualize the complex flow pattern of the blood flow as well as its transitional regime, we report in

Tab. 7.2: Details of the meshes used for the numerical simulations.

Fluid mesh		Structure mesh	
# Vertices	# Tetrahedra	# Vertices	# Tetrahedra
4,237,118	25,925,318	1,525,185	7,311,192

Tab. 7.3: Number of degrees of freedom (DoF) obtained after spatial discretization.

Fluid DoF	Structure DoF	Coupling DoF	Geometry DoF	Total
16,948,472	4,575,555	912,591	12,711,354	35,147,972

Tab. 7.4: PizDora Cray XC40 technical data.

Number of compute nodes	1256
Processor	64–bit Intel Xeon processor E5-2690v3-Haswell
Memory	64 GB per node in 1192 nodes 128GB per node in 64 nodes
Memory bandwidth	Up to 137 GB/s per node
Network	Dragonfly interconnect

(a) $t = 2.15$ s. (b) $t = 2.2$ s. (c) $t = 2.4$ s. (d) $t = 2.6$ s.

Fig. 7.6: Volume rendering of the magnitude of the fluid velocity at several time instances during the third heart beat simulated.

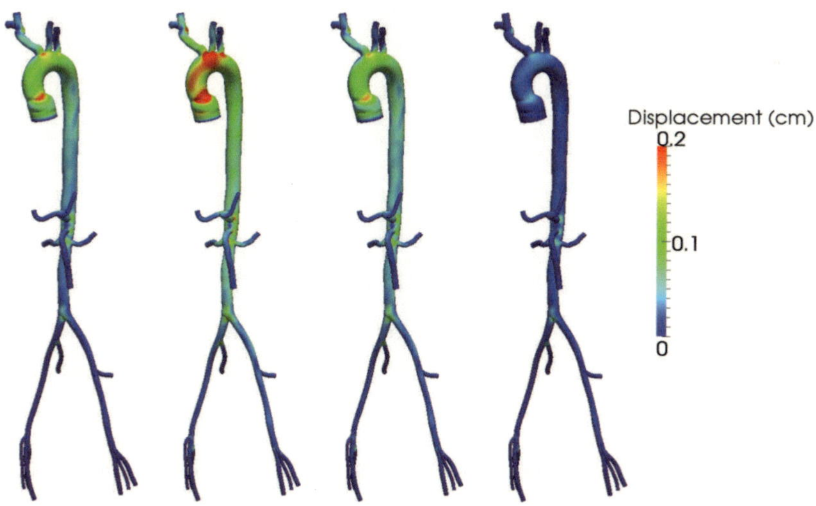

(a) $t = 2.15$ s. (b) $t = 2.2$ s. (c) $t = 2.4$ s. (d) $t = 2.6$ s.

Fig. 7.7: Solid displacement at several time instances during the third heart beat simulated.

Figure 7.9 the coherent vortex structures of the fluid flow identified by means of the Q criterion at different time frames during the third heart beat simulated.

The simulations were done with the finite element library LifeV [16]. The strong scalability results of the parallel solver are reported in Figure 7.10. In the left plot of

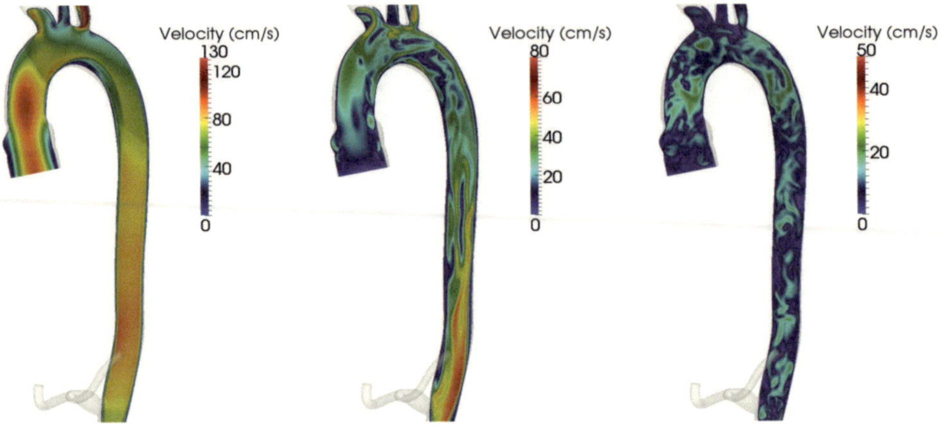

(a) Systolic peak. (b) End systole. (c) Diastole.

Fig. 7.8: Magnitude of the fluid velocity in the descending aorta at the systolic peak, at the end of systole and at diastole.

(a) $t = 2.2$ s. (b) $t = 2.4$ s. (c) $t = 2.6$ s.

Fig. 7.9: Coherent vortex structures identified by means of the Q criterion in the descending aorta.

Figure 7.10 we report the average number of GMRES iterations per linear solve over the simulation, while in the right plot the average time to compute a single time step. It includes: computation of the nonlinear residual, assembly of the Jacobian matrix, construction of the preconditioner and numerical solution of the Jacobian linear system. We remark that in our simulations, since the harmonic extension and the structure matrices are constant, the preconditioners $\mathcal{H}_S$ and $\mathcal{H}_G$ defined in Section 7.2.5 are built only once and stored at the beginning. As shown in Figure 7.10, we observe the robustness of the solver with respect to the number of cores utilized. Indeed, for the

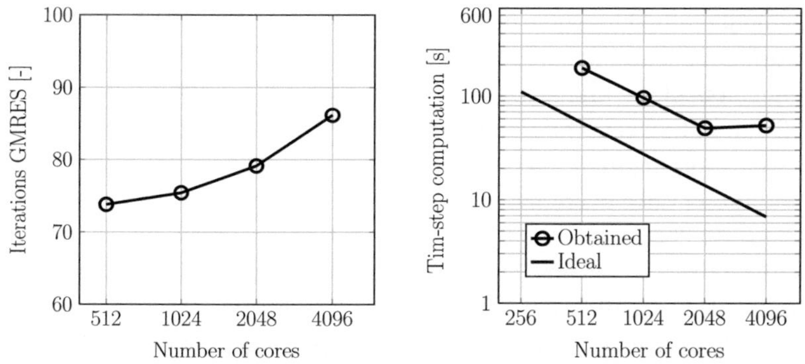

Fig. 7.10: Scalability results: average number of GMRES iterations (left) and average time to compute a single time step.

problem at hand, the number of linear solver iterations ranges from roughly 75 (with 512 cores) up to 85 iterations (with 4096 cores). The average time to compute a single time step is almost perfectly scalable up to 2048 cores; with this number of cores, a complete heartbeat therefore takes about 10 hours.

7.5.2 Numerical simulation setting for the RFSI

With the aim of performing a numerical reduction of the RFSI problem, we require *ad hoc* choices concerning the high-fidelity discretization to be taken. During the offline phase of the reduction algorithm several finite element vectors and matrices need to be stored and manipulated, we therefore prefer to lower the dimension of the finite element spaces with respect to the ones used for the three-dimensional FSI solver. We use a coarser grid for the discretization of the fluid domain with a thin boundary layer of elements near the interface Γ to maintain a good resolution towards the interface. We neglect the VMS modeling and use the stable P2-P1 finite elements pair for the discretization of the fluid system, such that no further stabilization terms are required in the formulation of the RFSI problem. This choice results in an approximation space with 8,903,292 degrees of freedom for the velocity field and 379,160 degrees of freedom for the pressure field. Also the discretization of the temporal derivatives differs between the FSI and the RFSI solver, which moreover do not take into account turbulence models. Note however that we choose to use an implicit Euler time integrator which is dissipative, with a time step equal to 0.001 s. Finally, regarding the physical modeling, we use a homogeneous thickness $h_s = 0.1$ and homogeneous coefficients $c_s = \kappa_s = 10^5$ for the surrounding tissue since they are both important for the stability estimates of the RFSI model [23].

We choose as model parameter the structural Young's modulus E. The numerical reduction is prepared with two different values of the parameters and two simulations with the finite element method for some heartbeats. We then apply twice the POD considering the two different sets of snapshots.

The set of snapshots of the first POD is computed fixing $E_1 := 10^6$ dyn/cm^2. The high-fidelity RFSI simulation is performed for three heartbeats in order to reach a periodic state. The snapshots are then selected among the finite element solutions of the third heartbeat sampling once every five time levels. The corresponding eigenvalue problem has dimension 200. A tolerance equal to $\delta = 0.001$ is fixed, which selects the first 48 velocity modes and the first 2 pressure ones. This remarkable difference in number of modes is due to the different behavior of eigenvalue decays as is shown in Figure 7.11. Adding the supremizers corresponding to the pressure modes, the reduced space resulting from the first POD produces 52 basis functions.

The second POD is performed considering Young's modulus to be equal to $E_2 := 4 \cdot 10^6$ dyn/cm^2. As before, we focus on the high-fidelity solutions of the third heartbeat that, this time, are sampled once every ten time levels. This choice results in an eigenvalue problem with a dimension equal to 100. With a tolerance $\delta = 0.01$, the POD selects 15 velocity eigenmodes and 2 pressure ones. We construct the corresponding 19 basis functions, which are orthonormalized with respect to the basis obtained with the first POD using a Gram-Schimdt algorithm.

The final reduced space has a dimension of 71 which is five orders of magnitude smaller than the dimension of the high-fidelity approximation space.

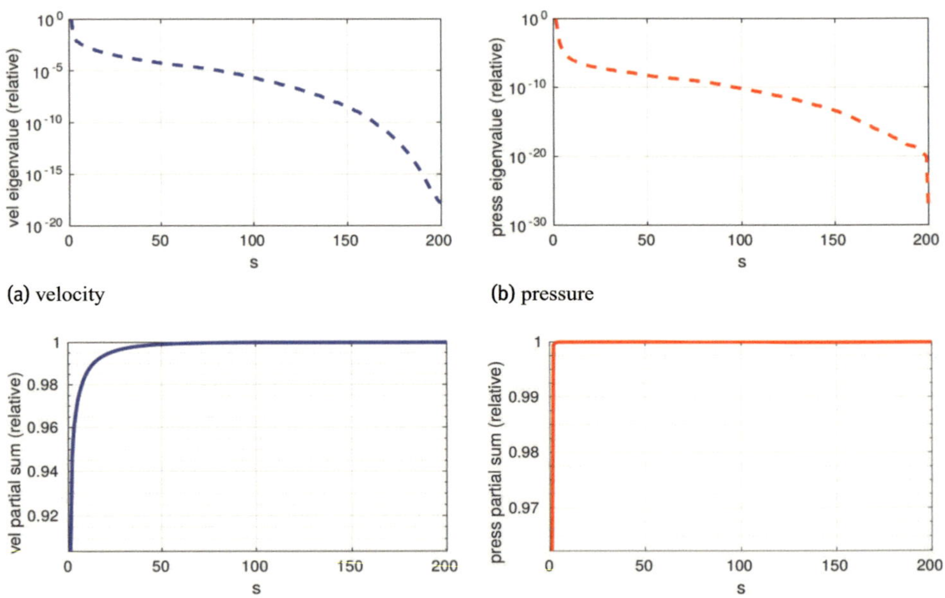

(a) velocity (b) pressure

(c) velocity (d) pressure

Fig. 7.11: Top: decay of POD eigenvalues computed with 200 snapshots with $E = 10^6$ dyn/cm^2. Bottom: partial sum of the POD eigenvalues computed with 200 snapshots with $E = 10^6$ dyn/cm^2.

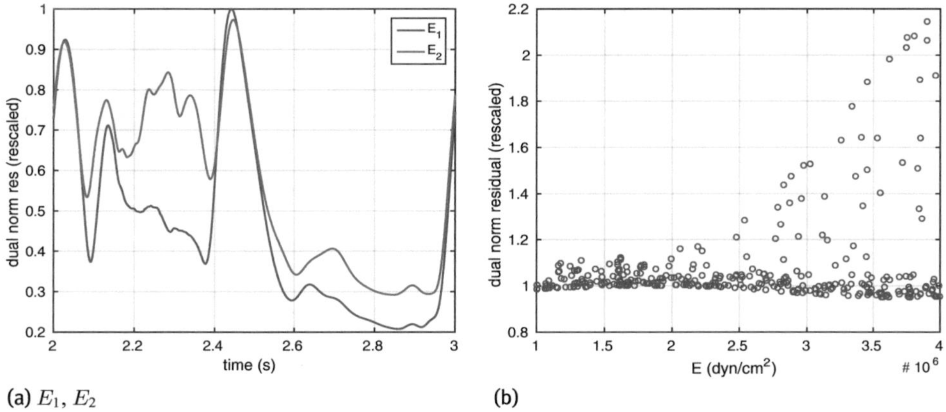

(a) E_1, E_2 (b)

Fig. 7.12: (a): Comparisons between dual norm of the residuals obtained solving the ROM-RFSI with E_1 and E_2. (b): Maximum of the dual norm of the residuals among the time levels of the third heartbeat computed for several values of Young's Modulus $E \in [10^6, 4 \cdot 10^6]$ dyn/cm^2.

We suppose to use as an *a-posteriori* error indicator the dual norm of the following residuals:

$$r_N^{n+1}((\mathbf{v}_h, q_h); \boldsymbol{\mu}) := f(\mathbf{v}_h; \mathbf{u}_N^n, \mathbf{d}_{s,N}^n) - a(\mathbf{u}_N^{n+1}, \mathbf{v}_h) - c(\mathbf{u}_N^n, \mathbf{u}_N^{n+1}, \mathbf{v}_h) - b(\mathbf{v}_h, p_N^{n+1}) - b(\mathbf{u}_N^{n+1}, q_h)\,,$$

being $n = 0, \dots, N_T$. In Figure 7.12a we show the dual norms rescaled by the $L^\infty([0,T], \widetilde{U}(\Omega_f))$ norm of the solution, for all the time levels of the third heartbeat, for both the values of E_1 and E_2 used to construct the reduced-order model. Even if the number of basis functions associated with E_2 is smaller than the one related to E_1, the values of the dual norms are comparable for both choices of Young's modulus.

We select 500 random values of $\boldsymbol{\mu}$ in the interval $[10^6, 4 \cdot 10^6]$ and for each of them we compute the maximum of the dual norm of the residuals among the time-levels $n = 0, \dots, N_T$. The values of the maximum residual dual norms shown in Figure 7.12b are of the same order as the ones computed for E_1 and E_2. Our reduced-order model can thus be used to simulate the entire interval of parameter values $E \in [10^6, 4 \cdot 10^6]$.

We focus now on the results obtained with $E = 4 \cdot 10^6$ dyn/cm^2 and we compare the high-fidelity solutions and the reduced-order ones at different time levels. In Figures 7.13 and 7.14 we notice the good agreement between the two discretizations for both the velocity and the pressure fields, especially during the systolic phase. In the diastolic phase the differences between the two discretizations are higher since the energy associated with these time levels is smaller. In Figure 7.15, we compare the evolution of the averaged magnitude of wall shear stresses in three different areas. Once again, we notice that the results obtained with the high-fidelity approximation and the reduced-order one are in good agreement.

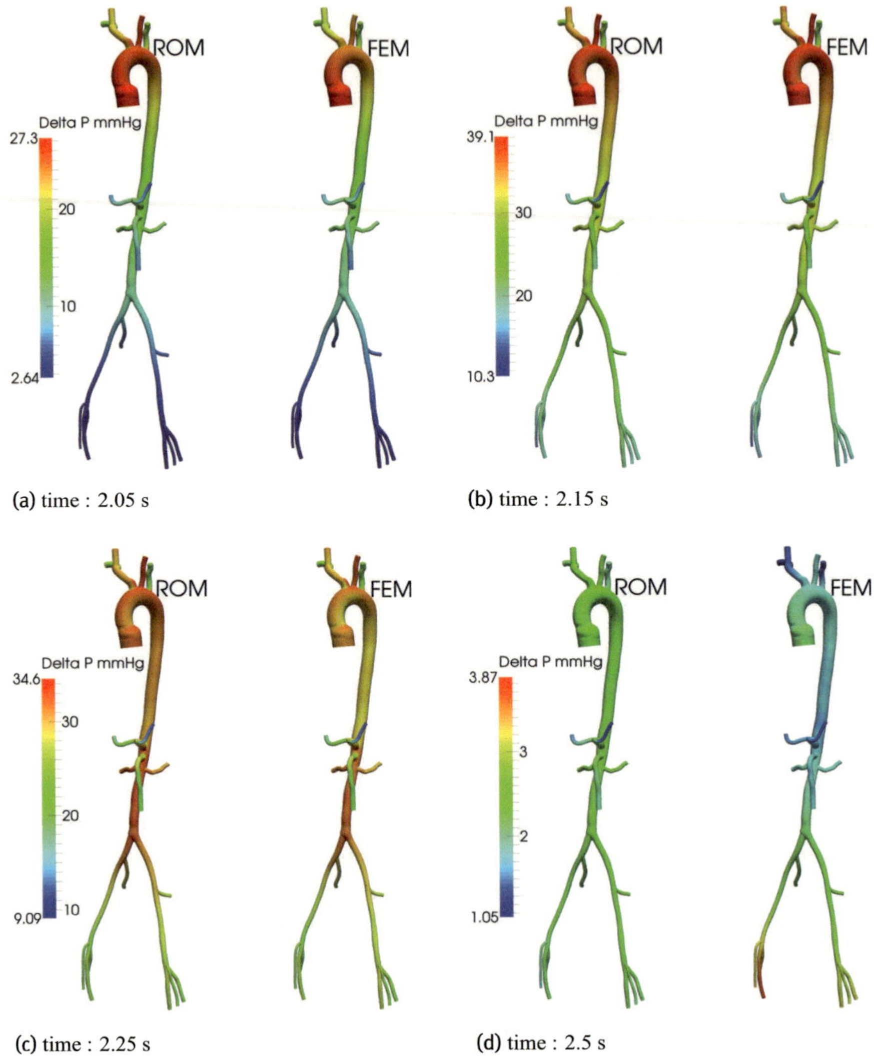

Fig. 7.13: Comparisons between pressure solutions computed for $E = 4 \cdot 10^6$ dyn/cm^2 using the reduced-order model with 71 basis functions (left) and the finite element one (right).

Concerning the computational costs, the solution of one single time instant of the RFSI solver takes 40 s on 720 processors, i.e., 11 hours for one heartbeat. The offline computation of the reduced-order model with 71 basis functions takes about 1 hour on 720 processors, plus, roughly, 7 hours to compute the reduced structures for the online evaluation of the dual norm of the residuals. The solution of the RFSI problem for three heartbeats using the reduced space of dimension 71 takes 3.4 s on one single processor of a laptop.

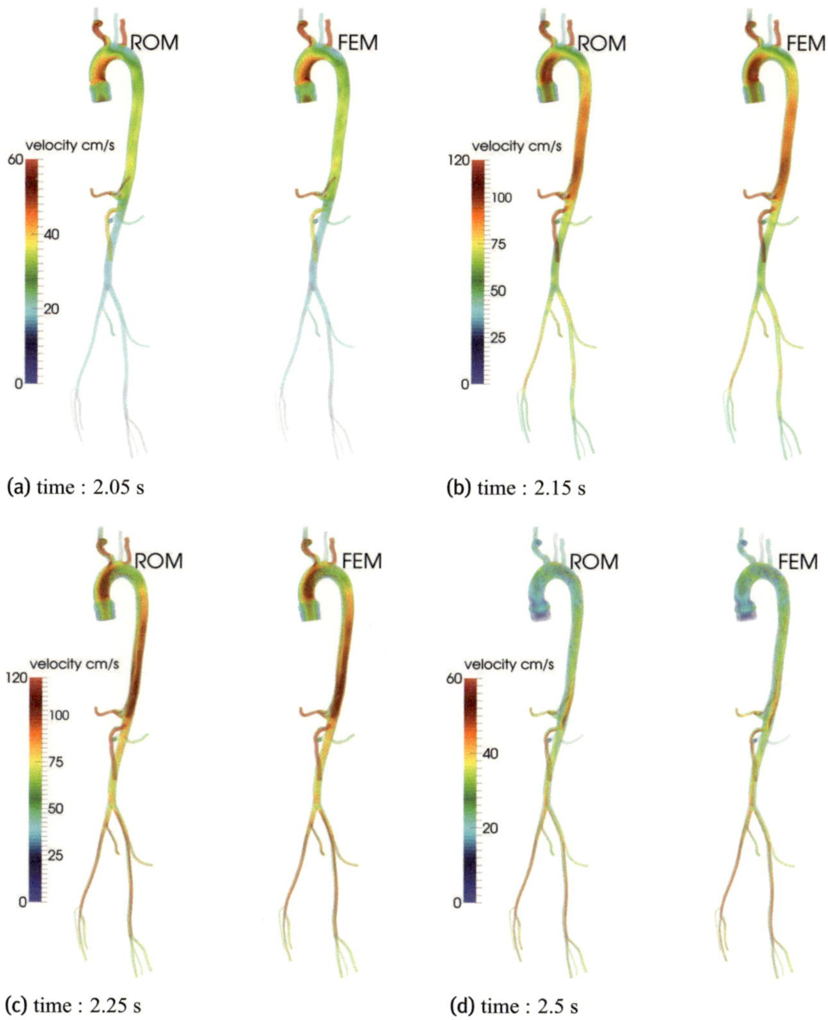

Fig. 7.14: Comparisons between velocity solutions computed for $E = 4 \cdot 10^6$ dyn/cm^2 using the reduced-order model with 71 basis functions (left) and the finite element one (right).

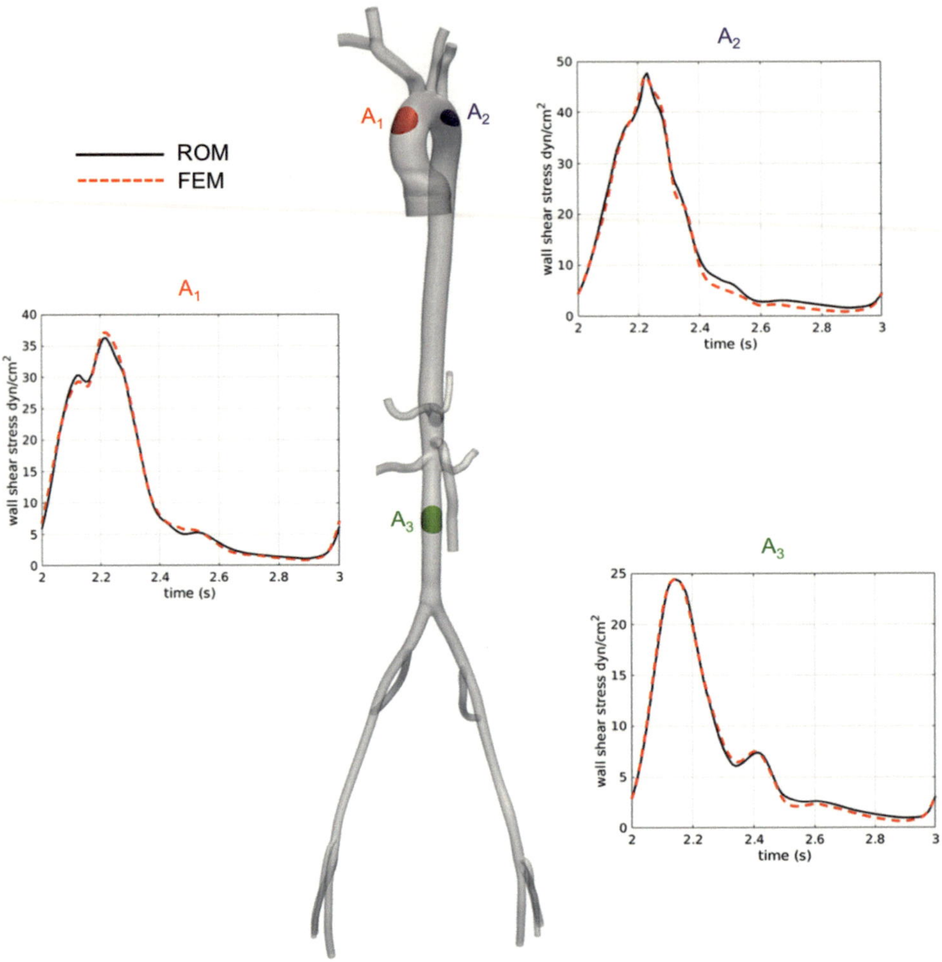

Fig. 7.15: Comparisons between WSS (Wall Shear Stresses) computed for $E = 4 \cdot 10^6$ dyn/cm^2 using the reduced-order model with 71 basis functions (left) and the finite element one (right).

7.5.3 Comparisons between FSI and RFSI

In this section we present a comparison of the results obtained using the 3D implicit FSI solver described in Section 2 against those generated by the RFSI one reported in Section 3. In particular, we focus on the outlet mean pressures, velocity profiles at several locations and wall shear stress quantities.

In Figure 7.16 we compare the time evolution of the fluid pressure at all the outflow sections. We notice that both solvers lead to comparable results in terms of pressure wave speed either in proximal or distal branches. By comparing the peak values of the pressure obtained we notice that they are in excellent agreement for those that concern the proximal arteries. Conversely, moving towards distal vessels, it is possible to observe a gap in the peak pressure values. We remark that these differences may be due to the dumping of the viscous term used in the RFSI solver for the Robin condition (7.34) used to model the surrounding tissue. Furthermore, such an effect is visible in the early diastolic phase wherein the pressure oscillations are dumped out in the RFSI solver with regard to the 3D FSI one.

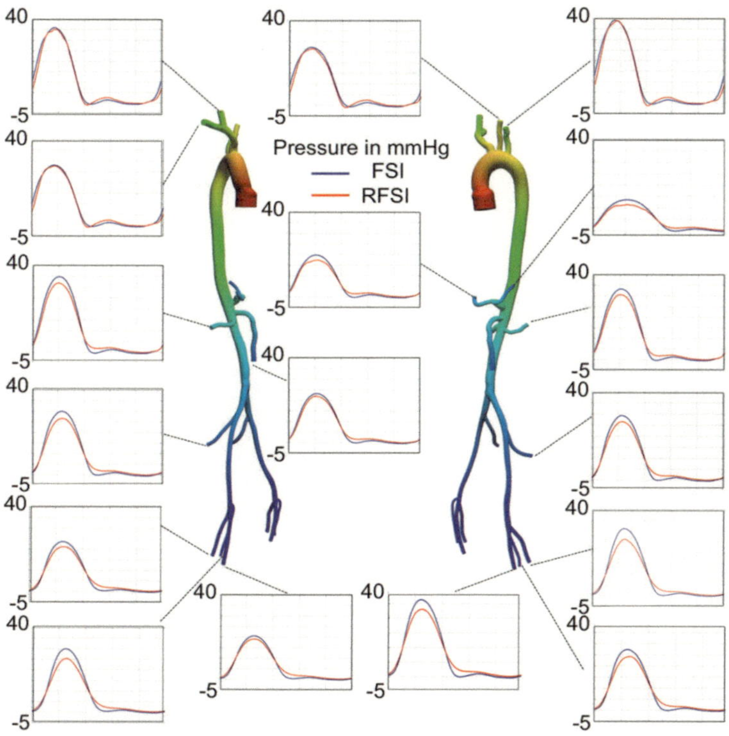

Fig. 7.16: Comparisons between outflow mean pressures computed in the third heartbeat obtained by solving the 3D FSI and the RFSI problems.

In Figures 7.17 and 7.18 we compare the velocity profiles at time 2.15 s and 2.4 s, respectively, at three locations of our domain. At time 2.15 s, which corresponds to the systolic peak of the third heartbeat, both solvers produce velocity profiles that are in good agreement. Of course, the results obtained with the full 3D FSI feature a much higher resolution thanks to the smaller mesh size used. At time 2.4 s, i.e., during the mid-deceleration phase, we notice that the velocity profiles are qualitatively similar although the FSI solver, wherein the SUPG-VMS stabilization is used, is able to capture better local features of the blood flow dynamic like small vortexes.

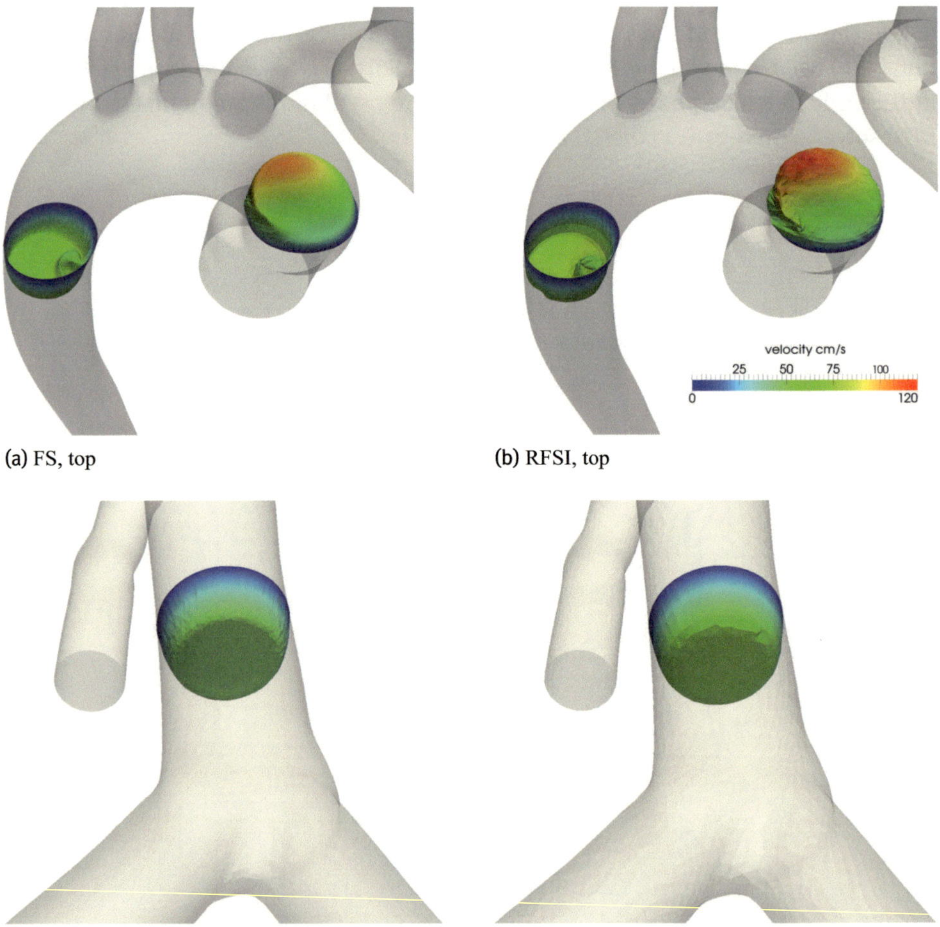

(a) FS, top

(b) RFSI, top

(c) FSI, bottom

(d) RFSI, bottom

Fig. 7.17: Comparisons between velocity profiles at time 2.15 s. On the right, the solution is P2 but the visualization is only P1.

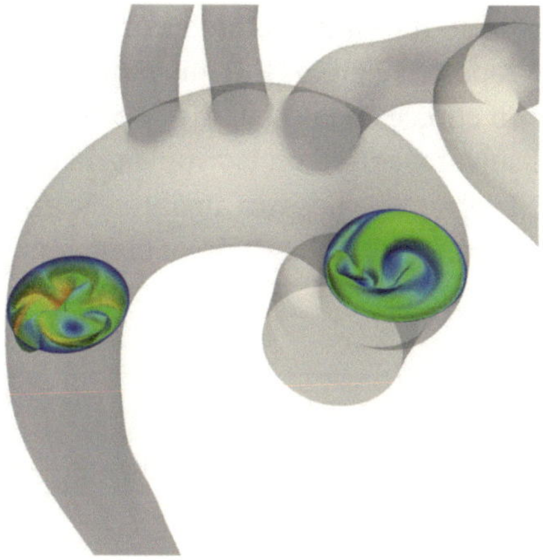

(a) FS, top

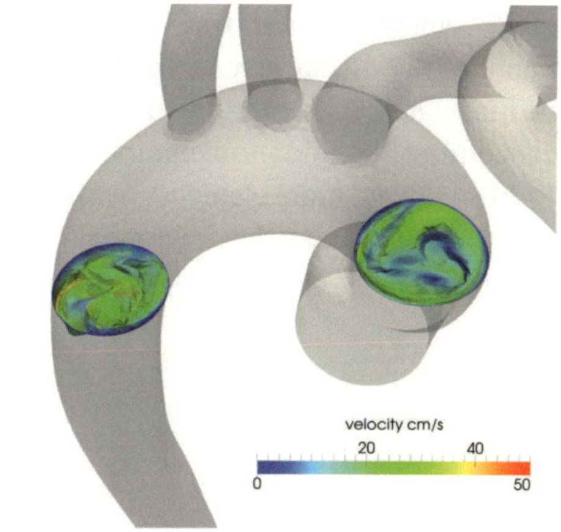

(b) RFSI, top

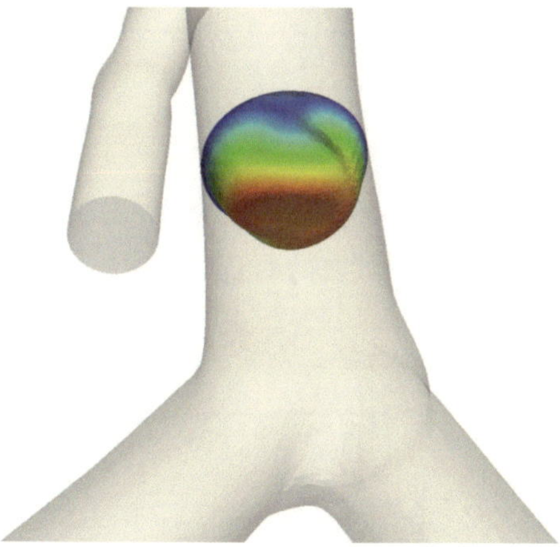

(c) FSI, bottom

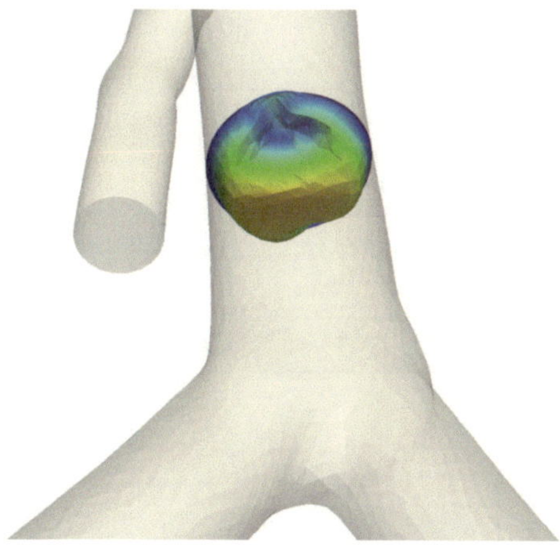

(d) RFSI, bottom

Fig. 7.18: Comparisons between velocity profiles at time 2.4 s. On the right, the solution is P2 but the visualization is only P1.

Figure 7.19 shows a comparison of the Wall Shear Stresses (WSS). Specifically, we focus on three different patches located at the fluid-structure interface where we evaluate the time evolution of the WSS magnitude averaged by the patch area and its normalization by the corresponding temporal mean. The WSS curves obtained by both solvers are in agreement in regard to their phase, especially in A_1 and A_2 areas. During the systolic phase the WSS values obtained by the two solvers feature a remarkable difference. Similar comparisons have already been carried out in [24] where the case of an aortic arch was considered. Nevertheless, in the same work a better agreement was reported for other geometries (femoropopliteal bypass). This difference may come from the use of transpiration conditions (7.25) in the RFSI solver in the presence of larger displacements. Indeed, transpiration conditions introduce a modeling error proportional to the displacement. In the comparisons reported herein, the mismatch may be associated also with the different discretization choices related to the two solvers concerning the fluid problem: in fact, in the FSI solver we used an isotropic grid and P1-P1 stabilized finite elements, while, in the RFSI solver we chose an anisotropic grid featuring a boundary layer on the fluid-structure interface and P2-P1 finite elements.

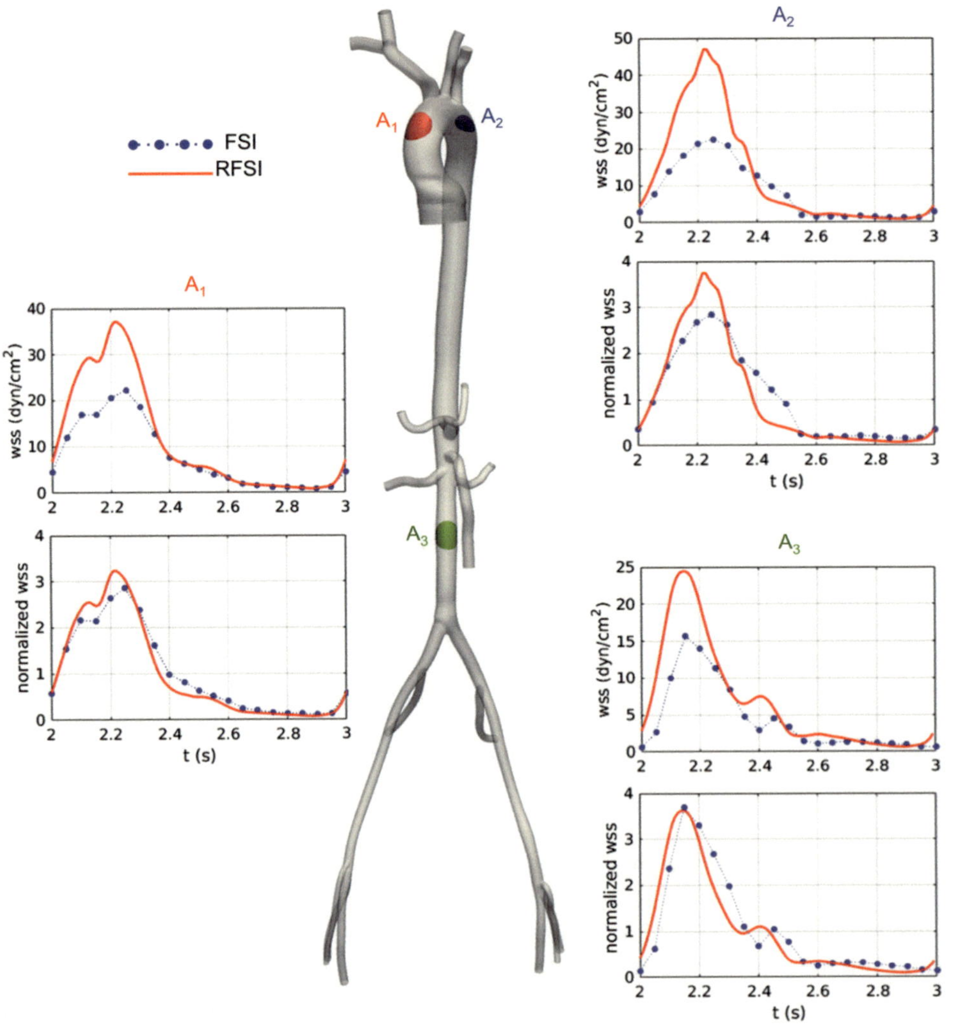

Fig. 7.19: Comparisons between the magnitude of the wall shear stress averaged on three different areas obtained by solving the 3D FSI and the RFSI problems.

7.6 Conclusions

When dealing with vascular flow simulation several modeling and numerical options are available. The choice depends on the targeted measures or indicators. In this work we have presented two possibilities, one relying on a detailed three-dimensional FSI model for both the solid and the fluid, and one, RFSI, which focuses on the fluid mechanics and allows for further numerical reduction. The FSI model is approximated by the finite element method and its resolution requires parallel methods, in particular we use a FaCSI preconditioner derived from domain decomposition and multigrid methods. RFSI relies on a fixed geometry and generalized Robin conditions to account for a membrane model. It is firstly approximated by the finite element method in an offline phase, then by POD and a Galerkin projection to a very small subspace in the online phase.

We have compared the two methods and showed that RFSI introduces modeling errors that have to be taken into account when interpreting the numerical results. However, it allows for drastic reductions of the computational times and runs online almost in real time on a simple laptop. It is important to notice that the online solver is specialized for the geometry at hand: changing the geometry, for example a new patient, requires recomputation of the offline phase.

Acknowledgment: The authors wish to thank Prof. Alfio Quarteroni for his mentoring on all the subjects of this chapter, Andrea Bartezzaghi for the realization of a very efficient online solver which was used for the presentations, and the whole LifeV community.

Bibliography

[1] Open source medical software corporation. http://www.vascularmodel.com/sandbox/doku.php.

[2] P. R. Amestoy, I. S. Duff, J. Koster, and J. Y. L'Excellent. A fully asynchronous multifrontal solver using distributed dynamic scheduling. *SIAM Journal on Matrix Analysis and Applications*, 23(1):15–41, 2001.

[3] P. R. Amestoy, A. Guermouche, J. Y. L'Excellent, and S. Pralet. Hybrid scheduling for the parallel solution of linear systems. *Parallel Computing*, 32(2):136–156, 2006.

[4] S. Badia, F. Nobile, and C. Vergara. Fluid-structure partitioned procedures based on Robin transmission conditions. *Journal of Computational Physics*, 227(14):7027–7051, 2008.

[5] S. Badia, F. Nobile, and C. Vergara. Robin-Robin preconditioned Krylov methods for fluid-structure interaction problems. *Computer Methods in Applied Mechanics and Engineering*, 198(33-36):2768–2784, 2009.

[6] S. Badia, A. Quaini, and A. Quarteroni. Modular vs. non-modular preconditioners for fluid-structure systems with large added-mass effect. *Computer Methods in Applied Mechanics and Engineering*, 197(49-50):4216–4232, 2008.

[7] S. Badia, A. Quaini, and A. Quarteroni. Modular vs. non-modular preconditioners for fluid-structure systems with large added-mass effect. *Computer Methods in Applied Mechanics and Engineering*, 197(49-50):4216–4232, 2008.

[8] S. Badia, A. Quaini, and A. Quarteroni. Splitting methods based on algebraic factorization for fluid-structure interaction. *SIAM Journal on Scientific Computing*, 30(4):1778–1805, 2008.

[9] F. Ballarin, E. Faggiano, S. Ippolito, A. Manzoni, A. Quarteroni, G. Rozza, and R. Scrofani. Fast simulations of patient-specific haemodynamics of coronary artery bypass grafts based on a pod-galerkin method and a vascular shape parametrization. *Journal of Computational Physics*, 315:609–628, 2016.

[10] F. Ballarin and G. Rozza. Pod-galerkin monolithic reduced order models for parametrized fluid-structure interaction problems. *International Journal for Numerical Methods in Fluids*, pages n/a–n/a, 2016. fld.4252.

[11] A. T. Barker. *Parallel monolithic fluid-structure interaction algorithms with application to blood flow simulation*. PhD thesis, University of Colorado, 2009.

[12] A. T. Barker and X.-C. Cai. Scalable parallel methods for monolithic coupling in fluid-structure interaction with application to blood flow modeling. *Journal of Computational Physics*, 229(3):642–659, 2010.

[13] Y. Bazilevs, V. M. Calo, J. A. Cottrell, T. J. R. Hughes, A. Reali, and G. Scovazzi. Variational multiscale residual-based turbulence modeling for large eddy simulation of incompressible flows. *Computer Methods in Applied Mechanics and Engineering*, 197(1):173–201, 2007.

[14] Y. Bazilevs, V. M. Calo, T. J. R. Hughes, and Y. Zhang. Isogeometric fluid-structure interaction: theory, algorithms, and computations. *Computational Mechanics*, 43:3–37, 2008.

[15] Y. Bazilevs, K. Takizawa, and T. E. Tezduyar. *Computational fluid-structure interaction: methods and applications*. John Wiley & Sons, 2013.

[16] L. Bertagna, L. Formaggia, S. Deparis, D. Forti, and A. Veneziani. The lifev library: engineering mathematics beyond the proof of concept. submitted, 2016.

[17] L. Bertagna and A. Veneziani. A model reduction approach for the variational estimation of vascular compliance by solving an inverse fluid-structure interaction problem. *Inverse Problems*, 30(5):055006, 2014.

[18] D. Boffi, N. Cavallini, and L. Gastaldi. Finite element approach to immersed boundary method with different fluid and solid densities. *Mathematical Models and Methods in Applied Sciences*, 21(12):2523–2550, 2011.

[19] L. Boussel, V. Rayz, C. McCulloch, A. Martin, G. Acevedo-Bolton, M. Lawton, R. Higashida, W. S. Smith, W. L. Young, and D. Saloner. Aneurysm growth occurs at region of low wall shear stress. *Stroke*, 39(11):2997–3002, 2008.

[20] J. Burkardt, M. Gunzburger, and H.-C. Lee. POD and CVT-based reduced-order modeling of Navier-Stokes flows. *Computer Methods in Applied Mechanics and Engineering*, 196(1–3):337–355, 2006.

[21] A. Caiazzo, R. Guibert, and I. E. Vignon-Clementel. A reduced-order modeling for efficient design study of artificial valve in enlarged ventricular outflow tracts. *Computer Methods in Biomechanics and Biomedical Engineering*, 19(12):1314–1318, 2016.

[22] P. Causin, J. F. Gerbeau, and F. Nobile. Added-mass effect in the design of partitioned algorithms for fluid-structure problems. *Computer Methods in Applied Mechanics and Engineering*, 194(42-44):4506–4527, 2005.

[23] C. M. Colciago. *Reduced Order Fluid-Structure Interaction Models for Haemodynamics Applications*. PhD thesis, SB, Lausanne, 2014.

[24] C. M. Colciago, S. Deparis, and A. Quarteroni. Comparisons between reduced order models and full 3D models for fluid-structure interaction problems in haemodynamics. *J. Comput.*

Appl. Math., 265(0):120–138, 2014. Current Trends and Progresses in Scientific Computation Dedicated to Professor Ben-yu Guo on His 70th Birthday.

[25] G.-H. Cottet, E. Maitre, and T. Milcent. Eulerian formulation and level set models for incompressible fluid-structure interaction. *Mathematical Modelling and Numerical Analysis*, 42(3):471–492, 2008.

[26] P. Crosetto, S. Deparis, G. Fourestey, and A. Quarteroni. Parallel algorithms for fluid-structure interaction problems in haemodynamics. *SIAM Journal on Scientific Computing*, 33(4):1598–1622, 2011.

[27] P. Crosetto, S. Deparis, G. Fourestey, and A. Quarteroni. Parallel algorithms for fluid-structure interaction problems in haemodynamics. *SIAM Journal on Scientific Computing*, 33(4):1598–1622, 2011.

[28] J. Degroote, K.-J. Bathe, and J. Vierendeels. Performance of a new partitioned procedure versus a monolithic procedure in fluid-structure interaction. *Computers & Structures*, 87:793–801, 2009.

[29] S. Deparis, M. Discacciati, G. Fourestey, and A. Quarteroni. Fluid-structure algorithms based on Steklov-Poincaré operators. *Computer Methods in Applied Mechanics and Engineering*, 195:5797–5812, 2006.

[30] S. Deparis, M. Discacciati, and A. Quarteroni. A domain decomposition framework for fluid-structure interaction problems. In C. Groth and D. Zingg, editors, *Computational Fluid Dynamics 2004*, pages 41–58. Springer Berlin Heidelberg, 2006.

[31] S. Deparis, M. A. Fernandez, and L. Formaggia. Acceleration of a fixed point algorithm for fluid-structure interaction using transpiration conditions. *M2AN Mathematical Modelling and Numerical Analysis.*, 37(4):601–616, 2003.

[32] S. Deparis, D. Forti, G. Grandperrin, and A. Quarteroni. FaCSI: A block parallel preconditioner for fluid–structure interaction in hemodynamics. Journal of Computational Physics (accepted), 2016.

[33] S. Deparis, D. Forti, A. Heinlein, A. Klawonn, A. Quarteroni, and O. Rheinbach. A comparison of preconditioners for the steklov-poincaré formulation of the fluid-structure coupling in hemodynamics. *Proceedings in Applied Mathematics and Mechanics*, 15:93–94, 2015.

[34] S. Deparis, G. Grandperrin, and A. Quarteroni. Parallel preconditioners for the unsteady Navier-Stokes equations and applications to hemodynamics simulations. *Computers & Fluids*, 92(0):253–273, 2014.

[35] W. G. Dettmer and D. Perić. A fully implicit computational strategy for strongly coupled fluid-solid interaction. *Archives of Computational Methods in Engineering*, 14(3):205–247, 2007.

[36] E. S. Di Martino and D. A. Vorp. Effect of variation in intraluminal thrombus constitutive properties on abdominal aortic aneurysm wall stress. *Annals of Biomedical Engineering*, 31(7):804–809, 2003.

[37] J. Donea, S. Giuliani, and J. Halleux. An arbitrary lagrangian-eulerian finite element method for transient dynamic fluid-structure interaction. *Computer Methods in Applied Mechanics and Engineering*, 33:689–723, 1982.

[38] T. Dunne. *Adaptive Finite Element Approximation of Fluid-Structure Interaction Based on Eulerian and Arbitrary Lagrangian-Eulerian Variational Formulations*. PhD thesis, Ruprechts-Karls Universität Heidelberg, 2007.

[39] E. Faggiano and L. Antiga. An open-source tool for patient-specific fluid-structure vessel mesh generation. 2015. In preparation.

[40] M. A. Fernandez, J. Gerbeau, and C. Grandmont. A projection semi-implicit scheme for the coupling of an elastic structure with an incompressible fluid. *International Journal for Numerical Methods in Engineering*, 64(4):794–821, 2007.

[41] M. A. Fernández and M. Moubachir. An exact block-Newton algorithm for solving fluid-structure interaction problems. *C. R. Math. Acad. Sci. Paris*, 336(8):681–686, 2003.

[42] M. A. Fernández, J. Mullaert, and M. Vidrascu. Generalized robin–neumann explicit coupling schemes for incompressible fluid-structure interaction: Stability analysis and numerics. *International Journal for Numerical Methods in Engineering*, 101(3):199–229, 2015.

[43] C. A. Figueroa, I. E. Vignon-Clementel, K. E. Jansen, T. J. Hughes, and C. A. Taylor. A coupled momentum method for modeling blood flow in three-dimensional deformable arteries. *Computer Methods in Applied Mechanics and Engineering*, 195(41-43):5685–5706, 2006.

[44] L. Formaggia, A. Quarteroni, and A. Veneziani. *Cardiovascular mathematics*. Springer, 2009.

[45] L. Formaggia, A. Quarteroni, and A. Veneziani, editors. *Cardiovascular mathematics*, volume 1 of *MS&A. Modeling, Simulation and Applications*. Springer-Verlag Italia, Milan, 2009. Modeling and simulation of the circulatory system.

[46] D. Forti and L. Dedè. Semi–implicit BDF time discretization of the Navier–Stokes equations with VMS–LES modeling in High Performance Computing framework. *Computers & Fluids*, 117:168–182, 2015.

[47] D. Forti, A. Quarteroni, and S. Deparis. A parallel algorithm for the solution of large-scale non-conforming fluid-structure interaction problems in hemodynamics. Submitted, 2015.

[48] T. C. Gasser, R. W. Ogden, and G. A. Holzapfel. Hyperelastic modelling of arterial layers with distributed collagen fibre orientations. *Journal of The Royal Society Interface*, 3(6):15–35, 2006.

[49] M. W. Gee, U. Küttler, and W. A. Wall. Truly monolithic algebraic multigrid for fluid-structure interaction. *International Journal for Numerical Methods in Engineering*, 26:52–72, 2010.

[50] J. F. Gerbeau and M. Vidrascu. A quasi-Newton algorithm based on a reduced model for fluid-structure interaction problems in blood flows. *Mathematical Modelling and Numerical Analysis*, 37(4):631–647, 2003.

[51] A.-L. Gerner, A. Reusken, and K. Veroy. Reduced basis a posteriori error bounds for the instationary Stokes equations. submitted, 2012.

[52] A.-L. Gerner and K. Veroy. Reduced basis a posteriori error bounds for the Stokes equations in parametrized domains: a penalty approach. *Mathematical Models and Methods in Applied Sciences*, 21(10):2103–2134, 2011.

[53] A. Gerstenberger and W. A. Wall. An extended finite element method/lagrange multiplier based approach for fluid–structure interaction. *Computer Methods in Applied Mechanics and Engineering*, 197(19):1699–1714, 2008.

[54] P. Gervasio, F. Saleri, and A. Veneziani. Algebraic fractional-step schemes with spectral methods for the incompressible navier-stokes equations. *Journal of Computational Physics*, 214(1):347–365, 2006.

[55] L. Grinberg and G. E. Karniadakis. Outflow boundary conditions for arterial networks with multiple outlets. *Annals of biomedical engineering*, 36(9):1496–1514, 2008.

[56] L. Grinberg, A. Yakhot, and G. Karniadakis. Analyzing transient turbulence in a stenosed carotid artery by proper orthogonal decomposition. *Annals of Biomedical Engineering*, 37(11):2200–2217, 2009.

[57] R. Guibert, K. McLeod, A. Caiazzo, T. Mansi, M. A. Fernéndez, M. Sermesant, X. Pennec, I. E. Vignon-Clementel, Y. Boudjemline, and J.-F. Gerbeau. Group-wise construction of reduced models for understanding and characterization of pulmonary blood flows from medical images. *Medical Image Analysis*, 18(1):63–82, 2014.

[58] M. D. Gunzburger, J. S. Peterson, and J. N. Shadid. Reduced-order modeling of time-dependent PDEs with multiple parameters in the boundary data. *Computer Methods in Applied Mechanics and Engineering*, 196(4–6):1030–1047, 2007.

[59] M. Heil. An efficient solver for the fully coupled solution of large-displacement fluid–structure interaction problems. *Computer Methods in Applied Mechanics and Engineering*, 193(1-2):1–23, 2004.

[60] M. Heil, A. L. Hazel, and J. Boyle. Solvers for large–displacement fluid–structure interaction problems: segregated versus monolithic approaches. *Computational Mechanics*, 43(1):91–101, 2008.

[61] J. S. Hesthaven, G. Rozza, and B. Stamm. *Certified Reduced Basis Methods for Parametrized Partial Differential Equations*. Springer International Publishing, The address, 1 edition, 2016.

[62] G. A. Holzapfel, T. C. Gasser, and R. W. Ogden. A new constitutive framework for arterial wall mechanics and a comparative study of material models. *Journal of elasticity and the physical science of solids*, 61(1):1–48, 2000.

[63] G. A. Holzapfel and R. W. Ogden. *Mechanics of biological tissue*. Springer Science & Business Media, 2006.

[64] J. Hron and S. Turek. A monolithic fem/multigrid solver for an ALE formulation of fluid-structure interaction with applications in biomechanics. In H.-J. Bungartz and M. Schäfer, editors, *Fluid-structure interaction*, volume 53 of *Lecture Notes in Computational Science and Engineering*, pages 146–170, 2006.

[65] T. J. Hughes, W. Liu, and T. Zimmermann. Lagrangian-eulerian finite element formulation for incompressible viscous flow. *Computer Methods in Applied Mechanics and Engineering*, 29:329–349, 1981.

[66] J. D. Humphrey and K. R. Rajagopal. A constrained mixture model for growth and remodeling of soft tissues. *Mathematical Models and Methods in Applied Sciences*, 12(03):407–430, 2002.

[67] L. Iapichino, A. Quarteroni, and G. Rozza. A reduced basis hybrid method for the coupling of parametrized domains represented by fluidic networks. *Computer Methods in Applied Mechanics and Engineering*, 221-222(0):63–82, 2012.

[68] B. Janssen and T. Wick. Block preconditioning with Schur complements for monolithic fluid-structure interactions. In *Proceedings of V European Conference on Computational Fluid Dynamics ECCOMAS CFD 2010*, Lisbon, Portugal,, June 14-17 2010.

[69] L.-D. Jou, D. H. Lee, H. Morsi, and M. E. Mawad. Wall shear stress on ruptured and unruptured intracranial aneurysms at the internal carotid artery. *AJNR Am J Neuroradiol*, 29:1761–1767, 2008.

[70] T. Kashiwabara, C. Colciago, L. Dedé, and A. Quarteroni. Well-posedness, regularity, and convergence analysis of the finite element approximation of a generalized robin boundary value problem. submitted, 2014.

[71] E. Kuhl, R. Maas, G. Himpel, and A. Menzel. Computational modeling of arterial wall growth. *Biomechanics and Modeling in Mechanobiology*, 6(5):321–331, 2007.

[72] K. Kunisch and S. Volkwein. Galerkin proper orthogonal decomposition methods for a general equation in fluid dynamics. *SIAM Journal on Numerical Analysis*, 40(2):492–515, 2003.

[73] U. Küttler, M. W. Gee, C. Förster, A. Comerford, and W. A. Wall. Coupling strategies for biomedical fluid-structure interaction problems. *International Journal for Numerical Methods in Biomedical Engineering*, 26(3-4):305–321, 2010.

[74] U. Küttler and W. A. Wall. The dilemma of domain decomposition approaches in fluid-structure interactions with fully enclosed incompressible fluids. In U. Langer, M. Discacciati, D. Keyes, O. Widlund, and W. Zulehner, editors, *Domain Decomposition Methods in Science and Engineering XVII*, volume 60 of *Lecture Notes in Computational Science and Engineering*, pages 575–582. Springer Berlin Heidelberg, 2008.

[75] U. Küttler and W. A. Wall. Fixed-point fluid–structure interaction solvers with dynamic relaxation. *Computational Mechanics*, 43(1):61–72, 2008.

[76] U. Langer and H. Yang. Domain decomposition solvers for some fluid-structure interaction problems. *Proceedings in Applied Mathematics and Mechanics*, 12(1):375–376, 2012.

[77] U. Langer and H. Yang. Partitioned solution algorithms for fluid-structure interaction problems with hyperelastic models. *Journal of Computational and Applied Mathematics*, 276:47–61, 2015.

[78] T. Lassila, A. Manzoni, A. Quarteroni, and G. Rozza. Boundary control and shape optimization for the robust design of bypass anastomoses under uncertainty. *ESAIM: Mathematical Modelling and Numerical Analysis*, 47:1107–1131, 7 2013.

[79] P. Le Tallec and J. Mouro. Fluid structure interaction with large structural displacements. *Computer Methods in Applied Mechanics and Engineering*, 190:3039–3067, 2001.

[80] F. Loth, P. F. Fischer, and H. S. Bassiouny. Blood flow in end-to-side anastomoses. *Annual Review of Fluid Mechanics*, 40:367–393, 2008.

[81] F. Loth, S. A. Jones, C. K. Zarins, D. P. Giddens, R. F. Nassar, S. Glagov, and H. S. Bassiouny. Relative contribution of wall shear stress and injury in experimental intimal thickening at PTFE end-to-side arterial anastomoses. *J Biomech Eng*, 124:44–51, 2002.

[82] M. A. M., A. S. L., and I. S. Hemodynamic shear stress and its role in atherosclerosis. *JAMA*, 282(21):2035–2042, 1999.

[83] A. Malossi and J. Bonnemain. Numerical comparison and calibration of geometrical multiscale models for the simulation of arterial flows. *Cardiovascular Engineering and Technology*, 4(4):440–463, 2013.

[84] A. Manzoni, A. Quarteroni, and G. Rozza. Model reduction techniques for fast blood flow simulation in parametrized geometries. *International Journal for Numerical Methods in Biomedical Engineering*, 28(6-7):604–625, 2012.

[85] A. Manzoni, A. Quarteroni, and G. Rozza. Shape optimization for viscous flows by reduced basis methods and free-form deformation. *International Journal for Numerical Methods in Fluids*, 70:646–670, 2012.

[86] H. G. Matthies, R. Niekamp, and J. Steindorf. Algorithms for strong coupling procedures. *Computer Methods in Applied Mechanics and Engineering*, 195(17-18):2028–2049, 2006.

[87] U. M. Mayer, A. Popp, A. Gerstenberger, and W. A. Wall. 3d fluid-structure-contact interaction based on a combined xfem fsi and dual mortar contact approach. *Computational Mechanics*, 46(1):53–67, 2010.

[88] C. Michler, S. J. Hulshoff, E. H. Van Brummelen, and R. De Borst. A monolithic approach to fluid–structure interaction. *Computers & fluids*, 33(5):839–848, 2004.

[89] R. Mittal and G. Iaccarino. Immersed boundary methods. In *Annual review of fluid mechanics. Vol. 37*, volume 37 of *Annu. Rev. Fluid Mech.*, pages 239–261. Annual Reviews, Palo Alto, CA, 2005.

[90] P. Moireau, N. Xiao, M. Astorino, C. A. Figueroa, D.Chapelle, C. A. Taylor, and J. Gerbeau. External tissue support and fluid-structure simulation in blood flows. *Biomechanics and Modeling in Mechanobiology*, 11(1-2):1–18, 2012.

[91] N. M. Newmark. A method of computation for structural dynamics. *Journal of the Engineering Mechanics Division*, 85(3):67–94, 1959.

[92] W. W. Nichols and M. F. O'Rourke. *McDonald's Blood Flow in Arteries - Theoretical, experimental and clinical principles -*. Arnold, 1998.

[93] F. Nobile. *Numerical approximation of fluid-structure interaction problems with application to haemodynamics*. PhD thesis, EPFL, 2001.

[94] F. Nobile. Coupling strategies for the numerical simulation of blood flow in deformable arteries by 3d and 1d models. *Mathematical and Computer Modelling*, 49(11–12):2152–2160, 2009. Trends in Application of Mathematics to Medicine Trends in Application of Mathematics to Medicine.

[95] F. Nobile, M. Pozzoli, and C. Vergara. Inexact accurate partitioned algorithms for fluid–structure interaction problems with finite elasticity in haemodynamics. *Journal of Computational Physics*, 273(0):598–617, 2014.

[96] R. W. Ogden. *Anisotropy and Nonlinear Elasticity in Arterial Wall Mechanics*, pages 179–258. Springer Vienna, Vienna, 2009.

[97] T. G. Papaioannou, E. N. Karatzis, M. Vavuranakis, J. P. Lekakis, and C. Stefanadis. Assessment of vascular wall shear stress and implications for atherosclerotic disease. *International Journal of Cardiology*, 113(1):12–18, 2006.

[98] C. S. Peskin. The immersed boundary method. *Acta Numerica*, 11:479–517, 2002.

[99] A. Quarteroni, A. Manzoni, and F. Negri. *Reduced Basis Methods for Partial Differential Equations*, volume 92. Springer International Publishing, 1 edition, 2016.

[100] A. Quarteroni and G. Rozza, editors. *Reduced Order Methods for Modeling and Computational Reduction*. MS&A, Modeling, Simulation and Applications. Springer, Milano, 2013. Book edited with selected contributions from CECAM workshop, EPFL, May 2012.

[101] R. Rannacher and T. Richter. An adaptive finite element method for fluid-structure interaction problems based on a fully eulerian formulation. In H.-J. B. et al., editor, *Fluid Structure Interaction II*, pages 159–192. Springer-Verlag Berlin Heidelberg, 2010. Lecture Notes in Computational Science and Engineering 73.

[102] T. Richter. A fully Eulerian formulation for fluid-structure-interaction problems with large deformations and free structure movement. In *Proceedings of the V European Conference on Computational Fluid Dynamics ECCOMAS CFD 2010*, Lisbon, Portugal, 14-17 June 2010, 2010.

[103] T. Richter. A fully eulerian formulation for fluid–structure-interaction problems. *Journal of Computational Physics*, 233:227–240, 2013.

[104] C. W. Rowley. Model reduction for fluids, using balanced Proper Orthogonal Decomposition. *International Journal of Bifurcation and Chaos (IJBC)*, 15(3):997–1013, 2005.

[105] G. Rozza, D. Huynh, and A. Manzoni. Reduced basis approximation and a posteriori error estimation for Stokes flows in parametrized geometries: roles of the inf-sup stability constants. *Numer. Math.*, 125(1):115–152, 2013.

[106] G. Rozza, A. Manzoni, and F. Negri. Reduction strategies for pde-constrained optimization probems in haemodynamics. In J. e. a. Eberhardsteiner, editor, *European Congress on Computational Methods in Applied Sciences and Engineering*, pages 1–20, September 10-14 2012.

[107] G. Rozza and K. Veroy. On the stability of the reduced basis method for Stokes equations in parametrized domains. *Computer Methods in Applied Mechanics and Engineering*, 196(7):1244–1260, 2007.

[108] M. Shojima, M. Oshima, K. Takagi, R. Torii, M. Hayakawa, K. Katada, A. Morita, and T. Kirino. Magnitude and role of wall shear stress on cerebral aneurysm: computational fluid dynamic study of 20 middle cerebral artery aneurysms. *Stroke*, 35(11):2500–5, 2004.

[109] M. Souli, A. Ouahsine, and L. Lewin. {ALE} formulation for fluid-structure interaction problems. *Computer Methods in Applied Mechanics and Engineering*, 190(5–7):659–675, 2000.

[110] K. Stein, T. Tezduyar, and R. Benney. Mesh moving techniques for fluid-structure interactions with large displacements. *Journal of Applied Mechanics*, 70(1):58–63, 2003.

[111] K. Takizawa and T. E. Tezduyar. Multiscale space–time fluid–structure interaction techniques. *Computational Mechanics*, 48(3):247–267, 2011.

[112] C. Taylor and J. Humphrey. Open problems in computational vascular biomechanics: Hemodynamics and arterial wall mechanics. *Computer Methods in Applied Mechanics and Engineering*, 198(45–46):3514–3523, 2009. Models and Methods in Computational Vascular and Cardiovascular Mechanics.

[113] T. E. Tezduyar and S. Sathe. Modelling of fluid–structure interactions with the space–time finite elements: solution techniques. *International Journal for Numerical Methods in Fluids*, 54(6-8):855–900, 2007.

[114] T. E. Tezduyar, S. Sathe, M. Schwaab, and B. S. Conklin. Arterial fluid mechanics modeling with the stabilized space–time fluid–structure interaction technique. *International Journal for Numerical Methods in Fluids*, 57(5):601–629, 2008.

[115] T. E. Tezduyar, S. Sathe, and K. Stein. Solution techniques for the fully discretized equations in computation of fluid-structure interactions with the space-time formulations. *Computer Methods in Applied Mechanics and Engineering*, 195(41-43):5743–5753, 2006.

[116] J. Tong, T. Cohnert, P. Regitnig, and G. Holzapfel. Effects of age on the elastic properties of the intraluminal thrombus and the thrombus-covered wall in abdominal aortic aneurysms: Biaxial extension behaviour and material modelling. *European Journal of Vascular and Endovascular Surgery*, 42(2):207–219, 2011.

[117] A. Toselli and O. Widlund. *Domain decomposition methods: algorithms and theory*. Springer, 2005.

[118] P. Tricerri, L. Dedè, S. Deparis, A. Quarteroni, A. M. Robertson, and A. Sequeira. Fluid-structure interaction simulations of cerebral arteries modeled by isotropic and anisotropic constitutive laws. *Computational Mechanics*, 55(3):479–498, 2015.

[119] A. Valencia and F. Solis. Blood flow dynamics and arterial wall interaction in a saccular aneurysm model of the basilar artery. *Computers & Structures*, 84(21):1326–1337, 2006.

[120] E. H. Van Brummelen. Added mass effects of compressible and incompressible flows in fluid-structure interaction. *Journal of Applied mechanics*, 76(2):021206–1–021206–7, 2009.

[121] E. H. Van Brummelen. Partitioned iterative solution methods for fluid-structure interaction. *International Journal for Numerical Methods in Fluids*, 65(1-3):3–27, 2011.

[122] W. A. Wall, A. Gerstenberger, P. Gamnitzer, C. Förster, and E. Ramm. Large deformation fluid-structure interaction–advances in ale methods and new fixed grid approaches. In *Fluid-structure interaction*, pages 195–232. Springer, 2006.

[123] W. A. Wall, A. Gerstenberger, U. Küttler, and U. M. Mayer. An XFEM based fixed-grid approach for 3d fluid-structure interaction. In H.-J. B. et al., editor, *Fluid Structure Interaction II*, pages 327–349. Springer-Verlag Berlin Heidelberg, 2010. Lecture Notes in Computational Science and Engineering 73.

[124] H. Wang, J. Chessa, W. K. Liu, and T. Belytschko. The immersed/fictitious element method for fluid-structure interaction: volumetric consistency, compressibility and thin members. *International Journal for Numerical Methods in Engineering*, 74(1):32–55, 2008.

[125] T. Wiesner. *Flexible Aggregation-based Algebraic Multigrid Methods for Contact and Flow Problems*. Phd thesis, Technische Universität München, 2015.

[126] K. Willcox and J. Peraire. Balanced model reduction via the proper orthogonal decomposition. *AIAA Journal*, pages 2323–2330, 2002.

[127] Y. Wu and X.-C. Cai. A fully implicit domain decomposition based ALE framework for three-dimensional fluid-structure interaction with application in blood flow computation. *Journal of Computational Physics*, 258:524–537, 2014.

E. Harald van Brummelen, Mahnaz Shokrpour Roudbari,
Görkem Şimşek, and Kristoffer G. van der Zee

8 Binary-fluid–solid interaction based on the Navier–Stokes–Cahn–Hilliard Equations

Abstract: We consider a model for binary-fluid–solid interaction based on a diffuse-interface model for the binary fluid and a hyperelastic-material model for the solid. The diffuse-interface binary-fluid model is described by the quasi-incompressible Navier–Stokes–Cahn–Hilliard equations with preferential-wetting boundary conditions at the fluid–solid interface. The fluid traction on the interface includes a capillary-stress contribution in addition to the regular viscous-stress and pressure contributions. The dynamic interface condition comprises the traction exerted by the nonuniform solid–fluid surface tension in accordance with the Young–Laplace law for the solid–fluid interface. The solid is modeled as a hyperelastic material. We present a weak formulation of the aggregated binary-fluid–solid interaction problem, based on an arbitrary Lagrangian–Eulerian formulation of the Navier–Stokes–Cahn–Hilliard equations and a proper weak evaluation of the binary-fluid traction and of the solid–fluid surface tension. We also present an analysis of the essential properties of the binary-fluid–solid interaction problem, including a dissipation relation for the complete fluid–solid interaction problem. To validate the presented binary-fluid–solid interaction model, we consider numerical simulations for the elasto-capillary interaction of a droplet with a soft solid substrate and present a comparison to corresponding experimental data.

Keywords: binary-fluid–solid interaction, elasto-capillarity, quasi-incompressibility, Navier–Stokes–Cahn–Hilliard equations

Classification: 65M60, 65M12, 65Z05, 35Q35, 35R37

8.1 Introduction

Binary fluids are fluids that comprise two constituents, viz. two phases of the same fluid (gas or liquid) or two distinct species (e.g., water and air). A distinctive feature of binary-fluids is the presence of a fluid–fluid interface that separates the two com-

E. Harald van Brummelen, Eindhoven University of Technology, P.O. Box 513, 5600 MB Eindhoven, The Netherlands, e.h.v.brummelen@tue.nl
Mahnaz Shokrpour Roudbari, Eindhoven University of Technology, P.O. Box 513, 5600 MB Eindhoven, The Netherlands, m.shokrpour.roudbari@tue.nl
Görkem Şimşek, Eindhoven University of Technology, P.O. Box 513, 5600 MB Eindhoven, The Netherlands, g.simsek@tue.nl
Kristoffer G. van der Zee, School of Mathematical Sciences, The University of Nottingham, University Park, Nottingham, NG7 2RD, United Kingdom, kg.vanderzee@nottingham.ac.uk

https://doi.org/10.1515/9783110494259-008

ponents. This interface generally carries surface energy and accordingly it introduces surface tension. The interaction of a binary fluid with a deformable solid engenders a variety of intricate physical phenomena. The solid–fluid interface also carries surface energy and, generally, this surface energy is distinct for the two components of the binary fluid. Consequently, the binary-fluid–solid problem will exhibit wetting behavior [18, 19]. The surface tension in the fluid–fluid interface in combination with the elasticity of the solid in addition leads to *elasto-capillary fluid–solid interaction* (or, concisely, *elasto-capillarity* [51]), viz. a deformation of the solid due to the capillary forces that are exerted by the binary fluid on the solid surface at the contact line. Because of the fact that the capillary traction is essentially supported on a line (in particular, a manifold of codimension 2) it leads to a singularity in the deformation of the solid [13, 39, 57]. Elasto-capillarity underlies miscellaneous complex physical phenomena such as *durotaxis* [58], i.e., seemingly spontaneous migration of liquid droplets on solid substrates with an elasticity gradient; *capillary origami* [49], i.e., large-scale solid deformations by capillary forces; and, in the case of visco-elastic solid substrates, *stick-slip* dynamics [38, 39]. Binary-fluid–solid interaction is moreover of fundamental technological relevance in a wide variety of high-tech industrial applications, such as inkjet printing and additive manufacturing.

Despite significant progress in computational models and simulation techniques for the interaction of solids and classical fluids (see [7, 63] for an overview) and for binary fluids separately (see, for instance, [1, 3, 31, 34, 36, 42, 55]), modeling and simulation of binary-fluid–solid interaction with capillary effects has remained essentially unexplored.

In this work we consider a model for binary-fluid–solid interaction based on a diffuse-interface model for the binary fluid and a hyperelastic solid model with a St. Venant–Kirchhoff stored energy function. Our motivation for selecting a diffuse-interface model for the binary fluid lies in the versatility of these models in describing wetting phenomena [37, 52, 67] and in their rigorous thermodynamic substructure [1, 32, 42, 55]. The diffuse-interface binary-fluid model that we consider is described by the quasi-incompressible Navier–Stokes–Cahn–Hilliard (NSCH) equations according to [55]. The interaction of the binary fluid with the solid substrate is represented by dynamic and kinematic interface conditions and a preferential-wetting boundary condition. The traction exerted by the binary fluid on the solid–fluid interface comprises a capillary-stress contribution in addition to the standard pressure and viscous-stress components. The dynamic condition imposes equilibrium of this binary-fluid traction, the traction exerted by the hyperelastic solid and the traction due to the nonuniform solid–fluid surface tension. We present a weak reformulation of the aggregated binary-fluid–solid interaction problem comprising an arbitrary Lagrangian–Eulerian (ALE) formulation of the NSCH system and a suitable weak representation of the binary-fluid traction and the solid–fluid surface tension. In addition, we investigate some basic attributes of the aggregated binary-fluid–solid interaction problem. In particular, we establish the following fundamental energy dissipation relation for the aggregated fluid-solid-interaction (FSI) problem:

$$d_t\mathscr{E}(\boldsymbol{\psi}, \boldsymbol{u}, \phi) = -\int_{\Omega_t^{\mathrm{F}}} \operatorname{grad} \boldsymbol{u} : \boldsymbol{\tau} - \int_{\Omega_t^{\mathrm{F}}} m|\operatorname{grad} \mu|^2 + \mathrm{bnd} \leq 0 + \mathrm{bnd}$$

with $m > 0$, where $\mathscr{E}$ indicates the energy functional of the binary-fluid–solid interaction problem, $\boldsymbol{\psi}$, $\boldsymbol{u}$, ϕ, $\boldsymbol{\tau}$ and μ refer to the deformation, fluid velocity, order parameter, viscous-stress tensor and chemical potential, respectively, and 'bnd' contains terms that are only supported on the external boundary of the fluid–solid system. Moreover, we present a formal analysis of the structure of the fluid–fluid interface and its relation to capillarity and wetting.

To assess the capability of the presented binary-fluid–solid interaction model to describe elasto-capillary phenomena, we conduct numerical computations for a configuration pertaining to a droplet on a soft solid substrate, and we present a comparison to experimental data from [57].

The remainder of this chapter is structured as follows. Section 8.2 presents the considered binary-fluid–solid interaction problem. In Section 8.3 we elaborate on the weak formulation of the aggregated fluid–solid interaction problem. In addition to establishing the weak formulation, we also present a formal consideration of the ambient function spaces. Section 8.4 is devoted to an analysis of some elementary properties of the aggregated binary-fluid–solid interaction problem. Section 8.5 is concerned with details of the numerical approximation of the FSI problem. This section introduces both the spatial and temporal discretization of the FSI problem and the partitioned iterative solution procedure to solve the discrete problem. Section 8.6 considers numerical experiments and results. Section 8.7 presents concluding remarks.

8.2 Problem statement

We consider a binary fluid composed of two distinct immiscible constituents separated by an interface. The fluid–fluid interface represents a thin but finite transition layer, in which the fluid consists of a mixture of the two species. The binary fluid comprises a liquid embedded in an ambient fluid; see Figure 8.1. The binary fluid is contiguous with a deformable elastic solid.

To furnish a setting for the binary-fluid–solid system, we consider a time interval $(0, t_{\mathrm{FIN}}) \subseteq \mathbb{R}_{>0}$ and two simply connected time-dependent open subsets $\Omega_t^{\mathrm{F}} \subset \mathbb{R}^d$ ($d = 2, 3$) and $\Omega_t^{\mathrm{S}} \subset \mathbb{R}^d$, which accommodate the binary fluid and solid, respectively. The fluid–solid interface $\Gamma_t := \partial\Omega_t^{\mathrm{F}} \cap \partial\Omega_t^{\mathrm{S}}$ is composed of the intersection between the boundaries of the fluid and solid domains. We assume that the time-dependent configuration $\Omega_t := \operatorname{int}(\operatorname{cl}\Omega_t^{\mathrm{F}} \cup \operatorname{cl}\Omega_t^{\mathrm{S}})$ can be conceived of as the image of a continuous transformation T_t acting on the initial configuration $\Omega_0 := \operatorname{int}(\operatorname{cl}\Omega_0^{\mathrm{F}} \cup \operatorname{cl}\Omega_0^{\mathrm{S}})$ such that $\Omega_t^{\mathrm{F}} = T_t\Omega_0^{\mathrm{F}}$ and $\Omega_t^{\mathrm{S}} = T_t\Omega_0^{\mathrm{S}}$; see Figure 8.1. Note that, in particular, the transformation T_t is continuous across the solid–fluid interface. Moreover, we assume that the restriction of T_t to both Ω_0^{F} and Ω_0^{S} is diffeomorphic.

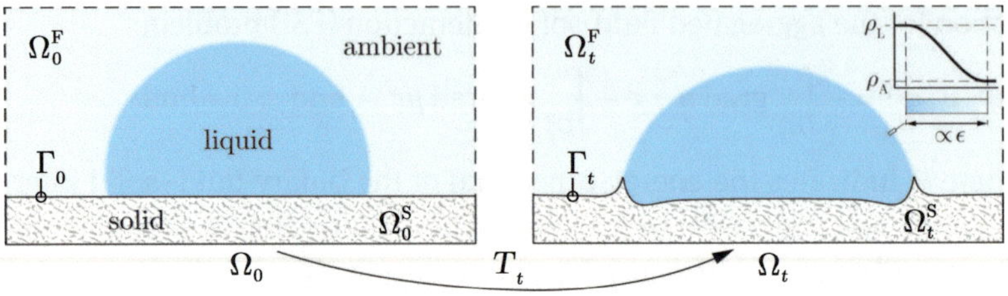

Fig. 8.1: Illustration of the considered binary-fluid–solid interaction problem. The binary fluid comprises a liquid embedded in an ambient fluid, separated by a diffuse interface of thickness $\propto \epsilon$, and is adjacent to a deformable solid. The fluid and solid domains, Ω_t^F and Ω_t^S, correspond to images of the time-dependent map T_t acting on the initial configurations Ω_0^F and Ω_0^S.

8.2.1 Navier–Stokes–Cahn–Hilliard binary-fluid model

We consider a quasi-incompressible Navier–Stokes–Cahn–Hilliard (NSCH) phase-field model for binary fluids composed of two immiscible incompressible constituents. NSCH models can generally be classified into incompressible [1, 27, 32, 34] and quasi-incompressible [31, 42, 53, 55] models. We select a variant of the quasi-incompressible model presented in [55], in view of its well-defined thermodynamic substructure and its consistent derivation from mixture theory, typical for phase-field models; see [29]. For a detailed derivation of this NSCH model in the context of mixture theory, we refer the reader to [55].

The two fluid constituents are equipresent in accordance with the volume fractions $\phi_L : (0, t_{FIN}) \times \Omega_t^F \to [0, 1]$ and $\phi_A : (0, t_{FIN}) \times \Omega_t^F \to [0, 1]$, which form a partition of unity:

$$\phi_L + \phi_A = 1 \tag{8.1}$$

The presence of each constituent, locally in space and time, is identified by the order parameter:

$$\phi := \phi_L - \phi_A : (0, t_{FIN}) \times \Omega_t^F \to [-1, 1] \tag{8.2}$$

and by virtue of (8.1) it holds that $\phi_L = (1 + \phi)/2$ and $\phi_A = (1 - \phi)/2$. One may note that $\phi = 1$ (resp. $\phi = -1$) pertains to a pure L-species (resp. A-species) composition of the binary fluid. Intermediate values $\phi \in (-1, 1)$ are indicative of a mixture. Denoting by ρ_L and ρ_A the specific densities of the two constituents, the mixture density $\rho : (0, t_{FIN}) \times \Omega_t^F \to \mathbb{R}_{>0}$ is related to the volume fractions by $\rho = \rho_L\phi_L + \rho_A\phi_A$. The mixture density is therefore related to the order parameter by the following algebraic relation:

$$\rho = \rho(\phi) := \frac{1+\phi}{2}\rho_L + \frac{1-\phi}{2}\rho_A \tag{8.3}$$

The mixture velocity $\boldsymbol{u} : (0, t_{FIN}) \times \Omega_t^F \to \mathbb{R}^d$ is defined as the barycentric (i.e., mass-averaged) velocity according to:

$$\boldsymbol{u} = \frac{1}{\rho}\left(\frac{1+\phi}{2}\rho_L\boldsymbol{u}_L + \frac{1-\phi}{2}\rho_A\boldsymbol{u}_A\right) \tag{8.4}$$

with $\boldsymbol{u}_{\mathrm{L}}$ and $\boldsymbol{u}_{\mathrm{A}}$ as the velocities of the liquid and ambient components, respectively.

We consider the quasi-incompressible NSCH phase-field model according to [55], in which the mixture density is defined by (8.3). The order parameter ϕ in (8.2), the volume-averaged mixture density ρ, and the mass-averaged mixture velocity $\boldsymbol{u}$ in (8.4) are subject to:

$$\left.\begin{aligned} \partial_t(\rho\boldsymbol{u}) + \operatorname{div}(\rho\boldsymbol{u}\otimes\boldsymbol{u}) + \operatorname{grad} p - \operatorname{div}\boldsymbol{\tau} - \operatorname{div}\boldsymbol{\zeta} &= 0 \\ \partial_t\phi + \operatorname{div}(\phi\boldsymbol{u} - m\operatorname{grad}\mu) &= 0 \\ \mu + \epsilon\sigma\,\Delta\phi - \frac{\sigma}{\epsilon}\Psi' + \frac{p}{\rho}\frac{d\rho}{d\phi} &= 0 \\ \operatorname{div}\boldsymbol{u} - \alpha\operatorname{div}(m\operatorname{grad}\mu) &= 0 \\ \rho - \rho_{\mathrm{L}}\frac{1+\phi}{2} - \rho_{\mathrm{A}}\frac{1-\phi}{2} &= 0 \end{aligned}\right\} \text{ in } \Omega_t^{\mathrm{F}} \tag{8.5}$$

where $\Psi' := \Psi'(\phi)$ indicates the derivate of the standard double-well potential that represents the energy density associated with mixing of the constituents:

$$\Psi := \Psi(\phi) = \frac{1}{4}(\phi^2 - 1)^2 \tag{8.6}$$

The auxiliary variables $p\colon (0, t_{\mathrm{FIN}}) \times \Omega_t^{\mathrm{F}} \to \mathbb{R}$ and $\mu\colon (0, t_{\mathrm{FIN}}) \times \Omega_t^{\mathrm{F}} \to \mathbb{R}$ in (8.5) represent the fluid pressure and chemical potential, respectively, and the constant $\alpha \in \mathbb{R}$ indicates the relative density difference of the ambient and liquid species:

$$\alpha = \frac{\rho_{\mathrm{A}} - \rho_{\mathrm{L}}}{\rho_{\mathrm{A}} + \rho_{\mathrm{L}}} \tag{8.7}$$

The fluid stress has been decomposed into a viscous component, $\boldsymbol{\tau}$, and a capillary component, $\boldsymbol{\zeta}$. The viscous stress tensor is related to the symmetric strain-rate tensor $\boldsymbol{\varepsilon} := \frac{1}{2}(\operatorname{grad}\boldsymbol{u} + (\operatorname{grad}\boldsymbol{u})^T)$ by the constitutive relation:

$$\boldsymbol{\tau} = \eta\Big(2\boldsymbol{\varepsilon} + \Lambda\,(\operatorname{tr}\boldsymbol{\varepsilon})\boldsymbol{I}\Big) \tag{8.8}$$

with $\operatorname{tr}\boldsymbol{\varepsilon} = \operatorname{div}\boldsymbol{u}$ the trace of the strain-rate tensor, η a viscosity parameter, viz. a strictly positive coefficient possibly dependent on the order parameter, $\boldsymbol{I}$ the identity tensor in $\mathbb{R}^d$ and $\Lambda \geq -2/d$ a constant. For mixtures, the viscosity is generally defined as the volumetric average of the species viscosities according to

$$\eta := \eta(\phi) = \frac{1+\phi}{2}\eta_{\mathrm{L}} + \frac{1-\phi}{2}\eta_{\mathrm{A}} \tag{8.9}$$

The lower bound on Λ arises from thermodynamic considerations; see Section 8.4.1 and e.g., [59, §7.3]. One can infer that, in particular, if $\Lambda = -2/d$ then $\boldsymbol{\tau}$ is traceless in accordance with Stokes' hypothesis. The capillary-stress tensor is defined by:

$$\boldsymbol{\zeta} = -\epsilon\sigma\operatorname{grad}\phi\otimes\operatorname{grad}\phi + \boldsymbol{I}\Big(\frac{\epsilon\sigma}{2}|\operatorname{grad}\phi|^2 + \frac{\sigma}{\epsilon}\Psi - \mu\phi\Big) \tag{8.10}$$

The parameter σ is related to the fluid–fluid (liquid–ambient) surface tension σ_{LA} by $2\sqrt{2}\,\sigma = 3\,\sigma_{\mathrm{LA}}$, and m designates a strictly positive (possibly order-parameter dependent) scalar referred to as mobility. The parameter $\epsilon > 0$ controls the thickness of the diffuse interface between the fluid constituents; see also Section 8.4.3.

Remark 8.1. The NSCH equations (8.5) imply the conservation-of-mass relation

$$\partial_t \rho + \operatorname{div} \rho \boldsymbol{u} = 0 \tag{8.11}$$

To derive (8.11), we first note that by (8.5$_5$) and the chain rule, it holds that

$$\begin{aligned}\partial_t \rho + \operatorname{div} \rho \boldsymbol{u} &= \tfrac{1}{2}(\rho_{\mathrm{L}} - \rho_{\mathrm{A}})(\partial_t \phi + \boldsymbol{u} \cdot \operatorname{grad} \phi) + \tfrac{1}{2}(\rho_{\mathrm{L}}(1+\phi) + \rho_{\mathrm{A}}(1-\phi)) \operatorname{div} \boldsymbol{u} \\ &= \tfrac{1}{2}(\rho_{\mathrm{L}} - \rho_{\mathrm{A}})(\partial_t \phi + \operatorname{div} \phi \boldsymbol{u}) + \tfrac{1}{2}(\rho_{\mathrm{L}} + \rho_{\mathrm{A}}) \operatorname{div} \boldsymbol{u}\end{aligned} \tag{8.12}$$

The ultimate expression in (8.12) vanishes by virtue of (8.5$_2$), (8.5$_4$) and (8.7).

Remark 8.2. The trace of the capillary stress tensor in (8.5$_1$) is

$$\operatorname{tr}(\boldsymbol{\zeta}) = \left(\frac{d}{2} - 1\right) \epsilon\sigma\,|\operatorname{grad} \phi|^2 + d\left(\frac{\sigma}{\epsilon}\Psi - \mu\phi\right) \tag{8.13}$$

and is generally nonzero. Hence, even if the viscous stress tensor is traceless in accordance with Stokes' hypothesis, the pressure p does not admit the usual interpretation as the trace of the stress tensor, on account of the capillary contribution.

The NSCH equations (8.5) must generally be furnished with initial and boundary conditions. A suitable initial condition for (8.5) is provided by a specification of the initial phase distribution and the initial mixture velocity according to

$$\phi(0,\,\cdot\,) = \phi_0, \qquad \boldsymbol{u}(0,\,\cdot\,) = \boldsymbol{u}_0 \tag{8.14}$$

where $\phi_0 : \Omega_0^{\mathrm{F}} \to [-1, 1]$ and $\boldsymbol{u}_0 : \Omega_0^{\mathrm{F}} \to \mathbb{R}^d$ correspond to exogenous data.

A comprehensive overview of boundary conditions for (8.5) is beyond the scope of this work. We restrict ourselves to a consideration of five elementary types of boundary conditions. Accordingly, we assume that the boundary $\partial\Omega_t^{\mathrm{F}}$ is composed of complementary disjoint open subsets $\Gamma_{\mathrm{W}}^{\mathrm{F}}$, $\Gamma_{0,\mathrm{S}}^{\mathrm{F}}$, $\Gamma_{0,\mathrm{N}}^{\mathrm{F}}$, Γ_{-}^{F} and Γ_{+}^{F}, each associated with a different type of boundary condition. In particular, $\Gamma_{\mathrm{W}}^{\mathrm{F}}$ represents a (possibly deformable) solid wall that exhibits preferential-wetting properties for the two fluid components, $\Gamma_{0,\mathrm{S}}^{\mathrm{F}}$ (resp. $\Gamma_{0,\mathrm{N}}^{\mathrm{F}}$) is a rigid impermeable free-slip (resp. no slip) boundary, neutral to wetting by the two fluid components, and Γ_{-}^{F} and Γ_{+}^{F} represent inflow and outflow boundaries, respectively. The inflow and outflow boundaries are characterized by:

$$\Gamma_{\pm}^{\mathrm{F}} = \left\{\boldsymbol{x} \in \partial\Omega_t^{\mathrm{F}} \setminus \Gamma_{\mathrm{W}}^{\mathrm{F}} : \boldsymbol{u}(\boldsymbol{x}) \cdot \boldsymbol{n}^{\mathrm{F}}(\boldsymbol{x}) \gtrless 0\right\} \tag{8.15}$$

with $\boldsymbol{n}^{\mathrm{F}}$ the exterior unit normal vector to $\partial\Omega_t^{\mathrm{F}}$. Let us note that $\Gamma_{\mathrm{W}}^{\mathrm{F}}$ includes the solid–fluid interface Γ_t. The wetting neutrality at $\Gamma_{0,\mathrm{S}}^{\mathrm{F}}$ and $\Gamma_{0,\mathrm{N}}^{\mathrm{F}}$ is represented by a homogeneous Neumann condition on the order parameter. The impermeability of these boundaries implies that both the normal component of the mixture velocity and the normal volume-fraction flux, $m\partial_n\mu$, vanish. Completion of the boundary conditions at $\Gamma^{\mathrm{F}}_{0,\mathrm{N}}$ and $\Gamma^{\mathrm{F}}_{0,\mathrm{s}}$ with the respective no-slip and free-slip conditions gives:

$$\left.\begin{aligned}\partial_n\mu &= 0\\ \partial_n\phi &= 0\\ u_n &= 0\\ \boldsymbol{u}_\Gamma &= 0\end{aligned}\right\}\ \text{on } \Gamma^{\mathrm{F}}_{0,\mathrm{N}} \qquad \left.\begin{aligned}\partial_n\mu &= 0\\ \partial_n\phi &= 0\\ u_n &= 0\\ (-\boldsymbol{\tau}\boldsymbol{n}^{\mathrm{F}} - \boldsymbol{\zeta}\boldsymbol{n}^{\mathrm{F}})_\Gamma &= 0\end{aligned}\right\}\ \text{on } \Gamma^{\mathrm{F}}_{0,\mathrm{s}} \tag{8.16}$$

where $(\,\cdot\,)_n$ and $(\,\cdot\,)_\Gamma$ denote the components of $(\,\cdot\,)$ in the normal and tangent bundles of $\partial\Omega^{\mathrm{F}}_t$, respectively. Appropriate boundary conditions at in- and outflow boundaries are:

$$\left.\begin{aligned}\partial_n\mu &= \overline{\partial_n\mu_-}\\ \phi &= \overline{\phi}_-\\ \boldsymbol{u} &= \overline{\boldsymbol{u}}_-\end{aligned}\right\}\ \text{on } \Gamma^{\mathrm{F}}_- \qquad \left.\begin{aligned}\partial_n\mu &= \overline{\partial_n\mu_+}\\ \partial_n\phi &= \overline{\partial_n\phi_+}\\ p\boldsymbol{n}^{\mathrm{F}} - \boldsymbol{\tau}\boldsymbol{n}^{\mathrm{F}} - \boldsymbol{\zeta}\boldsymbol{n}^{\mathrm{F}} &= \overline{\boldsymbol{t}}^{\mathrm{F}}_+\end{aligned}\right\}\ \text{on } \Gamma^{\mathrm{F}}_- \tag{8.17}$$

The right-hand sides in (8.17) represent exogenous data.

The preferential-wetting property of $\Gamma^{\mathrm{F}}_{\mathrm{W}}$ is encoded by the following boundary condition

$$\epsilon\sigma\,\partial_n\phi + \sigma'_{\mathrm{SF}}(\phi) = 0 \quad \text{on } \Gamma^{\mathrm{F}}_{\mathrm{W}} \tag{8.18}$$

where σ_{SF} represents the solid–fluid surface tension according to

$$\sigma_{\mathrm{SF}}(\phi) = \tfrac{1}{4}(\phi^3 - 3\phi)(\sigma_{\mathrm{SA}} - \sigma_{\mathrm{SL}}) + \tfrac{1}{2}(\sigma_{\mathrm{SL}} + \sigma_{\mathrm{SA}}) \tag{8.19}$$

with $\sigma_{\mathrm{SL}} \geq 0$ and $\sigma_{\mathrm{SA}} \geq 0$ the solid–fluid surface tensions of the liquid species and ambient species, respectively; see also [13, 37, 54, 67]. Boundary condition (8.18) imposes that in equilibrium the diffuse fluid–fluid interface intersects the solid–fluid interface at the static contact angle (interior to the liquid)

$$\theta = \arccos((\sigma_{\mathrm{SA}} - \sigma_{\mathrm{SL}})/\sigma_{\mathrm{LA}}) \tag{8.20}$$

See [54] and Section 8.4.3. Assuming that the solid wall is impermeable, the contiguity of the binary fluid and the solid wall at $\Gamma^{\mathrm{F}}_{\mathrm{W}}$ insists that the trace of the fluid velocity coincides with the solid velocity, and that the volume-fraction flux vanishes. Summarizing, we obtain:

$$\left.\begin{aligned}\partial_n\mu &= 0\\ \epsilon\sigma\,\partial_n\phi + \sigma'_{\mathrm{SF}}(\phi) &= 0\\ \boldsymbol{u} &= \overline{\boldsymbol{u}}_{\mathrm{W}}\end{aligned}\right\}\ \text{on } \Gamma^{\mathrm{F}}_{\mathrm{W}} \tag{8.21}$$

where the exogenous data $\overline{\boldsymbol{u}}_{\mathrm{W}}$ corresponds to the velocity of the contiguous solid.

Remark 8.3. The solid–binary-fluid surface tension $\sigma_{\mathrm{SF}}(\,\cdot\,)$ in (8.19) interpolates between the solid–fluid surface tensions of the components. In particular, recalling that $\phi = +1$ (resp. $\phi = -1$) characterizes the liquid (resp. ambient fluid), it must hold that

$\sigma_{SF}(1) = \sigma_{SL}$ (resp. $\sigma_{SF}(-1) = \sigma_{SA}$). In addition, to ensure that all line tensions vanish in the pure fluid states, it must hold that $\lim_{s\to\pm1} \sigma'_{SF}(s) = 0$. The simplest relation in compliance with the aforementioned conditions is the cubic-spline interpolation in (8.19).

8.2.2 Hyperelastic St. Venant–Kirchhoff solid model

We consider a hyperelastic solid with a St. Venant–Kirchhoff constitutive stress-strain relation. Our choice for this solid model is motivated by its well-defined energy conservation structure and its suitability for large deformations. Recalling that the reference configuration is identified with the initial configuration, and denoting the initial density of the solid by $\rho_S : \Omega_0^S \to \mathbb{R}_{>0}$, the solid deformation $\boldsymbol{\psi}^S : \Omega_0^S \to \Omega_t^S$ satisfies the equation of motion:

$$\rho_S \partial_t^2 \boldsymbol{\psi}^S - \operatorname{Div} \boldsymbol{P} = 0 \quad \text{in } \Omega_0^S \tag{8.22}$$

with $\boldsymbol{P}$ the first Piola–Kirchhoff stress tensor and Div(·) the divergence operator acting in the reference configuration, i.e., Div $\boldsymbol{P}$ is the vector with Cartesian components:

$$(\operatorname{Div} \boldsymbol{P})_i(X_1, \ldots, X_d) = \sum_{j=1}^{d} \frac{\partial P_{ij}(X_1, \ldots, X_d)}{\partial X_j} \tag{8.23}$$

For hyperelastic materials, the energy that is associated with a deformation of the material is represented by a stored energy function $W : \mathbb{M}_+^d \to \mathbb{R}$, where $\mathbb{M}_+^d$ denotes the class of all rank-2 tensors with strictly positive determinant; see, for example, [17]. Denoting by $\boldsymbol{F} := \boldsymbol{F}(\boldsymbol{\psi}^S)$ the deformation tensor, viz. the Fréchet derivative of the deformation $\boldsymbol{\psi}^S$, the strain energy associated with the deformation $\boldsymbol{\psi}^S$ for a material with stored-energy function W is represented by the functional

$$\boldsymbol{\psi}^S \mapsto \mathscr{W}(\boldsymbol{\psi}^S) := \int_{\Omega_0^S} W(\boldsymbol{F}) \tag{8.24}$$

The St. Venant–Kirchhoff constitutive relation defines the stored-energy function as $W(\boldsymbol{F}) = \frac{1}{2}\lambda_L(\operatorname{tr} \boldsymbol{E})^2 + \mu_L \operatorname{tr} \boldsymbol{E}^2$ with $\boldsymbol{E} := \boldsymbol{E}(\boldsymbol{F}) = \frac{1}{2}(\boldsymbol{F}^T\boldsymbol{F} - \boldsymbol{I})$ the Green–Lagrange strain tensor corresponding to $\boldsymbol{F}$ and λ_L and μ_L the Lamé parameters. The first Piola–Kirchhoff stress tensor corresponds to the derivative $\boldsymbol{P} = \partial W/\partial \boldsymbol{F}$. One can infer that $\boldsymbol{P} = \boldsymbol{F}\boldsymbol{S}$ with $\boldsymbol{S} = \lambda_L(\operatorname{tr} \boldsymbol{E})\boldsymbol{I} + 2\mu_L \boldsymbol{E}$ the second Piola–Kirchhoff stress tensor.

Remark 8.4. In this work, we restrict our considerations to a simple St. Venant–Kirchhoff stored energy function for the solid material. It is to be remarked that a setting more conducive to rigorous analysis is provided by polyconvex stored energy functions; see [4, 17].

The partial-differential equation (8.22) must be furnished with suitable initial and boundary conditions. The identification of the reference configuration and the initial configuration implies that the initial deformation coincides with the identity. Assuming in addition that the solid is initially stationary, the deformation complies with the initial conditions:

$$\left.\begin{aligned} \boldsymbol{\psi}^{\mathrm{S}}(0,\,\cdot\,) &= \mathrm{Id} \\ \partial_t \boldsymbol{\psi}^{\mathrm{S}}(0,\,\cdot\,) &= 0 \end{aligned}\right\} \quad \text{in } \Omega_0^{\mathrm{S}} \tag{8.25}$$

Equation (8.22) is generally furnished with Dirichlet (essential) or Neumann (natural) boundary conditions, which prescribe either the deformation or the traction at the boundary, respectively. We assume that the boundary $\partial\Omega_0^{\mathrm{S}}$ admits a decomposition into two complementary disjoint open subsets $\Gamma_{\mathrm{D}}^{\mathrm{S}}$ and $\Gamma_{\mathrm{N}}^{\mathrm{S}}$, each associated with one of these types of boundary conditions, and prescribe:

$$\begin{aligned} \boldsymbol{\psi}^{\mathrm{S}} &= \overline{\boldsymbol{\psi}}^{\mathrm{S}} && \text{on } \Gamma_{\mathrm{D}}^{\mathrm{S}} \\ \boldsymbol{P}\boldsymbol{n}_0^{\mathrm{S}} &= \overline{\boldsymbol{t}}^{\mathrm{S}} && \text{on } \Gamma_{\mathrm{N}}^{\mathrm{S}} \end{aligned} \tag{8.26}$$

with $\boldsymbol{n}_0^{\mathrm{S}}$ the exterior unit normal vector to $\partial\Omega_0^{\mathrm{S}}$ (i.e., to the boundary of the solid domain in the reference configuration) and with $\overline{\boldsymbol{\psi}}^{\mathrm{S}} : \Gamma_{\mathrm{D}}^{\mathrm{S}} \to \mathbb{R}^d$ and $\overline{\boldsymbol{t}}^{\mathrm{S}} : \Gamma_{\mathrm{N}}^{\mathrm{S}} \to \mathbb{R}^d$ exogenous deformation and traction data, respectively.

8.2.3 Interface conditions

The binary fluid (8.5) and solid (8.22) interact at the mutual interface Γ_t by kinematic and dynamic interface conditions, complemented by a preferential-wetting and impermeability condition in accordance with (8.21). The kinematic condition identifies the velocity of the binary fluid and the velocity of the solid at the interface. The dynamic condition imposes continuity of the tractions exerted at the interface. The preferential-wetting condition encodes the affinity of the solid substrate for the two fluid species. The impermeability condition insists that the phase flux into the solid substrate vanishes.

The kinematic condition can be conceived of as a Dirichlet boundary condition for fluid velocity on the moving boundary Γ_t, in accordance with

$$\boldsymbol{u} = \overline{\boldsymbol{u}}_{\mathrm{W}} = \partial_t \boldsymbol{\psi}^{\mathrm{S}} \circ T_t^{-1} \quad \text{on } \Gamma_t \subseteq \Gamma_{\mathrm{W}}^{\mathrm{F}} \tag{8.27}$$

The composition with the inverse transformation T_t^{-1} in the right member of (8.27) serves to transport the trace of the solid velocity, $\partial_t \boldsymbol{\psi}^{\mathrm{S}}$, from the reference configuration of the interface, Γ_0, to the current configuration, Γ_t. In combination with the impermeability condition and the preferential-wetting condition, we obtain a boundary condition for the NSCH equations (8.5) conforming to (8.21):

$$\left.\begin{aligned} \partial_n \mu &= 0 \\ \epsilon\sigma\, \partial_n \phi + \sigma_{\mathrm{SF}}'(\phi) &= 0 \\ \boldsymbol{u} &= \partial_t \boldsymbol{\psi}^{\mathrm{S}} \circ T_t^{-1} \end{aligned}\right\} \quad \text{on } \Gamma_t \tag{8.28}$$

The dynamic condition imposes equilibrium of the fluid and solid tractions and the traction exerted on the interface by the solid–fluid surface tension. In most conventional fluid–solid interaction problems, the latter contribution is negligible. However, in many binary-fluid–solid interaction problems, e.g., elasto-capillarity, the traction induced by solid–fluid surface tension yields a non-negligible effect. In fact, the traction induced by solid–fluid surface tension is proportional to the curvature of the solid–fluid interface and yields a regularization of the interface deformation. Without this regularization, the localized traction exerted by the fluid–fluid surface tension at the contact line where the fluid–fluid interface intersects with the solid–fluid interface, would lead to an inadmissible singularity in the sharp-interface limit, i.e., as the thickness of the diffuse-interface passes to zero ($\epsilon \to +0$); see also Remark 8.8 on page 298.

The traction due to the solid–fluid surface tension satisfies the Young–Laplace relation for a nonuniform surface tension, according to

$$\boldsymbol{t}_{\mathrm{SF}} = \sigma_{\mathrm{SF}} \kappa \boldsymbol{n}^{\mathrm{F}} + \operatorname{grad}_{\Gamma} \sigma_{\mathrm{SF}} \tag{8.29}$$

with $\sigma_{\mathrm{SF}} := \sigma_{\mathrm{SF}}(\phi)$ the solid–fluid surface tension in (8.19), $\kappa := \kappa(\boldsymbol{\psi}^{\mathrm{S}})$ as the additive curvature ($d-1$ times the mean curvature) of the solid–fluid interface Γ_t, and $\operatorname{grad}_{\Gamma}$ as the tangential gradient (or *Γ-gradient*) on Γ_t; see, e.g., [13, 30] and [64, §2.4.2]. We adopt the convention that curvature is negative if the osculating circle in the normal plane is located in the fluid domain.

Remark 8.5. In the sharp-interface limit $\epsilon \to +0$, the support of $\operatorname{grad}_{\Gamma} \sigma_{\mathrm{SF}}$ in (8.29) is essentially confined to the intersection of the solid–fluid interface Γ_t and the fluid–fluid interface. Accordingly, this contribution can be conceived of as a *contact-line force*. Specifically, $\operatorname{grad}_{\Gamma} \sigma_{\mathrm{SF}}$ represents a surface-tension component that emanates from variations of the solid–fluid surface energy corresponding to variations of the contract-line position; see also Section 8.4.3.

Remark 8.6. In the context of fluid–solid interaction problems, the tangential gradient in (8.29) represents a nonlinear operator on account of its implicit dependence on the deformation of the interface.

The binary fluid in (8.5) exerts a traction on the interface according to:

$$\boldsymbol{t}^{\mathrm{F}} := p\boldsymbol{n}^{\mathrm{F}} - \boldsymbol{\tau}\boldsymbol{n}^{\mathrm{F}} - \boldsymbol{\zeta}\boldsymbol{n}^{\mathrm{F}} \tag{8.30}$$

It is to be noted that the traction exerted by the binary fluid contains a capillary contribution, $\boldsymbol{\zeta}\boldsymbol{n}^{\mathrm{F}}$, in addition to the standard pressure and viscous traction. If the interface exerts a traction $\bar{\boldsymbol{t}}^{\mathrm{S}}$ on the solid in accordance with (8.26) then, by reciprocality, the solid exerts traction $-\boldsymbol{P}\boldsymbol{n}_0^{\mathrm{S}}$ on the interface.

The dynamic interface condition imposes that the tractions sustained by the interface are balanced. The traction induced by the solid–fluid surface tension in (8.29) and the traction exerted by the binary fluid in accordance with (8.30) are formulated in the current configuration. Conversely, the traction exerted by the solid is expressed in the reference configuration. To reconcile this configurational discrepancy in the traction contributions, we formulate the dynamic condition in distributional form:

$$\int_{\Gamma_0} \boldsymbol{\xi} \cdot \boldsymbol{P}\boldsymbol{n}_0^{\mathrm{S}} = \int_{\Gamma_t} (\boldsymbol{\xi} \circ T_t^{-1}) \cdot \boldsymbol{t}^{\mathrm{F}} + \int_{\Gamma_t} (\boldsymbol{\xi} \circ T_t^{-1}) \cdot \boldsymbol{t}_{\mathrm{SF}} \quad \forall \boldsymbol{\xi} \in \boldsymbol{C}_0^{\infty}(\Gamma_0) \tag{8.31}$$

where $\boldsymbol{C}_0^{\infty}(\Gamma_0)$ denotes the class of smooth compactly supported $\mathbb{R}^d$-valued functions on Γ_0. It is to be noted that in the left-hand side of (8.31), the integration measure is the surface measure carried by the interface in the reference configuration, and in the right-hand side, the integration measure is the surface measure carried by the interface in the actual configuration. A precise interpretation of the right member of (8.31) based on weak evaluation of the traction and the solid–fluid surface-tension contribution is presented in Section 8.3.2.

8.3 Weak formulation

This section presents a weak formulation of the binary-fluid–solid interaction problem in Section 8.2. To accommodate the motion of the fluid domain, we first introduce a weak arbitrary Lagrangian–Eulerian (ALE) formulation of the NSCH system (8.5). Section 8.3.2 presents a weak formulation of the interface conditions and of the traction exerted by the solid–fluid surface tension, including a weak formulation of the solid subsystem (8.22). Section 8.3.3 presents a formal consideration of the ambient function spaces for the binary-fluid–solid interaction problem. In Section 8.3.4 we introduce the aggregated weak formulation of the FSI problem.

8.3.1 ALE formulation of NSCH equations

To introduce the ALE formulation, we first recall the transformation T_t from the reference configuration to the actual configuration. On the subdomain associated with the solid, the transformation is generated by the solid deformation. On the fluid subdomain, the transformation can be constructed in various manners, e.g., by means of a pseudosolid [25] or by a harmonic map. We select the latter, and accordingly define T_t as the harmonic extension of the trace of the solid deformation at the interface onto the fluid domain. To this end, we consider a continuous extension $\boldsymbol{\psi}$ of the solid deformation $\boldsymbol{\psi}^{\mathrm{S}}$ to all of Ω_0, encompassing both the solid and the fluid reference configurations. The extension onto Ω_0^{F} is weakly harmonic, in compliance with:

$$\int_{\Omega_0^{\mathrm{F}}} \operatorname{grad} \boldsymbol{\vartheta} : \operatorname{grad} \boldsymbol{\psi} = 0 \qquad \forall \boldsymbol{\vartheta} \in \boldsymbol{C}_0^{\infty}(\Omega_0^{\mathrm{F}}) \tag{8.32}$$

Note that the gradient operator in (8.32) acts in the reference configuration. By virtue of the continuity of $\boldsymbol{\psi}$ across the solid–fluid interface, the restriction of $\boldsymbol{\psi}$ to the fluid domain subject to (8.32) corresponds to the (time-dependent) harmonic extension of the solid deformation, i.e., the restriction of $\boldsymbol{\psi}$ to the solid domain. We specify the domain map T_t via $T_t(\cdot) = \boldsymbol{\psi}(t, \cdot)$. The time-dependent transformation T_t associates a velocity $\boldsymbol{v}\colon \Omega_t \to \mathbb{R}^d$ to every point in the current configuration, according to:

$$\boldsymbol{v} = \partial_t \boldsymbol{\psi} \circ T_t^{-1} \tag{8.33}$$

The restriction $\boldsymbol{v}|_{\Omega_t^{\mathrm{S}}}$ corresponds to the material velocity of the solid and the restriction $\boldsymbol{v}|_{\Omega_t^{\mathrm{F}}}$ corresponds to the velocity of the fluid domain.

The motion of the fluid domain is accommodated by a weak formulation of (8.5) in ALE form. We consider a weak formulation that is set in the current configuration. To derive the weak ALE formulation, we first note that for any two sufficiently smooth functions $\chi\colon \Omega_0^{\mathrm{F}} \to \mathbb{R}$ and $\psi\colon (0, t_{\mathrm{FIN}}) \times \Omega_t^{\mathrm{F}} \to \mathbb{R}$, the following sequence of identities holds:

$$\begin{aligned}
\frac{d}{dt}\int_{\Omega_t^{\mathrm{F}}} (\chi \circ T_t^{-1})\,\psi &= \int_{\Omega_t^{\mathrm{F}}} \frac{\partial}{\partial t}((\chi \circ T_t^{-1})\,\psi) + \int_{\Omega_t^{\mathrm{F}}} \operatorname{div}(\boldsymbol{v}\,(\chi \circ T_t^{-1})\,\psi) \\
&= \int_{\Omega_t^{\mathrm{F}}} \psi\Big(\frac{\partial}{\partial t}(\chi \circ T_t^{-1}) + \boldsymbol{v}\cdot \operatorname{grad}(\chi \circ T_t^{-1})\Big) + \int_{\Omega_t^{\mathrm{F}}} (\chi \circ T_t^{-1})\Big(\frac{\partial \psi}{\partial t} + \operatorname{div}(\boldsymbol{v}\psi)\Big)
\end{aligned} \tag{8.34}$$

The first identity in (8.34) follows from the Reynolds–Leibniz theorem. The second identity results from the product rule. The first term in the ultimate expression in (8.34) vanishes by virtue of:

$$\Big(\frac{\partial}{\partial t}(\chi \circ T_t^{-1})\Big)(t, \boldsymbol{y}) = \frac{d}{ds}(\chi \circ T_{t+s}^{-1}(\boldsymbol{y}))\Big|_{s=0} = -\boldsymbol{v}(t, \boldsymbol{y})\cdot \operatorname{grad}(\chi \circ T_t^{-1}(\boldsymbol{y})) \tag{8.35}$$

for all $\boldsymbol{y} \in \Omega_t^{\mathrm{F}}$; see, for instance, [56]. From (8.34) and (8.35) it follows that

$$\int_{\Omega_t^{\mathrm{F}}} (\chi \circ T_t^{-1})\frac{\partial \psi}{\partial t} = \frac{d}{dt}\int_{\Omega_t^{\mathrm{F}}} (\chi \circ T_t^{-1})\,\psi - \int_{\Omega_t^{\mathrm{F}}} (\chi \circ T_t^{-1})\operatorname{div}(\boldsymbol{v}\psi) \tag{8.36}$$

From (8.5_1) and (8.36), we infer the following identity, valid for all $\boldsymbol{\theta} \in \boldsymbol{C}^\infty(\Omega_0^{\mathrm{F}})$:

$$\begin{aligned}
\frac{d}{dt}\int_{\Omega_t^{\mathrm{F}}} (\boldsymbol{\theta} \circ T_t^{-1})\cdot \rho \boldsymbol{u} &+ \int_{\Omega_t^{\mathrm{F}}} \operatorname{grad}(\boldsymbol{\theta} \circ T_t^{-1})\colon (\boldsymbol{\tau} + \boldsymbol{\zeta} - \rho \boldsymbol{u} \otimes (\boldsymbol{u} - \boldsymbol{v})) \\
&- \int_{\Omega_t^{\mathrm{F}}} p \operatorname{div}(\boldsymbol{\theta} \circ T_t^{-1}) = -\int_{\partial\Omega_t^{\mathrm{F}}} (\boldsymbol{\theta} \circ T_t^{-1})\cdot \boldsymbol{t}^{\mathrm{F}} - \int_{\partial\Omega_t^{\mathrm{F}}} (\boldsymbol{\theta} \circ T_t^{-1})\cdot \rho \boldsymbol{u}(u_n - v_n)
\end{aligned} \tag{8.37}$$

with $\boldsymbol{t}^{\mathrm{F}}$ the traction exerted by the binary fluid on the interface according to (8.30). Similarly, it follows from (8.5_2) and (8.36) that

$$\begin{aligned}
\frac{d}{dt}\int_{\Omega_t^{\mathrm{F}}} (\beta \circ T_t^{-1})\phi &+ \int_{\Omega_t^{\mathrm{F}}} \operatorname{grad}(\beta \circ T_t^{-1})\cdot (m \operatorname{grad}\mu - \phi(\boldsymbol{u} - \boldsymbol{v})) \\
&= \int_{\partial\Omega_t^{\mathrm{F}}} (\beta \circ T_t^{-1})\, m\, \partial_n \mu - \int_{\partial\Omega_t^{\mathrm{F}}} (\beta \circ T_t^{-1})\,\phi(u_n - v_n)
\end{aligned} \tag{8.38}$$

for all $\beta \in C^\infty(\Omega_0^{\mathrm{F}})$. The following identities follow straightforwardly from (8.5_3) and (8.5_4):

$$\int_{\Omega_t^{\mathrm{F}}} (\chi \circ T_t^{-1})\mu - \int_{\Omega_t^{\mathrm{F}}} \epsilon\sigma \operatorname{grad}(\chi \circ T_t^{-1}) \cdot \operatorname{grad}\phi - \int_{\Omega_t^{\mathrm{F}}} \frac{1}{\epsilon}(\chi \circ T_t^{-1})\,\Psi' + \int_{\Omega_t^{\mathrm{F}}} (\chi \circ T_t^{-1})\,\frac{p}{\rho}\frac{d\rho}{d\phi} = -\int_{\partial\Omega_t^{\mathrm{F}}} (\chi \circ T_t^{-1})\,\epsilon\sigma\,\partial_n\phi \tag{8.39}$$

for all $\chi \in C^\infty(\Omega_0^{\mathrm{F}})$ and

$$\int_{\Omega_t^{\mathrm{F}}} (q \circ T_t^{-1}) \operatorname{div}\boldsymbol{u} + \int_{\Omega_t^{\mathrm{F}}} \alpha m \operatorname{grad}(q \circ T_t^{-1}) \cdot \operatorname{grad}\mu = \int_{\partial\Omega_t^{\mathrm{F}}} q\,\alpha m\,\partial_n\mu \tag{8.40}$$

for all $q \in C^\infty(\Omega_0^{\mathrm{F}})$. By virtue of the wetting condition in (8.28), the term $\epsilon\sigma\,\partial_n\phi$ that appears in the right member of (8.39) can be replaced by $-\sigma'_{\mathrm{SF}}(\phi)$. Moreover, $\partial_n\mu$ in (8.38) and (8.40) vanishes on the interface due to (8.28). The harmonic extension according to (8.32) and the identities in (8.37)–(8.40) form the basis of the weak ALE formulation of the Navier–Stokes–Cahn–Hilliard equations in Section 8.3.4.

8.3.2 Interface conditions

In the weak formulation of the FSI problem, the kinematic condition (8.27) is enforced by means of a Dirichlet boundary condition that identifies the trace of the fluid velocity, $\boldsymbol{u}$, to the trace of the solid velocity, $\partial_t\boldsymbol{\psi}^{\mathrm{S}} \circ T_t^{-1}$, at the interface. To facilitate the presentation and without essential loss of generality, we assume that the boundary conditions for the NSCH subsystem on the complementary part of the boundary, $\partial\Omega_t^{\mathrm{F}} \setminus \Gamma_t$, are such that $u_n = v_n$. The continuity of the domain deformation, $\boldsymbol{\psi}$, implies that the trace of the fluid domain velocity at the interface coincides with the trace of the solid velocity at the interface, in accordance with $\boldsymbol{v} = \partial_t\boldsymbol{\psi} \circ T_t^{-1}$. The kinematic interface condition in turn implies $\boldsymbol{u} = \boldsymbol{v}$ and, hence, u_n and v_n cancel at Γ_t. The second terms in the right members of (8.37) and (8.38) therefore vanish.

To elucidate the weak formulation of the dynamic interface condition (8.31), we first note that the solid deformation according to (8.22) satisfies:

$$\frac{d^2}{dt^2}\int_{\Omega_0^{\mathrm{S}}} \rho_{\mathrm{S}}\boldsymbol{\psi}^{\mathrm{S}} \cdot \boldsymbol{\theta} + \mathscr{W}'(\boldsymbol{\psi}^{\mathrm{S}};\boldsymbol{\theta}) = \int_{\partial\Omega_0^{\mathrm{S}}} \boldsymbol{\theta} \cdot \boldsymbol{P}\boldsymbol{n}_0^{\mathrm{S}} \qquad \forall\boldsymbol{\theta} \in \boldsymbol{C}^\infty(\Omega_0^{\mathrm{S}}) \tag{8.41}$$

The right member of (8.41) corresponds to the contraction of the trace of $\boldsymbol{\theta}$ with the traction that is exerted on the solid at the boundary. The dynamic condition is enforced by replacing the traction on the solid at the interface by the fluid traction and the traction exerted by the solid–fluid surface tension, in accordance with (8.31). The specifications of the functions $\boldsymbol{t}_{\mathrm{SF}}$ and $\boldsymbol{t}^{\mathrm{F}}$ that appear in the right-hand side of (8.31), according to (8.29) and (8.30), respectively, should however be conceived of as distributional representations of *traction functionals*. Accordingly, the integrals in the right member of (8.31) require a reformulation into proper weak form. For the fluid traction, the corresponding reformulation is generally referred to as *flux extraction* or *variationally consistent flux evaluation*; see, for instance, [43, 65, 68].

The weak formulation of the fluid-traction functional derives directly from the weak ALE formulation of the momentum equation in the NSCH system according to (8.37). Recalling that the second term in the right-hand side of (8.37) vanishes under the standing assumptions on the boundary conditions, we infer the following identity from (8.37):

$$\int_{\Gamma_t} (\boldsymbol{\theta} \circ T_t^{-1}) \cdot \boldsymbol{t}^{\mathrm{F}} = \int_{\Omega_t^{\mathrm{F}}} p \operatorname{div} (\boldsymbol{\ell}_{\boldsymbol{\theta}} \circ T_t^{-1}) - \frac{d}{dt} \int_{\Omega_t^{\mathrm{F}}} (\boldsymbol{\ell}_{\boldsymbol{\theta}} \circ T_t^{-1}) \cdot \rho \boldsymbol{u} \\ - \int_{\Omega_t^{\mathrm{F}}} \operatorname{grad} (\boldsymbol{\ell}_{\boldsymbol{\theta}} \circ T_t^{-1}) : (\boldsymbol{\tau} + \boldsymbol{\zeta} - \rho \boldsymbol{u} \otimes (\boldsymbol{u} - \boldsymbol{v})) \tag{8.42}$$

where $\boldsymbol{\ell}_{\boldsymbol{\theta}}$ represents an arbitrary sufficiently smooth lifting of the trace of $\boldsymbol{\theta}$ from within Ω_0^{S} into the fluid domain in the reference configuration, Ω_0^{F}, i.e., any function $\boldsymbol{\ell}_{\boldsymbol{\theta}} : \Omega_0^{\mathrm{F}} \to \mathbb{R}^d$ such that the trace of $\boldsymbol{\ell}_{\boldsymbol{\theta}}$ at Γ_0 from within Ω_0^{F} coincides with the trace of $\boldsymbol{\theta}$ at Γ_0 from within Ω_0^{S}. It is to be noted that, in particular, if $\boldsymbol{\theta}$ is defined throughout Ω_0 and $\boldsymbol{\theta} \in \boldsymbol{C}^0(\Omega_0)$, then $\boldsymbol{\ell}_{\boldsymbol{\theta}}$ can be identified with $\boldsymbol{\theta}|_{\Omega_0^{\mathrm{F}}}$. This implies that in the aggregated weak formulation, the fluid traction on the solid in accordance with the dynamic interface condition can be accounted for by imposing continuity of the test functions for the momentum equations (8.5_1) and (8.22), pertaining to the binary fluid and solid, respectively. Alternatively, assuming that the test spaces for the binary-fluid and solid momentum equations are compatible, the fluid traction on the solid can be incorporated in the aggregated weak formulation by introducing a single global test function $\boldsymbol{\theta} : \Omega_0 \to \mathbb{R}^d$.

To recast the integral associated with the traction due to solid–fluid surface tension into an appropriate weak form, we note the following identities:

$$\int_{\Gamma_t} (\boldsymbol{\theta} \circ T_t^{-1}) \cdot \boldsymbol{t}_{\mathrm{SF}} = \int_{\Gamma_t} (\boldsymbol{\theta} \circ T_t^{-1}) \cdot (\sigma_{\mathrm{SF}}\, \kappa \boldsymbol{n}^{\mathrm{F}} + \operatorname{grad}_\Gamma \sigma_{\mathrm{SF}}) \\ = - \int_{\Gamma_t} \sigma_{\mathrm{SF}} \operatorname{grad}_\Gamma \mathrm{Id}_{\Gamma_t} : \operatorname{grad}_\Gamma (\boldsymbol{\theta} \circ T_t^{-1}) + \int_{\partial \Gamma_t} \sigma_{\mathrm{SF}} (\boldsymbol{\theta} \circ T_t^{-1}) \cdot \boldsymbol{\nu} \tag{8.43}$$

with Id_{Γ_t} the identity on Γ_t and $\boldsymbol{\nu}$ the exterior unit normal vector to $\partial \Gamma_t$ in the tangent bundle of Γ_t; see [5, 20]. The second term in the right-hand side of (8.43) can be conceived of as a line force that emanates from variations in the area of the solid boundary that is wetted by the binary fluid. We assume that this term can be suppressed by virtue of boundary conditions or auxiliary data of the problem. The first term in the right-hand side of (8.43) then provides a suitable reformulation of the traction induced by solid–fluid surface tension.

Remark 8.7. The functionals in the right members of (8.42) and (8.43) indeed correspond to weak formulations of the integrals in the corresponding left-hand sides, with $\boldsymbol{t}_{\mathrm{SF}}$ and $\boldsymbol{t}^{\mathrm{F}}$ according to (8.29) and (8.30). The identities (8.42) and (8.43) hold for all sufficiently smooth solutions of (8.5) and (8.22) for which the integrals in the left-hand sides exist, but the right-hand sides are defined for a larger class of solutions of (8.5) and (8.22) with weaker regularity. For instance, the integral in the left-hand side of (8.42) depends implicitly on the trace of p at Γ_t and, accordingly, it is not properly defined for $p \in L^2(\Omega_t^{\mathrm{F}})$, i.e., if p is square integrable on Ω_t^{F} but does not a priori satisfy any further regularity conditions. The right-hand side of (8.42) instead contains the L^2 inner product of p and $\operatorname{div}(\boldsymbol{\ell}_{\boldsymbol{\theta}} \circ T_t^{-1})$. Hence, this term is well defined for $p \in L^2(\Omega_t^{\mathrm{F}})$, provided that $\operatorname{div}(\boldsymbol{\ell}_{\boldsymbol{\theta}} \circ T_t^{-1}) \in L^2(\Omega_t^{\mathrm{F}})$. Similarly, the left-hand side of (8.43) depends implicitly on the additive curvature, κ, of the solid–fluid interface, which is defined in terms of second-order derivatives of the trace of the solid deformation at the interface. The right-hand side of (8.43) depends on the tangential gradient of the trace of the deformation only and it is well defined under much weaker conditions on the deformation.

8.3.3 Ambient function spaces

A complete specification of the ambient function spaces for the aggregated weak formulation of the binary-fluid–solid interaction problem in such a manner that existence and stability of its solutions can be established, is technical and beyond the scope of this work. We restrict ourselves here to identifying some elementary properties of these spaces.

We first introduce some general notational conventions. We denote by $L^2(\omega)$ the class of square-integrable functions on any $\omega \subset \mathbb{R}^d$, by $H^1(\omega)$ the Sobolev space of functions in $L^2(\omega)$ with weak derivatives in $L^2(\omega)$, and by $H^1_{0,\Xi}(\omega)$ the subspace of functions that vanish on $\Xi \subseteq \partial\Omega$ in the trace sense. For any vector space $V(\omega)$ of scalar-valued functions, $\boldsymbol{V}(\omega)$ is the extension to the corresponding vector space of $\mathbb{R}^d$-valued functions. Given a vector space V and a time interval $(0, t_{\mathrm{FIN}})$, we denote by $L(0, t_{\mathrm{FIN}}; V)$ the linear space of functions from $(0, t_{\mathrm{FIN}})$ into V.

The deformation $\boldsymbol{\psi}\colon (0, t_{\mathrm{FIN}}) \times \Omega_0 \to \mathbb{R}^d$ of the fluid and solid domains is continuous across the solid–fluid interface Γ_0. The restriction of the deformation to the solid domain Ω_0^{S} complies with the equations of motion (8.22). On the fluid domain, the deformation is given by the harmonic extension in accordance with (8.32). Standard arguments indicate the following setting of the deformation of the fluid domain:

$$\boldsymbol{\psi}^{\mathrm{F}} := \boldsymbol{\psi}|_{\Omega_0^{\mathrm{F}}} \in \mathrm{Id} + L(0, t_{\mathrm{FIN}}; \boldsymbol{H}^1_{0,\partial\Omega_0^{\mathrm{F}}\setminus\Gamma_0}(\Omega_0^{\mathrm{F}})) \tag{8.44}$$

Implicit to (8.44) is the assumption that the part of the fluid-domain boundary complementary to the interface is immobile, but this assumption is nonessential. The restriction $\boldsymbol{\psi}^{\mathrm{S}} := \boldsymbol{\psi}|_{\Omega_0^{\mathrm{S}}}$ of the deformation to the solid domain can be accommodated in $\mathrm{Id} + L(0, t_{\mathrm{FIN}}; \boldsymbol{H}^1(\Omega_0^{\mathrm{S}}))$. However, the solid deformation must satisfy additional requirements, to ensure that all terms in the formulation are continuous. We assume that the reference configuration of the interface, Γ_0, is a C^∞-manifold. To account for the solid–fluid surface tension in the Hilbert-space setting of the solid deformation, we introduce the principal part of (8.43) in the inner product. Accordingly, we define

$$\boldsymbol{H}(\mathrm{grad}_{\Gamma_0};\, \Omega_0^{\mathrm{s}}) = \{\boldsymbol{\theta} \in \boldsymbol{H}^1(\Omega_0^{\mathrm{s}}) \colon \|\mathrm{grad}_{\Gamma_0} \boldsymbol{\theta}\|_{L^2(\Gamma_0)} < \infty\} \tag{8.45}$$

which is equipped with a Hilbert structure corresponding to the inner product:

$$(\boldsymbol{\theta}, \boldsymbol{\zeta})_{\boldsymbol{H}(\mathrm{grad}_{\Gamma_0};\Omega_0^{\mathrm{s}})} = \int_{\Omega_0^{\mathrm{s}}} \boldsymbol{\theta} \cdot \boldsymbol{\zeta} + \int_{\Omega_0^{\mathrm{s}}} \mathrm{grad}\,\boldsymbol{\theta} \colon \mathrm{grad}\,\boldsymbol{\zeta} + \int_{\Gamma_0} \mathrm{grad}_{\Gamma_0} \boldsymbol{\theta} \colon \mathrm{grad}_{\Gamma_0} \boldsymbol{\zeta} \tag{8.46}$$

To provide a setting for the displacement in the nonlinear solid subproblem, we consider a suitable linear subspace $\boldsymbol{S}(\Omega_0^{\mathrm{s}}) \subset \boldsymbol{H}(\mathrm{grad}_{\Gamma_0}; \Omega_0^{\mathrm{s}})$, closed with respect to the norm $\|\cdot\|_{\boldsymbol{S}(\Omega_0^{\mathrm{s}})}$. We insist that the stored energy $W(\mathrm{grad}\,\boldsymbol{\psi}^{\mathrm{s}})$ corresponding to deformations $\boldsymbol{\psi}^{\mathrm{s}} \in \mathrm{Id} + \boldsymbol{S}(\Omega_0^{\mathrm{s}})$ is Lebesgue-measurable. In addition, we require that the following linear forms are continuous for all deformations $\boldsymbol{\psi}^{\mathrm{s}} \in \mathrm{Id} + \boldsymbol{S}(\Omega_0^{\mathrm{s}})$:

$$\begin{aligned} \boldsymbol{\theta} &\mapsto \int_{\Omega_0^{\mathrm{s}}} \frac{\partial W}{\partial \boldsymbol{F}}(\mathrm{grad}\,\boldsymbol{\psi}^{\mathrm{s}}) \colon \mathrm{grad}\,\boldsymbol{\theta} \\ \boldsymbol{\theta} &\mapsto \int_{\Gamma} \mathrm{grad}_{\Gamma}\,\mathrm{Id}_{\Gamma} \colon \mathrm{grad}_{\Gamma}\,(\boldsymbol{\theta} \circ T^{-1}) \end{aligned} \tag{8.47}$$

where $\Gamma := \boldsymbol{\psi}^{\mathrm{s}}(\Gamma_0)$ and the corresponding tangential gradient depend implicitly on $\boldsymbol{\psi}^{\mathrm{s}}$. Finally, it must hold that the linear forms in (8.47) admit an interpretation as Gâteaux derivatives of the strain energy and the solid–fluid interface energy. To this end, it must hold that

$$\begin{aligned} \int_{\Omega_0^{\mathrm{s}}} o(|\,\mathrm{grad}\,\boldsymbol{\theta}|) &= o(\|\boldsymbol{\theta}\|_{\boldsymbol{S}(\Omega_0^{\mathrm{s}})}) \\ \int_{\Gamma} o(|\,\mathrm{grad}_{\Gamma}\,\boldsymbol{\theta}|) &= o(\|\boldsymbol{\theta}\|_{\boldsymbol{S}(\Omega_0^{\mathrm{s}})}) \end{aligned} \tag{8.48}$$

as $\boldsymbol{\theta} \to 0$. Under the (nonessential) assumption that the solid is rigidly fixed at $\partial\Omega_0^{\mathrm{s}} \setminus \Gamma_0$ we arrive at the following ambient space for the deformation: $\boldsymbol{\psi} \in \mathrm{Id} + L(0, t_{\mathrm{FIN}}; \boldsymbol{W}(\Omega_0))$ with

$$\boldsymbol{W}(\Omega_0) := \{\boldsymbol{\theta} \in \boldsymbol{H}^1_{0,\partial\Omega_0}(\Omega_0) \colon \boldsymbol{\theta}|_{\Omega_0^{\mathrm{s}}} \in \boldsymbol{H}(\mathrm{grad}_{\Gamma_0}; \Omega_0^{\mathrm{s}}),\ \|\boldsymbol{\theta}|_{\Omega_0^{\mathrm{s}}}\|_{\boldsymbol{S}(\Omega_0^{\mathrm{s}})} < \infty\} \tag{8.49}$$

Remark 8.8. The restriction to the interface of the trace of functions in $\boldsymbol{H}(\mathrm{grad}_{\Gamma_0}; \Omega_0^{\mathrm{s}})$ according to (8.45) resides in $\boldsymbol{H}^1(\Gamma_0)$. The Sobolev embedding theorem (e.g., [22, Thm. B.40]) conveys that for $d = 2$ (resp. $d = 3$) the embedding of $\boldsymbol{H}^1(\Gamma_0)$ into $\boldsymbol{L}^\infty(\Gamma_0) \cap \boldsymbol{C}^0(\Gamma_0)$ (resp. $\boldsymbol{L}^q(\Gamma_0)$ for any $q < +\infty$) is continuous. Consequently, contact-line forces on the interface are continuous with respect to the norm induced by (8.46). In contrast, contact-line forces are not continuous with respect to the $\boldsymbol{H}^1(\Omega_0^{\mathrm{s}})$-norm.

In general, contact-line forces, i.e., surface point loads in $d = 2$ and surface line loads in $d = 3$, can fail to be continuous with respect to norms in which the strain energy is continuous. However, such localized loads on the solid surface are generally continuous with respect to norms in which the solid–fluid surface energy is continuous. This indicates that the configuration of the interface near a contact line is exclusively determined by solid–fluid surface tension, independent of elastic effects in the solid. Indeed, the solid–fluid interface exhibits a kink at the contact line, whose shape conforms to the so-called Neumann law [18, 48]. The Neumann law relates the angle of the kink to the surface tensions of the adjacent solid–liquid, solid–ambient and liquid–ambient interfaces.

The ambient spaces for the weak formulation of the NSCH system must be such that for all admissible $\boldsymbol{u}, \phi, \mu, p$ the mappings from $\boldsymbol{\theta}, \beta, \chi, q$ to the left members of (8.37)–(8.40) constitute continuous linear forms. Furthermore, in view of the definition of ϕ according to (8.2), the range of ϕ is (almost everywhere) confined to the interval $[-1, 1]$. Accordingly, it follows from (8.3) that the range of ρ is confined to $[\rho_{\mathrm{A}}, \rho_{\mathrm{L}}]$ and

$$\frac{1}{\rho}\frac{d\rho}{d\phi} \leq \frac{1}{\rho_{\mathrm{A}}}\frac{\rho_{\mathrm{L}} - \rho_{\mathrm{A}}}{2} \tag{8.50}$$

We assume that the transformation T_t and its inverse T_t^{-1} are sufficiently regular (in particular, Lipschitz continuous) to ensure that $H^1(\Omega_0^{\mathrm{F}})$ and $H^1(\Omega_t^{\mathrm{F}})$ are invariant under composition with T_t and T_t^{-1}, i.e.,

$$\begin{aligned} u \in H^1(\Omega_0^{\mathrm{F}}) &\Rightarrow (u \circ T_t^{-1}) \in H^1(\Omega_t^{\mathrm{F}}) \\ u \in H^1(\Omega_t^{\mathrm{F}}) &\Rightarrow (u \circ T_t) \in H^1(\Omega_0^{\mathrm{F}}) \end{aligned} \tag{8.51}$$

Taking the aforementioned considerations into account, a plausible setting for $\boldsymbol{u}, \phi, \mu, p$ after transportation to the reference domain is:

$$\boldsymbol{u} \circ T_t \in L(0, t_{\mathrm{FIN}}; \boldsymbol{H}^1(\Omega_0^{\mathrm{F}})) \qquad \mu \circ T_t \in L(0, t_{\mathrm{FIN}}; H^1(\Omega_0^{\mathrm{F}})) \tag{8.52}$$

$$\phi \circ T_t \in L(0, t_{\mathrm{FIN}}; V^{\phi}(\Omega_0^{\mathrm{F}})) \qquad p \circ T_t \in L(0, t_{\mathrm{FIN}}; L^2(\Omega_0^{\mathrm{F}})) \tag{8.53}$$

where $V^{\phi}(\Omega_0^{\mathrm{F}})$ is the convex space defined by

$$V^{\phi}(\Omega_0^{\mathrm{F}}) = \left\{\hat{\phi} \in H^1(\Omega_0^{\mathrm{F}}) \cap L^{\infty}(\Omega_0^{\mathrm{F}}) : |\hat{\phi}| \leq 1 \text{ a.e.}, \boldsymbol{\zeta}(\hat{\phi} \circ T_t^{-1}) \in L^2(\Omega_0^{\mathrm{F}}, \mathbb{M}^d)\right\} \tag{8.54}$$

with $\mathbb{M}^d$ the class of rank-2 tensors, equipped with the tensor-contraction inner product.

Remark 8.9. One can verify that for all $\boldsymbol{u}, \mu, \phi, p$ subject to (8.52), the mappings from

$$(\boldsymbol{\theta}, q, \beta, \chi) \in \boldsymbol{H}^1(\Omega_0^{\mathrm{F}}) \times L^2(\Omega_0^{\mathrm{F}}) \times H^1(\Omega_0^{\mathrm{F}}) \times V^{\phi}(\Omega_0^{\mathrm{F}}) \tag{8.55}$$

to the left members of (8.37)–(8.40) are continuous. For instance,

$$\left|\int_{\Omega_t^{\mathrm{F}}} \operatorname{grad}(\boldsymbol{\theta} \circ T_t^{-1}) : \boldsymbol{\zeta}(\phi)\right| \leq \|\boldsymbol{\theta} \circ T_t^{-1}\|_{\boldsymbol{H}^1(\Omega_t^{\mathrm{F}})} \|\boldsymbol{\zeta}(\phi)\|_{L^2(\Omega_t^{\mathrm{F}}, \mathbb{M}^d)} \leq \mathscr{C}\, \|\boldsymbol{\theta}\|_{\boldsymbol{H}^1(\Omega_0^{\mathrm{F}})} \tag{8.56}$$

for some constant[1] $\mathscr{C} > 0$ independent of $\boldsymbol{\theta}$. For the convection term in (8.37), we obtain

$$\left| \int_{\Omega_t^{\mathrm{F}}} \operatorname{grad}(\boldsymbol{\theta} \circ T_t^{-1}) : (\rho \boldsymbol{u} \otimes \boldsymbol{u}) \right| \leq \|\boldsymbol{\theta} \circ T_t^{-1}\|_{\boldsymbol{H}^1(\Omega_t^{\mathrm{F}})} \|\rho\|_{L^\infty(\Omega_t^{\mathrm{F}})} \|\boldsymbol{u}\|^2_{\boldsymbol{L}^4(\Omega_t^{\mathrm{F}})} \leq \mathscr{C} \|\boldsymbol{\theta}\|_{\boldsymbol{H}^1(\Omega_0^{\mathrm{F}})} . \tag{8.57}$$

The first inequality in (8.57) follows from Hölder's inequality. The Sobolev embedding theorem implies that for $d \in \{2, 3\}$ the embedding of $\boldsymbol{L}^4(\Omega_t^{\mathrm{F}})$ into $\boldsymbol{H}^1(\Omega_t^{\mathrm{F}})$ is continuous. Moreover, for $\phi \circ T_t \in V^\phi(\Omega_0^{\mathrm{F}})$ it holds that $\rho(\phi) \in L^\infty(\Omega_t^{\mathrm{F}})$. Continuity of all other terms in (8.37)–(8.40) can be established in a similar manner.

Remark 8.10. Arrangement of the phase field in a convex space instead of a general linear space, analogous to (8.54), has been investigated in the context of the Cahn–Hilliard equation with a nonsmooth free energy in [8]. For numerical approximation methods in this setting, we refer to [9, 33]. The L^∞-regularity of the phase field hypothesized in (8.54) has been established for the Cahn–Hilliard equation separately in [15].

The kinematic condition (8.27) is accounted for by identifying the trace of the fluid velocity at the interface to the trace of the time derivative of the solid deformation. We then arrive at the following ambient space for the aggregated binary-fluid–solid interaction problem:

$$\begin{aligned} \mathbb{U} := \{(\boldsymbol{\psi}, \boldsymbol{u}, p, \phi, \mu) :\ & \boldsymbol{\psi} \in \mathrm{Id} + L(0, t_{\mathrm{FIN}}; \boldsymbol{W}(\Omega_0)), \boldsymbol{u} \circ T_t \in L(0, t_{\mathrm{FIN}}; \boldsymbol{H}^1(\Omega_0^{\mathrm{F}})) , \\ & p \circ T_t \in L(0, t_{\mathrm{FIN}}; L^2(\Omega_0^{\mathrm{F}})), \phi \circ T_t \in L(0, t_{\mathrm{FIN}}; V^\phi(\Omega_0^{\mathrm{F}})) , \\ & \mu \circ T_t \in L(0, t_{\mathrm{FIN}}; H^1(\Omega_0^{\mathrm{F}})), (\boldsymbol{u} \circ T_t)|_{\Gamma_0} = (\partial_t \boldsymbol{\psi})|_{\Gamma_0} \ \text{ a.e. } t \in (0, t_{\mathrm{FIN}}) , \\ & \boldsymbol{\psi}|_{t=0} = \mathrm{Id}, \partial_t \boldsymbol{\psi}|_{t=0} = 0, \boldsymbol{u}|_{t=0} = \boldsymbol{u}_0, \phi|_{t=0} = \phi_0 \} \end{aligned} \tag{8.58}$$

Recalling that the dynamic condition is imposed by introducing a single sufficiently smooth global test function for the equations of motion of the solid and the fluid (see Sec. 8.3.2), a suitable setting for the test functions for the FSI problem is:

$$\begin{aligned} \mathbb{V} := \{(\boldsymbol{\theta}, \boldsymbol{\vartheta}, q, \beta, \chi) :\ & \boldsymbol{\theta} \in \boldsymbol{W}(\Omega_0), \boldsymbol{\vartheta} \in \boldsymbol{H}^1_{0,\partial\Omega_0^{\mathrm{F}}}(\Omega_0^{\mathrm{F}}), q \in L^2(\Omega_0^{\mathrm{F}}) , \\ & \beta \in H^1(\Omega_0^{\mathrm{F}}), \chi \in V^\phi(\Omega_0^{\mathrm{F}}) \} \end{aligned} \tag{8.59}$$

Let us note that the spaces $\mathbb{U}$ and $\mathbb{V}$ are generally subject to auxiliary boundary conditions on the boundary $\partial\Omega_0$ of the aggregated FSI domain. We will not consider these auxiliary conditions in further detail.

1 We use $\mathscr{C}$ to denote a generic positive constant, of which the value and connotation may change from one instance to the next, even within a single chain of expressions.

8.3.4 Aggregated weak formulation

The weak formulation of the aggregated binary-fluid–solid interaction problem is composed of the bilinear form in (8.32) associated with the harmonic extension of the solid deformation to the fluid domain, the semilinear forms in (8.37)–(8.40) pertaining to the ALE formulation of the NSCH equations in combination with the solid–fluid boundary conditions in (8.28), the semilinear form in (8.41) corresponding to the solid subsystem, the semilinear form in (8.42) corresponding to the fluid traction, and the semilinear form associated with the traction due to solid–fluid surface tension, in accordance with (8.43). The weak formulation of the aggregated FSI problem can be condensed into:

$$\begin{aligned}&\text{Find } (\boldsymbol{\psi}, \boldsymbol{u}, p, \phi, \mu) \in \mathbb{U} \text{ such that}\\ &\qquad \mathscr{A}(\boldsymbol{\psi}, \boldsymbol{u}, p, \phi, \mu; \boldsymbol{\theta}, \boldsymbol{\vartheta}, q, \beta, \chi) = \mathscr{L}(\boldsymbol{\theta}, \boldsymbol{\vartheta}, q, \beta, \chi)\\ &\qquad\qquad \forall(\boldsymbol{\theta}, \boldsymbol{\vartheta}, q, \beta, \chi) \in \mathbb{V}, \quad \text{a.e. } t \in (0, t_{\mathrm{FIN}})\end{aligned} \tag{8.60}$$

where $\mathscr{L}: \mathbb{V} \to \mathbb{R}$ denotes a linear form corresponding to exogenous data and the semilinear form $\mathscr{A}: \mathbb{U} \times \mathbb{V} \to \mathbb{R}$ is defined by:

$$\begin{aligned}&\mathscr{A}((\boldsymbol{\psi}, \boldsymbol{u}, p, \phi, \mu); (\boldsymbol{\theta}, \boldsymbol{\vartheta}, q, \beta, \chi))\\
&:= \frac{d^2}{dt^2}\int_{\Omega_0^{\mathrm{S}}} \rho_{\mathrm{S}}\boldsymbol{\psi}\cdot\boldsymbol{\theta} + \mathscr{W}'(\boldsymbol{\psi}|_{\Omega_0^{\mathrm{S}}}; \boldsymbol{\theta}) + \int_{\Gamma_t} \sigma_{\mathrm{SF}}(\phi)\,\mathrm{grad}_\Gamma \mathrm{Id}_{\Gamma_t} : \mathrm{grad}_\Gamma(\boldsymbol{\theta}\circ T_t^{-1})\\
&+ \int_{\Omega_0^{\mathrm{F}}} \mathrm{grad}\,\boldsymbol{\psi} : \mathrm{grad}\,\boldsymbol{\vartheta} + \frac{d}{dt}\int_{\Omega_t^{\mathrm{F}}} \rho\boldsymbol{u}\cdot(\boldsymbol{\theta}\circ T_t^{-1}) + \int_{\Omega_t^{\mathrm{F}}} (\boldsymbol{\tau}+\boldsymbol{\zeta}) : \mathrm{grad}\,(\boldsymbol{\theta}\circ T_t^{-1})\\
&- \int_{\Omega_t^{\mathrm{F}}} (\rho\boldsymbol{u}\otimes(\boldsymbol{u}-\boldsymbol{v})) : \mathrm{grad}\,(\boldsymbol{\theta}\circ T_t^{-1}) - \int_{\Omega_t^{\mathrm{F}}} p\,\mathrm{div}\,(\boldsymbol{\theta}\circ T_t^{-1})\\
&+ \frac{d}{dt}\int_{\Omega_t^{\mathrm{F}}} \phi(\beta\circ T_t^{-1}) + \int_{\Omega_t^{\mathrm{F}}} (m\,\mathrm{grad}\,\mu - \phi(\boldsymbol{u}-\boldsymbol{v}))\cdot\mathrm{grad}\,(\beta\circ T_t^{-1})\\
&+ \int_{\Omega_t^{\mathrm{F}}} \mu(\chi\circ T_t^{-1}) - \int_{\Omega_t^{\mathrm{F}}} \epsilon\sigma\,\mathrm{grad}\,\phi\cdot\mathrm{grad}\,(\chi\circ T_t^{-1}) - \int_{\Omega_t^{\mathrm{F}}} \frac{\sigma}{\epsilon}\Psi'\,(\chi\circ T_t^{-1})\\
&+ \int_{\Omega_t^{\mathrm{F}}} \frac{p(\rho_{\mathrm{L}}-\rho_{\mathrm{A}})}{2\rho}(\chi\circ T_t^{-1}) + \int_{\Omega_t^{\mathrm{F}}} \mathrm{div}\,\boldsymbol{u}\,(q\circ T_t^{-1})\\
&+ \int_{\Omega_t^{\mathrm{F}}} \alpha m\,\mathrm{grad}\,\mu\cdot\mathrm{grad}\,(q\circ T_t^{-1}) - \int_{\Gamma_t} \sigma'_{\mathrm{SF}}(\phi)(\chi\circ T_t^{-1})\end{aligned} \tag{8.61}$$

with $\boldsymbol{v} = \partial_t\boldsymbol{\psi}\circ T_t^{-1}$ in accordance with (8.33). Note that in (8.61), the term $\epsilon\sigma\,\partial_n\phi$ in the right-hand side of (8.39) has been replaced by $-\sigma'_{\mathrm{SF}}(\phi)$ in accordance with the wetting condition in (8.28).

8.4 Properties of the binary-fluid–solid interaction problem

In this section we determine some elementary properties of the binary-fluid–solid interaction problem (8.60). In Section 8.4.1 we establish the fundamental energy-dissipation structure of the binary-fluid–solid interaction problem. Section 8.4.2 is concerned with equilibrium solutions of the FSI problem. In Section 8.4.3, we look into the structure of the diffuse fluid–fluid interface in equilibrium in the specific case of a flat interface and we formally consider aspects related to wetting and capillarity in the sharp-interface limit.

8.4.1 Energy dissipation

The energy of the binary-fluid–solid interaction problem (8.60) comprises the kinetic energy of the fluid and solid subsystems, the free energy associated with mixing of the liquid and ambient fluid species, the solid–fluid surface energy, and the stored energy due to the elastic deformation of the solid. Accordingly, the energy functional of the aggregated FSI problem is given by:

$$\begin{aligned}\mathscr{E}(\boldsymbol{\psi}, \boldsymbol{u}, \phi) = \int_{\Omega_t^{\mathrm{F}}} \frac{1}{2}\rho|\boldsymbol{u}|^2 + \int_{\Omega_t^{\mathrm{F}}} \left(\frac{\epsilon\sigma}{2}|\operatorname{grad}\phi|^2 + \frac{\sigma}{\epsilon}\Psi\right) + \int_{\Gamma_t} \sigma_{\mathrm{SF}}(\phi) \\ + \int_{\Omega_0^{\mathrm{S}}} \frac{1}{2}\rho_{\mathrm{S}}|\partial_t\boldsymbol{\psi}|^2 + \mathscr{W}(\boldsymbol{\psi}|_{\Omega_0^{\mathrm{S}}})\end{aligned} \tag{8.62}$$

The first and fourth term in (8.62) represent the kinetic energy of the binary fluid and solid, respectively. The second term is a Ginzburg–Landau-type energy functional associated with mixing of the binary-fluid constituents. The third term represents an energetic contribution associated with solid–fluid surface tension. The fifth term corresponds to the strain energy of the solid.

We will establish the following fundamental energy-dissipation relation for the binary-fluid–solid interaction problem: If $(\boldsymbol{\psi}(t), \boldsymbol{u}(t), p(t), \phi(t), \mu(t))$ corresponds to a sufficiently smooth solution of (8.60) then:

$$d_t\mathscr{E}(\boldsymbol{\psi}(t), \boldsymbol{u}(t), \phi(t)) = -\int_{\Omega_t^{\mathrm{F}}} \operatorname{grad}\boldsymbol{u} : \boldsymbol{\tau} - \int_{\Omega_t^{\mathrm{F}}} m|\operatorname{grad}\mu|^2 + \mathrm{bnd} \leq 0 + \mathrm{bnd} \tag{8.63}$$

where 'bnd' serves as a placeholder for terms that are only supported on the external boundary $\partial\Omega_0$ of the aggregated fluid–solid system. To establish (8.63), we will first consider each of the energy contributions in (8.62) independently and then assemble the dissipation relation (8.63) of the aggregated FSI problem. For transparency, we will in most steps refer to the corresponding strong form of equations, rather than their weak form.

For the Ginzburg–Landau-type energy functional in (8.62), we obtain:

$$
\begin{aligned}
&\frac{d}{dt}\int_{\Omega_t^{\mathrm{F}}}\left(\frac{\epsilon\sigma}{2}|\operatorname{grad}\phi|^2+\frac{\sigma}{\epsilon}\Psi\right)\\
&=\int_{\Omega_t^{\mathrm{F}}}\left(\epsilon\sigma\operatorname{grad}\phi\cdot\operatorname{grad}\partial_t\phi+\frac{\sigma}{\epsilon}\Psi'\partial_t\phi\right)+\int_{\partial\Omega_t^{\mathrm{F}}}\left(\frac{\epsilon\sigma}{2}|\operatorname{grad}\phi|^2+\frac{\sigma}{\epsilon}\Psi\right)v_n\\
&=\underbrace{\int_{\Omega_t^{\mathrm{F}}}\left(\frac{\sigma}{\epsilon}\Psi'-\epsilon\sigma\,\Delta\phi\right)\partial_t\phi}_{T_1^{\mathrm{GL}}}+\underbrace{\int_{\Gamma_t}\left(\frac{\epsilon\sigma}{2}|\operatorname{grad}\phi|^2+\frac{\sigma}{\epsilon}\Psi\right)v_n}_{T_2^{\mathrm{GL}}}+\underbrace{\int_{\Gamma_t}\sigma\epsilon\,\partial_n\phi\,\partial_t\phi}_{T_3^{\mathrm{GL}}}+\mathrm{bnd}
\end{aligned}
\tag{8.64}
$$

Equations (8.5_3) and (8.5_4) convey

$$
\begin{aligned}
T_1^{\mathrm{GL}}&=\int_{\Omega_t^{\mathrm{F}}}\mu(\operatorname{div}m\operatorname{grad}\mu-\operatorname{div}\phi\boldsymbol{u})+\int_{\Omega_t^{\mathrm{F}}}\frac{p}{\rho}\frac{d\rho}{d\phi}\partial_t\phi\\
&=-\int_{\Omega_t^{\mathrm{F}}}m|\operatorname{grad}\mu|^2+\int_{\Gamma_t}m\mu\partial_n\mu-\int_{\Omega_t^{\mathrm{F}}}(\boldsymbol{u}\cdot\mu\operatorname{grad}\phi+\mu\phi\operatorname{div}\boldsymbol{u})\\
&\quad-\int_{\Omega_t^{\mathrm{F}}}p\operatorname{div}\boldsymbol{u}-\int_{\Omega_t^{\mathrm{F}}}\frac{p}{\rho}\frac{d\rho}{d\phi}\boldsymbol{u}\cdot\operatorname{grad}\phi+\mathrm{bnd}\,.
\end{aligned}
\tag{8.65}
$$

Let us note that in the second identity in (8.65), we applied the divergence theorem and conservation of mass (see Remark 8.1) to replace $d_\phi\rho\,\partial_t\phi$ by $\partial_t\rho=-\operatorname{div}\rho\boldsymbol{u}$. Moreover, by adding a suitable partition of zero to T_2^{GL}, we obtain

$$
\begin{aligned}
T_2^{\mathrm{GL}}&=\int_{\Gamma_t}\left(\frac{\epsilon\sigma}{2}|\operatorname{grad}\phi|^2+\frac{\sigma}{\epsilon}\Psi\right)(v_n-u_n)\\
&\quad+\int_{\Omega_t^{\mathrm{F}}}\operatorname{grad}\boldsymbol{u}:\left(\frac{\epsilon\sigma}{2}|\operatorname{grad}\phi|^2+\frac{\sigma}{\epsilon}\Psi\right)\boldsymbol{I}\\
&\quad+\int_{\Omega_t^{\mathrm{F}}}\boldsymbol{u}\cdot\operatorname{grad}\left(\frac{\epsilon\sigma}{2}|\operatorname{grad}\phi|^2+\frac{\sigma}{\epsilon}\Psi\right)+\mathrm{bnd}\,.
\end{aligned}
\tag{8.66}
$$

Elementary calculus in combination with the identity (8.5_4) yields

$$
\begin{aligned}
\operatorname{div}(\epsilon\sigma\operatorname{grad}\phi\otimes\operatorname{grad}\phi)&=\epsilon\sigma\Big(\Delta\phi\operatorname{grad}\phi+\frac{1}{2}\operatorname{grad}|\operatorname{grad}\phi|^2\Big)\\
&=\left(\frac{\sigma}{\epsilon}\Psi'-\mu-\frac{p}{\rho}\frac{d\rho}{d\phi}\right)\operatorname{grad}\phi+\frac{\epsilon\sigma}{2}\operatorname{grad}|\operatorname{grad}\phi|^2\\
&=\operatorname{grad}\left(\frac{\sigma}{\epsilon}\Psi+\frac{\epsilon\sigma}{2}|\operatorname{grad}\phi|^2\right)-\mu\operatorname{grad}\phi-\frac{p}{\rho}\frac{d\rho}{d\phi}\operatorname{grad}\phi\,.
\end{aligned}
\tag{8.67}
$$

Hence, by collecting the expansions of T_1^{GL} and T_2^{GL} and invoking the divergence theorem, we obtain:

$$
\begin{aligned}
T_1^{\mathrm{GL}}+T_2^{\mathrm{GL}}&=-\int_{\Omega_t^{\mathrm{F}}}m|\operatorname{grad}\mu|^2-\int_{\Omega_t^{\mathrm{F}}}p\operatorname{div}\boldsymbol{u}+\int_{\Omega_t^{\mathrm{F}}}\operatorname{grad}\boldsymbol{u}:\boldsymbol{\zeta}+\int_{\Gamma_t}m\mu\partial_n\mu\\
&\quad+\int_{\Gamma_t}\left(\frac{\epsilon\sigma}{2}|\operatorname{grad}\phi|^2+\frac{\sigma}{\epsilon}\Psi\right)(v_n-u_n)+\int_{\Gamma_t}\epsilon\sigma\,\partial_n\phi\,\boldsymbol{u}\cdot\operatorname{grad}\phi+\mathrm{bnd}
\end{aligned}
\tag{8.68}
$$

Considering the ultimate term in (8.68), it follows from the preferential-wetting boundary condition (8.18) that:

$$\begin{aligned}\int_{\Gamma_t} \epsilon\sigma\,\partial_n\phi\,\boldsymbol{u}\cdot\operatorname{grad}\phi &= -\int_{\Gamma_t} \sigma'_{\mathrm{SF}}(\phi)\,\boldsymbol{u}\cdot\operatorname{grad}\phi \\ &= -\int_{\Gamma_t} \sigma'_{\mathrm{SF}}(\phi)\,u_n\partial_n\phi - \int_{\Gamma_t} \boldsymbol{u}\cdot\operatorname{grad}_\Gamma \sigma_{\mathrm{SF}}(\phi)\end{aligned} \tag{8.69}$$

The second identity in (8.69) emanates from the decomposition of the velocity vector in its normal and tangential components and the chain rule. The preferential-wetting boundary condition (8.18) moreover yields the following identity for the contribution of the solid–fluid interface to the third term in (8.64):

$$T_3^{\mathrm{GL}} = -\int_{\Gamma_t} \sigma'_{\mathrm{SF}}(\phi)\partial_t\phi \tag{8.70}$$

Noting that the fourth and fifth term in (8.68) vanish on account of (8.33) and the interface conditions in (8.28), we collect the results in (8.64)–(8.70) to obtain:

$$\begin{aligned}\frac{d}{dt}\int_{\Omega_t^{\mathrm{F}}}\left(\frac{\epsilon\sigma}{2}|\operatorname{grad}\phi|^2 + \frac{\sigma}{\epsilon}\Psi\right) &= -\int_{\Omega_t^{\mathrm{F}}} m|\operatorname{grad}\mu|^2 - \int_{\Omega_t^{\mathrm{F}}} p\operatorname{div}\boldsymbol{u} + \int_{\Omega_t^{\mathrm{F}}} \operatorname{grad}\boldsymbol{u} : \boldsymbol{\zeta} \\ &\quad - \int_{\Gamma_t} \sigma'_{\mathrm{SF}}(\phi)\,u_n\partial_n\phi - \int_{\Gamma_t} \boldsymbol{u}\cdot\operatorname{grad}_\Gamma\sigma_{\mathrm{SF}}(\phi) - \int_{\Gamma_t}\sigma'_{\mathrm{SF}}(\phi)\partial_t\phi + \mathrm{bnd}\end{aligned} \tag{8.71}$$

To recast the third term in (8.62), i.e., the energetic contribution of the solid–fluid surface tension, in an appropriate form we require some elementary tangential calculus. We consider a general continuously differentiable one-parameter family of functions $f_{(\cdot)} : (0, t_{\mathrm{FIN}}) \to C^1(\Omega_t^{\mathrm{F}})$ and the map from $(0, t_{\mathrm{FIN}})$ into $\mathbb{R}$ according to $t \mapsto \int_{\Gamma_t} f_t$. Transportation of f_t to the reference configuration conveys that the derivative of this map satisfies:

$$\frac{d}{dt}\int_{\Gamma_t} f_t = \frac{d}{dt}\int_{\Gamma_0} f_t\circ T_t\,\jmath_t = \int_{\Gamma_0} d_t(f_t\circ T_t)\,\jmath_t + \int_{\Gamma_0} f_t\circ T_t\,d_t\jmath_t \tag{8.72}$$

with $\jmath_t := J_t|(T_t')^{-T}\boldsymbol{n}|$ the surface density (or *tangential Jacobian*), i.e., the ratio of surface measures in the actual configuration and in the reference configuration. For the first term in the right member it holds that $d_t(f_t\circ T_t) = (d_t f_t + \boldsymbol{v}\cdot\operatorname{grad} f_t)\circ T_t$ with $\boldsymbol{v} = (d_t T_t)\circ T_t^{-1}$ the domain velocity corresponding to T_t. For the second term we have $d_t\jmath_t = (\operatorname{div}_\Gamma \boldsymbol{v})\circ T_t$, where $\operatorname{div}_\Gamma(\cdot) = \operatorname{div}(\cdot) - \boldsymbol{n}\cdot\partial_n(\cdot)$ denotes the surface divergence. Invoking the decomposition of the gradient into its tangential and normal components according to $\operatorname{grad} = \boldsymbol{n}\partial_n + \operatorname{grad}_\Gamma$, it therefore follows from (8.72) that:

$$\frac{d}{dt}\int_{\Gamma_t} f_t = \int_{\Gamma_t}(d_t f_t + v_n\partial_n f_t) + \int_{\Gamma_t}(\operatorname{grad}_\Gamma f_t\cdot\boldsymbol{v} + f_t\operatorname{div}_\Gamma\boldsymbol{v}) \tag{8.73}$$

According to the tangential Green's identity (see, for instance, [56]), the ultimate term in (8.73) can be replaced by:

$$\int_{\Gamma_t}(\operatorname{grad}_\Gamma f_t\cdot\boldsymbol{v} + f_t\operatorname{div}_\Gamma\boldsymbol{v}) = -\int_{\Gamma_t} f_t\,\kappa\,v_n + \int_{\partial\Gamma_t} f_t\,\boldsymbol{v}\cdot\boldsymbol{\nu} \tag{8.74}$$

where $\boldsymbol{\nu}$ denotes the exterior unit normal vector to $\partial\Gamma_t$ in the tangent bundle of Γ_t. By applying (8.73) and (8.74) to the third term in the aggregate energy (8.62), we infer

$$\frac{d}{dt}\int_{\Gamma_t}\sigma_{\mathrm{SF}}(\phi)=\int_{\Gamma_t}\sigma'_{\mathrm{SF}}(\phi)(\partial_t\phi+v_n\partial_n\phi)-\int_{\Gamma_t}\sigma_{\mathrm{SF}}(\phi)\,\kappa\,v_n+\int_{\partial\Gamma_t}\sigma_{\mathrm{SF}}(\phi)\,\boldsymbol{v}\cdot\boldsymbol{\nu}\,. \tag{8.75}$$

Next, we consider the kinetic-energy contribution of the fluid subsystem to (8.62). The Reynolds–Leibniz theorem yields

$$\frac{d}{dt}\int_{\Omega_t^{\mathrm{F}}}\frac{1}{2}\rho|\boldsymbol{u}|^2=\int_{\Omega_t^{\mathrm{F}}}\frac{1}{2}|\boldsymbol{u}|^2\partial_t\rho+\int_{\Omega_t^{\mathrm{F}}}\boldsymbol{u}\cdot\rho\partial_t\boldsymbol{u}+\int_{\Gamma_t}\frac{1}{2}\rho|\boldsymbol{u}|^2v_n+\mathrm{bnd}\,. \tag{8.76}$$

From the conservation of momentum relation (8.5_1) and the conservation of mass relation (8.11) we deduce:

$$\begin{aligned}
&\int_{\Omega_t^{\mathrm{F}}}\frac{1}{2}|\boldsymbol{u}|^2\partial_t\rho+\int_{\Omega_t^{\mathrm{F}}}\boldsymbol{u}\cdot\rho\partial_t\boldsymbol{u}\\
&\quad=-\int_{\Omega_t^{\mathrm{F}}}\frac{1}{2}|\boldsymbol{u}|^2\,\mathrm{div}(\rho\boldsymbol{u})-\int_{\Omega_t^{\mathrm{F}}}\boldsymbol{u}\cdot(\rho\boldsymbol{u}\cdot\,\mathrm{grad}\,\boldsymbol{u}+\mathrm{grad}\,p-\mathrm{div}\,\boldsymbol{\tau}-\mathrm{div}\,\boldsymbol{\zeta})\\
&\quad=-\int_{\Gamma_t}\frac{1}{2}\rho|\boldsymbol{u}|^2u_n-\int_{\Gamma_t}\boldsymbol{u}\cdot(p\boldsymbol{n}-\boldsymbol{\tau}_n-\boldsymbol{\zeta}_n)-\int_{\Omega_t^{\mathrm{F}}}\mathrm{grad}\,\boldsymbol{u}:\boldsymbol{\tau}\\
&\qquad-\int_{\Omega_t^{\mathrm{F}}}\mathrm{grad}\,\boldsymbol{u}:\boldsymbol{\zeta}+\int_{\Omega_t^{\mathrm{F}}}p\,\mathrm{div}\,\boldsymbol{u}+\mathrm{bnd}
\end{aligned} \tag{8.77}$$

Finally, we consider the energetic contribution of the solid subsystem. Recalling the notational convention $\boldsymbol{\psi}|_{\Omega_0^{\mathrm{S}}}=\boldsymbol{\psi}^{\mathrm{S}}$, we infer from (8.41):

$$\begin{aligned}
\frac{d}{dt}\Big(\int_{\Omega_0^{\mathrm{S}}}\frac{1}{2}\rho_{\mathrm{S}}|\partial_t\boldsymbol{\psi}|^2+\mathscr{W}(\boldsymbol{\psi}|_{\Omega_0^{\mathrm{S}}})\Big)&=\int_{\Omega_0^{\mathrm{S}}}\partial_t\boldsymbol{\psi}^{\mathrm{S}}\cdot\rho_{\mathrm{S}}\partial_t^2\boldsymbol{\psi}^{\mathrm{S}}+\mathscr{W}'(\boldsymbol{\psi}^{\mathrm{S}};\partial_t\boldsymbol{\psi}^{\mathrm{S}})\\
&=\int_{\Gamma_0}\partial_t\boldsymbol{\psi}^{\mathrm{S}}\cdot\boldsymbol{P}\boldsymbol{n}_0^{\mathrm{S}}+\mathrm{bnd}
\end{aligned} \tag{8.78}$$

The kinematic condition (8.28_3) and the dynamic condition (8.31) in combination with (8.29) and (8.30) in turn convey that:

$$\begin{aligned}
&\frac{d}{dt}\Big(\int_{\Omega_0^{\mathrm{S}}}\frac{1}{2}\rho_{\mathrm{S}}|\partial_t\boldsymbol{\psi}|^2+\mathscr{W}(\boldsymbol{\psi}|_{\Omega_0^{\mathrm{S}}})\Big)\\
&\qquad=\int_{\Gamma_t}\boldsymbol{u}\cdot(p\boldsymbol{n}-\boldsymbol{\tau}_n-\boldsymbol{\zeta}_n)+\int_{\Gamma_t}\sigma_{\mathrm{sf}}\,\kappa\,\boldsymbol{u}\cdot\boldsymbol{n}+\int_{\Gamma_t}\boldsymbol{u}\cdot\,\mathrm{grad}_\Gamma\,\sigma_{\mathrm{sf}}+\mathrm{bnd}
\end{aligned} \tag{8.79}$$

We next identify the cancelations that occur among the various terms in the expansions of the energy contributions. The first term in the right member of (8.79) cancels versus the second term in the ultimate expression in (8.77). By virtue of $v_n=u_n=\boldsymbol{u}\cdot\boldsymbol{n}$, the second term in the right member of (8.79) cancels versus the second term in (8.75). The third term in (8.79) counterbalances the fifth term in the right-hand side of (8.71). Because $v_n=u_n$, the first term in the ultimate expression in (8.77) compensates the third term in (8.76). The fourth and fifth term in (8.77) cancel versus the second and third term in (8.71), respectively. What remains of the kinetic-energy contribution is the viscous term $-\int_{\Omega_t^{\mathrm{F}}}\mathrm{grad}\,\boldsymbol{u}:\boldsymbol{\tau}$. The first term in (8.75) counterbalances the fourth and sixth term in (8.71). The ultimate term in (8.75) will be ignored under the standing assumption that appropriate boundary conditions hold at $\partial\Gamma_t$. Finally, considering the contribution of the Ginzburg–Landau energy functional, we note that all terms in the right-hand side of (8.71) have been accounted for, except for the dissipation term $-\int_{\Omega_t^{\mathrm{F}}} m|\operatorname{grad}\mu|^2$. In conclusion, we have

$$d_t\mathscr{E}(\boldsymbol{\psi}(t),\boldsymbol{u}(t),\phi(t)) = -\int_{\Omega_t^{\mathrm{F}}} \operatorname{grad}\boldsymbol{u} : \boldsymbol{\tau} - \int_{\Omega_t^{\mathrm{F}}} m|\operatorname{grad}\mu|^2 + \text{bnd}\,. \tag{8.80}$$

To prove the inequality (8.63), it remains to show that $\int_{\Omega_t^{\mathrm{F}}} \operatorname{grad}\boldsymbol{u} : \boldsymbol{\tau}$ is non-negative. Recalling the definition of the viscous stress tensor in (8.8), we infer:

$$\begin{aligned}\int_{\Omega_t^{\mathrm{F}}} \operatorname{grad}\boldsymbol{u} : \boldsymbol{\tau} &= \int_{\Omega_t^{\mathrm{F}}} \eta \operatorname{grad}\boldsymbol{u} : \Big(2\boldsymbol{\varepsilon}(\boldsymbol{u}) + \Lambda\,(\operatorname{tr}\boldsymbol{\varepsilon}(\boldsymbol{u}))\boldsymbol{I}\Big)\\ &= \int_{\Omega_t^{\mathrm{F}}} \eta(\boldsymbol{\varepsilon}(\boldsymbol{u}) : \boldsymbol{\varepsilon}(\boldsymbol{u}) + \Lambda|\operatorname{tr}\boldsymbol{\varepsilon}(\boldsymbol{u})|^2)\end{aligned} \tag{8.81}$$

The term in parenthesis in the final expression in (8.81) is non-negative if and only if Λ complies with:

$$2(\varepsilon_{11}^2 + \cdots + \varepsilon_{dd}^2) + \Lambda(\varepsilon_{11} + \cdots + \varepsilon_{dd})^2 \geq 0 \tag{8.82}$$

for all $\boldsymbol{\varepsilon}$. Following a similar argumentation as in [59, §7.3], it can be shown that (8.82) holds under the standing assumption $d\Lambda + 2 \geq 0$; see Section 8.2.1.

8.4.2 Equilibrium solutions

An equilibrium solution of the aggregated binary-fluid–solid interaction problem (8.60) is any tuple $(\boldsymbol{\psi}_{\mathrm{E}}, \boldsymbol{u}_{\mathrm{E}}, p_{\mathrm{E}}, \phi_{\mathrm{E}}, \mu_{\mathrm{E}}) \in \mathbb{U}$ such that

$$d_t\mathscr{E}(\boldsymbol{\psi}_{\mathrm{E}}(t), \boldsymbol{u}_{\mathrm{E}}(t), \phi_{\mathrm{E}}(t)) = 0 \tag{8.83}$$

and that complies with the auxiliary boundary conditions. We assume that the auxiliary boundary conditions impose $\boldsymbol{u} = 0$ on some non-negligible part of the boundary $\partial\Omega_t^{\mathrm{F}}$. From the dissipation inequality (8.63), one can deduce directly that $\mu_{\mathrm{E}} = \mathscr{C}$. In addition, by minimizing (8.82) with respect to $(\varepsilon_{11}, \ldots, \varepsilon_{dd})$, we find that equality holds in (8.82) if and only if

$$\varepsilon_{11} = \varepsilon_{22} = \cdots = \varepsilon_{dd} \tag{8.84}$$

and

$$(d\Lambda + 2)\,\varepsilon_{11}^2 = 0\,. \tag{8.85}$$

For $d\Lambda + 2 > 0$, it follows directly from (8.85) that $\varepsilon_{11} = 0$ and one can infer from the dissipation inequality (8.63) that $\boldsymbol{u}_{\mathrm{E}} = \mathscr{C}$. The aforementioned assumption on $\boldsymbol{u}$ then leads to the conclusion $\boldsymbol{u}_{\mathrm{E}} = 0$. For the limit case $\Lambda = -2/d$, Equation (8.85) does not allow the inference $\varepsilon_{11} = 0$. However, the identities in (8.84) in combination with $\mu_{\mathrm{E}} = \mathscr{C}$ and (8.5$_4$) convey

$$\operatorname{div} \boldsymbol{u}_{\mathrm{E}} - \alpha \operatorname{div}(\operatorname{grad} \mu_{\mathrm{E}}) = \operatorname{tr} \boldsymbol{\varepsilon}(\boldsymbol{u}_{\mathrm{E}}) - 0 = d\, \varepsilon_{11}(\boldsymbol{u}_{\mathrm{E}}) = 0 \,. \tag{8.86}$$

Hence, it holds that $\varepsilon_{11}(\boldsymbol{u}_{\mathrm{E}}) = 0$ and the dissipation inequality (8.63) again implies that $\boldsymbol{u}_{\mathrm{E}} = 0$.

From $\boldsymbol{u}_{\mathrm{E}} = 0$ and $\mu_{\mathrm{E}} = \mathscr{C}$ in combination with (8.5$_2$) we deduce that $\partial_t \phi_{\mathrm{E}} = 0$. The equilibrium condition for the mixture velocity implies that $\partial_t \boldsymbol{u}_{\mathrm{E}} = 0$ and, therefore,

$$\partial_t(\rho_{\mathrm{E}} \boldsymbol{u}_{\mathrm{E}}) = \boldsymbol{u}_{\mathrm{E}} d_\phi \rho\, \partial_t \phi_{\mathrm{E}} + \rho_{\mathrm{E}} \partial_t \boldsymbol{u}_{\mathrm{E}} = 0 \,. \tag{8.87}$$

Hence, for any equilibrium solution $(\boldsymbol{\psi}_{\mathrm{E}}, \boldsymbol{u}_{\mathrm{E}}, p_{\mathrm{E}}, \phi_{\mathrm{E}}, \mu_{\mathrm{E}})$ of the binary-fluid–solid-interaction problem, the binary-fluid tuple $(\boldsymbol{u}_{\mathrm{E}}, p_{\mathrm{E}}, \phi_{\mathrm{E}}, \mu_{\mathrm{E}})$ corresponds to a stationary solution of the NSCH equations (8.5). A similar deduction cannot be made for the solid subsystem in (8.60) on account of the absence of a dissipative mechanism in the solid.

8.4.3 Equilibrium fluid–fluid interface profile

The derivation in Section 8.4.2 conveys that equilibrium solutions of the binary-fluid–solid interaction problem satisfy the stationary NSCH equations (8.5) with vanishing mixture velocity and constant chemical potential:

$$\left.\begin{aligned} \operatorname{grad} p_{\mathrm{E}} - \operatorname{div} \boldsymbol{\zeta}_{\mathrm{E}} &= 0 \\ \mu_{\mathrm{E}} + \sigma\Big(\epsilon \Delta \phi_{\mathrm{E}} - \frac{1}{\epsilon} \Psi'(\phi_{\mathrm{E}})\Big) + \frac{p_{\mathrm{E}}}{\rho(\phi_{\mathrm{E}})} \frac{d\rho}{d\phi} &= 0 \end{aligned}\right\} \quad \text{in } \Omega^{\mathrm{F}}_{\mathrm{E}} \tag{8.88}$$

To elucidate the manner in which the fluid–fluid interface is represented in the Navier–Stokes–Cahn–Hilliard model (8.5), we consider the particular case of a planar fluid–fluid interface. Denoting by s a coordinate normal to the interface and centered at the interface, and expanding the capillary-stress tensor in accordance with its definition in (8.10), equations (8.88) reduce to:

$$\left.\begin{aligned} \frac{dp_{\mathrm{E}}}{ds} - \frac{d}{ds}\Big(- \frac{\epsilon\sigma}{2}\Big|\frac{d\phi_{\mathrm{E}}}{ds}\Big|^2 + \frac{\sigma}{\epsilon}\Psi(\phi_{\mathrm{E}}) - \mu_{\mathrm{E}}\phi_{\mathrm{E}}\Big) &= 0 \\ \mu_{\mathrm{E}} + \epsilon\sigma \frac{d^2\phi_{\mathrm{E}}}{ds^2} - \frac{\sigma}{\epsilon}\Psi'(\phi_{\mathrm{E}}) + \frac{p_{\mathrm{E}}}{\rho(\phi_{\mathrm{E}})}\frac{d\rho}{d\phi} &= 0 \end{aligned}\right\} \quad \text{in } \Omega^{\mathrm{F}}_{\mathrm{E}} \tag{8.89}$$

We insist that the mixture reduces to the pure species and exhibits a uniform pressure away from the interface:

$$\lim_{s\to\pm\infty} \phi_{\mathrm{E}}(s) = \pm 1 \qquad \lim_{s\to\pm\infty} (dp_{\mathrm{E}} - \operatorname{tr} \boldsymbol{\zeta}_{\mathrm{E}})(s) = dp_\infty \tag{8.90}$$

Note that the *pressure* in the mixture in fact corresponds to $1/d$ times the trace of the compound stress tensor, which for vanishing velocity comprises $-\operatorname{tr}\boldsymbol{\zeta}_{\mathrm{E}}$ in addition to dp_{E}; see Remark 8.2 on page 288.

Recalling that μ_{E} is constant, it follows from (8.89) and the chain rule that

$$\frac{dp_{\mathrm{E}}}{ds} = \frac{d\phi_{\mathrm{E}}}{ds}\left(-\sigma\epsilon\frac{d^2\phi_{\mathrm{E}}}{ds^2} + \frac{\sigma}{\epsilon}\Psi'(\phi_{\mathrm{E}}) - \mu_{\mathrm{E}}\right) = \frac{d\phi_{\mathrm{E}}}{ds}\left(\frac{p_{\mathrm{E}}}{\rho(\phi_{\mathrm{E}})}\frac{d\rho}{d\phi}\right). \tag{8.91}$$

Equation (8.91) can be solved via separation of variables to obtain the general solution $p_{\mathrm{E}} = \mathscr{C}\rho(\phi_{\mathrm{E}})$. It can then be verified by substitution that

$$\phi_{\mathrm{E}} = \tanh\left(\frac{s}{\sqrt{2}\epsilon}\right) \qquad p_{\mathrm{E}} = p_\infty + p_\infty\phi_{\mathrm{E}}\left(\frac{\rho_{\mathrm{L}}-\rho_{\mathrm{A}}}{\rho_{\mathrm{L}}+\rho_{\mathrm{A}}}\right) \qquad \mu_{\mathrm{E}} = -p_\infty\left(\frac{\rho_{\mathrm{L}}-\rho_{\mathrm{A}}}{\rho_{\mathrm{L}}+\rho_{\mathrm{A}}}\right) \tag{8.92}$$

solves (8.89)–(8.90). It is to be noted that, in particular, in equilibrium the order parameter ϕ_{E} assumes the typical tanh-form; see also [36, 37].

Considering the traction $p_{\mathrm{E}}\boldsymbol{n} - \boldsymbol{\zeta}_{\mathrm{E}}\boldsymbol{n}$ that the binary fluid exerts in equilibrium on a surface normal to the fluid–fluid interface, and noting that $\partial_n\phi$ vanishes, we infer from (8.10) and (8.92) that:

$$p_{\mathrm{E}}\boldsymbol{n} - \boldsymbol{\zeta}_{\mathrm{E}}\boldsymbol{n} = \left(p_\infty - \frac{\sigma}{2\epsilon}\operatorname{sech}^4\left(\frac{s}{\sqrt{2}\epsilon}\right)\right)\boldsymbol{n} \tag{8.93}$$

with sech(·) the hyperbolic-secant function. The first term in parentheses in the right member of (8.93) represents the ambient pressure level. The second term represents the fluid–fluid surface tension. The second term constitutes a strictly negative function that is essentially localized in a neighborhood of the fluid–fluid interface, and that collapses onto the interface in the sharp-interface limit $\epsilon \to +0$; see Figure 8.2 and Remark 8.11. The integral

$$-\sigma\int_{-\infty}^{\infty}\left(\frac{1}{2\epsilon}\right)\operatorname{sech}^4\left(\frac{s}{\sqrt{2}\epsilon}\right)ds = -\sigma\frac{2\sqrt{2}}{3} = -\sigma_{\mathrm{LA}} \tag{8.94}$$

represents the magnitude of the force that is exerted by the fluid–fluid surface tension on the surface per unit length. Equation (8.94) confirms that the force per unit length indeed coincides with the surface tension σ_{LA}. Alternatively, the fluid–fluid surface tension can be conceived of as the increase in the free energy that accompanies an increase in surface area of the meniscus [18]. Accordingly, the surface tension associated with the equilibrium solution (8.92) of the order parameter can be derived as:

$$\int_{-\infty}^{+\infty}\left(\frac{\epsilon\sigma}{2}\left|\frac{d\phi_{\mathrm{E}}}{ds}\right|^2 + \frac{\sigma}{\epsilon}\Psi(\phi_{\mathrm{E}})\right)ds = \sigma\int_{-\infty}^{\infty}\left(\frac{1}{2\epsilon}\right)\operatorname{sech}^4\left(\frac{s}{\sqrt{2}\epsilon}\right)ds = \frac{2\sqrt{2}}{3}\sigma = \sigma_{\mathrm{LA}} \tag{8.95}$$

cf. Equation (8.62). Note that the second expression in (8.95) coincides with (8.94) modulo its sign.

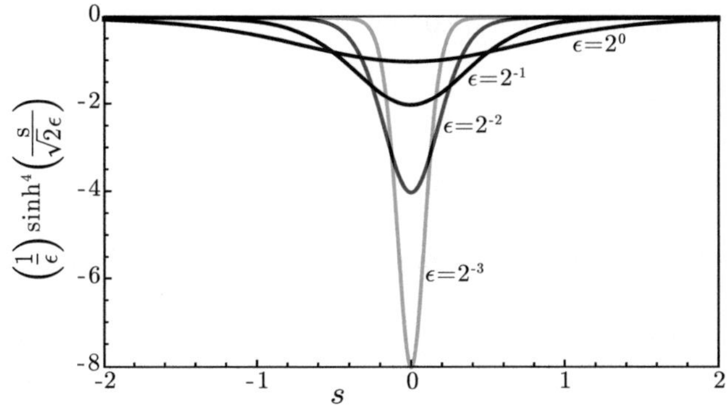

Fig. 8.2: Illustration of the traction on a surface normal to the fluid–fluid interface in accordance with (8.93) for $\epsilon \in 2^{\{-3,-2,-1,0\}}$.

Remark 8.11. In the sharp interface limit $\epsilon \to +0$, the integrands in (8.94) and (8.95) converge to a constant multiple of the Dirac delta distribution. In particular, defining

$$\delta_\epsilon(s) = \left(\frac{3}{4\sqrt{2}\epsilon}\right)\operatorname{sech}^4\left(\frac{s}{\sqrt{2}\epsilon}\right) \tag{8.96}$$

it holds that

$$\lim_{\epsilon\to+0}\int_{-\infty}^{\infty} f(s)\,\delta_\epsilon(s)\,ds = f(0) \qquad \forall f \in C_0^\infty(\mathbb{R})\,. \tag{8.97}$$

To establish (8.97), we first note that

$$\lim_{\epsilon\to+0}\int_{-\sqrt{\epsilon}}^{\sqrt{\epsilon}} \delta_\epsilon(s)\,ds = \lim_{\epsilon\to+0}\left(\frac{3}{4}\sinh\left(\frac{1}{\sqrt{2\epsilon}}\right) + \frac{1}{4}\sinh\left(\frac{3}{\sqrt{2\epsilon}}\right)\right)\operatorname{sech}^3\left(\frac{1}{\sqrt{2\epsilon}}\right) = 1\,. \tag{8.98}$$

Hence, in the limit $\epsilon \to +0$ the entire mass of δ_ϵ is confined to the interval $(-\sqrt{\epsilon}, \sqrt{\epsilon})$. By virtue of their regularity, functions in $C_0^\infty(\mathbb{R})$ can be represented by a Maclaurin series. Considering terms of order $k \geq 1$ in this series, it holds that

$$\left|\int_{-\sqrt{\epsilon}}^{\sqrt{\epsilon}} s^k\,\delta_\epsilon(s)\,ds\right| \leq |\epsilon|^{k/2}\int_{-\sqrt{\epsilon}}^{\sqrt{\epsilon}} \delta_\epsilon(s)\,ds \leq |\epsilon|^{k/2}\,. \tag{8.99}$$

The contributions of terms of order $k \geq 1$ in the Maclaurin series to the left member of (8.97) therefore vanish in the limit $\epsilon \to +0$ and only the constant term remains.

To establish that the equilibrium solution of the order parameter in (8.92) is compatible with the wetting boundary condition (8.18) with solid–fluid surface energy (8.19), we consider the intersection of the fluid–fluid interface with a flat surface at the static contact angle according to (8.20). Denoting the unit normal vector to the surface interior to the fluid by $\boldsymbol{n}^{\mathrm{S}}$ and the normal to the fluid–fluid meniscus exterior to the liquid by $\boldsymbol{n}^{\mathrm{I}}$ (illustrated in Figure 8.3 on page 310), one can infer that $\partial_n\phi_{\mathrm{E}} = \boldsymbol{n}^{\mathrm{S}}\cdot\boldsymbol{n}^{\mathrm{I}}\phi_{\mathrm{E}}'$. Inverting the relation between $\phi_{\rm E}$ and s in (8.92), we obtain the identities:

$$\epsilon\partial_n\phi = \boldsymbol{n}^{\rm S}\cdot\boldsymbol{n}^{\rm I}\,\epsilon\phi_{\rm E}' = \frac{\boldsymbol{n}^{\rm S}\cdot\boldsymbol{n}^{\rm I}}{\sqrt{2}}\operatorname{sech}^2\Big(\frac{s}{\sqrt{2}\epsilon}\Big)$$
$$= \frac{\boldsymbol{n}^{\rm S}\cdot\boldsymbol{n}^{\rm I}}{\sqrt{2}}\operatorname{sech}^2(\operatorname{arctanh}(\phi_{\rm E})) = \frac{\boldsymbol{n}^{\rm S}\cdot\boldsymbol{n}^{\rm I}}{\sqrt{2}}(1-\phi_{\rm E}^2) \quad (8.100)$$

Therefore, the equilibrium solution $\phi_{\rm E}$ in (8.92) is compatible with (8.18) and (8.19) if

$$\sigma\frac{\boldsymbol{n}^{\rm S}\cdot\boldsymbol{n}^{\rm I}}{\sqrt{2}}(1-\phi^2) - \frac{3(\sigma_{\rm SA}-\sigma_{\rm SL})}{4}(1-\phi^2) = 0 \quad (8.101)$$

or, equivalently,

$$\boldsymbol{n}^{\rm S}\cdot\boldsymbol{n}^{\rm I} = \frac{3(\sigma_{\rm SA}-\sigma_{\rm SL})}{2\sqrt{2}\,\sigma} = \frac{\sigma_{\rm SA}-\sigma_{\rm SL}}{\sigma_{\rm LA}} = \cos(\theta)\,. \quad (8.102)$$

Noting that $\boldsymbol{n}^{\rm S}\cdot\boldsymbol{n}^{\rm I}$ corresponds to the cosine of the angle between the meniscus and the surface, interior to the liquid, Equation (8.102) corroborates the compatibility of $\phi_{\rm E}$ according to (8.92) and the wetting condition (8.18) with solid–fluid surface energy (8.19).

Remark 8.12. We have restricted ourselves here to a consideration of a flat fluid–fluid interface. An investigation of curved diffuse interfaces in the context of the Cahn–Hilliard equation separately with a degenerate mobility, as well of the corresponding sharp-interface limits, is presented in [41].

To elucidate the behavior of the contact-line-force term $\operatorname{grad}_\Gamma \sigma_{\rm SF}$ in (8.29), we consider the sharp-interface limit of this term in the setting of an intersection between essentially flat fluid–fluid and solid–fluid interfaces at the static contact angle θ according to (8.20). The analysis above conveys that the order parameter exhibits the equilibrium profile in (8.92). Let $\boldsymbol{v}^{\rm S}$ denote the tangent vector to the solid–fluid interface normal to the contact line and interior with respect to the liquid component; see Figure 8.3. Invoking the chain rule, we obtain:

$$\operatorname{grad}_\Gamma \sigma_{\rm SF}(\boldsymbol{v}^{\rm S}x) = \boldsymbol{v}^{\rm S}\frac{\sigma_{\rm SF}'(\phi_{\rm E})\phi_{\rm E}'(x\cos\theta)}{\cos\theta} \quad (8.103)$$

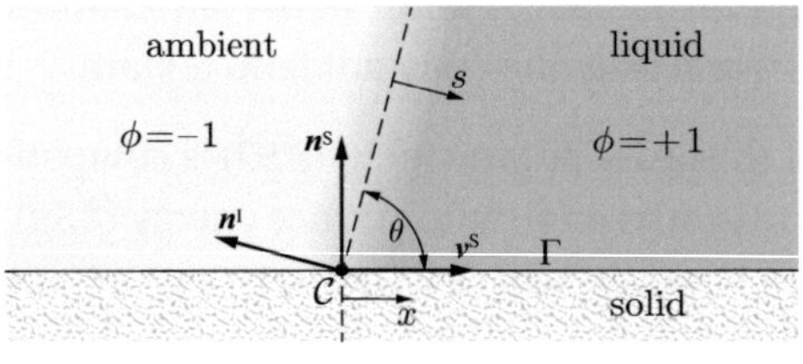

Fig. 8.3: Illustration of the contact line in the diffuse-interface model. The contact line $\mathcal{C}$ is the intersection of the solid–fluid interface Γ and the zero level-set of the phase field. The diffuse fluid–fluid interface intersects the solid–fluid interface at the static contact angle θ according to (8.20).

with x an arc-length coordinate along $\boldsymbol{v}^{\mathrm{S}}$. Introducing the expression for the equilibrium profile in (8.92) and expanding the right member of (8.103), it follows that:

$$\sigma'_{\mathrm{SF}}(\phi_{\mathrm{E}})\phi'_{\mathrm{E}}(s) = \left(\frac{3}{4\sqrt{2}\epsilon}\right)(\sigma_{\mathrm{SL}} - \sigma_{\mathrm{SA}})\,\mathrm{sech}^4\left(\frac{s}{\sqrt{2}\epsilon}\right) = (\sigma_{\mathrm{SL}} - \sigma_{\mathrm{SA}})\delta_\epsilon(s) \tag{8.104}$$

with δ_ϵ according to (8.97). The line measure dx in the solid–fluid interface is related to the line measure ds by $dx\ \cos\theta = ds$. The analysis in Remark 8.11 then conveys that in the sharp-interface limit, $\delta_\epsilon(x\cos\theta)/\cos\theta$ converges to a Dirac distribution supported on the contact line with respect to the line measure in the solid–fluid interface. In particular, denoting the solid–fluid interface by Γ and the contact line by $\mathcal{C}$, it holds that

$$\lim_{\epsilon\to+0}\int_\Gamma \boldsymbol{f}\cdot\mathrm{grad}_\Gamma\,\sigma_{\mathrm{SF}} = (\sigma_{\mathrm{SL}} - \sigma_{\mathrm{SA}})\int_{\mathcal{C}} \boldsymbol{f}\cdot\boldsymbol{v}^{\mathrm{S}} \qquad \forall \boldsymbol{f} \in \boldsymbol{C}_0^\infty(\Gamma)\,. \tag{8.105}$$

Hence, the term $\mathrm{grad}_\Gamma\,\sigma_{\mathrm{SF}}$ can indeed be conceived of as a line traction proportional to the surface-energy difference $\sigma_{\mathrm{SL}} - \sigma_{\mathrm{SA}}$ of the liquid and the ambient fluid, and which is concentrated on the contact line and acts in the direction tangential to the solid–fluid boundary Γ and normal to the contact line. From a variational vantage point, this line traction represents the variation in the energy due to a shift in the position of the contact line.

8.5 Numerical approximation methods

8.5.1 Spatial and temporal discretization

We consider discrete approximations of (8.61) based on finite-element discretization in the spatial dependence. We restrict ourselves to three-dimensional configurations with rotational symmetry, and we exploit this symmetry in the finite-element approximation; see Figure 8.4 for an illustration. The temporal dependence is discretized by means of an implicit backward Euler approximation, in conjunction with a standard decomposition of the double-well potential into contractive and expansive components [21, 23, 24, 55, 66] to enhance stability.

To furnish a setting for the finite-element discretization of (8.61) in cylindrical coordinates, we first introduce the transformation $\Phi\colon \tilde{\Omega}_0\times(0,2\pi] \to \Omega_0$ from cylindrical coordinates to Cartesian coordinates according to:

$$\Phi\colon (y_1, y_2, y_3) \mapsto (y_1\cos y_3, y_2, y_1\sin y_3) \tag{8.106}$$

The domain $\tilde{\Omega}_0 \subset \mathbb{R}^2$ corresponds to the intersection of Ω_0 with the radial–longitudinal half plane $\{y_3 = 0, y_1 \geq 0\}$. Similarly, $\tilde{\Omega}_0^{\mathrm{F}}$ and $\tilde{\Omega}_0^{\mathrm{S}}$ indicate the intersections of the fluid and solid reference domains with the radial–longitudinal half plane, respectively. We denote by $\mathcal{S}^h$ a finite-element mesh on $\tilde{\Omega}_0$ characterized by a mesh parameter $h > 0$, viz. a cover of the reference domain $\tilde{\Omega}_0$ by nonoverlapping open element domains $E_{\mathrm{Y}} \subset \tilde{\Omega}_0$; see Figure 8.4. We impose on $\mathcal{S}^h$ the standard conditions of regularity

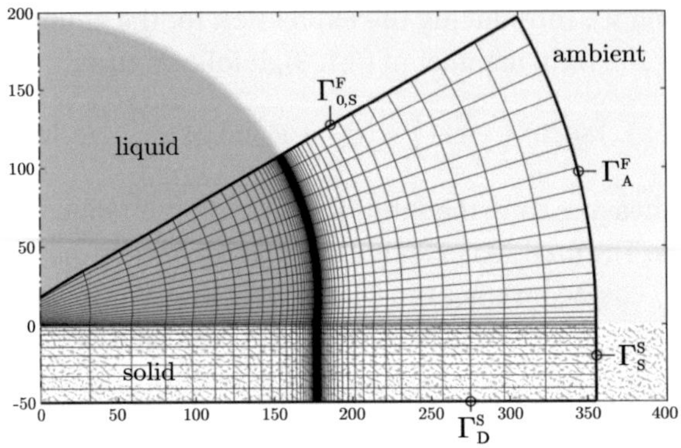

Fig. 8.4: Illustration of the considered rotationally symmetric liquid–ambient–solid configuration and of the corresponding finite-element mesh (units: μm). The partitioning of the boundary $\partial\Omega$ serves to specify the boundary conditions in Section 8.6.

and shape-regularity; see, for instance, [22] for further details. In addition, we insist that the considered mesh conforms to the fluid and solid domains, i.e., each element is exclusively contained in $\tilde{\Omega}^{\mathrm{F}}_0$ or $\tilde{\Omega}^{\mathrm{S}}_0$. To each element E_{Y} we associate the cylinder $e = E_{\mathrm{Y}} \times (0, 2\pi]$. We introduce the following finite-element approximation spaces, subordinate to the mesh $\mathcal{S}^h$:

$$\begin{aligned}
\boldsymbol{W}^h(\Omega_0) &= \{\boldsymbol{\psi} \in \boldsymbol{C}^0(\Omega_0) \colon (\boldsymbol{\psi} \circ \Phi)|_e \in \boldsymbol{\mathcal{P}}^2(E_{\mathrm{Y}}),\ E_{\mathrm{Y}} \in \mathcal{S}^h\}\\
\boldsymbol{U}^h(\Omega^{\mathrm{F}}_0) &= \{\boldsymbol{u} \in \boldsymbol{C}^0(\Omega^{\mathrm{F}}_0) \colon (\boldsymbol{u} \circ \Phi)|_e \in \boldsymbol{\mathcal{P}}^2(E_{\mathrm{Y}}),\ E_{\mathrm{Y}} \in \mathcal{S}^h\}\\
P^h(\Omega^{\mathrm{F}}_0) &= \{p \in C^0(\Omega^{\mathrm{F}}_0) \colon (p \circ \Phi)|_e \in \mathcal{P}^1(E_{\mathrm{Y}}),\ E_{\mathrm{Y}} \in \mathcal{S}^h\}\\
V^h(\Omega^{\mathrm{F}}_0) &= \{v \in C^1(\Omega^{\mathrm{F}}_0) \colon (v \circ \Phi)|_e \in \mathcal{P}^2(E_{\mathrm{Y}}),\ E_{\mathrm{Y}} \in \mathcal{S}^h\}
\end{aligned} \tag{8.107}$$

where $\mathcal{P}^k(E_{\mathrm{Y}})$ and $\boldsymbol{\mathcal{P}}^k(E_{\mathrm{Y}})$ denote the classes of bivariate polynomials of degree k from E_{Y} into $\mathbb{R}$ and $\mathbb{R}^d$, respectively. Note that the finite-element approximation functions in (8.107) depend on the radial and longitudinal coordinates only and, hence, they are constant in the azimuthal direction.

To discretize (8.61) in the temporal dependence, we regard a natural number N and we introduce the index set $\mathbb{J} := \{j \in \mathbb{N} : j \leq N\}$ and a uniform partition $\tau\mathbb{J} + (-\tau, 0)$ of the time interval $[0, t_{\mathrm{FIN}}]$ into N subintervals of length $\tau = t_{\mathrm{FIN}}/N$. Given a vector space V and the index set $\mathbb{J}$, we denote by $L(\mathbb{J}; V)$ the linear space of functions from $\mathbb{J}$ into V. We consider the following ambient trial space for the discrete approximation of (8.61):

$$\begin{aligned}
\mathbb{U}^{h,\tau} := \{(\boldsymbol{\psi}^{h,\tau}, \boldsymbol{u}^{h,\tau}, p^{h,\tau}, \phi^{h,\tau}, \mu^{h,\tau}) \colon &\boldsymbol{\psi}^{h,\tau} \in \mathrm{Id} + L(\mathbb{J}; \boldsymbol{W}^h(\Omega_0)),\\
&\boldsymbol{u}^{h,\tau} \circ T_n \in L(\mathbb{J}; \boldsymbol{U}^h(\Omega^{\mathrm{F}}_0)), p^{h,\tau} \circ T_n \in L(\mathbb{J}; P^h(\Omega^{\mathrm{F}}_0)),\\
&\phi^{h,\tau} \circ T_n \in L(\mathbb{J}; V^h(\Omega^{\mathrm{F}}_0)), \mu^{h,\tau} \circ T_n \in L(\mathbb{J}; V^h(\Omega^{\mathrm{F}}_0)),\\
&(\boldsymbol{u}^{h,\tau}_n \circ T_n)|_{\Gamma_0} = (\boldsymbol{\psi}^{h,\tau}_n|_{\Gamma_0} - \boldsymbol{\psi}^{h,\tau}_{n-1}|_{\Gamma_0})/\tau \ \text{ for all } \ n \in \mathbb{J}\}
\end{aligned} \tag{8.108}$$

with $T_n\colon \Omega_0 \to \Omega_n$ the transformation corresponding to the deformation $\boldsymbol{\psi}_n^{h,\tau}$, i.e., $T_n(\cdot) = \boldsymbol{\psi}_n^{h,\tau}(\cdot)$; cf. (8.58). The initial configuration $\boldsymbol{\psi}_0^{h,\tau}$ that appears in the discrete approximation of the kinematic condition in (8.108) for $n = 1$ is identified with the identity, in accordance with $T_0(\cdot) = (\cdot)$ and the initial condition in (8.25). We consider the following test space for the discrete approximation of (8.61):

$$\begin{aligned}\mathbb{V}^h := \{(\boldsymbol{\theta}^h, \boldsymbol{\vartheta}^h, q^h, \beta^h, \chi^h)\colon \boldsymbol{\theta}^h \in \boldsymbol{W}^h(\Omega_0), \boldsymbol{\vartheta}^h \in \boldsymbol{U}^h_{0,\partial\Omega_0^{\mathrm{F}}}(\Omega_0^{\mathrm{F}}),\\ q^h \in P^h(\Omega_0^{\mathrm{F}}), \beta^h \in V^h(\Omega_0^{\mathrm{F}}), \chi^h \in V^h(\Omega_0^{\mathrm{F}})\}\end{aligned} \tag{8.109}$$

with $\boldsymbol{U}^h_{0,\partial\Omega_0^{\mathrm{F}}}(\Omega_0^{\mathrm{F}}) = \boldsymbol{U}^h(\Omega_0^{\mathrm{F}}) \cap \boldsymbol{H}^1_{0,\partial\Omega_0^{\mathrm{F}}}(\Omega_0^{\mathrm{F}})$ the subspace of $\boldsymbol{U}^h(\Omega_0^{\mathrm{F}})$ of functions that vanish on the boundary.

The approximation of the aggregated FSI problem can then be condensed into:

$$\begin{aligned}&\text{Find } (\boldsymbol{\psi}^{h,\tau}, \boldsymbol{u}^{h,\tau}, p^{h,\tau}, \phi^{h,\tau}, \mu^{h,\tau}) \in \mathbb{U}^{h,\tau} \text{ such that}\\ &\mathscr{A}_n^{h,\tau}(\boldsymbol{\psi}^{h,\tau}, \boldsymbol{u}^{h,\tau}, p^{h,\tau}, \phi^{h,\tau}, \mu^{h,\tau}; \boldsymbol{\theta}^h, \boldsymbol{\vartheta}^h, q^h, \beta^h, \chi^h) = \mathscr{L}_n^{h,\tau}(\boldsymbol{\theta}^h, \boldsymbol{\vartheta}^h, q^h, \beta^h, \chi^h)\\ &\qquad\qquad \forall(\boldsymbol{\theta}^h, \boldsymbol{\vartheta}^h, q^h, \beta^h, \chi^h) \in \mathbb{V}^h,\ \forall n \in \mathbb{J}\end{aligned} \tag{8.110}$$

with $\{\mathscr{A}_n^{h,\tau}\}_{n\in\mathbb{J}}\colon \mathbb{U}^{h,\tau} \times \mathbb{V}^h \to \mathbb{R}$ a sequence of semilinear forms according to:

$$\begin{aligned}&\mathscr{A}_n^{h,\tau}(\boldsymbol{\psi}^{h,\tau}, \boldsymbol{u}^{h,\tau}, p^{h,\tau}, \phi^{h,\tau}, \mu^{h,\tau}; \boldsymbol{\theta}^h, \boldsymbol{\vartheta}^h, q^h, \beta^h, \chi^h)\\ &:= \frac{1}{\tau}\int_{\Omega_0^{\mathrm{S}}} \rho_{\mathrm{S}} \boldsymbol{v}_n^{h,\tau} \cdot \boldsymbol{\theta}^h + \mathscr{W}'(\boldsymbol{\psi}_n^{h,\tau}|_{\Omega_0^{\mathrm{S}}}; \boldsymbol{\theta}^h)\\ &+ \int_{\Gamma_n} \sigma_{\mathrm{SF}}(\phi_n^{h,\tau}) \operatorname{grad}_\Gamma \mathrm{Id}_{\Gamma_n} : \operatorname{grad}_\Gamma (\boldsymbol{\theta}^h \circ T_n^{-1}) + \int_{\Omega_0^{\mathrm{F}}} \operatorname{grad} \boldsymbol{\psi}_n^{h,\tau} : \operatorname{grad} \boldsymbol{\vartheta}^h\\ &+ \frac{1}{\tau}\int_{\Omega_n^{\mathrm{F}}} \rho_n^{h,\tau} \boldsymbol{u}_n^{h,\tau} \cdot (\boldsymbol{\theta}^h \circ T_n^{-1}) + \int_{\Omega_n^{\mathrm{F}}} (\boldsymbol{\tau}_n^{h,\tau} + \boldsymbol{\zeta}_n^{h,\tau}) : \operatorname{grad}(\boldsymbol{\theta}^h \circ T_n^{-1})\\ &- \int_{\Omega_n^{\mathrm{F}}} (\rho \boldsymbol{u}_n^{h,\tau} \otimes (\boldsymbol{u}_n^{h,\tau} - \boldsymbol{v}_n^{h,\tau})) : \operatorname{grad}(\boldsymbol{\theta}^h \circ T_n^{-1}) - \int_{\Omega_t^{\mathrm{F}}} p_n^{h,\tau} \operatorname{div}(\boldsymbol{\theta}^h \circ T_n^{-1})\\ &+ \frac{1}{\tau}\int_{\Omega_n^{\mathrm{F}}} \phi_n^{h,\tau} (\beta^h \circ T_n^{-1}) + \int_{\Omega_n^{\mathrm{F}}} m \operatorname{grad} \mu_n^{h,\tau} \cdot \operatorname{grad}(\beta^h \circ T_n^{-1})\\ &- \int_{\Omega_n^{\mathrm{F}}} \phi_n^{h,\tau} (\boldsymbol{u}_n^{h,\tau} - \boldsymbol{v}_n^{h,\tau}) \cdot \operatorname{grad}(\beta^h \circ T_n^{-1}) + \int_{\Omega_n^{\mathrm{F}}} \mu_n^{h,\tau} (\chi^h \circ T_n^{-1})\\ &- \int_{\Omega_n^{\mathrm{F}}} \epsilon\sigma \operatorname{grad} \phi_n^{h,\tau} \cdot \operatorname{grad}(\chi^h \circ T_n^{-1}) - \int_{\Omega_n^{\mathrm{F}}} \frac{\sigma}{\epsilon} (\Psi'_{c,n} - \Psi'_{e,n-1})(\chi^h \circ T_n^{-1})\\ &+ \int_{\Omega_n^{\mathrm{F}}} \frac{p_n^{h,\tau}(\rho_{\mathrm{L}} - \rho_{\mathrm{A}})}{2\rho_n^{h,\tau}} (\chi^h \circ T_n^{-1}) + \int_{\Omega_n^{\mathrm{F}}} \operatorname{div} \boldsymbol{u}_n^{h,\tau} (q^h \circ T_n^{-1})\\ &+ \int_{\Omega_n^{\mathrm{F}}} \alpha m \operatorname{grad} \mu_n^{h,\tau} \cdot \operatorname{grad}(q^h \circ T_n^{-1}) - \int_{\Gamma_n} \sigma'_{\mathrm{SF}}(\phi_n^{h,\tau})(\chi^h \circ T_n^{-1})\end{aligned} \tag{8.111}$$

where $\boldsymbol{v}_n^{h,\tau} \circ T_n = (\boldsymbol{\psi}_n^{h,\tau} - \boldsymbol{\psi}_{n-1}^{h,\tau})/\tau$ with $\boldsymbol{\psi}_0^{h,\tau} = \mathrm{Id}$ in accordance with (8.25). The functions Ψ_{c} and Ψ_{e} in (8.111) correspond to contractive and expansive parts of the double-well potential:

$$\Psi_{\mathrm{c}}(\phi) = \phi^2 + \frac{1}{4} \qquad \Psi_{\mathrm{e}}(\phi) = \frac{3}{2}\phi^2 - \frac{1}{4}\phi^4 \tag{8.113}$$

The decomposition $\Psi = \Psi_c - \Psi_e$ of the double-well potential in combination with the implicit (resp. explicit) treatment of Ψ_c (resp. Ψ_e) serves to enhance the stability of the formulation; see also [21, 23, 24, 55, 66]. The linear forms $\{\mathscr{L}_n^{h,\tau}\}_{n\in\mathbb{J}} : \mathbb{V}^h \to \mathbb{R}$ in (8.110) are given by:

$$\mathscr{L}_n^{h,\tau}(\boldsymbol{\theta}^h, \boldsymbol{\vartheta}^h, q^h, \beta^h, \chi^h) = \mathscr{L}(\boldsymbol{\theta}^h, \boldsymbol{\vartheta}^h, q^h, \beta^h, \chi^h) + \frac{1}{\tau}\int_{\Omega_0^S} \rho_S \boldsymbol{v}_{n-1}^{h,\tau} \cdot \boldsymbol{\theta}^h$$
$$+ \frac{1}{\tau}\int_{\Omega_{n-1}^F} \rho_{n-1}^{h,\tau} \boldsymbol{u}_{n-1}^{h,\tau} \cdot (\boldsymbol{\theta}^h \circ T_{n-1}^{-1}) + \frac{1}{\tau}\int_{\Omega_{n-1}^F} \phi_{n-1}^{h,\tau} (\beta^h \circ T_{n-1}^{-1}) \qquad (8.114)$$

with $\boldsymbol{u}_0^{h,\tau} = 0$ and $\phi_0^{h,\tau} = \phi_0$ and $\boldsymbol{v}_0^{h,\tau} = 0$ in accordance with the initial conditions (8.14) and (8.25). The linear form $\mathscr{L}$ corresponds to the exogenous data; cf. Section 8.3.4. To condense the notation in (8.111) and (8.114), we have suppressed the arguments of ρ, $\boldsymbol{\tau}$, $\boldsymbol{\zeta}$, Ψ_c and Ψ_e and appended the affixes of these arguments instead, for instance $\rho_n^{h,\tau} := \rho(\phi_n^{h,\tau})$.

Remark 8.13. For each $n \in \mathbb{J}$, the discrete approximation $\phi_n^{h,\tau} \circ T_n$ of the order parameter is set in the finite-element space $V^h(\Omega_0^F)$. As opposed to functions in the ambient space $V^\phi(\Omega_0^F)$ in (8.54), the range of functions in $V^h(\Omega_0^F)$ is not restricted to the interval $[-1, 1]$. This implies that the discrete approximation of the order parameter can exceed the bounds ± 1. In practice, one observes that the discrete approximation of the order parameter can violate the bounds in the vicinity of the diffuse interface if the interface is not adequately resolved by the finite-element space. Although minor under- or overshoots need not be detrimental, in the case of disparate component densities or viscosities, excessive under- or overshoots can lead to terminal negative mixture density in (8.3) or negative mixture viscosity in (8.9).

8.5.2 Partitioned iterative solution method

The discretized binary-fluid–solid interaction problem in (8.110) corresponds to a sequence of nonlinear algebraic equations for the degrees of freedom associated with the finite-element approximations of the deformation, mixture velocity, pressure, order parameter and chemical potential at time levels $n \in \mathbb{J}$. Various solution procedures have been developed in the context of conventional FSI problems. These solution procedures can generally be categorized into monolithic methods and partitioned methods. In monolithic procedures, the fluid and solid subsystems are solved simultaneously in a fully coupled manner; see, for instance, [6, 28, 35, 46, 62]. Partitioned procedures introduce a decomposition of the aggregated FSI problem into the fluid and solid subsystems. These subsystems are then solved alternatingly until convergence up to a prescribed tolerance is reached. This basic *subiteration* method can be enhanced by means of under-relaxation or modified-mass methods [11, 60, 61], Aitken's method [2, 40], interface-GMRES (generalized minimal residual) and interface-Newton–Krylov methods [44, 45, 47], or multigrid methods [14, 28, 50]; see [11] for an overview. For conventional FSI problems, the stability and efficiency of partitioned solution methods depends on the *added-mass* of the fluid subsystem relative to the mass of the solid subsystem [10, 11, 16, 26].

We opt to solve the binary-fluid–solid interaction problem in (8.110) by means of a basic subiteration procedure with under-relaxation. Our motivation for applying subiteration is twofold. First, the subiteration procedure engenders a separate treatment of the NSCH equations and of the deformation of the fluid and solid domains and hence it significantly simplifies the solution process. Second, the subiteration procedure enables us to investigate the mechanisms that underly the coupling strength in elasto-capillary binary-fluid–solid interaction processes.

To facilitate the presentation of the subiteration process, we introduce the direct-sum decomposition $\boldsymbol{W}^h(\Omega_0) = \boldsymbol{W}^h_{\mathrm{F}}(\Omega_0) \oplus \boldsymbol{W}^h_{\mathrm{S}}(\Omega_0)$ of the finite-element space $\boldsymbol{W}^h(\Omega_0)$ in (8.107), where $\boldsymbol{W}^h_{\mathrm{F}}(\Omega_0)$ contains all finite-element functions of which the support is confined to the fluid reference domain:

$$\boldsymbol{W}^h_{\mathrm{F}}(\Omega_0) = \{\boldsymbol{\theta} \in \boldsymbol{W}^h(\Omega_0) : \ \mathrm{supp}(\boldsymbol{\theta}) \subset \Omega^{\mathrm{F}}_0\} \tag{8.116}$$

It is to be noted that the support of functions in the complementary subspace $\boldsymbol{W}^h_{\mathrm{S}}(\Omega_0)$ is not necessarily restricted to Ω^{S}_0. All functions that have support in Ω^{S}_0 belong to $\boldsymbol{W}^h_{\mathrm{S}}(\Omega_0)$, including functions that are supported across the solid–fluid interface; see Figure 8.5 for an illustration. If the test function $\boldsymbol{\theta}^h$ in (8.110) is restricted to $\boldsymbol{W}^h_{\mathrm{F}}(\Omega_0)$ and $\boldsymbol{\vartheta}^h$ is set to zero, then the terms associated with the solid subproblem and the harmonic extension of the solid deformation onto the fluid domain vanish. Conversely, if $\boldsymbol{\theta}^h$ is restricted to $\boldsymbol{W}^h_{\mathrm{S}}(\Omega_0)$ and $\boldsymbol{\vartheta}^h \in \boldsymbol{U}^h_{0,\partial\Omega^{\mathrm{F}}_0}(\Omega^{\mathrm{F}}_0)$, and q^h, β^h and χ^h are set to zero, then (8.110) reduces to a weak formulation of the solid subproblem with a proper weak formulation of the traction induced by the binary fluid (Section 8.3.2) and the solid–fluid surface tension, and of the harmonic extension of the solid displacement onto the fluid domain. Hence, the partitioning of the test function $\boldsymbol{\theta}^h$ in (8.110) implied by (8.116) yields a decomposition of the aggregated FSI problem into subproblems associated with the solid and binary-fluid.

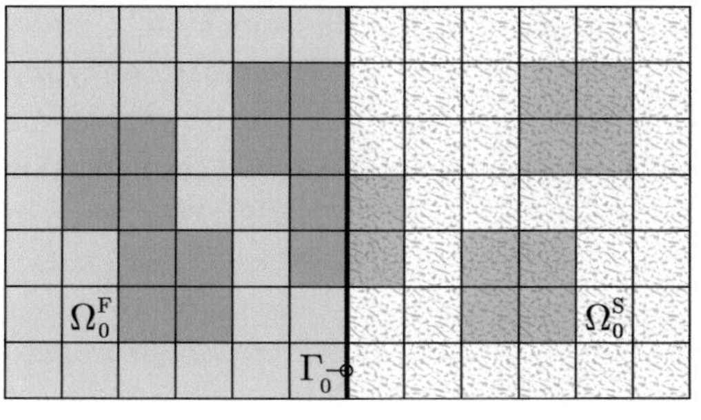

Fig. 8.5: Illustration of the decomposition $\boldsymbol{W}^h(\Omega_0) = \boldsymbol{W}^h_{\mathrm{F}}(\Omega_0) \oplus \boldsymbol{W}^h_{\mathrm{S}}(\Omega_0)$ with $\boldsymbol{W}^h_{\mathrm{F}}(\Omega_0)$ according to (8.116). Finite-element shape functions with support restricted to Ω^{F}_0 (*red*) belong to $\boldsymbol{W}^h_{\mathrm{F}}(\Omega_0)$. Shape functions of which the support intersects Ω^{S}_0 (*green*) belong to $\boldsymbol{W}^h_{\mathrm{S}}(\Omega_0)$.

Algorithm 8.1 Subiteration process with under-relaxation.

1: $\boldsymbol{\psi}_0^{h,\tau} = \mathrm{Id}, \boldsymbol{v}_0^{h,\tau} = 0, \boldsymbol{u}_0^{h,\tau} = 0, \phi_0^{h,\tau} = \phi_0, n = 1$ ▷ initialize time integration
2: **while** $n \le N$ **do**
3: $\boldsymbol{\psi}_{n,0}^{h,\tau} = \boldsymbol{\psi}_{n-1}^{h,\tau}, \boldsymbol{v}_{n,0}^{h,\tau} = 0, k = 0$ ▷ initialize subiteration
4: **repeat**
5: $k \leftarrow k+1$
6: solve ▷ solve NSCH problem

$$(\boldsymbol{u}_{n,k}^{h,\tau}, p_{n,k}^{h,\tau}, \phi_{n,k}^{h,\tau}, \mu_{n,k}^{h,\tau}) \circ T_{n,k-1} \in \boldsymbol{U}^h(\Omega_0^{\mathrm{F}}) \times P^h(\Omega_0^{\mathrm{F}}) \times V^h(\Omega_0^{\mathrm{F}}) \times V^h(\Omega_0^{\mathrm{F}}):$$
$$\mathscr{A}_n^{h,\tau}(\boldsymbol{\psi}_{n,k-1}^{h,\tau}, \boldsymbol{u}_{n,k}^{h,\tau}, p_{n,k}^{h,\tau}, \phi_{n,k}^{h,\tau}, \mu_{n,k}^{h,\tau}; \boldsymbol{\theta}^h, 0, q^h, \beta^h, \chi^h) = \mathscr{L}_n^{h,\tau}(\boldsymbol{\theta}^h, 0, q^h, \beta^h, \chi^h)$$
$$\forall(\boldsymbol{\theta}^h, q^h, \beta^h, \chi^h) \in \boldsymbol{W}_{\mathrm{F}}^h(\Omega_0^{\mathrm{F}}) \times P^h(\Omega_0^{\mathrm{F}}) \times V^h(\Omega_0^{\mathrm{F}}) \times V^h(\Omega_0^{\mathrm{F}})$$
$$\text{subject to } (\boldsymbol{u}_{n,k}^{h,\tau} \circ T_{n,k-1})|_{\Gamma_0} = (\boldsymbol{v}_{n,k-1}^{h,\tau} \circ T_{n,k-1})|_{\Gamma_0}$$

7: solve ▷ solve domain deformation

$$\boldsymbol{\psi}_{n,k}^{h,\tau} \in \mathrm{Id} + \boldsymbol{W}^h(\Omega_0):$$
$$\mathscr{A}_n^{h,\tau}(\boldsymbol{\psi}_{n,k}^{h,\tau}, \boldsymbol{u}_{n,k}^{h,\tau}, p_{n,k}^{h,\tau}, \phi_{n,k}^{h,\tau}, \mu_{n,k}^{h,\tau}; \boldsymbol{\theta}^h, \boldsymbol{\vartheta}^h, 0, 0, 0) = \mathscr{L}_n^{h,\tau}(\boldsymbol{\theta}^h, \boldsymbol{\vartheta}^h, 0, 0, 0)$$
$$\forall(\boldsymbol{\theta}^h, \boldsymbol{\vartheta}^h) \in \boldsymbol{W}_{\mathrm{S}}^h(\Omega_0^{\mathrm{F}}) \times \boldsymbol{U}_{0,\partial\Omega_0^{\mathrm{F}}}^h(\Omega_0^{\mathrm{F}})$$

8: $\boldsymbol{\psi}_{n,k}^{h,\tau} \leftarrow \zeta \boldsymbol{\psi}_{n,k}^{h,\tau} + (1-\zeta)\boldsymbol{\psi}_{n,k-1}^{h,\tau}$ ▷ introduce under-relaxation
9: **until** $\|\boldsymbol{\psi}_{n,k}^{h,\tau} - \boldsymbol{\psi}_{n,k-1}^{h,\tau}\|_{\boldsymbol{H}^1(\Omega_0^{\mathrm{S}})} < \mathrm{tol}$
10: $\boldsymbol{\psi}_n^{h,\tau} = \boldsymbol{\psi}_{n,k}^{h,\tau}, \boldsymbol{u}_n^{h,\tau} = \boldsymbol{u}_{n,k}^{h,\tau}, p_n^{h,\tau} = p_{n,k}^{h,\tau}, \phi_n^{h,\tau} = \phi_{n,k}^{h,\tau}, \mu_n^{h,\tau} = \mu_{n,k}^{h,\tau}$
11: $n \leftarrow n+1$
12: **end while**

On the basis of decomposition in (8.116), Algorithm 8.1 describes the subiteration procedure with under-relaxation. The NSCH subsystem and domain-deformation subsystem are solved in lines 6 and 7 of Algorithm 8.1, and these steps are repeated within each time step until the update between consecutive approximations complies with a prescribed tolerance. The solution of the domain deformation in line 7 comprises the solid deformation and its harmonic extension onto the fluid domain simultaneously. However, in practice these deformations can be determined consecutively. Line 8 with $\zeta \in (0,1]$ corresponds to the under-relaxation procedure. It is to be noted that the transportation map $T_{n,k}$ depends implicitly on the deformation $\boldsymbol{\psi}_{n,k}^{h,\tau}$. Implicit to the solution of the domain deformation in line 7 is that the mixture velocity $\boldsymbol{u}_{n,k}^{h,\tau}$, order parameter $\phi_{n,k}^{h,\tau}$ and chemical potential $\mu_{n,k}^{h,\tau}$ are transported to the updated configuration via $T_{n,k}$. However, in an actual implementation of Algorithm 8.1 all variables are defined in the reference configuration and a modification of the actual configuration is hence automatically accounted for.

8.6 Numerical experiments

To assess the properties of the presented binary-fluid–solid interaction model (8.60) and of the subiteration process in Algorithm 8.1, we consider numerical approximations according to (8.110) for the experimental set-up in [57]. The test case corresponds to a sessile droplet on a deformable substrate; see Figure 8.4 on page 312. Section 8.6.1 elaborates the set-up of the considered numerical experiments. Section 8.6.2 presents the numerical results.

8.6.1 Experimental set-up

We consider a liquid droplet of 13.8 nL set in an ambient fluid on a 50 μm-thick soft solid substrate. This droplet volume has been selected such that the radius of the circular footprint, i.e., the contact area of the droplet with the solid substrate, agrees closely with the value of 176.7 μm in the experiments in [57]. The substrate is modeled as a nearly incompressible solid with a St. Venant–Kirchhoff stored energy function. The Young's modulus is set to $E_{\mathrm{Y}} = 3$ kPa in accordance with the experimental set-up in [57] and the Poisson ratio is selected as $\nu_{\mathrm{P}} = 0.499$ to describe the nearly incompressible behavior of the solid. The Lamé parameters are derived accordingly as $\mu_{\mathrm{L}} = E_{\mathrm{Y}}/(2 + 2\nu_{\mathrm{P}})$ and $\lambda_{\mathrm{L}} = \nu_{\mathrm{P}} E_{\mathrm{Y}}/(1 + \nu_{\mathrm{P}})(1 - 2\nu_{\mathrm{P}})$. The surface tension of the fluid–fluid interface is $\sigma_{\mathrm{LA}} = 46\,\mathrm{mJ/m^2}$. The solid–fluid surface tension of the liquid (resp. ambient fluid) is $\sigma_{\mathrm{SL}} = 36\,\mathrm{mJ/m^2}$ (resp. $\sigma_{\mathrm{SA}} = 31\,\mathrm{mJ/m^2}$).

We are primarily interested in equilibrium configurations of the binary-fluid–solid interaction problem. The densities of the liquid, ambient fluid and solid, ρ_{L}, ρ_{A} and ρ_{S}, respectively, the viscosities of the liquid and ambient fluid, η_{L} and η_{A}, and the mobility, m, only affect the dynamic behavior and do not influence the equilibrium configuration. The liquid considered in [57] is Glycerol, which has density and viscosity $1.26\,\mathrm{pg/\mu m^3}$ and 1412 mPas, respectively. The solid substrate is composed of silicone gel, which has a density of approximately $1.00\,\mathrm{pg/\mu m^3}$. The ambient fluid, air, exhibits a density of approximately $10^{-3}\,\mathrm{pg/\mu m^3}$ and viscosity of approximately 10 μPas. The density (resp. viscosity) ratio of the liquid and ambient fluid hence exceeds 10^3 (resp. 10^5), which inhibits stability of the numerical approximation for small values of ϵ; see also Remark 8.13. To enhance the stability of the computation, the density and viscosity of the ambient fluid are increased to $10^{-1}\,\mathrm{pg/\mu m^3}$ and 100 mPas, respectively. To examine the effect of the solid–fluid density ratio on the convergence behavior of the subiteration procedure, we will also consider different solid and ambient-fluid densities. An overview of the parameter settings for the considered test cases is presented in Table 8.1.

To complete the specification of the considered binary-fluid–solid interaction problem, we define the boundary conditions on the external boundary of the system; see Figure 8.4. The boundary $\Gamma^{\mathrm{F}}_{0,\mathrm{s}}$ of the fluid domain corresponds to a slip boundary and the corresponding boundary conditions are specified in (8.16). On the lateral

Tab. 8.1: Overview of the liquid and ambient densities and viscosities, solid density, time step and relaxation parameters in test cases 1 and 2.

TC#	ρ_l pg/μm³	η_l mPas	ρ_a pg/μm³	η_a mPas	ρ_s pg/μm³	τ μs	ζ –
1	1.26	1412	0.1	100	1.00	10	0.01
2	1.26	1412	1.26	1412	12.6	1000	0.1

boundary $\Gamma^{\mathrm{F}}_{\mathrm{A}}$, we impose

$$\left.\begin{aligned} p\boldsymbol{n}^{\mathrm{F}} - \boldsymbol{\tau}\boldsymbol{n}^{\mathrm{F}} - \boldsymbol{\zeta}\boldsymbol{n}^{\mathrm{F}} &= 0 \\ \phi &= -1 \\ \partial_n \mu &= 0 \end{aligned}\right\} \quad \text{in } \Gamma^{\mathrm{F}}_{\mathrm{A}}\,. \tag{8.117}$$

The deformation of the solid complies with the following boundary conditions:

$$\left.\begin{aligned} \psi_n &= \mathrm{Id}_n \\ (\boldsymbol{P}\boldsymbol{n}^{\mathrm{S}})_{\Gamma} &= 0 \end{aligned}\right\} \quad \text{in } \Gamma^{\mathrm{S}}_{\mathrm{S}} \qquad \boldsymbol{\psi} = \mathrm{Id} \;\; \text{in } \Gamma^{\mathrm{S}}_{\mathrm{D}} \tag{8.118}$$

where Id_n denotes the normal component of the identity on $\Gamma^{\mathrm{S}}_{\mathrm{S}}$, i.e., $\mathrm{Id}_n(\boldsymbol{x}) = \boldsymbol{x} \cdot \boldsymbol{n}^{\mathrm{S}}$ for all $\boldsymbol{x} \in \Gamma^{\mathrm{S}}_{\mathrm{S}}$, and $(\boldsymbol{P}\boldsymbol{n}^{\mathrm{S}})_{\Gamma}$ stands for the tangential component of the solid traction. The first boundary condition in (8.118) implies that at the lateral boundary $\Gamma^{\mathrm{S}}_{\mathrm{S}}$ the substrate is constrained in the normal direction and is allowed to slide in the tangential direction.

8.6.2 Numerical results

We first consider the dynamics of the aggregated fluid–solid system. To this end, we regard the evolution of the binary-fluid–solid interaction problem for parameter values TC1 in Table 8.1. The initial configuration corresponds to an undeformed flat solid–fluid interface. Figure 8.6 presents the evolution of the deformation of the solid–fluid interface in the time interval $t \in [0, 300]\,\mu\mathrm{s}$. The thick solid curve in Figure 8.6 indicates the stationary deformation of the solid–fluid interface obtained in TC2; cf. Figure 8.9. The surface tension of the fluid–fluid interface exerts a localized load on the solid–fluid interface in the vicinity of the contact line; cf. Figure 8.2. This contact-line traction leads to instantaneous formation of a kink in the surface of the soft substrate. Because of the finite width of the diffuse fluid–fluid interface corresponding to $\epsilon = 1\,\mu\mathrm{m}$, the kink exhibits finite curvature. The kink develops into a so-called *wetting ridge* of approximately 7 μm height. By extrapolation, we estimate that for the considered solid, liquid and ambient densities, the wetting ridge below the contact line develops in approximately 1 sm. The fluid–fluid surface tension causes an elevated pressure in the droplet relative to the pressure in the ambient fluid; see also

Figure 8.11. In accordance with the Young–Laplace relation, this capillary pressure difference is $\sigma_{\mathrm{LA}}\varkappa$ with $\varkappa$ the additive curvature of the fluid–fluid interface. The elevated pressure in the droplet induces an indentation of the substrate below the wetted surface. The evolution of this indentation occurs on a time scale that is several orders of magnitude larger than the development of the wetting ridge. The considered elasto-capillary binary-fluid–solid interaction problem hence exhibits marked spatial and temporal multiscale behavior.

Figure 8.7 displays the temporal variation of the various components of the energy and of the aggregate energy in (8.62) corresponding to test case 1, viz.

$$\begin{aligned} \Delta\mathscr{E}^{\mathrm{F}}_{\mathrm{kin}}(t) &= \int_{\Omega^{\mathrm{F}}_t} \frac{1}{2}\rho(t)|\boldsymbol{u}(t)|^2 & \Delta\mathscr{E}^{\Gamma}(t) &= \int_{\Gamma_t} \sigma_{\mathrm{SF}}(\phi(t)) - \int_{\Gamma_0} \sigma_{\mathrm{SF}}(\phi_0) \\ \Delta\mathscr{E}^{\mathrm{S}}_{\mathrm{kin}}(t) &= \int_{\Omega^{\mathrm{S}}_0} \frac{1}{2}\rho_{\mathrm{S}}|\partial_t\boldsymbol{\psi}(t)|^2 & \Delta\mathscr{E}^{\mathrm{S}}_{\mathrm{str}}(t) &= \mathscr{W}(\boldsymbol{\psi}(t)|_{\Omega^{\mathrm{s}}_0}) \end{aligned} \tag{8.119a}$$

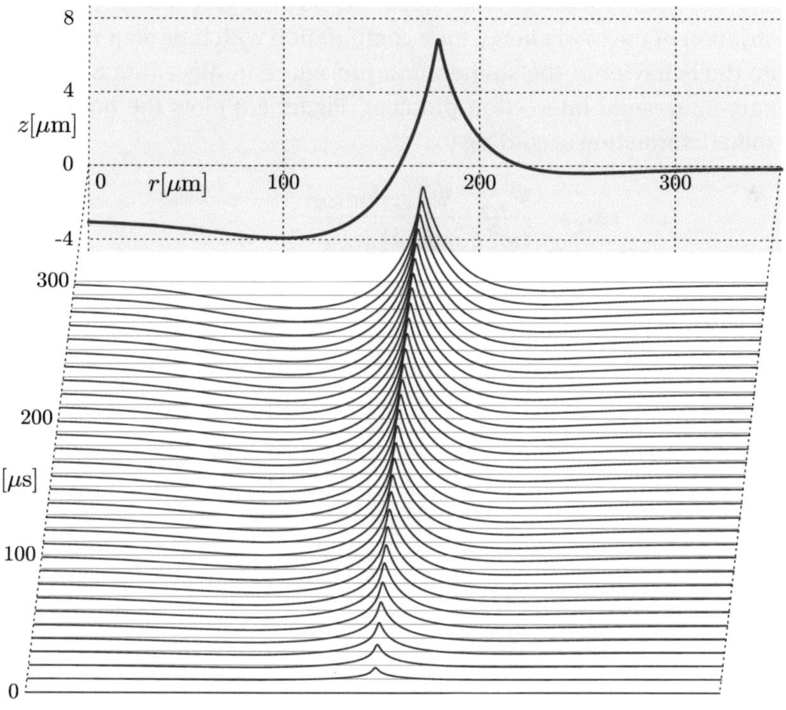

Fig. 8.6: Illustration of the evolution of the solid–fluid interface deformation for test case 1: vertical position, z, of the interface versus the radial distance from the center line, r, for $t \in [0, 300]\,\mu s$. The thick solid curve represents the stationary deformation obtained in test case 2.

and

$$\Delta\mathscr{E}^{\mathrm{GL}}(t) = \int_{\Omega_t^{\mathrm{F}}} \left(\frac{\epsilon\sigma}{2}|\operatorname{grad}\phi(t)|^2 + \frac{\sigma}{\epsilon}\Psi(\phi(t))\right) - \int_{\Omega_0^{\mathrm{F}}} \left(\frac{\epsilon\sigma}{2}|\operatorname{grad}\phi_0|^2 + \frac{\sigma}{\epsilon}\Psi(\phi_0)\right)$$

$$\Delta\mathscr{E}^{\mathrm{tot}}(t) = \Delta\mathscr{E}^{\mathrm{F}}_{\mathrm{kin}}(t) + \Delta\mathscr{E}^{\mathrm{GL}}(t) + \Delta\mathscr{E}^{\Gamma}(t) + \Delta\mathscr{E}^{\mathrm{S}}_{\mathrm{kin}}(t) + \Delta\mathscr{E}^{\mathrm{S}}_{\mathrm{str}}(t) \tag{8.119b}$$

By virtue of the initial conditions $\boldsymbol{u}(0,\cdot) = 0$, $\boldsymbol{\psi}^{\mathrm{S}}(0,\cdot) = \mathrm{Id}$ and $\partial_t\boldsymbol{\psi}^{\mathrm{S}}(0,\cdot) = 0$, the kinetic energy of the fluid and the strain energy and kinetic energy of the solid vanish initially. One can observe that the kinetic-energy contributions in the fluid and solid remain negligible. During the evolution of the binary-fluid–solid interaction problem, the mixture energy decays, while the energy carried by the solid–fluid interface and the strain energy of the solid increase. The aggregated energy decays in accordance with the fundamental energy-dissipation relation (8.63). It is to be noted that the discrete approximation according to (8.110) retains the energy-dissipation property (8.63) only for sufficiently small time steps. To illustrate that violations of the energy-dissipation property can occur for larger time steps, the dashed gray line in Figure 8.7 indicates the variation of the total energy for a computation with time step $\tau = 50\ \mu\mathrm{s}$.

To illustrate the behavior of the subiteration procedure in Algorithm 8.1 for the considered binary-fluid–solid interaction problem, Figure 8.8 plots the normalized update in the solid deformation according to

$$\mathrm{res}_k = \frac{\|\boldsymbol{\psi}^{h,\tau}_{n,k} - \boldsymbol{\psi}^{h,\tau}_{n,k-1}\|_{\boldsymbol{H}^1(\Omega_0^{\mathrm{S}})}}{\|\boldsymbol{\psi}^{h,\tau}_{n,1} - \boldsymbol{\psi}^{h,\tau}_{n,0}\|_{\boldsymbol{H}^1(\Omega_0^{\mathrm{S}})}} \tag{8.120}$$

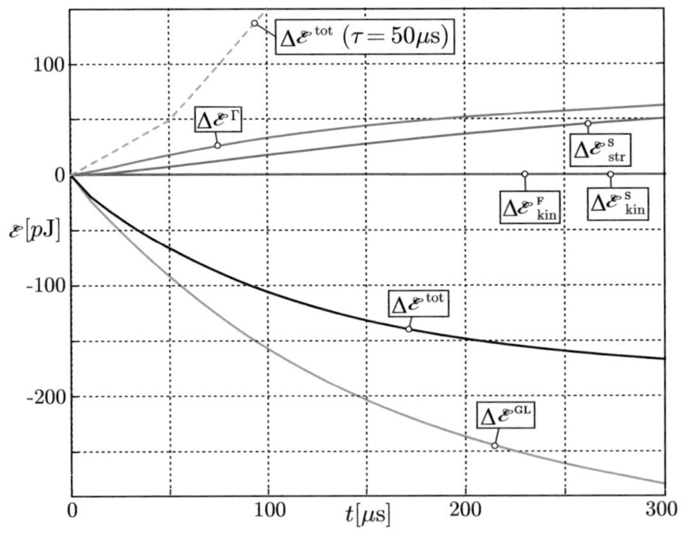

Fig. 8.7: Temporal variation of the components of the energy according to (8.119) for test case 1. The dashed line indicates the variation in the total energy for a computation with time step $\tau = 50\ \mu\mathrm{s}$ instead of $\tau = 10\ \mu\mathrm{s}$.

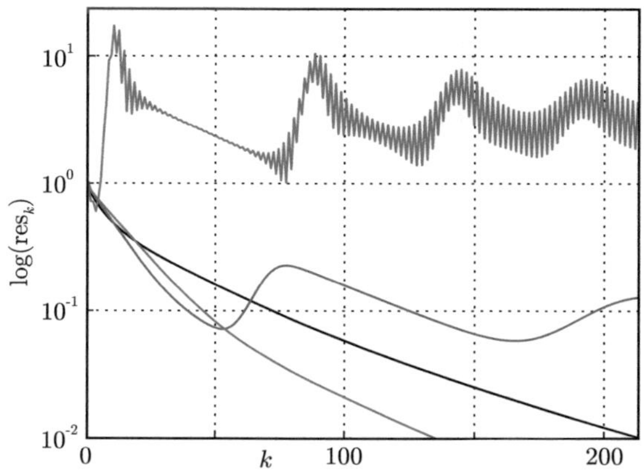

Fig. 8.8: Convergence of the subiteration procedure with under-relaxation: Normalized update (8.120) of the solid deformation in Algorithm 8.1 versus the iteration counter k for TC1-settings $\zeta = 0.01$ and $\rho_S = 1\,\text{pg}/\mu\text{m}^3$ (black), for $\zeta = 0.02$ and $\rho_S = 1\,\text{pg}/\mu\text{m}^3$ (green), for $\zeta = 0.02$ and $\rho_S = 10\,\text{pg}/\mu\text{m}^3$ (red) and for $\zeta = 0.02$ and $\rho_S = 20\,\text{pg}/\mu\text{m}^3$ (blue).

versus the iteration counter k in the first time step ($n = 1$) for the settings TC1 in Table 8.1 (*black* curve) and for twice larger relaxation parameter $\zeta = 0.02$ (*green*) and twice larger relaxation parameter and increased solid densities, viz. $\rho_S = 10\,\text{pg}/\mu\text{m}^3$ (*red*) and $\rho_S = 20\,\text{pg}/\mu\text{m}^3$ (*blue*). The results in Figure 8.8 convey that severe under-relaxation ($\zeta = 0.01$) is required in the subiteration procedure to attain convergence. A relative reduction in the update in the subiteration process by 10^{-2} then requires more than 200 iterations. An increase of the relaxation parameter to $\zeta = 0.02$ preempts convergence of the subiteration process. Although an increase of the solid density does improve the convergence behavior, in accordance with a corresponding reduction of the added-mass effect [10, 11, 16, 26], for $\zeta = 0.02$ a tenfold increase of the solid density to $\rho_S = 10\,\text{pg}/\mu\text{m}^3$ still does not yield convergence. If the solid density is further increased to $\rho_S = 20\,\text{pg}/\mu\text{m}^3$ then convergence of the subiteration procedure is restored for $\zeta = 0.02$. It is to be mentioned that for parameter settings for which the subiteration process does yield convergence, convergence can be nonmonotonous (results not displayed). Such nonmonotonous convergence can be attributed to non-normality of the subiteration procedure; see [12].

The second test case that we consider pertains to computation of the stationary equilibrium interface configuration, to enable a comparison to the experimental data in [57]. To enhance the stability of the computation, we select matching liquid and ambient-fluid densities and viscosities and we increase the density of the solid substrate. The corresponding parameters are designated TC2 in Table 8.1. At $t = 128\,\text{ms}$ the solid–fluid interface has essentially settled into its stationary equilibrium configuration. Figure 8.9 presents a comparison of the corresponding configuration of the solid–fluid interface and experimental data from [57]. Juxtaposition of the experimen-

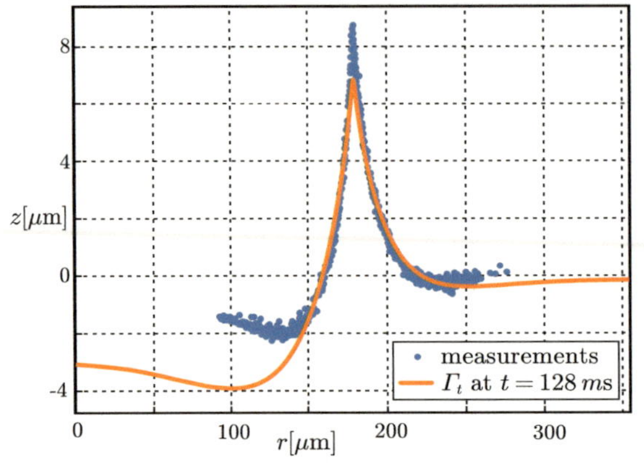

Fig. 8.9: Comparison of the computed fluid–solid interface configuration Γ_t for test case 2 at t = 128 ms and rendering of experimental data from [57].

tal and computed results shows that the elevation of the solid–fluid interface at the contact line is underestimated by approximately 25%. The computed shape of the wetting ridge is however overall in very good agreement with the experimental data. Figure 8.9 conveys a marked deviation between the computed and experimental results with respect to the indentation of the soft substrate below the droplet. We conjecture that this discrepancy is related to compressibility effects in the pseudo-incompressible solid model. Because of the radial symmetry of the considered configuration, the volumetric deviation associated with the difference in the indentation is less than 1%.

To further illustrate the properties of the NSCH binary-fluid model and the ALE formulation, Figure 8.10 presents a magnification of the contact-line region for test

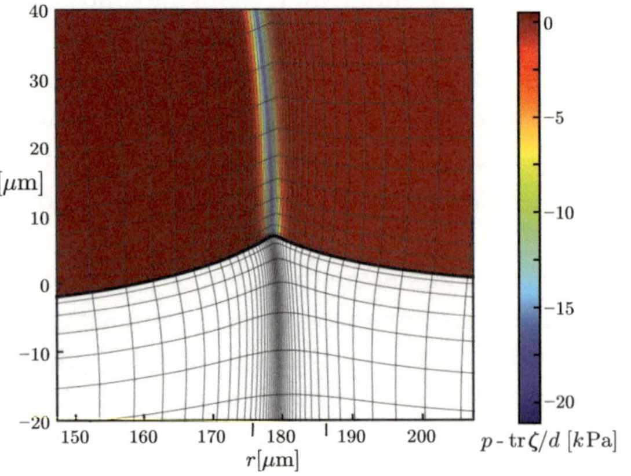

Fig. 8.10: Magnification of the contact-line region for test case 2 at t = 128 ms with deformed fluid and solid meshes and the computed pressure $p - \operatorname{tr}\boldsymbol{\zeta}/d$.

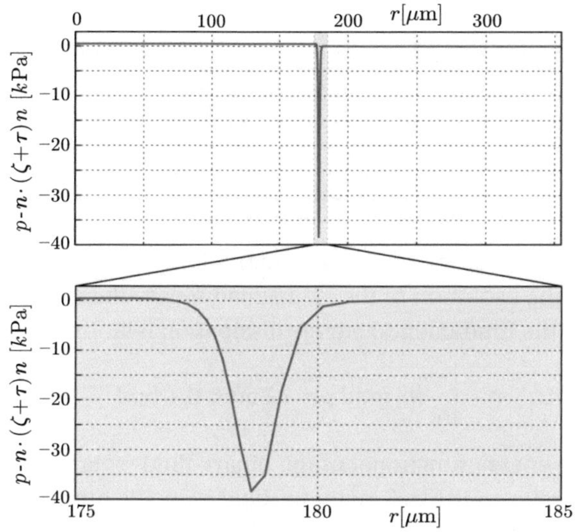

Fig. 8.11: Normal component of the binary-fluid traction $p - \boldsymbol{n}\cdot(\boldsymbol{\zeta} + \boldsymbol{\tau})\boldsymbol{n}$ on the solid–fluid interface for test case 2 at $t = 128$ ms (top) and an expanded view in the interval $r \in [175, 185]$ µm (bottom).

case 2 at $t = 128$ ms with the fluid and solid meshes in the actual configuration and the computed pressure (viz. $p - \operatorname{tr}\boldsymbol{\zeta}/d$) distribution in the binary fluid. In addition, Figure 8.11 plots the traction that the binary fluid exerts on the solid–fluid interface, including an expanded view in the interval $r \in [175, 185]$ µm. Figure 8.10 conveys that the pressure in the binary fluid exhibits a localized minimum at the diffuse fluid–fluid interface. From Figure 8.11 one can observe that the normal traction below the droplet is virtually uniform at an elevated level of approximately 520 Pa relative to the ambient pressure. This elevated normal traction is close to the theoretical Laplace pressure $2\sigma_{\mathrm{LA}}/R$ in a droplet on a rigid substrate with radius $R = 178$ µm and surface tension $\sigma_{\mathrm{LA}} = 46$ mN/m. Let us note that the expanded view in Figure 8.11 provides a magnification of approximately 36× along the radial axis.

8.7 Conclusion

We presented a model for the elasto-capillary interaction of a binary fluid with an elastic solid, based on a quasi-incompressible Navier–Stokes–Cahn–Hilliard diffuse-interface model for the binary fluid and a hyperelastic material model with a St. Venant–Kirchhoff stored-energy function for the solid. The interaction between the fluid and the solid at their mutual interface is described by a preferential-wetting condition in addition to the usual kinematic and dynamic interface conditions. In the diffuse-interface model, the fluid traction on the interface includes a capillary-stress contribution in addition to the regular viscous-stress and pressure contributions. The dynamic interface condition comprises the traction exerted by the nonuniform solid–fluid surface tension in accordance with the Young–Laplace law for the solid–fluid interface. We derived a weak formulation of the aggregated binary-fluid–solid interaction problem, based on an arbitrary Lagrangian–Eulerian formulation of the Navier–Stokes–Cahn–Hilliard equations and a proper weak evaluation of the binary-fluid traction and of the solid–fluid surface tension.

We conducted an analysis of the main properties of the binary-fluid–solid interaction problem. We established that the evolution of the aggregated binary-fluid–solid interaction problem complies with the fundamental energy dissipation relation:

$$d_t\mathscr{E}(\boldsymbol{\psi}, \boldsymbol{u}, \phi) = -\int_{\Omega_t^{\mathrm{F}}} \operatorname{grad} \boldsymbol{u} : \boldsymbol{\tau} - \int_{\Omega_t^{\mathrm{F}}} m|\operatorname{grad}\mu|^2 + \mathrm{bnd} \leq 0 + \mathrm{bnd}$$

with $m > 0$, where $\mathscr{E}$ indicates the energy functional of the binary-fluid–solid interaction problem, $\boldsymbol{\psi}, \boldsymbol{u}, \phi, \boldsymbol{\tau}$ and μ refer to the deformation, fluid velocity, order parameter, viscous-stress tensor and chemical potential, respectively, and 'bnd' contains terms that are only supported on the external boundary of the fluid–solid system.

By means of a formal analysis of the structure of the fluid–fluid interface, we established that the diffuse-interface binary-fluid model exhibits the usual capillary and wetting effects in the sharp-interface limit, i.e., as the diffuse-interface thickness passes to zero. In particular, the capillary traction exerted by the binary fluid on the solid–fluid interface converges to a contact-line force in the sharp-interface limit. We argued that because contact-line forces are not generally continuous with respect to norms in which the strain energy of the solid is continuous, but are continuous with respect to norms in which the solid–fluid surface energy is continuous, the configuration of the solid–fluid interface near the contact line is exclusively determined by solid–fluid surface tension, and thus exhibits a kink in accordance with the Neumann law.

Numerical results were presented for a sessile droplet on a soft solid substrate, based on a finite-element approximation of the weak formulation of the FSI problem, and basic subiteration with under-relaxation to solve the aggregated FSI problem in each time step. The numerical results conveyed that the considered elasto-capillary fluid–solid interaction problem exhibits marked spatial and temporal multiscale behavior. The contact-line traction leads to instantaneous formation of a kink in the solid–fluid interface and noticeable deformation on a μs-scale, followed by the formation of a wetting ridge at the contact line on a ms-scale. The elevated Young–Laplace pressure in the droplet relative to the ambient pressure induces an indentation in the soft substrate, which develops on a time scale that is again several orders of magnitude larger.

The numerical experiments indicate that the subiteration process is generally unsuitable as a solution procedure for elasto-capillary fluid–solid interaction problems. For relevant settings of the problem parameters, severe under-relaxation was required to attain convergence, leading to excessive iteration counts per time step. We observed that the convergence behavior of the subiteration process does improve with increasing solid–fluid mass ratios. However, even for increased mass ratios the convergence behavior of the subiteration procedure remains sensitive to the relaxation parameter, and criteria for selecting the relaxation parameter are nonobvious.

Comparison of the computed results with experimental data for the considered test case exhibited very good agreement in the wetting ridge region. The indentation of the substrate below the droplet was noticeably overestimated in the computational results relative to the available experimental data. In view of the close agreement between the computed pressure in the droplet and the theoretical Laplace pressure, it appears that the discrepancy between the computed and observed depression is to be attributed to corresponding differences in the constitutive behavior of the solid substrate. The overall good agreement between the computed results and experimental data corroborates the potential of computational binary-fluid–solid interaction models based on the quasi-incompressible NSCH equations for describing elasto-capillary FSI phenomena.

Acknowledgment: The authors would like to express their gratitude to Gertjan van Zwieten (Evalf Computing) for his invaluable support in relation to the software implementations in the Nutils library (Nutils.org) and for executing the numerical experiments in Section 8.6. The support of this work by NanoNextNL, a micro- and nanotechnology consortium of the Government of the Netherlands and 130 partners, is gratefully acknowledged.

Bibliography

[1] H. Abels, H. Garcke, and G. Grün. Thermodynamically consistent, frame indifferent diffuse interface models for incompressible two-phase flows with different densities. *Math. Mod. Meth. Appl. Sci.*, 22:1150013, 2012.

[2] A. Aitken. On Bernoulli's numerical solution of algebraic equations. *Proceedings of the Royal Society of Edinburgh*, 46:289–305, 1926.

[3] S. Aland and A. Voigt. Benchmark computations of diffuse interface models for two-dimensional bubble dynamics. *Int. J. Numer. Meth. Fluids*, 69:747–761, 2012.

[4] J. Ball. Convexity conditions and existence theorems in nonlinear elasticity. *Arch. Rational Mech. Anal.*, 63:337–403, 1977.

[5] E. Bänsch. Finite element discretization of the navier–stokes equations with a free capillary surface. *Numer. Math.*, 88:203–235, 2001.

[6] Y. Bazilevs, V. M. Calo, T. J. R. Hughes, and Y. Zhang. Isogeometric fluid-structure interaction: theory, algorithms, and computations. *Comput. Mech.*, 43:3–37, 2008.

[7] Y. Bazilevs, K. Takizawa, and T. Tezduyar. *Computational Fluid-Structure Interaction: Methods and Applications*. Wiley, 2013.

[8] J. F. Blowey and C. M. Elliott. The cahn–hilliard gradient theory for phase separation with non-smooth free energy part i: Mathematical analysis. 2(3):233–280, 1991.

[9] J. Bosch, M. Stoll, and P. Benner. Fast solution of Cahn–Hilliard variational inequalities using implicit time discretization and finite elements. *Journal of Computational Physics*, 262:38–57, 2014.

[10] E. van Brummelen. Added mass effects of compressible and incompressible flows in fluid-structure interaction. *J. Appl. Mech.*, 76:021206–7, 03 2009.

[11] E. van Brummelen. Partitioned iterative solution methods for fluid-structure interaction. *Int. J. Numer. Meth. Fluids*, 65:3–27, 2011.

[12] E. van Brummelen and R. de Borst. On the nonnormality of subiteration for a fluid-structure interaction problem. *SIAM J. Sci. Comput.*, 27:599–621, 2005.

[13] E. van Brummelen, H. Shokrpour Roudbari, and G. van Zwieten. Elasto-capillarity simulations based on the Navier-Stokes-Cahn-Hilliard equations. In *Advances in Computational Fluid-Structure Interaction and Flow Simulation*, Modeling and Simulation in Science, Engineering and Technology, pages 451–462. Birkhäuser, 2016.

[14] E. van Brummelen, K. van der Zee, and R. de Borst. Space/time multigrid for a fluid-structure-interaction problem. *Appl. Num. Math.*, 58(12):1951–1971, 2008.

[15] L. Caffarelli and N. Muler. An l^{∞} bound for solutions of the Cahn–Hilliard equation. *Arch. Rational Mech. Anal.*, 133:129–144, 1995.

[16] P. Causin, J. Gerbeau, and F. Nobile. Added-mass effect in the design of partitioned algorithms for fluid-structure problems. *Comput. Methods Appl. Mech. Engrg.*, 194:4506–4527, 2005.

[17] P. Ciarlet. *Mathematical Elasticity*. North-Holland, 1988.

[18] P.-G. de Gennes, F. Brochard-Wyart, and D. Queré. *Capillarity and Wetting Phenomena: Drops, Bubbles, Pearls, Waves*. Springer, 2004.

[19] P. de Gennes. Wetting: statics and dynamics. *Rev. Mod. Phys.*, 57:827–863, 1985.

[20] G. Dziuk. An algorithm for evolutionary surfaces. *Numer. Math.*, 58:603–611, 1991.

[21] C. Elliott and A. Stuart. The global dynamics of discrete semilinear parabolic equations. *SIAM Journal on Numerical Analysis*, 30(6):1622–1663, 2014/03/04 1993.

[22] A. Ern and J.-L. Guermond. *Theory and Practice of Finite Elements*. Number 159 in Applied Mathematical Sciences. Springer-Verlag, New York, 2004.

[23] D. Eyre. An unconditionally stable one step scheme for gradient systems. unpublished, 1997.

[24] D. Eyre. Unconditionally gradient stable time marching the Cahn–Hilliard equation. In J. Bullard, R. Kalia, M. Stoneham, and L.-Q. Chen, editors, *Computational and mathematical models of microstructural evolution*, volume 529 of *The Materials Research Society Proceedings*, pages 39–46, 1998.

[25] C. Farhat. CFD-based nonlinear computational aeroelasticity. In E. Stein, R. de Borst, and T. Hughes, editors, *Encyclopedia of Computational Mechanics*, volume 3: Fluids, chapter 13, pages 459–480. John Wiley & Sons, Ltd., 2004.

[26] C. Förster, W. Wall, and E. Ramm. Artificial added mass instabilities in sequential staggered coupling of nonlinear structures and incompressible viscous flows. *Comput. Methods Appl. Mech. Engrg.*, 196:1278–1293, 2007.

[27] H. Garcke, M. Hinze, and C. Kahle. A stable and linear time discretization for a thermodynamically consistent model for two-phase incompressible flow. *Appl. Num. Math.*, 99:151–171, 1 2016.

[28] M. W. Gee, U. Küttler, and W. A. Wall. Truly monolithic algebraic multigrid for fluid–structure interaction. *Int. J. Numer. Meth. Engng.*, 85:987–1016, 2011.

[29] H. Gomez and K. van der Zee. Computational phase-field modeling. In E. Stein, R. de Borst, and T. Hughes, editors, *Encyclopedia of Computational Mechanics*. John Wiley & Sons, Ltd., 2nd edition, 2017.

[30] H. Gouin. Interfaces endowed with nonconstant surface energies revisited with the d'Alembert–Lagrange principle. *Mathematics and Mechanics of Complex Systems*, 2:23–43, 2014.
[31] Z. Guo, P. Lin, and J. Lowengrub. A numerical method for the quasi-incompressible cahn–hilliard–navier–stokes equations for variable density flows with a discrete energy law. *J. Comput. Phys.*, 276:486–507, 11 2014.
[32] M. Gurtin, D. Polignone, and J. Vinals. Two-phase binary fluids and immiscible fluids described by an order parameter. *Math. Models Meth. Appl. Sci.*, 6:815–831, 1996.
[33] M. Hintermüller, M. Hinze, and M. H. Tber. An adaptive finite-element moreau–yosida-based solver for a non-smooth cahn–hilliard problem. *Optim. Methods Softw.*, 26:777–811, 2011.
[34] P. Hohenberg and B. Halperin. Theory of dynamic critical phenomena. *Rev. Mod. Phys.*, 49:435–479, 1977.
[35] J. Hron and S. Turek. *Fluid-Structure Interaction - Modelling, Simulation, Optimisation*, chapter A Monolithic FEM/Multigrid Solver for an ALE Formulation of Fluid-Structure Interaction with Applications in Biomechanics, pages 146–170. Number 53 in Computational Science and Engineering. Springer, Berlin, 2006.
[36] D. Jacqmin. Calculation of two-phase Navier-Stokes flows using phase-field modeling. *J. Comput. Phys.*, 155:96–127, 10 1999.
[37] D. Jacqmin. Contact-line dynamics of a diffuse fluid interface. *J. Fluid Mech.*, 402:57–88, 2000.
[38] T. Kajiya, A. Daer, T. Narita, L. Royon, F. Lequex, and L. Limat. Advancing liquid contact line on visco-elastic gel substrates: Stick-slip vs. continuous motions. *Soft Matter*, 9:454–461, 2013.
[39] S. Karpitschka, S. Das, M. van Gorcum, H. Perrin, B. Andreotti, and J. H. Snoeijer. Droplets move over viscoelastic substrates by surfing a ridge. *Nat Commun*, 6, 08 2015.
[40] U. Küttler and W. Wall. Fixed-point fluid–structure interaction solvers with dynamic relaxation. *Comput. Mech.*, 43(1):61–72, 12 2008.
[41] A. Lee, A. Münch, and E. Süli. Sharp-interface limits of the Cahn–Hilliard equation with degenerate mobility. *SIAM J. Appl. Math.*, 76:433–456, 2016/10/02 2016.
[42] J. Lowengrub and L. Truskinovsky. Quasi-incompressible Cahn-Hilliard fluids and topological transitions. *Proceedings of the Royal Society of London A: Mathematical, Physical and Engineering Sciences*, 454:2617–2654, 1998.
[43] H. Melbø and T. Kvamsdal. Goal oriented error estimators for Stokes equations based on variationally consistent postprocessing. *Comput. Methods Appl. Mech. Engrg.*, 192:613–633, 2003.
[44] C. Michler, E. van Brummelen, and R. de Borst. An interface Newton–Krylov solver for fluid-structure interaction. *Int. J. Numer. Meth. Fluids*, 47:1189–1195, 2005.
[45] C. Michler, E. van Brummelen, and R. de Borst. Error-amplification analysis of subiteration-preconditioned GMRES for fluid-structure interaction. *Comput. Methods Appl. Mech. Engrg.*, 195:2124–2148, 2006.
[46] C. Michler, S. Hulshoff, E. van Brummelen, and R. de Borst. A monolithic approach to fluid-structure interaction. *Computers and Fluids*, 33:839–848, 2004.
[47] C. Michler, H. van Brummelen, and R. de Borst. An investigation of interface-GMRES(R) for fluid–structure interaction problems with flutter and divergence. *Comput. Mech.*, 47:17–29, 2011.
[48] F. Neumann. *Vorlesungen über die Theorie der Kapillarität*. Teubner, Leipzig, 1894.
[49] C. Py, P. Reverdy, L. Doppler, J. Bico, B. Roman, and C. Baroud. Capillary origami: Spontaneous wrapping of a droplet with an elastic sheet. *Phys. Rev. Lett.*, 98:156103, Apr 2007.
[50] T. Richter. A monolithic geometric multigrid solver for fluid-structure interactions in Ale formulation. *Int. J. Num. Meth. Engng.*, 104:372–390, 2015.
[51] B. Roman and J. Bico. Elasto-capillarity: Deforming an elastic structure with a liquid droplet. *J. Phys.: Condens. Matter*, 22, 2010.

[52] P. Seppecher. Moving contact lines in the Cahn-Hilliard theory. *Int. J. Engng. Sci.*, 34:977–992, 7 1996.

[53] J. Shen, X. Yang, and Q. Wang. Mass and volume conservation in phase field models for binary fluids. *Commun. Comput. Phys.*, 13:1045–1065, 2013.

[54] M. Shokrpour Roudbari, E. van Brummelen, and C. Verhoosel. A multiscale diffuse-interface model for two-phase flow in porous media. *Computers & Fluids*, 141:212–222, 2016.

[55] G. Simsek, K. van der Zee, M. Shokrpour Roudbari, and E. van Brummelen. Linear and energy dissipative scheme for two-phase flows of quasi-incompressible Navier-Stokes Cahn-Hilliard models. *ArXiv [physics.flu-dyn]*, page arXiv:1603.06475v2, 2016.

[56] J. Sokołowski and J.-P. Zolésio. *Introduction to Shape Optimization: Shape Sensitivity Analysis*, volume 16 of *Springer Series in Computational Mathematics*. Springer-Verlag, 1992.

[57] R. Style, R. Boltyanskiy, Y. Che, J. Wettlaufer, L. Wilen, and E. Dufresne. Universal deformation of soft substrates near a contact line and the direct measurement of solid surface stresses. *Phys. Rev. Lett.*, 110:066103, Feb 2013.

[58] R. Style et al. Patterning droplets with durotaxis. *PNAS*, 110(31):12541–12544, 2013.

[59] R. Temam and A. Mirainville. *Mathematical Modeling in Continuum Mechanics*. Cambridge University Press, 2001.

[60] T. Tezduyar. Finite element methods for fluid dynamics with moving boundaries and interfaces. In E. Stein, R. de Borst, and T. Hughes, editors, *Encyclopedia of Computational Mechanics*, volume 3, chapter 17. John Wiley & Sons, 2004.

[61] T. Tezduyar, S. Sathe, R. Keedy, and K. Stein. Space-time finite element techniques for computation of fluid-structure interactions. *Comput. Methods Appl. Mech. Engrg.*, 195:2002–2027, 2006.

[62] T. Tezduyar, S. Sathe, and K. Stein. Solution techniques for the fully discretized equations in computation of fluid-structure interactions with the space-time formulation. *Comput. Methods Appl. Mech. Engrg.*, 195:5743–5753, 2006.

[63] T. Tezduyar, K. Takizawa, C. Moorman, S. Wright, and J. Christopher. Space–time finite element computation of complex fluid–structure interactions. *Int. J. Numer. Meth. Fluids*, 64:1201–1218, 2010.

[64] G. Tryggvason, R. Scardovelli, and S. Zaleski. *Direct Numerical Simulations of Gas-Liquid Multiphase Flows*. Cambridge University Press, 2011.

[65] E. van Brummelen, K. van der Zee, V. Garg, and S. Prudhomme. Flux evaluation in primal and dual boundary-coupled problems. *J. Appl. Mech.*, 79:010904–8, 2012.

[66] X. Wu, G. J. van Zwieten, and K. G. van der Zee. Stabilized second-order convex splitting schemes for cahn-hilliard models with application to diffuse-interface tumor-growth models. 30(2):180–203, 2014.

[67] P. Yue and J. Feng. Wall energy relaxation in the cahn–hilliard model for moving contact lines. *Phys. Fluids*, 23:012106, 2011.

[68] K. van der Zee, E. van Brummelen, I. Akkerman, and R. de Borst. Goal-oriented error estimation and adaptivity for fluid-structure interaction using exact linearized adjoints. *Comput. Methods Appl. Mech. Engrg.*, 200:2738–2757, 2011.

Thomas Wick

9 Coupling fluid-structure interaction with phase-field fracture: Algorithmic details

Abstract: In this chapter, further extensions, algorithmic motivations and descriptions of a framework for coupling arbitrary Lagrangian–Eulerian fluid-structure interaction with phase-field fracture are addressed. This setting is formulated via variational-monolithic coupling and has four unknowns: velocities, displacements, pressure, and a phase-field variable. Moreover, the phase-field variable is subject to a crack irreversibility constraint. The resulting formulation is a highly nonlinear multiphysics variational inequality system. Here, the irreversibility constraint is imposed through penalization using an augmented Lagrangian algorithm. This formulation is treated with a variational-monolithic arbitrary Lagrangian–Eulerian approach, which allows one to use implicit time-stepping methods for a robust discretization in time. The first aim is a revisit of our concept to couple fluid-structure and phase-field fracture including lists of major challenges. Our second focus is on adaptive time-step control of goal functionals based on a heuristic estimator. Third, algorithmic details on Newton's method for solving the nonlinear problem are provided. The proposed concepts are substantiated with numerical test cases.

Keywords: fluid-structure interaction, arbitrary Lagrangian–Eulerian technique, phase-field fracture, Newton's method, adaptive time-step control, numerical algorithms

Classification: 74R10; 74F10; 65M60; 49M15; 35Q74. See www.ams.org/msc

9.1 Introduction

In this contribution we revisit and further develop our recent efforts [111, 112] to couple fluid-structure interaction (FSI) and phase-field fracture (PFF) modeling. Both FSI and fracture propagation are difficult, but fascinating, research fields with numerous applications in applied mathematics and engineering. For FSI literature we refer the reader to the books [9, 18, 19, 38, 42] and also to the other chapters of this book. For fracture mechanics we refer to [5, 14, 49, 91, 95, 97, 120], including references cited therein.

Note: This work has been supported by the Austrian Academy of Sciences and Ecole Polytechnique in France.

Thomas Wick, Centre de Mathématiques Appliquées, Ecole Polytechnique, Université Paris-Saclay, 91128 Palaiseau, France, thomas.wick@polytechnique.edu

https://doi.org/10.1515/9783110494259-009

It is of interest to derive a common framework for FSI and PFF for applications in geomechanics, geophysics, pipelines, nuclear power plants, and porous media, which are possibly filled with fluids or coupled to surrounding flow, e.g. [1, 85, 99, 103, 121]. At first view, the principle idea is simple – take an FSI code and add a fracture into the solid part – but it turns out that the realization is significantly nontrivial; see also [103, 121] with a similar goal using extended finite elements for fracture modeling or an embedded interface finite element method, respectively, rather than phase-field modeling. Several challenges will be provided in this introduction and a list of open questions is given in the conclusions at the end. In the following two subsections, we first recapitulate known aspects and set-up the basic ideas. Section 9.1.1 is rather short since all contributions in this book deal with fluid-structure interaction such that enough additional information is provided. Conversely, Section 9.1.2 introduces the concept of phase-field fracture, which is not addressed in the other book chapters, so more detailed information will be provided. In the third subsection 9.1.3, the novel FSI-PFF approach will be explained.

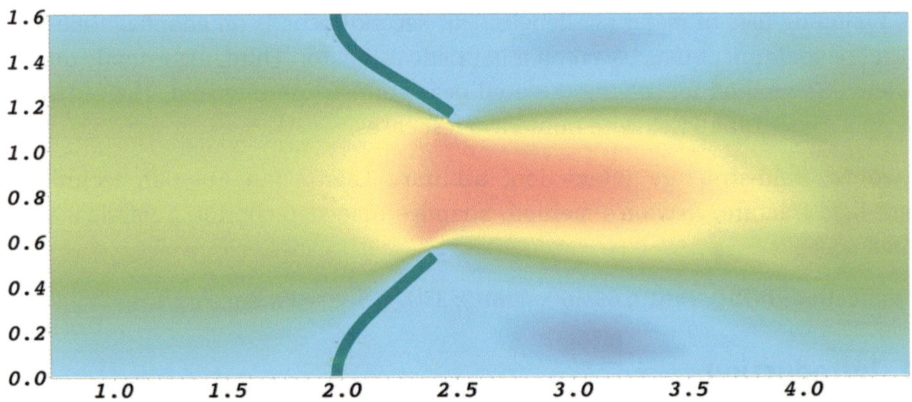

Fig. 9.1: Fluid-structure interaction (FSI). Two elastic solid flaps interact with a fluid modeled by the Navier–Stokes equations. Figure based on computations done in [109].

9.1.1 Fluid-structure interaction (FSI)

In classical FSI, the isothermal, incompressible Navier–Stokes equations are coupled with elastodynamics. The constitutive law in the solid is based on the geometrically nonlinear Saint-Venant Kirchhoff (STVK) model, see e.g. [22]. Here, three unknowns are sought: velocities, pressure, and displacements. The FSI coupling technique is based on an interface-tracking method; namely the now standard arbitrary Lagrangian–Eulerian (ALE) technique [28, 29, 37, 55, 57, 87, 98, 100]. Here, the flow equations are re-written such that their coordinate system matches the Lagrangian framework of the solid. The resulting formulation using variational-monolithic coupling in the reference configuration is outlined in [56, 93, 106]. The key feature of the ALE approach is that the interface aligns with mesh edges and therefore interface terms such as traction forces can be computed with high accuracy. In addition, it allows for many settings up to large solid deformations as long as the ALE mapping is a C^1-diffeomorphism. This last condition is of importance and finally limits standard ALE for very, very large deformations, topology changes, and contact of multiple solids. However, choosing a good mesh motion model to construct the ALE mapping still allows us to investigate a great variety of different applications.

Summarizing this FSI part we obtain:

Challenge 9.1. With regard to the previous descriptions, we classify several great challenges in FSI, which are important in particular from an algorithmic point of view:
- Dealing and coupling of different classes of partial differential equations (PDEs): elliptic, parabolic, hyperbolic that require different mathematical analysis and numerical tools;
- Nonlinearities in various equations and nonlinear coupling terms;
- Multidomain character and interface coupling conditions;
- Combining different coordinate systems: Eulerian and Lagrangian;
- Moving boundaries, i.e. moving interfaces;
- Designing robust and efficient numerical methods.

9.1.2 Phase-field fracture (PFF)

Conversely, at first glance, the concept of treating fractures and damage with a phase-field approach seems much easier because the coupled PDE system is of elliptic type for quasistatic brittle fracture. Moreover, the very basic problem of a fracture (namely a discontinuity) in a solid consists only of one equation: the solid problem. But for fracture modeling and its numerical treatment, phase-field fracture (in contrast to other methods such as extended/generalized finite elements [41, 82]), introduces a second equation, which is nothing else than a regularized indicator function and very similar to a level-set function. However, this additional function leads to a quasilinear PDE system. From an application viewpoint, two situations are of interest for given fracture(s): first, observing the variation of the fracture width (crack opening displacement) and second, the evolution of the fracture length. The latter situation is significantly more complicated. However, both configurations are of importance and variational fracture techniques as discussed in the following can be used for both of them.

Investigating in more detail, there are surprisingly many traps, which are – in all variety – totally different to the difficulties listed in Challenge 9.1. For instance, the most striking feature of FSI is its multidomain character and the accurate information on the interface. In basic PFF, the PDE system is coupled through volume terms, and the two equations (solid and phase-field) are both described in the entire domain – which is in general easier than multidomain problems.

In the following we revisit the most important features. Brittle fracture propagation in elasticity using variational techniques has attracted attention in recent years since the pioneering work in [13, 40]. Since in FSI the constitutive stress tensor is generally nonlinear, we also refer to [24] who formulated quasistatic fracture growth using a variational setting employing nonlinear elasticity. The numerical approach [13] is based on Ambrosio-Tortorelli elliptic functionals [3, 4]. Here, discontinuities in the displacement field across the lower dimensional crack surface are approximated by an auxiliary function φ; both features are visualized in Figure 9.2. In the fracture, $\varphi = 0$ and outside $\varphi = 1$. Therefore, this function can be viewed as an indicator function, which introduces a diffusive transition zone, a smeared interface, (of width ε) where $0 < \varphi < 1$ between the broken and the unbroken material. The number ε is a regularization parameter from a mathematical viewpoint and known as a length-scale parameter in physics and engineering-oriented research [75, 96].

An important modification of [40] towards a thermodynamically consistent phase-field fracture (PFF) model has been accomplished in [74, 75]. This approach has been extended to pressurized fractures in [79, 80]. Here, the crack irreversibility constraint has been imposed through penalization. In phase-field fracture, two unknowns are sought: displacements and a phase-field function that determines the crack loca-

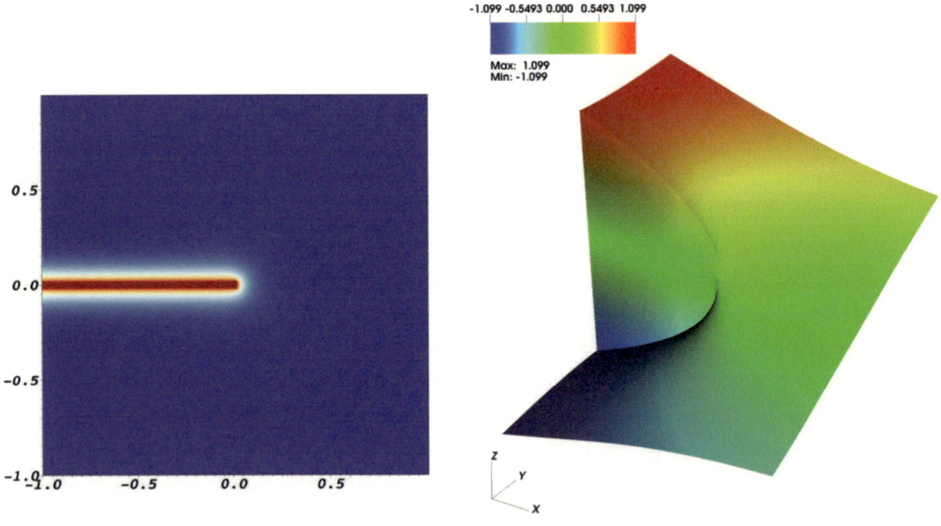

Fig. 9.2: Left: the phase-field variable φ acts as a smoothed indicator function which is 0 in the fracture (here red) and 1 outside (in blue). In between, we have a diffusive transition zone of width ε and where $0 < \varphi < 1$. Right: a 3D surface plot of the displacement variable is shown. Across the fracture, the displacements are discontinuous resulting in a jump of the displacement field. Both figures are taken from [115]. It is clear that an accurate description of the fracture path and the related displacement jump require (very) small ε for which local mesh refinement becomes indispensable as briefly described in the main text.

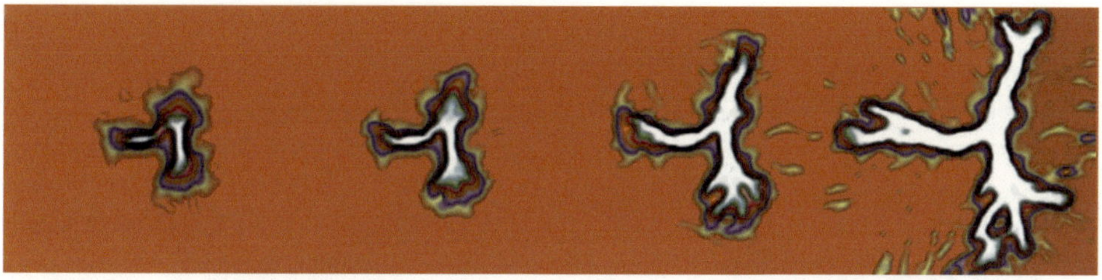

Fig. 9.3: Phase-field fracture (PFF) propagation at four times t_0, t_1, t_2, t_3. The white zone displays the fracture, $\varphi = 0$. The brown region is the unbroken material, $\varphi = 1$. In between (displayed in rainbow isolines) the smooth transition zone $0 < \varphi < 1$ of width ε is visualized. These four subfigures are results of a numerical simulation in a heterogeneous solid material and graphically display attractive features of the PFF method: fractures join, may propagate in arbitrary curvilinear path-directions, and finally branch. The underlying mesh (not displayed here) is regular and not aligned with the fracture path. Thus, specific remeshing is not necessary. The results shown in this figure are based on the code developed in [104] based on [80](Section 5.4).

tion. Recent advances and numerical studies towards pressurized and fluid-filled fracture and other multiphysics applications including thermo-elastic-plastic solids and coupling with a reservoir simulator have been considered in [73, 77, 78, 80, 104, 117]. However, to the best of our knowledge coupling with classical FSI and the need to work with different coordinate systems has not yet been considered, which constitutes a major novelty of [111, 112] and the present work.

Fracture evolutions satisfy a crack irreversibility constraint $\partial_t \varphi \leq 0$ (the crack can never heal) such that the resulting system can be characterized as a variational inequality. Our motivation for employing such a variational approach is that fracture nucleation, propagation, kinking, and curvilinear paths (see Figure 9.3) are automatically included in the model. In addition, explicit remeshing or reconstruction of the crack path is not necessary. The underlying equations are based on continuum mechanics principles that can be treated with (adaptive) Galerkin finite elements. Conversely, the underlying energy functional is not simultaneously convex in both solution variables and a crucial difficulty as discussed in several papers [2, 12, 14, 20, 44, 72] and constitutes a serious difficulty in developing efficient numerical methods.

Another challenge is the resolution of ϵ in relation to the spatial discretization parameter $h < \epsilon$, which requires ideally local mesh adaptivity around the crack zone [52] (see also Figure 9.4) when very small ε are of interest as well as a posteriori error estimation [6, 20] and goal functional evaluations [115]; otherwise the computational cost becomes prohibitive.

A further obstacle using a phase-field approach (i.e. with a smeared interface) is associated with the computation of the displacement jump across the crack resulting in the fracture width (also called aperture or crack opening displacement). An accurate calculation is indispensable since the fracture width describes the radius of the fracture channel and therefore defines the fracture domain. This is in partic-

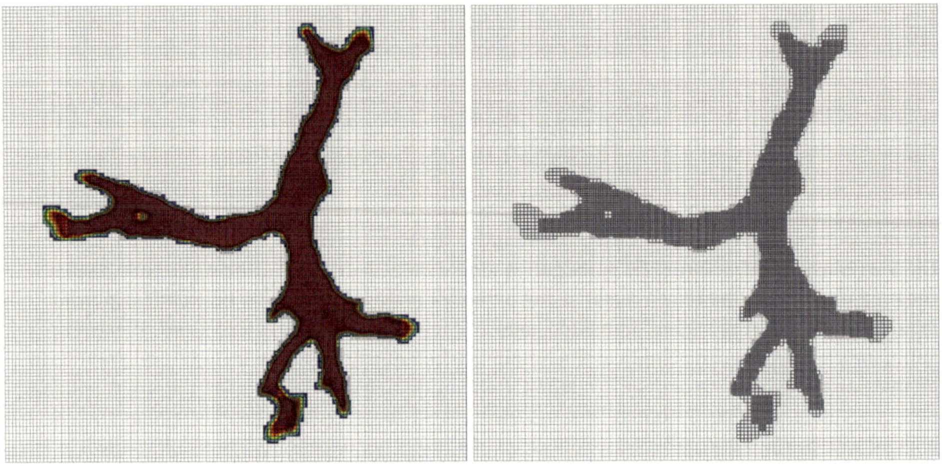

Fig. 9.4: Local mesh refinement near and inside the fracture. Left: the fracture path (in red) is shown. Right: here the focus is on the locally refined mesh. In this computation, three mesh levels are present. The results shown in this figure are based on the code developed in [104] based on [80](Section 5.4).

ular important when other equations are prescribed inside the fracture such as fluid flow [51, 67, 76, 78, 118], or chemical reactions [66]. To date, only a few studies considering accurate width computations have been published. A first approach was presented in [102] using an additional variable. Another method using two-stage level sets with applications in microstructures was introduced in [84]. This last technique has been slightly simplified by translating the phase-field function directly into a level-set function in [68].

As previously described in Section 9.1.1, the solid part of FSI is based on elastodynamics, and therefore we accentuate the work of [11, 15, 63, 64] who extended variational quasistatic brittle fracture to dynamic brittle fracture taking into account the solid acceleration term. Moreover, the authors of [63] introduced an elastic dissipation term, which is for wave propagation problems known as strong solid damping. This term improves the regularity of the solid velocity, see e.g. [43]. In the analysis for dynamic fracture this term was crucial, see [63][Remark 2.2]. It turns out that in our numerical simulations we also need such a term.

Challenge 9.2. The great challenges in phase-field fracture (PFF) are:
- The coupled system is derived from an energy functional that is nonconvex and is therefore extremely difficult to solve using monolithic formulations;
- Line-search or trust region techniques inside Newton's method that allow for a temporary increase of the residual rather than pure monotone convergence;

- Fine spatial meshes (i.e. small h) around the fracture zone in order to allow one to work with a reasonably small regularization parameter ϵ for the crack representation;
- Local mesh adaptivity to accomplish the previous aspect, but which is able to follow a propagating fracture with an a priori unknown fracture path;
- Inequality constraint in time to impose crack irreversibility;
- Nontrivial formulations when conditions (e.g. traction conditions) on the (smeared) crack boundary are described;
- Calculation of the fracture width.

9.1.3 Combining FSI and PFF

Having two frameworks that address, first of all, the challenges of each single subproblem is still ambitious and still a matter of current research as can be seen in the other chapters of this book. However, the combination of Challenge 9.1 (FSI) and 9.2 (PFF) seems a big adventure – and we do by no means claim that we can develop a final approach at this moment. It is rather a starting point. Major difficulties are related to the choice of an appropriate coordinate system and boundary/interface conditions for the PFF-system. To be more specific, an FSI-PFF model using ALE coordinates was first proposed in [112] and then extensively discussed in [111].

Challenge 9.3. To create a common FSI-PFF framework, the crucial steps can be summarized as follows:
- Developing a model for dynamic pressurized fractures including a crack irreversibility constraint.
- Rewriting the PFF model in a fixed reference domain $\widehat{\Omega}$.
- Formulating a common framework for ALE-FSI coupled with PFF.
- Deriving modified dynamic FSI interface conditions for the solid/fluid part. Using monolithic coupling, all coupling conditions are satisfied in a variational exact fashion on the continuous level. This formulation can be written in terms of a Galerkin form.

Moreover, in addition to the computational challenges highlighted in Challenge 9.1 and 9.2, the following aspects must be taken into account as we saw in [111]:
- The solid velocity of the pressurized PFF model is too irregular (a theoretical justification is however missing and our observations are from purely computational experiences) causing the nonlinear solver to break. Consequently, adding solid damping terms was necessary to obtain higher regularity.
- Here we found that changing the time-stepping scheme to dissipative techniques (i.e. backward Euler) would not help.
- However, we also found that damping in the initial steps was sufficient (inspired by Rannacher's time stepping: Crank–Nicolson damped with backward Euler [53, 69, 89]).

- Finally, even PFF alone has not yet been tested in detail for higher order finite elements and (higher order) time-stepping schemes other than backward Euler. Detailed numerical analysis including corresponding computations remain to be done. In [111], however, higher order discretization in space was necessary for the Navier–Stokes part. Less dissipative time stepping (e.g. Crank–Nicolson) is in general preferable for transient FSI because of the wave equation character of the solid equation. Possible consequences (of positive or negative nature) have not yet been investigated in more detail.

All these aspects led to a final model in which the numerical part is (in principle) straightforward under the condition that one has a modular software (such as the C++ open source finite element library deal.II [7, 8]) that allows one to easily add additional equations to a nonlinear multiphysics problem, and additionally allows one freedom for:
- a flexible spatial discretization in terms of the choice of finite elements (and their order) and local mesh adaptivity;
- a hand on all terms to implement a user-defined time-stepping scheme;
- and a nonlinear solver that allows access to all algorithmic steps.

The principle rule of our discretization follows the Rothe method (first time, then space). The discretized, nonlinear coupled problem is solved with Newton's method. Here, the convergence criterion must in particular allow for an increase of the residual. This can be achieved either with an error-oriented line-search globalization [27] or simply using a classical residual-based approach that sometimes violates the convergence criterion. Some convergence results of such Newton methods are provided in [27, 58].

We leave out two things for future development: spatial mesh adaptivity and development of linear solvers. The latter are extensively discussed in two chapters [23, 62] of this book (including references therein) for pure FSI problems. For PFF, linear iterative solvers with high-performance computing (HPC) have been developed in [34, 52, 67]. Spatial mesh adaptivity (in particular with respect to goal functional evaluations) has been developed in FSI for some years now, e.g. [48, 59, 92, 107, 122]. For PFF only recently a posteriori error estimation and mesh adaptivity have been developed [6, 20, 52, 67, 116] and for goal functionals newly in [115]. Thus, all ingredients to implement iterative solvers for FSI-PFF as well as a scheme for local space adaptivity are given, but implementation has yet to be done.

Apart from this overview including the outline of current challenges, we shall address a topic that in general receives a little less attention:
- Adaptive time-step control with respect to a goal functional.

Secondly, we focus on more understanding of
- solid damping as it is currently necessary

and provide
- details on the nonlinear Newton solver.

This chapter is organized as follows: in Section 9.2, the notation, spaces, and necessary variables are gathered. Next, in Section 9.3 the entire problem is stated with a few comments. Then, in our main Section 9.4, we provide descriptions of temporal discretization, adaptive time-step control, spatial discretization, the augmented Lagrangian iteration to impose the crack irreversibility constraint, and the nonlinear Newton solver. Then in Section 9.5 two numerical examples substantiate the algorithmic developments. We finish with concluding remarks and open questions in Section 9.6.

9.2 Notation, spaces, variables, constitutive laws

9.2.1 Philosophy

Our philosophy is to transform all equations from current (i.e. physical) domains $\Omega := \Omega(t)$ to a fixed, arbitrary, reference configuration $\widehat{\Omega}$ and to solve the coupled problem there. Thus, all spaces are formulated with respect to the reference configuration. Moreover, the moving domains are never explicitly computed but represented in terms of mappings (the ALE transformation $\hat{\mathcal{A}}$) that result into the deformation gradient $\widehat{F}$ and its determinant $\hat{J}$. The first quantity measures spatial changes of derivative information, whereas the determinant measures volume changes between the current and the reference configuration. In phase-field fracture, it is usually assumed that we work with small deformations and displacements resulting in the assumption that $\Omega \simeq \widehat{\Omega}$ (which is however questionable in the case where a material is completely damaged). However in FSI, due to large deformations, it is important to decide which coordinate system we want to work with. In this work (i.e. in [111, 112]) we decide to assume that PFF is given in Eulerian coordinates and therefore, we need to transform into. $\widehat{\Omega}$.

9.2.2 Notation

We denote by $\Omega := \Omega(t) \subset \mathbb{R}^d$, $d = 2$, the domain of the FSI phase-field fracture problem; see Figure 9.5. The domain consists of three time-dependent subdomains $\Omega_\mathrm{f}(t)$, $\Omega_\mathrm{s}(t)$, and $\mathcal{C}(t)$. We assume $\mathcal{C}(t) \subset\subset \Omega_\mathrm{s}(t)$. The FSI-interface between $\Omega_\mathrm{f}(t)$ and $\Omega_\mathrm{s}(t)$ is denoted by $\Gamma_i(t) = \overline{\partial\Omega_\mathrm{f}}(t) \cap \overline{\partial\Omega_\mathrm{s}}(t)$. The initial (or later reference) domains are denoted by $\widehat{\Omega}$, $\widehat{\Omega}_\mathrm{f}$, and $\widehat{\Omega}_\mathrm{s}$, respectively, with the interface $\widehat{\Gamma}_i = \overline{\partial\widehat{\Omega}_\mathrm{f}} \cap \overline{\partial\widehat{\Omega}_\mathrm{s}}$. Furthermore, we denote the outer boundary by $\partial\widehat{\Omega} = \widehat{\Gamma} = \widehat{\Gamma}_\mathrm{in} \cup \widehat{\Gamma}_\mathrm{D} \cup \widehat{\Gamma}_\mathrm{out}$ where $\widehat{\Gamma}_\mathrm{D}$ and $\widehat{\Gamma}_\mathrm{in}$ are Dirichlet boundaries (for the velocities and displacements) and $\widehat{\Gamma}_\mathrm{out}$ denotes a fluid outflow Neumann boundary, respectively. The displacements are set to zero on $\widehat{\Gamma}_\mathrm{out}$.

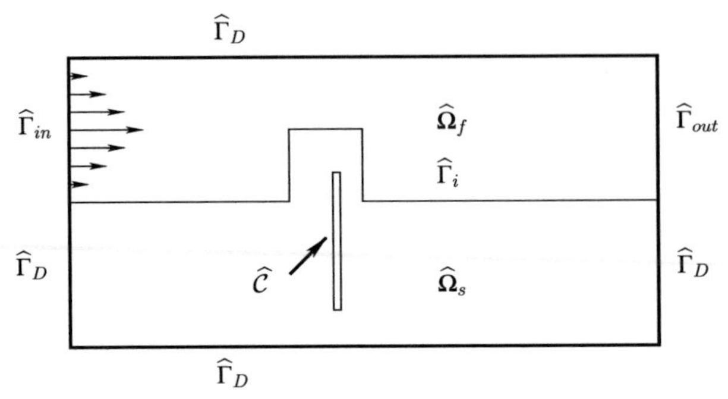

Fig. 9.5: Notation of domains and boundaries.

As frequently used in the literature, we denote the L^2 scalar product in Ω with $(a, b) := (a, b)_\Omega := \int_\Omega a \cdot b\, dx$ for vectors a, b. For (2nd order) tensor-valued functions A, B, it yields $(A, B) := (A, B)_\Omega := \int_\Omega A : B\, dx$, where $A : B = \sum_{ij=1}^{d} A_{ij}B_{ij}$ and A_{ij} and B_{ij} are the entries of A and B.

9.2.3 Function spaces

For the function spaces in the (fixed) reference domains $\widehat{\Omega}, \widehat{\Omega}_\mathrm{f}, \widehat{\Omega}_\mathrm{s}, \widehat{C}$, we define spaces for spatial discretization only. Rather than employing Bochner-spaces [25, 119] for space-time functions, the time t is later explicitly accounted for, e.g. [32] (Section 7.1). Here, let $I := [0, T]$ be the time interval and T the end time value. In the fluid subdomain, we define the following function spaces:

$$
\begin{aligned}
\hat{L}_\mathrm{f} &:= L^2(\widehat{\Omega}_\mathrm{f}) \,, \\
\hat{L}_\mathrm{f}^0 &:= L^2(\widehat{\Omega}_\mathrm{f})/\mathbb{R} \,, \\
\hat{V}_\mathrm{f}^0 &:= \{\hat{v}_\mathrm{f} \in H^1(\widehat{\Omega}_\mathrm{f})^d : \ \hat{v}_\mathrm{f} = 0 \ \text{ on } \widehat{\Gamma}_\mathrm{in} \cup \widehat{\Gamma}_\mathrm{D}\} \,, \\
\hat{V}_{f,\hat{u}}^0 &:= \{\hat{u}_\mathrm{f} \in H^1(\widehat{\Omega}_\mathrm{f})^d : \ \hat{u}_\mathrm{f} = \hat{u}_\mathrm{s} \ \text{ on } \widehat{\Gamma}_i \,, \quad \hat{u}_\mathrm{f} = 0 \ \text{ on } \widehat{\Gamma}_\mathrm{in} \cup \widehat{\Gamma}_\mathrm{D} \cup \widehat{\Gamma}_\mathrm{out}\} \,, \\
\hat{V}_{f,\hat{u},\widehat{\Gamma}_i}^0 &:= \{\hat{\psi}_\mathrm{f} \in H^1(\widehat{\Omega}_\mathrm{f})^d : \ \hat{\psi}_\mathrm{f} = 0 \ \text{ on } \widehat{\Gamma}_i \cup \widehat{\Gamma}_\mathrm{in} \cup \widehat{\Gamma}_\mathrm{D} \cup \widehat{\Gamma}_\mathrm{out}\} \,.
\end{aligned}
$$

The spaces for the solid part read

$$
\begin{aligned}
\hat{L}_\mathrm{s} &:= L^2(\widehat{\Omega}_\mathrm{s})^d \,, \\
\hat{V}_\mathrm{s}^0 &:= \{\hat{u}_\mathrm{s} \in H^1(\widehat{\Omega}_\mathrm{s})^d : \ \hat{u}_\mathrm{s} = 0 \ \text{ on } \widehat{\Gamma}_\mathrm{D}\} \,, \\
\hat{W} &:= \{\hat{\varphi} \in H^1(\widehat{\Omega}_\mathrm{s} \cup \widehat{C}) : \ \hat{\varphi} \le \hat{\varphi}^{old} \ \text{ a.e. on } \widehat{\Omega}_\mathrm{s} \cup \widehat{C}\} \,.
\end{aligned}
$$

In the last space $\hat{\varphi}^{old}$ will later represent the previous time-step solution of the phase-field variable. Moreover, in the presence of a fracture $\widehat{C}$ inside the solid domain $\widehat{\Omega}_\mathrm{s}$ the solid function spaces are extended to hold for functions defined in $\widehat{\Omega}_\mathrm{s} \cup \widehat{C}$.

For the FSI problem using variational-monolithic coupling [30, 31, 56] the velocity spaces are extended from $\widehat{\Omega}_{\mathrm{f}}$ and $\widehat{\Omega}_{\mathrm{s}}$ to the entire domain $\widehat{\Omega}$ such that we can work with global H^1 functions. Thus, we define:

$$\hat{V}^0 := \{\hat{v} \in H^1(\widehat{\Omega})^d : \ \hat{v} = 0 \ \text{ on } \hat{\Gamma}_{\mathrm{in}} \cup \hat{\Gamma}_{\mathrm{D}}\} . \tag{9.1}$$

By this choice, the kinematic and dynamic coupling conditions are automatically satisfied in a variational sense. Alternative definitions with a similar aim can be found, for example, in [17, 35, 36].

Finally, we notice that the spaces on the current domains Ω, Ω_{f}, Ω_{s}, $\mathcal{C}$ are defined correspondingly without the 'hat' notation.

9.2.4 Parameters, variables, vectors, tensors, and constitutive laws

We now provide all the notation for the following.

Definition 9.4. We define:
- $\hat{v} : \widehat{\Omega} \to \mathbb{R}^d$: global velocity for fluid and solid;
- $\hat{u}_{\mathrm{f}} : \widehat{\Omega}_{\mathrm{f}} \to \mathbb{R}^d$: fluid displacement for mesh motion;
- $\hat{u}_{\mathrm{s}} : \widehat{\Omega}_{\mathrm{s}} \to \mathbb{R}^d$: solid displacement;
- $\hat{p}_{\mathrm{f}} : \widehat{\Omega}_{\mathrm{f}} \to \mathbb{R}^d$: fluid pressure;
- $\hat{\varphi} : \widehat{\Omega}_{\mathrm{s}} \to \mathbb{R}^d$: phase-field function to determine fracture;
- $\hat{p}_F : \widehat{\Omega}_{\mathrm{f}} \to \mathbb{R}^d$: given, fixed, fracture pressure (not a solution variable);
- $\hat{\mathcal{A}}(\hat{x}, t) = \hat{x} + \hat{u}_{\mathrm{f}}(\hat{x}, t)$: ALE mapping;
- $\partial_t := \frac{\partial}{\partial t}$: partial time derivative;
- $\hat{w} := \partial_t \hat{\mathcal{A}}$: mesh velocity;
- $\hat{\nabla} = (\hat{\partial}_x, \hat{\partial}_y)$: nabla-operator, gradient in the reference configuration;
- $\hat{I}$: identity matrix;
- $\widehat{F} := \widehat{F}_{\mathrm{f}} := \hat{\nabla}\hat{\mathcal{A}} = \hat{I} + \hat{\nabla}\hat{u}_{\mathrm{f}}$: fluid deformation gradient;
- $\widehat{F} := \widehat{F}_{\mathrm{s}} := \hat{I} + \hat{\nabla}\hat{u}_{\mathrm{s}}$: solid deformation gradient;
- $\hat{J} := \det(\widehat{F})$: determinant of deformation gradient;
- $\hat{\rho}_{\mathrm{f}}, \hat{\rho}_{\mathrm{s}}$: fluid and solid density, respectively;
- ν_{f}: kinematic fluid viscosity;
- $\hat{n}_{\mathrm{f}}$: normal vector;
- $\hat{\sigma}_{\mathrm{f}} = -\hat{p}_{\mathrm{f}}\hat{I} + 2\hat{\rho}_{\mathrm{f}}\nu_{\mathrm{f}}(\hat{\nabla}\hat{v}_{\mathrm{f}}\widehat{F}^{-1} + \widehat{F}^{-T}\hat{\nabla}\hat{v}_{\mathrm{f}}^T)$: fluid Cauchy stress tensor in ALE format;
- μ, λ: Lamé coefficients;
- $tr(A) = A_{11} + A_{22}$: trace operator applied to a matrix A;
- $\widehat{E} = \frac{1}{2}(\widehat{F}^T\widehat{F} - I)$: Green–Lagrange strain tensor;
- $\widehat{\Sigma} = \widehat{\Sigma}_{\mathrm{s}}(\hat{u}_{\mathrm{s}}) = 2\mu\widehat{E} + \lambda tr(\widehat{E})I$: solid stress tensor, STVK material;
- $\widehat{E}_{\mathrm{lin}} = \frac{1}{2}(\hat{\nabla}\hat{u} + \hat{\nabla}\hat{u}^T)$: linearized Green–Lagrange strain tensor;
- μ_v, λ_v: coefficients of the (strong) damping term in the solid;
- $\widehat{\Sigma}_v = \widehat{\Sigma}_v(\hat{u}_{\mathrm{s}}) = 2\mu_v\widehat{E}_{\mathrm{lin}} + \lambda_v tr(\widehat{E}_{\mathrm{lin}})I$: constitutive linearized tensor for solid damping;

- $\widehat{\sigma}_{\text{mesh}} = \hat{J}^{-1}\alpha_u \widehat{\nabla}\hat{u}_{\text{f}}$, $\hat{J}^{-1}\alpha_u > 0$: constitutive law for mesh motion;
- h: mesh cell diameter, spatial discretization parameter;
- $k = t^n - t^{n-1}$: time-step size, temporal discretization parameter;
- ϵ: phase-field regularization parameter, $\epsilon < h$;
- κ: phase-field regularization of the bulk energy, $\kappa \approx 0$;
- Ξ: augmented Lagrangian penalization function;
- γ: augmented Lagrangian penalization parameter.

9.3 The fluid-structure interaction phase-field fracture (FSI-PFF) model

9.3.1 The FSI-PFF system

In [111][Formulation 8] a very detailed derivation of the following system has been performed:

Formulation 9.5 (A variational-monolithic ALE FSI-PFF framework in $\widehat{\Omega}$). Let a fracture pressure $\hat{p}_F \in H^1(\widehat{\Omega}_{\text{s}} \cup \widehat{C})$ be given. Find a global vector-valued velocity, vector-valued displacements, a scalar-valued fluid pressure, and a scalar-valued phase-field function, that is to say that $\{\hat{v}, \hat{u}_{\text{f}}, \hat{u}_{\text{s}}, \hat{p}_{\text{f}}, \hat{\varphi}\} \in \{\hat{v}^D + \hat{V}^0\} \times \{\hat{u}_{\text{f}}^D + \hat{V}^0_{f,\hat{u}}\} \times \hat{V}^0_{\text{s}} \times \hat{L}^0_{\text{f}} \times H^1(\widehat{\Omega}_{\text{s}} \cup \widehat{\mathcal{C}})$, such that $\hat{v}(0) = \hat{v}^0$, $\hat{u}_{\text{f}}(0) = \hat{u}^0_{\text{f}}$, $\hat{u}_{\text{s}}(0) = \hat{u}^0_{\text{s}}$, and $\hat{\varphi}(0) = \hat{\varphi}^0$ are satisfied, and for almost all times $t \in I$ holds:

$$\text{Fluid/solid momentum} \begin{cases} (\hat{J}\hat{\rho}_{\text{f}}\partial_t\hat{v}, \hat{\psi}^v)_{\widehat{\Omega}_{\text{f}}} + (\hat{\rho}_{\text{f}}\hat{J}(\widehat{F}^{-1}(\hat{v} - \hat{w}) \cdot \widehat{\nabla})\hat{v}), \hat{\psi}^v)_{\widehat{\Omega}_{\text{f}}} \\ +(\hat{J}\widehat{\sigma}_{\text{f}}\widehat{F}^{-T}, \widehat{\nabla}\hat{\psi}^v)_{\widehat{\Omega}_{\text{f}}} + \langle \rho_{\text{f}}\nu_{\text{f}}\hat{J}(\widehat{F}^{-T}\widehat{\nabla}\hat{v}^T\hat{n}_{\text{f}})\widehat{F}^{-T}, \hat{\psi}^v\rangle_{\widehat{\Gamma}_{\text{out}}} \\ +(\hat{\rho}_{\text{s}}\partial_t\hat{v}, \hat{\psi}^v)_{\widehat{\Omega}_{\text{s}}} + (E(\hat{\varphi})\widehat{F}\widehat{\Sigma}, \widehat{\nabla}\hat{\psi}^v)_{\widehat{\Omega}_{\text{s}}} + (E(\hat{\varphi})\widehat{\Sigma}_v(\hat{v}), \widehat{\nabla}\hat{\psi}^v)_{\widehat{\Omega}_{\text{s}}} \\ +(\hat{\varphi}^2\hat{p}_F, \widehat{\nabla}\cdot(\hat{J}\widehat{F}^{-1}\hat{\psi}^v)) \quad = 0 \quad \forall\hat{\psi}^v \in \hat{V}^0 , \end{cases}$$

$$\text{Fluid mesh motion} \left\{ (\widehat{\sigma}_{\text{mesh}}, \widehat{\nabla}\hat{\psi}^u_{\text{f}})_{\widehat{\Omega}_{\text{f}}} \quad = 0 \quad \forall\hat{\psi}^u_{\text{f}} \in \hat{V}^0_{f,\hat{u},\widehat{\Gamma}_i} , \right.$$

$$\text{Solid momentum, 2nd eq.} \left\{ \hat{\rho}_{\text{s}}(\partial_t\hat{u}_{\text{s}} - \hat{v}|_{\widehat{\Omega}_{\text{s}}}, \hat{\psi}^u_{\text{s}})_{\widehat{\Omega}_{\text{s}}} \quad = 0 \quad \forall\hat{\psi}^u_{\text{s}} \in \hat{L}_{\text{s}} , \right.$$

$$\text{Fluid mass conservation} \left\{ (\widehat{\operatorname{div}}\,(\hat{J}\widehat{F}^{-1}\hat{v}), \hat{\psi}^p_{\text{f}})_{\widehat{\Omega}_{\text{f}}} \quad = 0 \quad \forall\hat{\psi}^p_{\text{f}} \in \hat{L}^0_{\text{f}} , \right.$$

$$\text{Phase-field} \begin{cases} (1-\kappa)(\hat{J}\hat{\varphi}\,(\widehat{\Sigma} + \widehat{\Sigma}_v) : \widehat{E}, \hat{\psi}^\varphi_{\text{s}})_{\widehat{\Omega}_{\text{s}}} + 2(\hat{\varphi}\hat{p}_F\,\widehat{\nabla}\cdot(\hat{J}\widehat{F}^{-1}\hat{u}_{\text{s}}), \hat{\psi}^\varphi_{\text{s}})_{\widehat{\Omega}_{\text{s}}} \\ -\langle 2\hat{J}\hat{\varphi}\hat{p}_F\widehat{F}^{-T}\hat{n}_{\text{s}}\hat{u}_{\text{s}}, \hat{\psi}^\varphi_{\text{s}}\rangle_{\widehat{\Gamma}_i} \\ +G_c\Big(-\frac{1}{\epsilon}(\hat{J}(1-\hat{\varphi}), \hat{\psi}^\varphi_{\text{s}}) + \epsilon(\hat{J}(\widehat{\nabla}\hat{\varphi}\widehat{F}^{-1})\widehat{F}^{-T}, \widehat{\nabla}\hat{\psi}^\varphi_{\text{s}})\Big)_{\widehat{\Omega}_{\text{s}}} \\ +(\hat{J}[\Xi + \gamma(\hat{\varphi} - \hat{\varphi}^{old})]^+, \hat{\psi}^\varphi_{\text{s}})_{\widehat{\Omega}_{\text{s}}} \quad = 0 \quad \forall\hat{\psi}^\varphi_{\text{s}} \in H^1(\widehat{\Omega}_{\text{s}} \cup \widehat{\mathcal{C}}) . \end{cases}$$

In the following we make further comments on the
- crack irreversibility constraint;
- FSI interface conditions;
- fluid and fracture pressure.

9.3.2 Treatment of the crack irreversibility constraint

We apply the augmented Lagrangian technique to relax the inequality constraint $\partial_t \hat{\varphi} \le 0$ (the crack irreversibility condition) in terms of a variational equality [104]. Similar penalization-type approaches have been suggested in [80, 81, 83]. Other approaches are based on a direct Dirichlet-boundary-like enforcement of the irreversibility constraint [12, 13, 61], via a strain-history variable [11, 74], or with a semismooth Newton method [52, 67].

In more detail, we first apply a backward difference quotient to $\partial_t \varphi \le 0$ in order to obtain an incremental version of the inequality constraint, first in Eulerian coordinates, and then in the transformed coordinate system:

$$\partial_t \varphi \le 0 \quad \Rightarrow \frac{\varphi - \varphi^{old}}{k} \le 0 \quad \Rightarrow \varphi \le \varphi^{old} \quad \Rightarrow \hat{\varphi} \le \hat{\varphi}^{old}\,, \tag{9.2}$$

here $\hat{\varphi}^{old}$ represents later the previous time-step solution, i.e. $\hat{\varphi}^{old} := \hat{\varphi}^{n-1}$. Furthermore, k denotes the time-step size, but does not play a role in this section. Following [39, 46] and our previous work [104], we obtain as augmented Lagrangian penalization:

$$\left[\Xi + \gamma(\hat{\varphi} - \hat{\varphi}^{old})\right]^{+}, \tag{9.3}$$

where $\Xi \in L^2$ and $\gamma > 0$. Moreover, $[x]^{+} := max(x, 0)$. The key idea is to iterate the function Ξ in order to realize a reliable approximation to the constraint.

9.3.3 FSI interface coupling conditions

The coupling of a fluid with a solid must satisfy two physical conditions: namely continuity of velocities and continuity of normal stresses. A third condition of geometric nature is necessary when working with the ALE framework: continuity of displacements, which couples the physical solid $\hat{u}_s$ and the fluid mesh motion $\hat{u}_f$. Mathematically, the first and third conditions can be classified as (nonhomogeneous) Dirichlet conditions and the second condition is a (nonhomogeneous) Neumann condition.

In variational-monolithic coupling these Dirichlet conditions are built into the corresponding function space by employing a globalized Sobolev space $\hat{V}^0$; see (9.1). Neumann-type conditions are weakly incorporated through interface integrals, which actually cancel out in the later models because of their weak continuity thanks to working with the space $\hat{V}^0$. For the fluid problem, continuity of velocities is required (i.e. a kinematic coupling condition):

$$v_f = v_s \quad \text{on } \Gamma_i\,. \tag{9.4}$$

For the geometric problem we have

$$u_f = u_s \quad \text{on } \Gamma_i\,. \tag{9.5}$$

Finally, the Neumann coupling conditions (dynamic coupling conditions) on $\widehat{\Gamma}_i$ read:

$$\langle \hat{J}\hat{\sigma}_f\widehat{F}^{-T}\hat{n}_f, \hat{\psi}^v\rangle_{\widehat{\Gamma}_i} + \langle E(\hat{\varphi})[\widehat{F}\widehat{\Sigma} + \hat{\Sigma}_v(\hat{v}) - \hat{J}\hat{p}_F\widehat{F}^{-T}]\hat{n}_s, \hat{\psi}^v\rangle_{\widehat{\Gamma}_i} \tag{9.6}$$

$$= \langle \hat{J}\hat{\sigma}_f\widehat{F}^{-T}\hat{n}_f, \hat{\psi}^v\rangle_{\widehat{\Gamma}_i} + \langle E(\hat{\varphi})\widehat{F}\widehat{\Sigma}\hat{n}_s, \hat{\psi}^v\rangle_{\widehat{\Gamma}_i} \tag{9.7}$$

$$+ \langle E(\hat{\varphi})\hat{\Sigma}_v(\hat{v})\hat{n}_s, \hat{\psi}^v\rangle_{\widehat{\Gamma}_i} - \langle E(\hat{\varphi})\hat{J}\hat{p}_F\widehat{F}^{-T}\hat{n}_s, \hat{\psi}^v\rangle_{\widehat{\Gamma}_i} \tag{9.8}$$

$$= 0 \quad \forall \hat{\psi}^v \in \hat{V}^0 . \tag{9.9}$$

Here, the two terms in (9.7) are the standard FSI interface Neumann conditions. The other two terms in (9.8) are nonstandard. The first of these new terms comes from the solid damping term, see for example also in [105]. The second new term is obtained from the PFF system by Gauss' divergence theorem. A detailed derivation can be found in [111].

9.3.4 Remarks on pressures

Extensive remarks in regard to modeling that is currently possible are made in [111]. Namely, therein the (nonexisting direct) relationship between the Navier–Stokes pressure $\hat{p}_f$ and the fracture pressure $\hat{p}_F$ is explained. However, there is an indirect relationship between $\hat{p}_f$ and $\hat{p}_F$ already present. If the flow velocity is too high (thus $\hat{p}_f \gg \hat{p}_F$) then the fracture will be compressed and will finally collapse. Conversely, if $\hat{p}_f \ll \hat{p}_F$, then the obstacle will burst because the solid can no longer resist against the too high fracture pressure. Both phenomena are physically absolutely possible. Therefore, $\hat{p}_f$ and $\hat{p}_F$ should be in a similar range if one is interested in (academic) test cases in order to study the approach for reasonably long computation times.

9.4 Discretization

Formulation 9.5 is first formulated in terms of a compact single semilinear form $\hat{A}(\cdot)(\cdot)$: Find $\hat{U}$ from a space $\widehat{X}$ such that

$$\int_0^T \hat{A}(\hat{U})(\hat{\Psi}) = 0 \quad \forall \hat{\Psi} \in \widehat{X} .$$

This system is tackled with the Rothe method in which temporal discretization is carried out first, and subsequently spatial discretization. This procedure leaves the flexibility to vary the spatial discretization during the time-stepping process. Next, two iteration loops are required: as outer loop an augmented Lagrangian iteration to impose the temporal inequality constraint. As inner solver, actually the heart of treating the nonlinear problem, a Newton method has been implemented.

9.4.1 Time stepping

In the domain $\widehat{\Omega}$ and the time interval $I = [0, T]$, we consider Formulation 9.5. Our goal is to apply A-stable finite differences in time. Specifically, time discretization is based on a One-step-θ scheme as presented for the pure FSI problem [56, 106]. Computational stability of these schemes has been investigated in [94]. With regard to general phase-field models discretized with second-order time-stepping schemes we refer to [47].

In more detail, semidiscretization in time yields a sequence of generalized steady-state problems that are completed by appropriate boundary values at every time step. Let

$$I = \{0\} \cup I_1 \cup \ldots \cup I_N$$

be a partition of the time interval I into half-open subintervals $I_n := (t_{n-1}, t_n]$ of (time step) size $k := k_n := t_n - t_{n-1}$ with

$$0 = t_0 < \cdots < t_N = T .$$

Time derivatives are discretized with a backward difference quotient such that

$$\partial_t \hat{\varphi} \approx \frac{\hat{\varphi} - \hat{\varphi}^{n-1}}{k}, \quad \partial_t \hat{u} \approx \frac{\hat{u} - \hat{u}^{n-1}}{k}, \quad \partial_t \hat{v} \approx \frac{\hat{v} - \hat{v}^{n-1}}{k},$$

where $\hat{\varphi} := \hat{\varphi}^n := \hat{\varphi}(t_n)$, $\hat{u} := \hat{u}^n := \hat{u}(t_n)$, $\hat{v} := \hat{v}^n := \hat{v}(t_n)$, $\hat{\varphi}^{n-1} := \hat{\varphi}(t_{n-1})$, $\hat{u}^{n-1} := \hat{u}(t_{n-1})$, $\hat{v}^{n-1} := \hat{v}(t_{n-1})$. Furthermore, the mesh velocity $\partial_t \hat{\mathcal{A}} = \hat{w}$ is numerically realized as $\hat{w} = k^{-1}(\hat{u}_{\mathrm{f}} - \hat{u}_{\mathrm{f}}^{n-1})$.

Formulation 9.6 (The time-discretized abstract problem). We seek $\hat{U}^n = \{\hat{v}^n, \hat{u}_{\mathrm{f}}^n, \hat{u}_{\mathrm{s}}^n, \hat{p}_{\mathrm{f}}^n, \hat{\varphi}_{\mathrm{s}}^n\} \in \widehat{X}_{\mathrm{D}}^0$, where $\widehat{X}_{\mathrm{D}}^0 := \{\hat{v}^D + \hat{V}^0\} \times \{\hat{u}_{\mathrm{f}}^D + \hat{V}_{f,\hat{u}}^0\} \times \hat{V}_{\mathrm{s}}^0 \times \hat{L}_{\mathrm{f}}^0 \times H^1(\widehat{\Omega}_{\mathrm{s}} \cup \widehat{C})$ and $\widehat{X} = \hat{V}^0 \times \hat{V}_{f,\hat{u},\widehat{\Gamma}_i}^0 \times \hat{V}_{\mathrm{s}}^0 \times \hat{L}_{\mathrm{f}}^0 \times H^1(\widehat{\Omega}_{\mathrm{s}} \cup \widehat{C})$, for all $n = 1, 2, \ldots, N$ such that

$$\hat{A}(\hat{U}^n)(\hat{\Psi}) = 0 \quad \forall \hat{\Psi} \in \widehat{X}, \tag{9.10}$$

where the semilinear form $\hat{A}(\cdot)(\cdot)$ is split into

$$\hat{A}(\hat{U}^n)(\hat{\Psi}) := \hat{A}_T(\hat{U}^{n,t})(\hat{\Psi}) + \hat{A}_I(\hat{U}^n)(\hat{\Psi}) + \hat{A}_E(\hat{U}^n)(\hat{\Psi}) + \hat{A}_P(\hat{U}^n)(\hat{\Psi}) .$$

The previous splitting has four categories: time equation terms (including the time derivatives); implicit terms (such as the fluid incompressibility and also the phase-field equation); pressure terms; and finally all 'standard' terms (stress terms, fluid convection, and elastic dissipation). In detail we have the decomposition:

$$\begin{aligned} \hat{A}_T(\hat{U})(\hat{\Psi}) = {} & (\hat{J}\hat{\rho}_{\mathrm{f}}\partial_t\hat{v}, \hat{\psi}^v)_{\widehat{\Omega}_{\mathrm{f}}} - (\hat{\rho}_{\mathrm{f}}\hat{J}(\hat{F}^{-1}\hat{w} \cdot \hat{\nabla})\hat{v}, \hat{\psi}^v)_{\widehat{\Omega}_{\mathrm{f}}} \\ & + (\hat{\rho}_{\mathrm{s}}\partial_t\hat{v}, \hat{\psi}^v)_{\widehat{\Omega}_{\mathrm{s}}} + (\hat{\rho}_{\mathrm{s}}\partial_t\hat{u}_{\mathrm{s}}, \hat{\psi}_{\mathrm{s}}^u)_{\widehat{\Omega}_{\mathrm{s}}}, \end{aligned}$$

$$
\begin{aligned}
\hat{A}_I(\hat{U})(\hat{\Psi}) &= (\widehat{\sigma}_{\mathrm{mesh}}, \widehat{\nabla}\hat{\psi}^u_{\mathrm{f}})_{\widehat{\Omega}_{\mathrm{f}}} + (\widehat{\mathrm{div}}\,(\hat{J}\widehat{F}^{-1}\hat{v}), \hat{\psi}^p_{\mathrm{f}})_{\widehat{\Omega}_{\mathrm{f}}} \\
&\quad + (1-\kappa)(\hat{J}\hat{\varphi}\,(\widehat{\Sigma} + \widehat{\Sigma}_v)\colon \widehat{E}, \hat{\psi}^\varphi_{\mathrm{s}})_{\widehat{\Omega}_{\mathrm{s}}} + 2(\hat{\varphi}\hat{p}_F\,\widehat{\nabla}\cdot(\hat{J}\widehat{F}^{-1}\hat{u}_{\mathrm{s}}), \hat{\psi}^\varphi_{\mathrm{s}})_{\widehat{\Omega}_{\mathrm{s}}} \\
&\quad - \langle 2\hat{J}\hat{\varphi}\hat{p}_F\widehat{F}^{-T}\hat{n}_{\mathrm{s}}\hat{u}_{\mathrm{s}}, \hat{\psi}^\varphi_{\mathrm{s}}\rangle_{\widehat{\Gamma}_i} \\
&\quad + G_c\Big(-\frac{1}{\epsilon}(\hat{J}(1-\hat{\varphi}), \hat{\psi}^\varphi_{\mathrm{s}}) + \epsilon(\hat{J}(\widehat{\nabla}\hat{\varphi}\widehat{F}^{-1})\widehat{F}^{-T}, \widehat{\nabla}\hat{\psi}^\varphi_{\mathrm{s}})\Big)_{\widehat{\Omega}_{\mathrm{s}}} \\
&\quad + (\hat{J}[\Xi + \gamma(\hat{\varphi} - \hat{\varphi}^{n-1})]^+, \hat{\psi}^\varphi_{\mathrm{s}})_{\widehat{\Omega}_{\mathrm{s}}}, \\
\hat{A}_P(\hat{U})(\hat{\Psi}) &= (\hat{J}\widehat{\sigma}_{f,p}\widehat{F}^{-T}, \widehat{\nabla}\hat{\psi}^v)_{\widehat{\Omega}_{\mathrm{f}}}, \\
\hat{A}_E(\hat{U})(\hat{\Psi}) &= (\hat{\rho}_{\mathrm{f}}\hat{J}(\widehat{F}^{-1}\hat{v}\cdot\widehat{\nabla})\hat{v}), \hat{\psi}^v)_{\widehat{\Omega}_{\mathrm{f}}} + (\hat{J}\widehat{\sigma}_{f,vu}\widehat{F}^{-T}, \widehat{\nabla}\hat{\psi}^v)_{\widehat{\Omega}_{\mathrm{f}}} \\
&\quad + \langle \rho_{\mathrm{f}}\nu\hat{J}(\widehat{F}^{-T}\widehat{\nabla}v_{\mathrm{f}}^T)\widehat{F}^{-T}\hat{n}, \hat{\psi}^v\rangle_{\widehat{\Gamma}_{\mathrm{out}}} \\
&\quad + (E(\hat{\varphi})\widehat{F}\widehat{\Sigma}, \widehat{\nabla}\hat{\psi}^v)_{\widehat{\Omega}_{\mathrm{s}}} \\
&\quad + (E(\hat{\varphi})\widehat{\Sigma}_v(\hat{v}), \widehat{\nabla}\hat{\psi}^v)_{\widehat{\Omega}_{\mathrm{s}}} + (\hat{\varphi}^2\hat{p}_F, \widehat{\nabla}\cdot(\hat{J}\widehat{F}^{-1}\hat{\psi}^v))_{\widehat{\Omega}_{\mathrm{s}}} - (\hat{\rho}_{\mathrm{s}}\hat{v}, \hat{\psi}^u_{\mathrm{s}})_{\widehat{\Omega}_{\mathrm{s}}},
\end{aligned}
$$

where the fluid stress tensor $\widehat{\sigma}_{\mathrm{f}}$ is further split into $\widehat{\sigma}_{f,vu}, \widehat{\sigma}_{f,p}$:

$$
\widehat{\sigma}_{f,p} = -\hat{p}_{\mathrm{f}}\hat{I}, \quad \widehat{\sigma}_{f,vu} = \rho_{\mathrm{f}}\nu_{\mathrm{f}}(\widehat{\nabla}\hat{v}\widehat{F}^{-1} + \widehat{F}^{-T}\widehat{\nabla}\hat{v}^T).
$$

The (nonlinear) time derivative in $\hat{A}_T(\hat{U})(\hat{\Psi})$ is approximated by a backward difference quotient. For the time-step $t_n \in I$, for $n = 1, 2, \ldots, N(N \in \mathbb{R})$, we compute $\hat{v} := \hat{v}^n, \hat{u}_i := \hat{u}_i^n\ (i = f, s)$ via

$$
\begin{aligned}
\hat{A}_T(\hat{U}^{n,t})(\hat{\Psi}) &\approx \frac{1}{k}(\hat{\rho}_{\mathrm{f}}\hat{J}^{n,\theta}(\hat{v} - \hat{v}^{n-1}), \hat{\psi}^v)_{\widehat{\Omega}_{\mathrm{f}}} - \frac{1}{k}(\hat{\rho}_{\mathrm{f}}(\hat{J}\widehat{F}^{-1}(\hat{u}_{\mathrm{f}} - \hat{u}_{\mathrm{f}}^{n-1})\cdot\widehat{\nabla})\hat{v}, \hat{\psi}^v)_{\widehat{\Omega}_{\mathrm{f}}} \\
&\quad + \frac{1}{k}(\hat{\rho}_{\mathrm{s}}(\hat{v} - \hat{v}^{n-1}), \hat{\psi}^v)_{\widehat{\Omega}_{\mathrm{s}}} + (\hat{u}_{\mathrm{s}} - \hat{u}_{\mathrm{s}}^{n-1}, \hat{\psi}^u)_{\widehat{\Omega}_{\mathrm{s}}},
\end{aligned}
\tag{9.11}
$$

where we introduce the parameter $\theta \in [0, 1]$. Furthermore, we use

$$
\hat{J}^{n,\theta} = \theta\hat{J}^n + (1-\theta)\hat{J}^{n-1},
$$

and $\hat{u}_i^n := \hat{u}_i(t_n)$, $\hat{v}^n := \hat{v}(t_n)$, and $\hat{J} := \hat{J}^n := \hat{J}(t_n)$. The former time step is given by $\hat{v}^{n-1}$, etc., for $i = f, s$.

Remark 9.7 (Convenience of the splitting). The previous splitting has no advantage except a clean notation. In more detail, it allows for a very compact notation when applying One-Step-θ and Fractional-Step-θ schemes [16] to nonlinear coupled PDE systems and multiphysics problems in which some terms are treated explicitly and other terms fully implicitly. For instance, for temporal discretization of the Navier–Stokes equations, the pressure should be treated fully implicitly, thus no terms such as $(1-\theta)\hat{p}_{\mathrm{f}}^{n-1}$ should appear. Separating all terms in the above way yields immediately the correct temporal scheme as used for instance in [106]. Using this notational idea was one key part that greatly facilitated the implementation for the further extension to FSI plus PFF. Of course, it makes no difference in the code itself (except its readability) if such a splitting is used or not. However, the readability and sustainability of programming code is extremely important. The goal should be to have a code that can be immediately used again by yourself and extended after some time (let's say after three years of nonusage), or a code that is immediately usable by others; in this regard the author has had very positive experiences with the code in [108]. For general comments on code development and its treatment we also refer the reader to [65].

Formulation 9.8. Let the previous time-step solution $\hat{U}^{n-1} = \{\hat{v}^{n-1}, \hat{u}_f^{n-1}, \hat{u}_s^{n-1}, \hat{p}_f^{n-1}, \hat{\varphi}_s^{n-1}\}$ and the time-step $k := k_n = t_n - t_{n-1}$ be given. In order to solve (9.10), we seek $\hat{U}^n = \{\hat{v}^n, \hat{u}_f^n, \hat{u}_s^n, \hat{p}_f^n, \hat{\varphi}_s^n\} \in \hat{X}_D^0$ by employing One-step-θ splitting:

$$\begin{aligned}&\hat{A}_T(\hat{U}^{n,t})(\hat{\Psi}) + \theta\hat{A}_E(\hat{U}^n)(\hat{\Psi}) + \hat{A}_P(\hat{U}^n)(\hat{\Psi}) + \hat{A}_I(\hat{U}^n)(\hat{\Psi}) \\ &\quad = -(1-\theta)\hat{A}_E(\hat{U}^{n-1})(\hat{\Psi}) \quad \forall \hat{\Psi} \in \hat{X}\,.\end{aligned} \tag{9.12}$$

The concrete scheme depends on the choice of the parameter $\theta \in [0,1]$. For $\theta = 1$ we obtain the strongly A-stable backward Euler scheme (BE). For $\theta = 0.5 + k_{\text{shift}}$, we obtain the 2nd-order (shown for linear parabolic problems in [53, 90]), A-stable, globally stabilized, Crank–Nicolson scheme (CNs), which is shifted in this paper by $k_{\text{shift}} = 0.01\,s$ towards the implicit side. Ideally, one could use $k_{\text{shift}} = k_n$, which however requires the computation of a characteristic time-step size with $k_n \ll 0.5$ because the choice $k_{\text{shift}} = k_n \geq 0.5$ is useless since $\theta \geq 1$ would follow.

Remark 9.9. Formulation 9.8 is still nonlinear and continuous on the spatial level.

9.4.2 Adaptive time-step control with respect to goal functionals

Our aim is to control the accuracy of a certain physical quantity of interest $J(\hat{U})$ during the computation. To achieve a desired accuracy, we employ adaptive time-step control. The principle idea, e.g. [88], is based on ODE (ordinary differential equations) theory and utilizes the local truncation error. Thus, the only assumptions we have to make is enough regularity of the truncation error and discrete stability of the underlying difference scheme used for discretizing the continuous time derivatives such as $\partial_t \hat{v}_f$.

The following description has been adapted from [33] (based on [70, 71]) in which time-step control has been developed for the Fractional-Step-θ scheme for FSI computations. For simplicity, we are only interested in the heuristic estimator of [33] and apply this idea now to One-Step-θ schemes, namely the shifted Crank–Nicolson scheme. We also want to mention [10] who considered time-adaptivity for fluid-structure interaction coupled to thermal problems.

The entire approach is a combination of known results from the literature, which we briefly recapitulate to acknowledge the original authors. We follow an idea proposed by [101] but extend his criterion with additional terms, which are inspired by [50, 60]. The result is a (very) simple method but the price to pay is additional solves of the problem per time step. Specifically, per time step, we compute once the problem with $2k$, compute the goal functional $J(\hat{U})$, and then compute two times with time-step size k again. Afterwards, we compare $J(\hat{U}_k)$ and $J(\hat{U}_{2k})$ by evaluating the absolute error $abs_{\text{err}} := |J(\hat{U}_{2k}) - J(\hat{U}_k)|$.

Proposition 9.10. *Let the error tolerance TOL_{TS} be given. We evaluate:*

$$abs_{err} := |J(\hat{U}_{2k}) - J(\hat{U}_k)| \tag{9.13}$$

and then

$$\theta = \gamma * \left(\frac{TOL_{TS}}{abs_{err}}\right)^K , \tag{9.14}$$

where $\gamma \approx 1$ is a safety factor (in [50], $\gamma = 0.9$), $K = \frac{1}{15}$ as in [50]. We then compute the new time-step size as

$$k_{new} = \begin{cases} k_{old} & 1 \le \theta \le 1.2\,, \\ k_{old} * \theta & \text{otherwise}\,. \end{cases} \tag{9.15}$$

Moreover, we check whether $k_{min} \le k_{new} \le k_{max}$. If $k_{new} < 0.5k_{old}$, i.e. the new time step would be much smaller than the old one, we redo the current time step [101].

Remark 9.11. The check (9.15) prevents high time-step oscillations and just keeps the old time step if the difference is not too large. Several illustrative computations have been presented in [50].

Remark 9.12. The last check $k_{\text{new}} < 0.5k_{\text{old}}$ is very expensive since the entire calculation of the solution of this time step has to be repeated. However, in practice this happens rarely, but is necessary since a dramatic decrease of the time-step size indicates that the current solution is no where near accurate enough.

Algorithm 9.13. The adaptive time-step control for One-Step-θ schemes reads at t^n:
1. Set $k_{\text{old}} := k_{n-1}$
2. Perform
   - one computation with $2k_{\text{old}}$, evaluate $J(\hat{u}_{2k})$ at $t^n + 2k_{\text{old}}$;
   - compute two steps with k_{old}, evaluate at the end of the 2nd step $J(\hat{u}_k)$ at $t^n + k_{\text{old}} + k_{\text{old}}$;
3. Evaluate (9.13) and compute (9.14);
4. Compute the new k_{new} with (9.15);
5. If $k_{\text{new}} < 0.5k_{\text{old}}$, set $k_{\text{old}} := k_{\text{new}}$ and go to Step (ii);
6. Otherwise accept k_{new} and set $k_n = k_{\text{new}}$;
7. Increment $n \to n + 1$ and go to Step (i).

9.4.3 Spatial discretization

The time-discretized Formulation 9.8 is the starting point for the Galerkin discretization in space. To this end, we construct finite dimensional subspaces $\widehat{X}_h \subset \widehat{X}$ to find an approximate solution to the continuous problem. As previously explained, in the context of our variational-monolithic formulation, the computations are done in the reference configuration $\widehat{\Omega}$. We use two-dimensional shape-regular meshes. A mesh consists of quadrilateral cells $\hat{K}$. They perform a nonoverlapping cover of the computation domain $\hat{\Omega} \subset \mathbb{R}^d$, $d = 2$. The corresponding mesh is given by $\hat{\mathfrak{T}}_h = \{\hat{K}\}$. The discretization parameter in the reference configuration is denoted by $\hat{h}$ and is a cellwise constant that is given by the diameter $\hat{h}_{\hat{K}}$ of the cell $\hat{K}$.

On $\hat{\mathfrak{T}}_h$, the conforming finite element space for $\{\hat{v}_h, \hat{u}_{f,h}, \hat{u}_{s,h}, \hat{p}_{f,h}, \hat{\varphi}_h\}$ is denoted by the space $\widehat{X}_h \subset \widehat{X}$ and $\widehat{X}_h^0 \subset \widehat{X}_D^0$, respectively. For Navier–Stokes flow, we prefer the biquadratic, discontinuous-linear Q_2^c/P_1^{dc} element. The definitions of the spaces for the velocities, displacements, pressure, and phase-field unknowns on a time interval I_m read:

$$
\begin{aligned}
\hat{V}_h &:= \{\hat{v}_h \in [C(\hat{\Omega}_h)]^d,\ \hat{v}_h|_{\hat{K}} \in [Q_2^c(\hat{K})]^d \quad \forall \hat{K} \in \hat{\mathfrak{T}}_h, \hat{v}_h|_{\hat{\Gamma}\setminus\hat{\Gamma}_i} = 0\}\,,\\
\hat{U}_{f,h} &:= \{\hat{u}_h \in [C(\hat{\Omega}_h)]^d, \hat{u}_h|_{\hat{K}} \in [Q_2^c(\hat{K})]^d \quad \forall \hat{K} \in \hat{\mathfrak{T}}_h, \hat{u}_h|_{\hat{\Gamma}\setminus\hat{\Gamma}_i} = 0\}\,,\\
\hat{U}_{s,h} &:= \{\hat{u}_h \in [C(\hat{\Omega}_h)]^d, \hat{u}_h|_{\hat{K}} \in [Q_2^c(\hat{K})]^d \quad \forall \hat{K} \in \hat{\mathfrak{T}}_h, \hat{u}_h|_{\hat{\Gamma}\cup\hat{\Gamma}_i} = 0\}\,,\\
\hat{P}_{f,h} &:= \{\hat{p}_h \in [\hat{L}^2(\hat{\Omega}_h)],\ \hat{p}_h|_{\hat{K}} \in [P_1^{dc}(\hat{K})] \quad \forall \hat{K} \in \hat{\mathfrak{T}}_h\}\,,\\
\hat{\Phi}_{s,h} &:= \{\hat{\varphi}_h \in [C(\hat{\Omega}_h)]^d, \hat{\varphi}_h|_{\hat{K}} \in [Q_1^c(\hat{K})]^d \quad \forall \hat{K} \in \hat{\mathfrak{T}}_h, \hat{\varphi}_h|_{\hat{\Gamma}\setminus\hat{\Gamma}_i} = 0\}\,.
\end{aligned}
$$

Thus, the discrete space is defined as

$$
\widehat{X}_h := \hat{V}_h \times \hat{U}_{f,h} \times \hat{U}_{s,h} \times \hat{P}_{f,h} \times \hat{\Phi}_{s,h}\,.
$$

And correspondingly the space $\widehat{X}_h^0$:

$$
\widehat{X}_h^0 := \{\hat{v}_h^D + \hat{V}_h\} \times \{\hat{u}_{f,h}^D + \hat{U}_{f,h}\} \times \hat{U}_{s,h} \times \hat{P}_{f,h} \times \hat{\Phi}_{s,h}\,.
$$

We consider for each $\hat{K} \in \hat{\mathfrak{T}}_h$ the bilinear transformation $\hat{\sigma}_K \colon \hat{K}_{\text{unit}} \to \hat{K}$, where $\hat{K}_{\text{unit}}$ denotes the master cell. Then, the Q_1^c and Q_2^c elements are defined by

$$
\begin{aligned}
Q_1^c(\hat{K}) &= \{q \circ \hat{\sigma}_K^{-1} : q \in \operatorname{span} < 1, x, y, xy >\}\,,\\
Q_2^c(\hat{K}) &= \{q \circ \hat{\sigma}_K^{-1} : q \in \operatorname{span} < 1, x, y, xy, x^2, y^2, x^2y, y^2x, x^2y^2 >\}
\end{aligned}
$$

with $\dim Q_1^c = 4$ and $\dim Q_2^c = 9$. The P_1^{dc} element consists of linear functions defined by

$$
P_1^{dc}(\hat{K}) = \{q \circ \hat{\sigma}_K^{-1} : q \in \operatorname{span} < 1, x, y >\}
$$

with $\dim P_1^{dc}(\hat{K}) = 3$.

The property of the Q_2^c/P_1^{dc} element is continuity of the velocity values across different mesh cells [45]. However, the pressure is defined by discontinuous test functions. Therefore, this element preserves local mass conservation, is of low order, gains the *inf-sup stability*, and is an optimal choice for both fluid problems and fluid-structure interaction problems.

Formulation 9.14. Let the previous, spatially discretized, time-step solution $\hat{U}_h^{n-1} = \{\hat{v}_h^{n-1}, \hat{u}_{f,h}^{n-1}, \hat{u}_{s,h}^{n-1}, \hat{p}_{f,h}^{n-1}, \hat{\varphi}_{s,h}^{n-1}\}$ be given.
We seek $\hat{U}_h^n = \{\hat{v}_h^n, \hat{u}_{f,h}^n, \hat{u}_{s,h}^n, \hat{p}_{f,h}^n, \hat{\varphi}_{s,h}^n\} \in \hat{X}_h^0$ via

$$\begin{aligned}&\hat{A}_T(\hat{U}_h^{n,t})(\hat{\Psi}_h) + \theta\hat{A}_E(\hat{U}_h^n)(\hat{\Psi}_h) + \hat{A}_P(\hat{U}_h^n)(\hat{\Psi}_h) + \hat{A}_I(\hat{U}_h^n)(\hat{\Psi}_h) \\ &\quad = -(1-\theta)\hat{A}_E(\hat{U}_h^{n-1})(\hat{\Psi}_h) \quad \forall \hat{\Psi} \in \hat{X}_h \,.\end{aligned} \tag{9.16}$$

Remark 9.15. The finite element polynomial degrees have been chosen with respect to the FSI part such that the Navier–Stokes solver is stable with respect to the inf-sup stability. For phase-field fracture most studies are conducted with lowest order elements, some reasons are provided in [13]. Herein, we work with Q_2^c for the solid part and Q_1^c for the phase-field function. In practice we did not observe difficulties, but a detailed numerical analysis is missing to date in the literature.

9.4.4 Solution algorithms

At each time-step t_n, we design as outer loop an augmented Lagrangian iteration (indexed by k in the following) in order to enforce the crack irreversibility constraint. At each augmented Lagrangian step, in the inner loop, the nonlinear, spatially discretized coupled forward Formulation 9.14 is solved with Newton's method. At each Newton iteration a linear equation system is solved.

Outer loop: An augmented Lagrangian iteration

Per time step, we have to satisfy the irreversibility constraint (9.2). As mentioned earlier we work with an augmented Lagrangian formulation [39, 46] resulting in the penalization (9.3). Here, a penalization function Ξ needs to be computed, which is realized in terms of a discrete finite element function Ξ_h. The key idea in the augmented Lagrangian method is to iterate in order to obtain the correct penalization. Thus, for the iteration steps $k = 1, 2, 3, \ldots$ we have

$$\Xi_{h,k+1} = (\Xi_{h,k} + \gamma(\hat{\varphi}_{h,k+1} - \hat{\varphi}_h^{n-1}))_+ \,, \quad k = 0, 1, 2, \ldots \,.$$

Algorithm 9.16 (Augmented Lagrangian loop). At t^0 let $\Xi_{h,0}$ be given, e.g. $\Xi_{h,0} = 0$. Moreover, let $\gamma > 0$ be fixed and given for all t_n. For $k = 0, 1, 2, \ldots$

1. Given $\Xi_{h,k}$ and $\hat{\varphi}_h^{n-1}$; we seek $\hat{U}_{h,k+1}^n$ by solving Formulation 9.14 with Newton's method via Algorithm 9.17.
2. Update
$$\Xi_{h,k+1} = (\Xi_{h,k} + \gamma(\hat{\varphi}_{h,k+1} - \hat{\varphi}_h^{n-1}))_+$$
3. Check the stopping criterion
$$\|\Xi_{h,k+1} - \Xi_{h,k}\| \leq TOL_{\mathrm{AL}} \,, \quad TOL_{\mathrm{AL}} > 0 \,.$$

4. Set: $\hat{U}_h^n := \hat{U}_{h,k^*}^n$ where k^* is the k which satisfies the stopping criterion.
5. Increment $t_n \to t_{n+1}$.

In this study we modify Algorithm 9.16 and apply all five steps (i)–(v) only in the first time-steps $n = 1, 2$. For all remaining time steps, augmented Lagrangian iterations are no longer performed. Namely, for $n \geq 2$, we set $k = 0$ and never increment k while leaving out step (iii) in Algorithm 9.16. Thus, we update Ξ_h but do not check the convergence criterion and do not iterate $k > 0$ until step (iii) is fulfilled. This procedure significantly reduces the computational cost and still yields good satisfaction of the constraint.

Inner loop: Nonlinear iterations using Newton's method

For the nonlinear problem that arises in this paper, we cannot always expect a monotone decrease of Newton's residual. In particular, the monolithic formulation of the phase-field subproblem already requires line search/trust region versions that allow for a temporary increase of the residuals [44, 113, 114]. Consequently, the coupled FSI-PFF problem inherits the bad characteristics of the phase-field subproblem – an example of the performance of Newton's method is provided in Section 9.5.2.

To this end, temporal (indexed by n) and spatial (indexed by h) discretization results per augmented Lagrangian step (indexed by k) within each single time step in a nonlinear quasistationary problem

$$\hat{A}(\hat{U}_{h,k}^n)(\hat{\Psi}) = 0 \quad \forall \hat{\Psi} \in \hat{X}_h ,$$

which is solved by Newton's method. Still, the semilinear form $\hat{A}(\cdot)(\cdot)$ is decomposed as explained in Formulation 9.8 or 9.14, respectively.

Algorithm 9.17. For fixed n, h, k, let an initial Newton guess $\hat{U}_{h,k}^{n,0}$ be given. Find for $j = 0, 1, 2, \dots$ the update $\delta\hat{U}_{h,k}^{n,j}$ of the linear defect-correction problem

$$\begin{aligned} \hat{A}'(\hat{U}_{h,k}^{n,j})(\delta\hat{U}_{h,k}^{n,j}, \hat{\Psi}) &= -\hat{A}(\hat{U}_{h,k}^{n,j})(\hat{\Psi}) , \\ \hat{U}_{h,k}^{n,j+1} &= \hat{U}_{h,k}^{n,j} + \omega\delta\hat{U}_{h,k}^{n,j} . \end{aligned} \tag{9.17}$$

Here, $\omega \in (0, 1]$ is a damping parameter which is determined through backtracking line search. The convergence criterion is usually based on a decrease of the residuals:

$$\|A(U_{h,k}^{n,j+1})(\Psi_h)\|_{l^2} < \|A(U_{h,k}^{n,j})(\Psi_h)\|_{l^2} . \tag{9.18}$$

Definition 9.18 (Classical backtracking line search). A crucial role for (highly) nonlinear problems includes the appropriate determination of ω. A simple strategy is to modify the update step in (9.17) as follows: For given $\omega \in (0, 1)$ determine the minimal $l^* \in \mathbb{N}$ via $l = 0, 1, \dots, N_l$, such that

$$\begin{aligned} R(\hat{U}_{h,k,l}^{n,j+1}) &< R(\hat{U}_{h,k,l}^{n,j}) , \\ \hat{U}_{h,k,l}^{n,j+1} &= \hat{U}_h^{n,j} + \omega^l \delta\hat{U}_h^n . \end{aligned}$$

For the minimal l, we set

$$\hat{U}^{n,j+1}_{h,k} := \hat{U}^{n,j+1}_{h,k,l^*} .$$

In this context, the nonlinear residual $R(\cdot)$ is defined as

$$R(\hat{U}^{n,j}_{h,k}) := \|\hat{A}(\hat{U}^{n,j}_{h,k})(\hat{\Psi}_h)\|_{l^2} \quad \text{for } \hat{\Psi}_h \in \widehat{X}_h .$$

Definition 9.19 (Backtracking line search allowing for increasing residual). A heuristic modification of Definition 9.18 is as follows. For given $\omega \in (0,1)$ compute $l = 0, 1, \ldots, N_l$ for very few steps, e.g. $N_l = 3$. If in the final step $N_l = 3$ the condition

$$R(\hat{U}^{n,j+1}_{h,k,l}) < R(\hat{U}^{n,j}_{h,k,l})$$

is violated, take nevertheless this violating iterate $\hat{U}^{n,j+1}_{h,l}$, increment $j \to j+1$ and go to the next Newton step. Of course, we have no guarantee that the next Newton step will not again violate. If we obtain during the Newton loop at later iteration steps again decreasing residuals

$$R(\hat{U}^{n,j+1}_{h,k,l}) < R(\hat{U}^{n,j}_{h,k,l})$$

this procedure can be adopted. In our examples (see Figure 9.16), this idea worked well, but there is however no theoretical justification except for some special cases [58].

It remains to construct the Jacobian $A'(\hat{U})(\delta\hat{U}, \hat{\Psi})$ of Newton's method. The entries are given by evaluating the directional derivatives at a point $\hat{U}$. Specifically, the directional derivative (omitting all indices) $\hat{A}'(\hat{U})(\hat{\delta}\hat{U}, \hat{\Psi})$ for the Jacobian is defined by

$$\hat{A}'(\hat{U})(\delta\hat{U}, \hat{\Psi}) := \lim_{\varepsilon\to 0} \frac{1}{\varepsilon}\left\{\hat{A}(\hat{U} + \varepsilon\delta\hat{U})(\hat{\Psi}) - \hat{A}(\hat{U})(\hat{\Psi})\right\} = \frac{d}{d\varepsilon}\, \hat{A}(\hat{U} + \varepsilon\delta\hat{U})(\hat{\Psi})\Big|_{\varepsilon=0} .$$

As the residual, the Jacobian is split into four categories and is then evaluated analytically and most details are omitted for the convenience of the reader. Here, we present one representative example, namely the probably most tricky term: the ALE fluid convection:

$$\hat{A}_{\text{conv}}(\hat{U})(\hat{\psi}^v) := (\hat{\rho}_f\hat{J}(\hat{F}^{-1}(\hat{v} - \hat{w})\cdot\hat{\nabla})\hat{v}, \hat{\psi}^v) \tag{9.19}$$

and decompose the first argument (using the relation $\hat{u}_f \cdot \hat{\nabla}\hat{v} = \hat{\nabla}\hat{v}\,\hat{u}_f$) in the scalar product as

$$\hat{\rho}_f\hat{J}(\hat{F}^{-1}(\hat{v} - \hat{w})\cdot\hat{\nabla})\hat{v} = \hat{\rho}_f\hat{J}(\hat{\nabla}\hat{v}\widehat{F}^{-1})\hat{v} - \hat{\rho}_f\hat{J}(\hat{\nabla}\hat{w}\widehat{F}^{-1})\hat{v} . \tag{9.20}$$

The derivative of the first part on the right-hand side of (9.20) reads:

$$\begin{aligned}\partial_U(\hat{\rho}_f\hat{J}(\hat{\nabla}\hat{v}\widehat{F}^{-1})\hat{v})(\delta\hat{U}) = {}& \hat{\rho}_f\hat{J}\,\mathrm{tr}(\widehat{F}^{-1}\hat{\nabla}\delta\hat{u}_f)(\hat{\nabla}\hat{v}\widehat{F}^{-1})\hat{v}\\ &+ \hat{\rho}_f\hat{J}(\hat{\nabla}\delta\hat{v}\widehat{F}^{-1})\hat{v}\\ &+ \hat{\rho}_f\hat{J}\hat{\nabla}\hat{v}(-\widehat{F}^{-1}\hat{\nabla}\delta\hat{u}_f\widehat{F}^{-1})\hat{v}\\ &+ \hat{\rho}_f\hat{J}(\hat{\nabla}\hat{v}\widehat{F}^{-1})\delta\hat{v} .\end{aligned} \tag{9.21}$$

Here $\partial_U(\cdot)(\delta\hat{U})$ denotes the directional derivative into direction $\delta\hat{U}$.

In the second part of (9.20), we cannot directly differentiate the fluid domain velocity $\hat{w}$ since it is not an unknown itself. However, $\hat{w}$ can be represented in terms of $\hat{u}_f$ because of $\partial_t\hat{\mathcal{A}} = \hat{w}$. Here, we need to decide how to construct $\hat{w}$ in time. The simplest possibility is a linear construction in terms of a backward difference quotient:

$$\hat{w} = \partial_t\hat{\mathcal{A}} \approx \frac{1}{k}(\hat{u}_f - \hat{u}_f^{n-1}), \tag{9.22}$$

where $\hat{u}_f^{n-1}$ denotes the solution of the previous time step.

Remark 9.20. For higher order constructions of $\hat{w}$ and its impact we refer the reader to [37, 86].

Next, we readily get

$$\partial_u\hat{w}(\delta\hat{u}_f) := \partial_u\frac{1}{k}(\hat{u}_f - \hat{u}_f^{n-1})(\delta\hat{u}_f) = \frac{1}{k}\delta\hat{u}_f. \tag{9.23}$$

With this, we obtain for the second equation on the right-hand-side in (9.21)

$$\frac{1}{k}\hat{\rho}_f\hat{J}(\hat{\nabla}\delta\hat{u}_f\hat{F}^{-1})\hat{v}.$$

In the remaining terms of (9.21), we formally replace $\hat{\nabla}\hat{v}$ with $\hat{\nabla}\hat{w}$. Thus we have shown:

Proposition 9.21. *Let the ALE convection term $\hat{A}_{conv}(\hat{U})(\hat{\psi}^v)$ be given by (9.19) and let $\hat{w}$ be linearly constructed in time as given in (9.22). Then, the directional derivative in direction $\delta\hat{U} := \{\delta\hat{v}, \delta\hat{u}_f\}$ is given by:*

$$\begin{aligned}
\hat{A}_{conv}(\hat{U})'(\delta\hat{U}, \hat{\psi}^v) &= (\hat{\rho}_f\hat{J}\,\mathrm{tr}(\hat{F}^{-1}\hat{\nabla}\delta\hat{u}_f)(\hat{\nabla}\hat{v}\hat{F}^{-1})\hat{v}, \hat{\psi}^v) \\
&+ (\hat{\rho}_f\hat{J}(\hat{\nabla}\delta\hat{v}\hat{F}^{-1})\hat{v}, \hat{\psi}^v) + (\hat{\rho}_f\hat{J}\hat{\nabla}\hat{v}(-\hat{F}^{-1}\hat{\nabla}\delta\hat{u}_f\hat{F}^{-1})\hat{v}, \hat{\psi}^v) \\
&+ (\hat{\rho}_f\hat{J}(\hat{\nabla}\hat{v}\hat{F}^{-1})\delta\hat{v}, \hat{\psi}^v) \\
&+ (\hat{\rho}_f\hat{J}\,\mathrm{tr}(\hat{F}^{-1}\hat{\nabla}\delta\hat{u}_f)(\hat{\nabla}\hat{w}\hat{F}^{-1})\hat{v}, \hat{\psi}^v) + (\hat{\rho}_f\hat{J}(\hat{\nabla}k^{-1}\delta\hat{u}_f\hat{F}^{-1})\hat{v}, \hat{\psi}^v) \\
&+ (\hat{\rho}_f\hat{J}\hat{\nabla}\hat{w}(-\hat{F}^{-1}\hat{\nabla}\delta\hat{u}_f\hat{F}^{-1})\hat{v}, \hat{\psi}^v) + (\hat{\rho}_f\hat{J}(\hat{\nabla}\hat{w}\hat{F}^{-1})\delta\hat{v}, \hat{\psi}^v).
\end{aligned}$$

All details for pure variational-monolithic ALE-FSI in $\widehat{\Omega}$ can be found in [106, 110] with the respective programming code in [108].

Remark 9.22 (Quasi-Newton steps). In order to reduce computational cost, quasi-Newton steps are applied if the decrease of two subsequent Newton iterates is sufficiently good (here a factor of $\rho < 0.1$). Then the Jacobian matrix from the previous step is kept and only the right-hand side in (9.17) is updated.

Remark 9.23 (Dirichlet boundary conditions). In Newton's method, nonhomogeneous Dirichlet boundary conditions are only prescribed on the initial guess $\hat{U}_{h,k}^{n,0}$ and in all further updates nonhomogeneous Dirichlet conditions are replaced by homogeneous Dirichlet conditions.

Our final Newton algorithm reads as follows:

Algorithm 9.24. Choose an initial guess $\hat{U}_{h,k}^{n,0}$ and perform for $j = 0, 1, 2, 3, \ldots$:
1. Stop if $R(\hat{U}_{h,k}^{n,j}) < TOL_N$, solution found;
2. Compute $\rho := \frac{R(\hat{U}_{h,k}^{n,j})}{R(\hat{U}_{h,k}^{n,j-1})}$. If $\rho > 0.1$ assemble Jacobian $\hat{A}'(\hat{U}_{h,k}^{n,j})(\delta\hat{U}_{h,k}^{n,j}, \hat{\Psi})$; otherwise keep old Jacobian from step $j-1$ (quasi-Newton step);
3. Compute Newton update $\delta\hat{U}_{h,k}^{n,j}$ via solving (9.17)[a] using the Jacobian from step (ii);
4. Check if (9.18) is satisfied; if not perform line search in Definition 9.19 and obtain ω;
5. $\hat{U}_{h,k}^{n,j+1} = \hat{U}_{h,k}^{n,j} + \omega\delta\hat{U}_{h,k}^{n,j}$;
6. Increment $j \to j+1$ and go to Step (i).

Remark 9.25. Within Newton's method the linear equation systems are solved with the direct solver UMFPACK [26].

9.5 Numerical examples

The numerical examples have two goals. First, we conduct further studies on the solid damping parameter. As extensively discussed in [111] we are currently not able to avoid solid damping completely for an FSI-PFF setting. Second, we apply adaptivity in time for the accurate measurement of the u_x-displacement of the right tip of the fractured solid obstacle. As discussed in the introduction, the examples are computed with the open source finite element package deal.II [7, 8] by implementing the method from [104] in the open source FSI code [108]. We finally remark that in this work we 'only' consider nonpropagating fractures that vary in their width but not in their length. Even this situation is difficult enough in a coupled FSI-PFF system and for the sake of a clear understanding of the algorithmic developments we believe it is worth restricting our focus.

Test cases We conduct four studies. In Example 1 (a,b) a constant inflow profile is used. In Example 2 (a,b) a sinus-inflow profile is prescribed. In the a-test cases, a fixed time-step size is chosen, whereas the solid damping parameter varies. Specifically we choose $\mu_v = 0, 10^2, 10^3$. We notice that for the largest choice it holds that $\mu_v = \mu = 10^3$. For $\mu_v = 0$ the very first time step is computed with $\mu_v = 10^3$ (for a motivation we refer to [111]). In the b-test cases, $\mu_v = 0$ is fixed and adaptive time-step control is applied.

Configuration The configuration of both examples is displayed in Figure 9.6. The initial geometry is globally refined four times resulting in 8192 elements and 165,573 degrees of freedom. The smallest element diameter is $h = 0.018$.

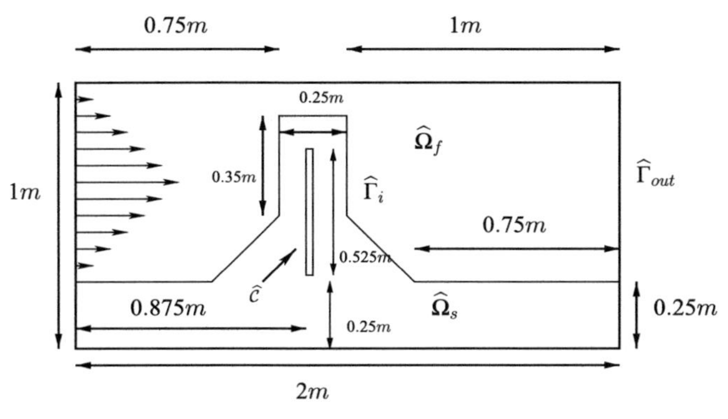

Fig. 9.6: Example 1 and 2: Notation of domains and boundaries.

Boundary and initial conditions For the upper, lower, and left boundaries, 'no-slip' conditions for velocity and zero displacements for the solid are prescribed. At the fluid outlet $\widehat{\Gamma}_{\text{out}}$, the 'do-nothing' outflow condition [54] is imposed. Moreover, $\hat{u} = 0$ on $\widehat{\Gamma}_{\text{out}}$. For the phase-field variable $\hat{\varphi}$ 'natural' boundary conditions are of homogeneous Neumann type. However, on the left and right boundaries as well as on $\widehat{\Gamma}_i$, we set $\hat{\varphi} = 1$ (no fracture).

The boundary condition on the inflow boundary $\widehat{\Gamma}_{\text{in}}$ is specified in each example separately. The maximal velocity in both examples is about $v_{\max} \approx 1.5$ m/s. The lowest channel width is $D = 0.15\ m$. Thus, the Reynolds number is $Re = \frac{v_{\max} D}{\nu_{\text{f}}} = 22.5$, resulting in a laminar flow regime.

Parameters For the fluid we use $\varrho_{\text{f}} = 1\,\text{kg}\,\text{m}^{-3}$, $\nu_{\text{f}} = 10^{-2} m^2 s^{-1}$. The elastic solid is characterized by $\varrho_{\text{s}} = 10\,\text{kg}\,\text{m}^{-3}$, $\nu_{\text{s}} = 0.2$, $\mu = 1\,\text{kg}\,\text{m}^{-1}\text{s}^{-2}$. The fracture pressure is $\hat{p}_F = 10\,Pa$. The model parameter $\varepsilon = 0.044 = 2\,h$ is fixed in all computations as well as $\kappa = 10^{-10}$. Furthermore, $\gamma = 100$ and $G_c = 1$ N/m. The (absolute) Newton tolerance is chosen as 10^{-10}. In the first time step, we determine an appropriate penalization parameter with the help of the augmented Lagrangian algorithm. In the remaining time steps, we keep this penalization parameter fixed. The time-step size is $k = 1$ s (in Example 1a) and $k = 0.25$ s (in Example 2a) and the total times are $T = 20$ s and $T = 50$ s, respectively. As a time-stepping scheme, we use the shifted Crank–Nicolson scheme. The solid damping parameters are specified below.

Quantities of interest We compute various goal functionals that cover different parts of the model:
- Evaluation of $\hat{u}_x(1, 0.85)$ and $\hat{u}_y(1, 0.85)$, which are the horizontal and vertical displacements of the tip of the solid in the point $(1, 0.85)$, respectively;
- Evaluation of the drag force on the FSI interface;
- Evaluation of min(J);

- Observing the time-step sizes using time-step control;
- Observing the absolute error using time-step control.
- Observing behavior of Newton's method at two selected time points.

9.5.1 Example 1: Constant inflow

In this first numerical test, a constant inflow is prescribed resulting in a visible deformation of the solid obstacle and finally in a steady-state solution.

Inflow boundary condition A parabolic inflow velocity profile is given on $\widehat{\Gamma}_{in}$ by

$$v_f(0, y) = \bar{U}(y - 0.25)(y - 1)\,, \qquad \bar{U} = 3.0\,\mathrm{ms}^{-1}\,.$$

For $t < 2.0\,s$, $v_f(0, y)$ is scaled with $\frac{1-\cos(\frac{\pi}{2}t)}{2}$ in order to have a smooth inflow profile.

Discussion of our findings The solution is illustrated in Figure 9.7 and 9.8. In the latter, a zoom-in is provided showing that the mesh in the fracture is squeezed. A discussion of why we emphasize this observation is provided in Section 9.3.4. The four quantities of interest are provided in Figure 9.9. Here we obtain the expected results that solid damping has clearly an influence on the findings. In Figure 9.10, our results to test case (b) are displayed. Here, when the tolerances $TOL_{TS} = 10^{-2}$ and $TOL_{TS} = 10^{-3}$ are used similar behavior is shown (mainly because the maximum step-size k_{max} is mostly taken), but $TOL_{TS} = 10^{-4}$ yields a significant reduction of the absolute error of the goal functional $J(\hat{U}) = \hat{u}_x(1, 0.85)$. In conclusion, the results in (b) are not yet entirely convincing and therefore we refer the reader forward to Example 2b.

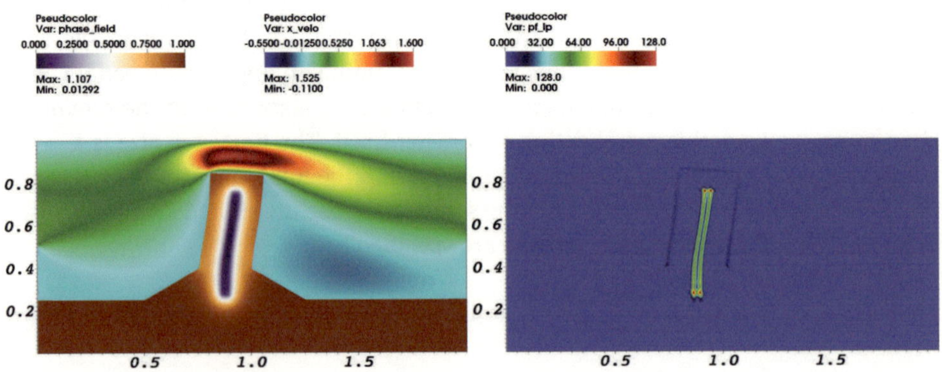

Fig. 9.7: Example 1: In the left figure, the velocity field and the fracture in terms of the phase-field variable are displayed. In the right figure, the augmented Lagrangian function Ξ is shown. Here, we see that penalization mainly acts on the fracture boundary in the transition zone.

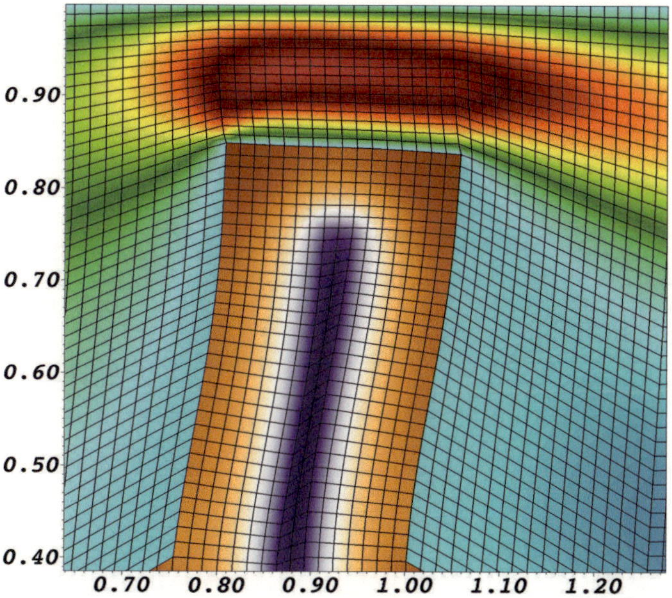

Fig. 9.8: Example 1: In this figure a zoom-in is provided, which shows as well the finite element mesh. In particular a critical part is a squeezed mesh inside the fracture; a detailed motivation for this observation is provided in Section 9.3.4. The white zone around the fracture has thickness ε and thus visualizes the effect of the phase-field regularization parameter.

9.5.2 Example 2: Sinus inflow

In this example, the inflow profile is a sinus curve and in particular we construct the inflow in such a way that also backflow is generated; see Figure 9.11.

Inflow boundary condition A parabolic time-dependent inflow velocity profile is given on $\widehat{\Gamma}_{\text{in}}$ by

$$v_f(0, y) = \bar{U}(y - 0.25)(y - 1) , \quad \bar{U} = [2 \sin(\frac{2t}{\pi} - 0.15\pi) + 1] ms^{-1} .$$

Discussion of our findings The augmented Lagrangian penalization yields very similar values as in the first example as observed in Figure 9.12. The next result in Figure 9.13 shows the oscillating obstacle in terms of $\hat{u}_x$ and $\hat{u}_y$. With respect to solid damping we again meet our expectations. In Figure 9.14 our findings of test case (b) are displayed. Here, we now have a significant difference of time-step sizes depending on the chosen tolerance TOL_{TS}. Finally, the corresponding absolute error is shown in Figure 9.15. We observe nice results that clearly show that the absolute error is concentrated around the chosen tolerance indicating that the developed formula in Proposition 9.10 is reliable. Finally in Figure 9.16, we display two examples that illustrate the convergence of Newton's method and shows the necessity of working with Def-

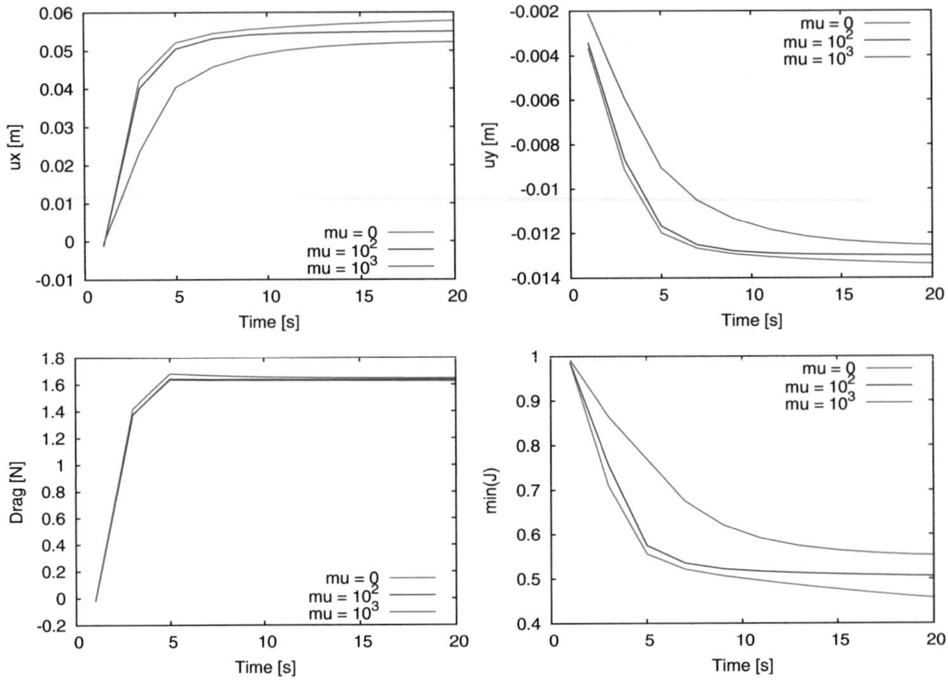

Fig. 9.9: Example 1a: Evaluation of four quantities of interest. Here, the drag shows less influence on the different μ_v than the other three functionals. For all four functionals, already working with $\mu_v = 10^2$ shows a very close behavior to the undamped setting $\mu_v = 0$.

inition 9.19. Specifically, the behavior of Newton's method is almost independent of μ_v. However, we note (not shown here) that in most time steps Newton's residual decreases monotonically.

9.6 Conclusions and future questions

In this chapter, fluid-structure interaction (FSI) has been coupled with phase-field fracture (PFF). Having both frameworks available is significantly nontrivial and therefore the biggest endeavor in this work. Together with part I [112] and part II [111], a coupled FSI-PFF system has not yet been presented in the existing literature. In this part III, algorithmic details and descriptions have been the focus. In extension to part II [111], adaptive time-step control for goal functionals has been worked out and implemented. Secondly, many details on the nonlinear Newton solver have been provided. Thirdly, more studies in relation to solid damping have been carried out, in particular the impact of μ_v on the Newton solver. However, we have to admit that this work

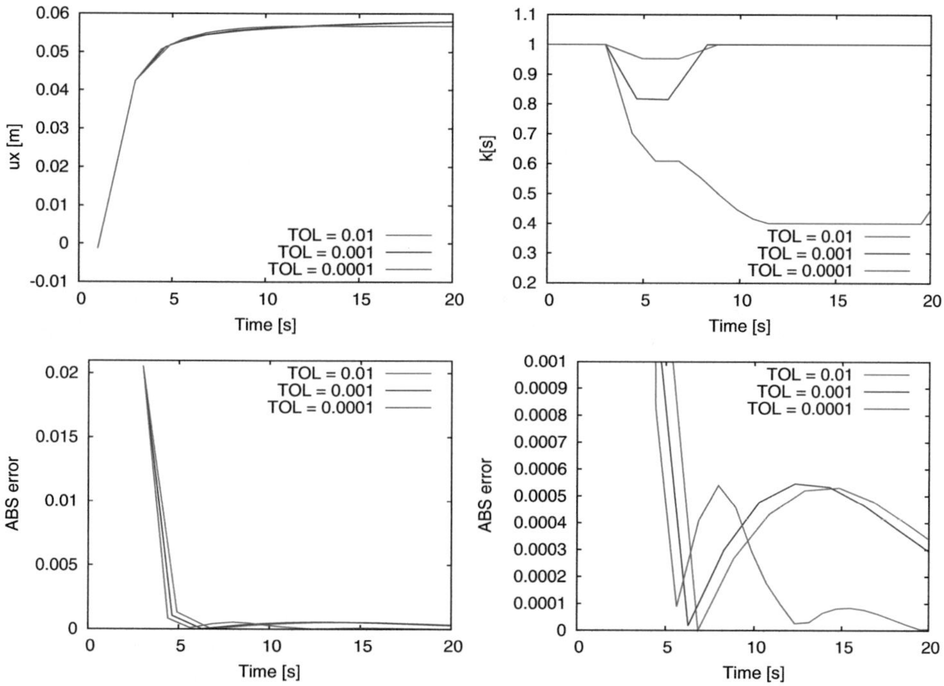

Fig. 9.10: Example 1b: Adaptive time-step control for when goal functional $J(\hat{U}) = \hat{u}_x(1, 0.85)$ (top left) is taken. In the top right figure the development of the time-step sizes are shown. Here, the results for $TOL_{TS} = 0.01$ and $TOL_{TS} = 0.001$ are very similar, which might be justified that we use as upper bound $k_{max} = 1\,s$, which is selected as time step size in most of the time steps. For the lowest tolerance $TOL_{TS} = 0.0001$ (best seen in the two bottom figures), the results are different and also lead to a significant reduction of the absolute error.

raises more questions than providing satisfactory answers. We provide a (definitely incomplete!) list of questions and concerns:

- Further understanding of the impact of the interface term $\langle 2\hat{J}\hat{\varphi}\hat{p}_F\widehat{F}^{-T}\hat{n}_s\hat{u}_s, \hat{\psi}_s^{\varphi}\rangle_{\hat{\Gamma}_i}$ that is due to Gauss divergence in order to impose the fracture pressure $\hat{p}_F$;
- The impact, in general, of boundary conditions on the FSI-PFF system, and in particular on crack growth: how do boundary conditions influence the crack path?
- Modeling a fluid equation inside the fracture from which a nonconstant $\hat{p}_F$ can be computed;
- Modeling the phase transition/topology changes when the fluid-filled fracture reaches the interface and couples directly with the Navier–Stokes flow. Thus, coupling of the Navier–Stokes pressure $\hat{p}_f$ with the fracture pressure $\hat{p}_F$;
- Further understanding of solid damping, which requires deeper analysis of the regularity of the solid velocity when working with elastodynamics coupled with a phase-field fracture model;
- Dealing with large solid deformations in which ALE reaches its limits;

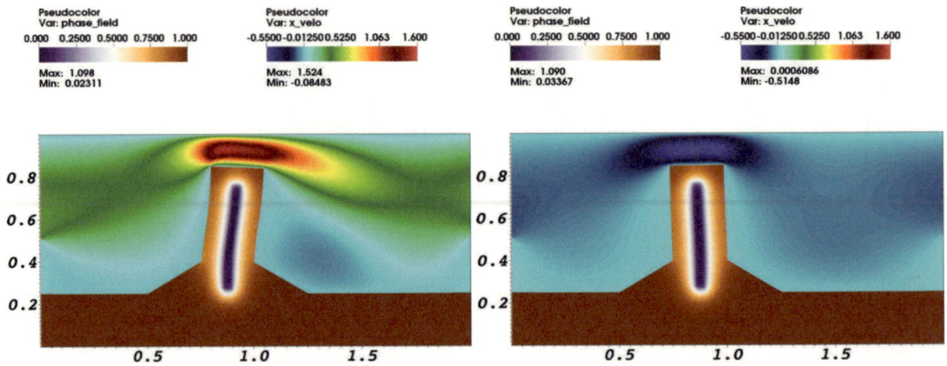

Fig. 9.11: Example 2: Display of the velocity field at two different times in which the highest forward flow and the highest backward flow (negative *x*-velocity) are obtained.

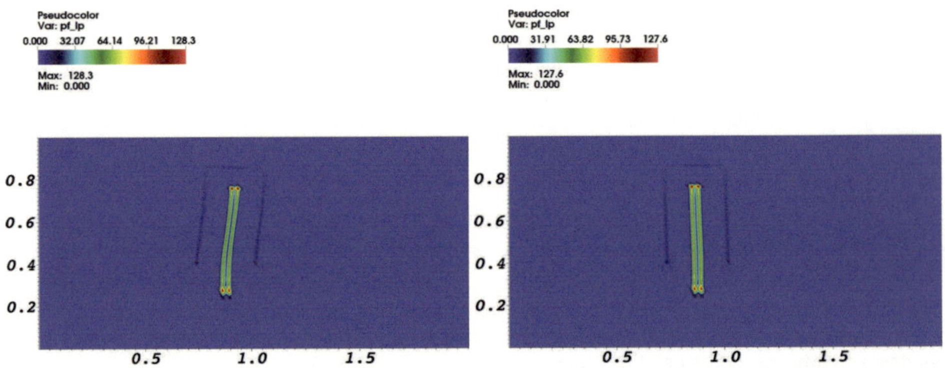

Fig. 9.12: Example 2: Display of the augmented Lagrangian function Ξ at two different times in which the highest forward flow and the highest backward flow are obtained. The values are similar to the first example.

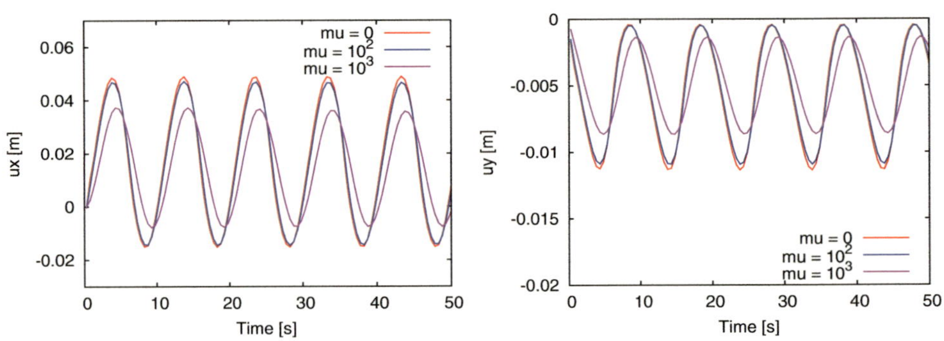

Fig. 9.13: Example 2a: Display of the displacement at the obstacle tip for various μ_v.

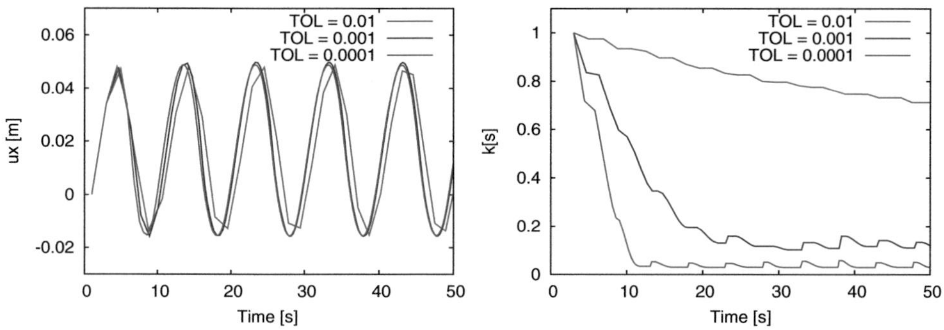

Fig. 9.14: Example 2b: Display of the goal functional (left) and the adaptively obtained time-step sizes. In contrast to Example 1b, we observe significantly reduced time-step sizes for each tolerance.

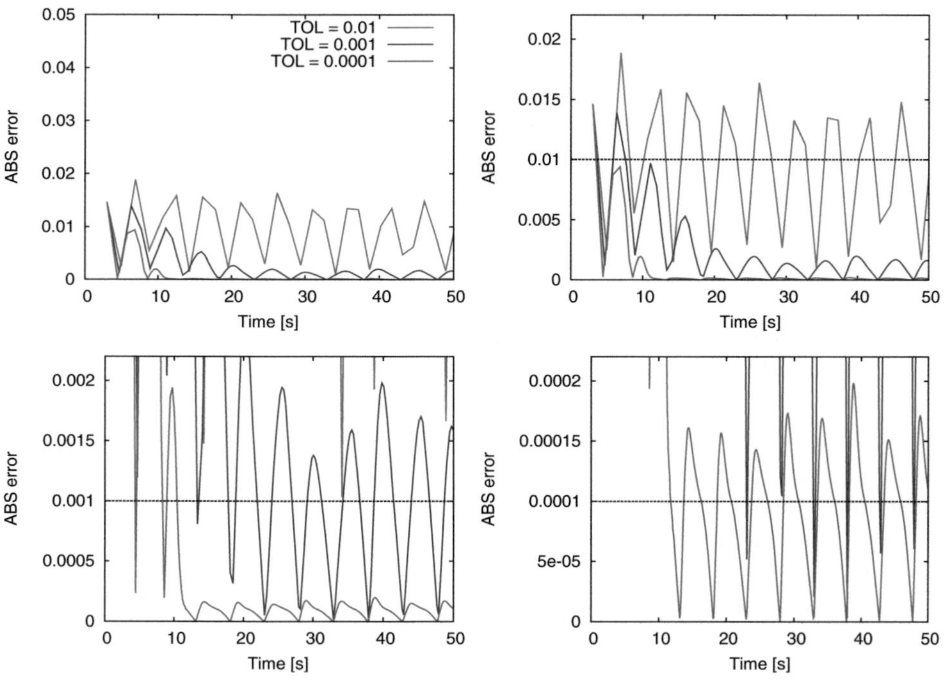

Fig. 9.15: Example 2b: The evolution of the absolute error is shown. The entire evolution is shown in the top left figure. Then, in the other three figures, the respective tolerance is shown as a black dotted line. It can be very well observed that the absolute error evolution centers around the given tolerance. Thus, the error estimator chooses the time steps in such a way that indeed (as expected) the absolute error is reduced according to the prescribed tolerance. This finding yields the computational justification that our time-step control formula works well.

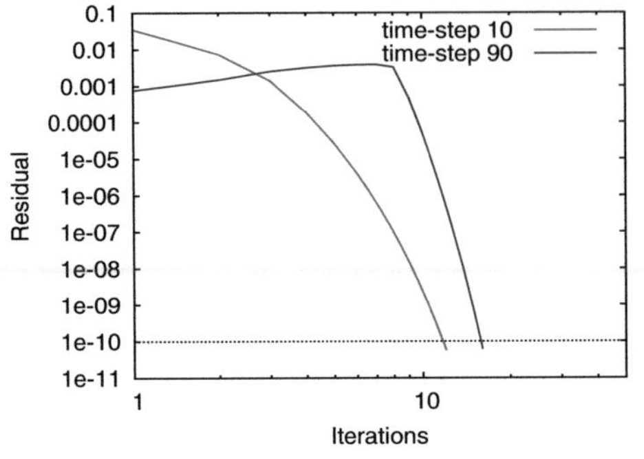

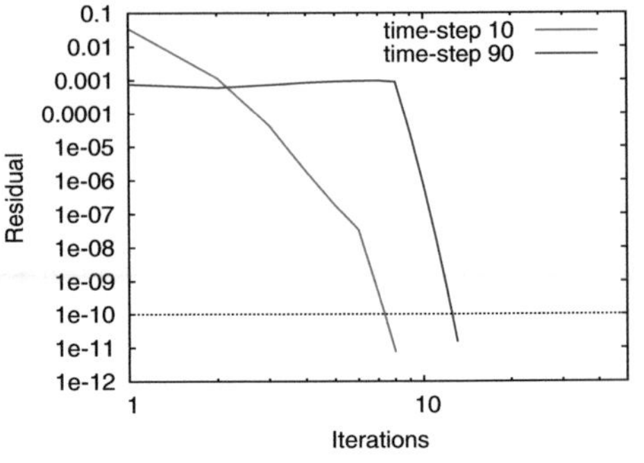

Fig. 9.16: Example 2a: Display of the Newton residual using $\mu_v = 0$ (left) and $\mu_v = 10^3$ (right) at two specific times $t = 4.75\,s$ (time-step 10) and $t = 44.75\,s$ (time-step 90). At $44.75\,s$, we see the necessity of working with a Newton method that allows for a temporary increase of the residual. Here, it can be observed that Newton's method is not greatly affected by the different μ_v.

- Efficient numerical solution, i.e. development of a (parallel) iterative linear solver;
- Mesh adaptivity with respect to residual and goal-oriented error estimators;
- Mesh adaptivity around the fracture in order to work with a very small $\epsilon < h$. Here, a predictor-corrector concept has been proposed in [52] (2D), and extended to 3D [67, 116], but the question raised in [6] remains as to whether the crack follows the mesh or vice versa as well as the combination with residual-based/goal-oriented mesh refinement;
- Numerical analysis of this system; but in particular, even for PFF alone (first studies for adaptive finite elements have been presented in [20, 21]), there are no conclusive numerical studies on the ϵ-h-relationship yet;
- 3D and further validation of the entire framework.

There is hope some of the above questions will be addressed in future work.

Acknowledgment: I wish to express my deepest gratitude to Professor Ulrich Langer for supporting this fascinating piece of research at RICAM Linz (Austria) before I joined École Polytechnique in France. Moreover, I express my great thankfulness to Marie-Cécile, Elisa, Céline, Henri, and Hélène for their patience since I had to finalize this work in the turbulent times of our moving to France.

Bibliography

[1] Adler PM, Thovert JF and Mourzenko VV, Fractured Porous Media, Oxford University Press, 2012.

[2] Allaire G, Jouve F and N. Van Goethem, Damage and fracture evolution in brittle materials by shape optimization methods, Journal of Computational Physics 230 (2011), 5010–5044.

[3] Ambrosio L and Tortorelli VM, Approximation of functionals depending on jumps by elliptic functionals via Γ-convergence, Comm. Pure Appl. Math. 43 (1990), 999–1036.
[4] Ambrosio L and Tortorelli VM, On the approximation of free discontinuity problems, Boll. Un. Mat. Ital. B 6 (1992), 105–123.
[5] Anderson TL, Fracture mechanics: Fundamental and Applications, 3rd edition, CRC Press, Boca Raton, London, New York, Singapore, 2005.
[6] Artina M, Fornasier M, Micheletti S and Perotto S, Anisotropic mesh adaptation for crack detection in brittle materials, SIAM J. Sci. Comput. 37 (2015), B633–B659.
[7] Bangerth W, Hartmann R and Kanschat G, deal.II – a General Purpose Object Oriented Finite Element Library, ACM Trans. Math. Softw. 33 (2007), 24/1–24/27.
[8] Bangerth W, Davydov D, Heister T, Heltai L, Kanschat G, Kronbichler M, Maier M, Turcksin B and Wells D, The deal.II Library, Version 8.4, Journal of Numerical Mathematics 24 (2016), 135–141.
[9] Bazilevs Y, Takizawa K and Tezduyar TE, Computational Fluid-Structure Interaction: Methods and Applications, Wiley, 2013.
[10] Birken P, Quint KJ, Hartmann S and Meister A, A time-adaptive fluid-structure interaction method for thermal coupling, Computing and Visualization in Science 13 (2010), 331–340.
[11] Borden MJ, Verhoosel CV, Scott MA, Hughes TJR and Landis CM, A phase-field description of dynamic brittle fracture, Comput. Meth. Appl. Mech. Engrg. 217 (2012), 77–95.
[12] Bourdin B, Numerical implementation of the variational formulation for quasi-static brittle fracture, Interfaces and free boundaries 9 (2007), 411–430.
[13] Bourdin B, Francfort GA and Marigo JJ, Numerical experiments in revisited brittle fracture, J. Mech. Phys. Solids 48 (2000), 797–826.
[14] Bourdin B, Francfort GA and Marigo JJ, The variational approach to fracture, J. Elasticity 91 (2008), 1–148.
[15] Bourdin B, Larsen CJ and Richardson CL, A time-discrete model for dynamic fracture based on crack regularization, Int. J. Frac. 168 (2011), 133–143.
[16] Bristeau MO, Glowinski R and Periaux J, Numerical methods for the Navier-Stokes equations, Comput. Phys. Rep. 6 (1987), 73–187.
[17] Bukac M, Canic S, Glowinski R, Muha B and Quaini A, A modular, operator-splitting scheme for fluid-structure interaction problems with thick structures, International Journal for Numerical Methods in Fluids 74 (2014), 577–604.
[18] Bungartz HJ, Miriam Mehl and Schäfer M, Fluid-Structure Interaction II: Modelling, Simulation, Optimization, Lecture Notes in Computational Science and Engineering, Springer, 2010.
[19] Bungartz HJ and Schäfer M, Fluid-Structure Interaction: Modelling, Simulation, Optimization, Lecture Notes in Computational Science and Engineering 53, Springer, 2006.
[20] Burke S, Ortner C and Süli E, An adaptive finite element approximation of a variational model of brittle fracture, SIAM J. Numer. Anal. 48 (2010), 980–1012.
[21] Burke S, An adaptive finite element approximation of a generalized Ambrosio-Tortorelli functional, M3AS 23 (2013), 1663–1697.
[22] Ciarlet PG, Mathematical Elasticity. Volume 1: Three Dimensional Elasticity, North-Holland, 1984.
[23] Wick T Jodlbauer D, A monolithic FSI solver applied to the FSI 1,2,3 benchmarks, Fluid-Structure Interaction: Modeling, Adaptive Discretization and Solvers (Frei S, Holm B, Richter T, Wick T and Yang H, eds.), Radon Series on Computational and Applied Mathematics, Walter de Gruyter, Berlin, 2017.
[24] G. dal Maso, Francfort GA and Toader R, Quasistatic crack growth in nonlinear elasticity, Arch. Ration. Mech. Anal. 176 (2005), 165–225.

[25] Dautray R and Lions JL, Mathematical Analysis and Numerical Methods for Science and Technology, 5, Springer, Berlin Heidelberg, 2000.

[26] Davis TA and Duff IS, An unsymmetric-pattern multifrontal method for sparse LU factorization, SIAM J. Matrix Anal. Appl. 18 (1997), 140–158.

[27] Deuflhard P, Newton Methods for Nonlinear Problems, Springer Series in Computational Mathematics 35, Springer Berlin Heidelberg, 2011.

[28] Donéa J, Fasoli-Stella P and Giuliani S, Lagrangian and Eulerian finite element techniques for transient fluid-structure interaction problems, in: Trans. 4th Int. Conf. on Structural Mechanics in Reactor Technology, p. Paper B1/2, 1977.

[29] Donea J, Huerta A, Ponthot PJ and Rodriguez-Ferran A, Arbitrary Lagrangian-Eulerian methods, Encyclopedia of Computational Mechanics, pp. 1–25, John Wiley and Sons, 2004.

[30] Dunne T, Adaptive Finite Element Approximation of Fluid-Structure Interaction Based on Eulerian and Arbitrary Lagrangian-Eulerian Variational Formulations, Ph.D. thesis, University of Heidelberg, 2007.

[31] Dunne T, Thomas Richter and Rolf Rannacher, Numerical simulation of fluid-structure interaction based on monolithic variational formulations, Comtemporary Challenges in Mathematical Fluid Mechanics, pp. 1–75, Springer, World Scientific, Singapore, 2010.

[32] Evans LC, Partial differential equations, American Mathematical Society, 2000.

[33] Failer L and Wick T, Adaptive Time-Step Control for Nonlinear Fluid-Structure Interaction, submitted for publication, Aug 2017.

[34] Farrell PE and Maurini C, Linear and nonlinear solvers for variational phase-field models of brittle fracture, arXiv:1511.08463, 2015.

[35] Fernández MA and Gerbeau JF, Algorithms for fluid-structure interaction problems, pp. 307–346, vol. 1 of Formaggia et al. [38], 2009.

[36] Fernandez MA and Mullaert J, Displacement-velocity correction schemes for incompressible fluid-structure interaction, Comptes Rendus Mathematique 349 (2011), 1011–1015.

[37] Formaggia L and Nobile F, A stability analysis for the Arbitrary Lagrangian Eulerian Formulation with Finite Elements, East-West Journal of Numerical Mathematics 7 (1999), 105–132.

[38] Formaggia L, Quarteroni A and Veneziani A, Cardiovascular Mathematics: Modeling and simulation of the circulatory system, Springer, Italia, Milano, 2009.

[39] Fortin M and Glowinski R, Augmented Lagrangian Methods: Applications to the Numerical Solution of Boundary Value Problems, Stud. Math. Appl. 15, North Holland, Amsterdam, 1983.

[40] Francfort GA and Marigo JJ, Revisiting brittle fracture as an energy minimization problem, J. Mech. Phys. Solids 46 (1998), 1319–1342.

[41] Fries TP and Belytschko T, The extended/generalized finite element method: An overview of the method and its applications, Int. J. Numer. Meth. Engrg. 84 (2010), 253–304.

[42] Galdi G and Rannacher R, Fundamental Trends in Fluid-Structure Interaction, World Scientific, 2010.

[43] Gazzola F and Squassina M, Global solutions and finite time blow up for damped semilinear wave equations, Ann. I. H. Poincaré 23 (2006), 185–207.

[44] Gerasimov T and De Lorenzis L, A line search assisted monolithic approach for phase-field computing of brittle fracture, Computer Methods in Applied Mechanics and Engineering 312 (2016), 276–303, Phase Field Approaches to Fracture.

[45] Girault V and Raviart PA, Finite Element method for the Navier-Stokes equations, Number 5 in Computer Series in Computational Mathematics, Springer, 1986.

[46] Glowinski R and Le Tallec P, Augmented Lagrangian and Operator-Splitting Methods in Nonlinear Mechanics, SIAM Stud. Appl. Math. 9, SIAM, Philadelphia, 1989.

[47] Gomez H and Hughes TJR, Provably unconditionally stable, second-order time accurate, mixed variational methods for phase-field models, Journal of Computational Physics 230 (2011), 5310–5327.

[48] Grätsch T and Bathe KJ, Goal-oriented error estimation in the analysis of fluid flows with structural interactions, Comp. Methods Appl. Mech. Engrg. 195 (2006), 5673–5684.

[49] Griffith AA, The phenomena of rupture and flow in solids., Philos. Trans. R. Soc. Lond. 221 (1921), 163–198.

[50] Gustafsson K, Lundh M and Soederlind G, A PI stepsize control for the numerical solution of ordinary differential equations, BIT 28 (1988), 270–287.

[51] Heider Y and Markert B, A phase-field modeling approach of hydraulic fracture in saturated porous media, Mechanics Research Communications (2016).

[52] Heister T, Wheeler MF and Wick T, A primal-dual active set method and predictor-corrector mesh adaptivity for computing fracture propagation using a phase-field approach, Comp. Meth. Appl. Mech. Engrg. 290 (2015), 466–495.

[53] Heywood JG and Rannacher R, Finite-Element Approximation of the Nonstationary Navier-Stokes Problem Part IV: Error Analysis for Second-Order Time Discretization, SIAM Journal on Numerical Analysis 27 (1990), 353–384.

[54] Heywood JG, Rannacher R and Turek S, Artificial boundaries and flux and pressure conditions for the incompressible Navier-Stokes Equations, International Journal of Numerical Methods in Fluids 22 (1996), 325–352.

[55] Hirt C, Amsden AA and Cook J, An Arbitrary Lagrangian-Eulerian computing method for all flow speeds, J. Comput. Phys. 14 (1974), 227–253.

[56] Hron J and Turek S, A monolithic FEM/Multigrid solver for ALE formulation of fluid structure with application in biomechanics, 53, pp. 146–170, Springer, 2006.

[57] Hughes TJR, Liu WK and Zimmermann T, Lagrangian-Eulerian finite element formulation for incompressible viscous flows, Comput. Methods Appl. Mech. Engrg. 29 (1981), 329–349.

[58] Dennis JJE and Schnabel RB, Numerical Methods for Unconstrained Optimization and Nonlinear Equations, Classics in Applied Mathematics, SIAM, 1996.

[59] Jansson J, Degirmenci NC and Hoffman J, Adaptive Unified Continuum FEM Modeling of a 3D FSI Benchmark Problem, International Journal for Numerical Methods in Biomedical Engineering, n/a–n/a, cnm.2851.

[60] John V and Rang J, Adaptive time step control for the incompressible Navier–Stokes equations, Computer Methods in Applied Mechanics and Engineering 199 (2010), 514–524.

[61] Kuhn C and Müller R, A continuum phase field model for fracture, Engineering Fracture Mechanics 77 (2010), 3625–3634, Computational Mechanics in Fracture and Damage: A Special Issue in Honor of Prof. Gross.

[62] Langer U and Yang H, Recent development of robust monolithic fluid-structure interaction solvers, Fluid-Structure Interaction: Modeling, Adaptive Discretization and Solvers (Frei S, Holm B, Richter T, Wick T and Yang H, eds.), Radon Series on Computational and Applied Mathematics, Walter de Gruyter, Berlin, 2017.

[63] Larsen CJ, Ortner C and Süli E, Existence of solutions to a regularized model of dynamics fracture, Methods in Applied Sciences 20 (2010), 1021–1048.

[64] Larsen CJ, IUTAM Symposium on Variational Concepts with Applications to the Mechanics of Materials: Proceedings of the IUTAM Symposium on Variational Concepts with Applications to the Mechanics of Materials, Bochum, Germany, September 22–26, 2008, ch. Models for Dynamic Fracture Based on Griffith's Criterion, pp. 131–140, Springer Netherlands, Dordrecht, 2010.

[65] R. le Veque, Top Ten Reasons to Not Share Your Code (and why you should anyway), http://faculty.washington.edu/rjl/pubs/topten/topten.pdf, Last Access: Jan 15, 2017.

[66] Lee S, Mikelic A, Wheeler MF and Wick T, Phase-field modeling of proppant-filled fractures in a poroelastic medium, accepted for publication in Comp. Meth. Appl. Mech. Engrg., doi: 10.1016/j.cma.2016.02.008, ICES-report 16–03, Feb 2016.
[67] Lee S, Wheeler MF and Wick T, Pressure and fluid-driven fracture propagation in porous media using an adaptive finite element phase field model, Computer Methods in Applied Mechanics and Engineering 305 (2016), 111–132.
[68] Lee S, Wheeler MF and Wick T, Iterative coupling of flow, geomechanics and adaptive phase-field fracture including level-set crack width approaches, Journal of Computational and Applied Mathematics 314 (2017), 40–60.
[69] Luskin M and Rannacher R, On the soothing property of the Crank–Nicolson scheme, Applicable Analysis 14 (1982), 117–135.
[70] Meidner D and Richter T, Goal-oriented error estimation for the fractional step theta scheme, Comput. Methods Appl. Math. 14 (2014), 203–230.
[71] Meidner D and Richter T, A posteriori error estimation for the fractional step theta discretization of the incompressible Navier-Stokes equations, Comput. Methods Appl. Mech. Engrg. 288 (2015), 45–59.
[72] Mesgarnejad A, Bourdin B and Khonsari MM, Validation simulations for the variational approach to fracture, Computer Methods in Applied Mechanics and Engineering 290 (2015), 420–437.
[73] Miehe C, Hofacker M, Schaenzel LM and Aldakheel F, Phase field modeling of fracture in multi-physics problems. Part II. Coupled brittle-to-ductile failure criteria and crack propagation in thermo-elastic–plastic solids, Computer Methods in Applied Mechanics and Engineering 294 (2015), 486–522.
[74] Miehe C, Hofacker M and Welschinger F, A phase field model for rate-independent crack propagation: Robust algorithmic implementation based on operator splits, Comput. Meth. Appl. Mech. Engrg. 199 (2010), 2765–2778.
[75] Miehe C, Welschinger F and Hofacker M, Thermodynamically consistent phase-field models of fracture: variational principles and multi-field FE implementations, Int. J. Numer. Methods Engrg. 83 (2010), 1273–1311.
[76] Miehe C, Mauthe S and Teichtmeister S, Minimization principles for the coupled problem of Darcy–Biot-type fluid transport in porous media linked to phase field modeling of fracture, Journal of the Mechanics and Physics of Solids 82 (2015), 186–217.
[77] Miehe C, Schaenzel LM and Ulmer H, Phase field modeling of fracture in multi-physics problems. Part I. Balance of crack surface and failure criteria for brittle crack propagation in thermo-elastic solids, Computer Methods in Applied Mechanics and Engineering 294 (2015), 449–485.
[78] Mikelić A, Wheeler MF and Wick T, A phase-field method for propagating fluid-filled fractures coupled to a surrounding porous medium, SIAM Multiscale Model. Simul. 13 (2015), 367–398.
[79] Mikelić A, Wheeler MF and Wick T, Phase-field modeling of a fluid-driven fracture in a poroelastic medium, Computational Geosciences 19 (2015), 1171–1195.
[80] Mikelić A, Wheeler MF and Wick T, A quasi-static phase-field approach to pressurized fractures, Nonlinearity 28 (2015), 1371–1399.
[81] Mikelić A, Wheeler MF and Wick T, A phase-field approach to the fluid filled fracture surrounded by a poroelastic medium, ICES Report 13–15, Jun 2013.
[82] Moes N, Dolbow J and Belytschko T, A finite element method for crack growth without remeshing, Int. J. Numer. Meth. Engrg. 46 (1999), 131–150.
[83] Neitzel I, Wick T and Wollner W, An Optimal Control Problem Governed by a Regularized Phase-Field Fracture Propagation Model, Hamburger Beiträge zur Angewandten Mathematik, Preprint no. 2015–2018, 2015.

[84] Nguyen TT, Yvonnet J, Zhu QZ, Bornert M and Chateau C, A phase-field method for computational modeling of interfacial damage interacting with crack propagation in realistic microstructures obtained by microtomography, Computer Methods in Applied Mechanics and Engineering (2015).
[85] Niklasch C and Herrmann N, Nonlinear fluid-structure interaction calculation of the leakage behaviour of cracked concrete walls, Nuclear Engineering and Design 239 (2009), 1628–1640, 15th International Conference on Nuclear Engineering (ICONE 15).
[86] Nobile F, Numerical Approximation of Fluid-Structure Interaction Problems with Applications to Haemodynamics, Ph.D. thesis, École Polytechnique Fédérale de Lausanne, 2001.
[87] Noh W, A time-dependent two-space-dimensional coupled Eulerian-Lagrangian code, Methods Comput. Phys. 3, pp. 117–179, Academic Press, New York, 1964.
[88] Rannacher R, Numerische Methoden für gewöhnliche Differentialgleichungen (Numerische Mathematik 1), Vorlesungsskriptum, 2002.
[89] Rannacher R, Finite element solution of diffusion problems with irregular data, Numer. Math. 43 (1984), 309–327.
[90] Rannacher R, On the stabilization of the Crank–Nicolson scheme for long time calculations, Preprint, August 1986.
[91] Rice JR, Mathematical analysis in the mechanics of fracture, chapter 3 of fracture: an advanced treatise ed, pp. 191–311, Academic Press New York, 1968.
[92] Richter T, Goal-oriented error estimation for fluid-structure interaction problems, Comp. Methods Appl. Mech. Engrg. 223–224 (2012), 38–42.
[93] Richter T and Wick T, Finite elements for fluid-structure interaction in ALE and fully Eulerian coordinates, Comp. Methods Appl. Mech. Engrg. 199 (2010), 2633–2642.
[94] Richter T and Wick T, On time discretizations of fluid-structure interactions, in: Multiple Shooting and Time Domain Decomposition Methods (Th. Carraro, Geiger M, Körkel S and Rannacher R, eds.), pp. 377–400, Contributions in Mathematical and Computational Science, 2015.
[95] Schreurs PJG, Fracture mechanics, Lecture notes course 4A780, Eindhoven University of Technology, 2012.
[96] Sicsic P and Marigo JJ, From Gradient Damage Laws to Griffith's Theory of Crack Propagation, Journal of Elasticity 113 (2013), 55–74.
[97] Sneddon IN and Lowengrub M, Crack problems in the classical theory of elasticity, SIAM series in Applied Mathematics, John Wiley and Sons, Philadelphia, 1969.
[98] Stein K, Tezduyar T and Benney R, Mesh moving techniques for fluid-structure interactions with large displacements, J. Appl. Mech. 70 (2003), 58–63.
[99] Talemi RH, Brown S, Martynov S and Mahgerefteh H, Hybrid fluid-structure interaction modelling of dynamic brittle fracture in steel pipelines transporting CO2 streams, International Journal of Greenhouse Gas Control (2016).
[100] Tezduyar T, Behr M and Liou J, A new strategy for finite element computations involving moving boundaries and interfaces – the deforming-spatial-domain/space-time procedure: I. The concept and the preliminary numerical tests, Comp. Methods Appl. Mech. Engrg. 94 (1992), 339–351.
[101] Turek S, Efficient solvers for incompressible flow problems, Springer, 1999.
[102] Verhoosel CV and R. de Borst, A phase-field model for cohesive fracture, International Journal for Numerical Methods in Engineering 96 (2013), 43–62.
[103] Wang KG, Lea P and Farhat C, A computational framework for the simulation of high-speed multi-material fluid–structure interaction problems with dynamic fracture, International Journal for Numerical Methods in Engineering 104 (2015), 585–623.

[104] Wheeler MF, Wick T and Wollner W, An augmented-Lagangrian method for the phase-field approach for pressurized fractures, Comp. Meth. Appl. Mech. Engrg. 271 (2014), 69–85.
[105] Wick T, Adaptive Finite Element Simulation of Fluid-Structure Interaction with Application to Heart-Valve Dynamics, Ph.D. thesis, University of Heidelberg, 2011.
[106] Wick T, Fluid-Structure Interactions using Different Mesh Motion Techniques, Computers and Structures 89 (2011), 1456–1467.
[107] Wick T, Goal-Oriented Mesh Adaptivity for Fluid-Structure Interaction with Application to Heart-Valve Settings, Arch. Mech. Engrg. 59 (2012), 73–99.
[108] Wick T, Solving Monolithic Fluid-Structure Interaction Problems in Arbitrary Lagrangian Eulerian Coordinates with the deal.II Library, Archive of Numerical Software 1 (2013), 1–19.
[109] Wick T, Flapping and Contact FSI Computations with the Fluid-Solid Interface-Tracking/Interface-Capturing Technique and Mesh Adaptivity, Computational Mechanics 53 (2014), 29–43.
[110] Wick T, Modeling, Discretization, Optimization, and Simulation of Fluid-Structure Interaction, Lecture notes at Heidelberg University, TU Munich, and JKU Linz available on https://www-m17.ma.tum.de/Lehrstuhl/LehreSoSe15NMFSIEn, Last access: Feb 01, 2017.
[111] Wick T, Coupling fluid-structure interaction with phase-field fracture, Journal of Computational Physics 327 (2016), 67–96.
[112] Wick T, Coupling fluid-structure interaction with phase-field fracture: modeling and a numerical example, RICAM preprint No. 2015–2022, accepted for publication in Numerical Mathematics and Advanced Applications, Proceedings of ENUMATH 2015, Feb 2016.
[113] Wick T, An error-oriented Newton/inexact augmented Lagrangian approach for fully monolithic phase-field fracture propagation, SIAM J. Sci. Comput 39 (2017), B589–B617.
[114] Wick T, Modified Newton methods for solving fully monolithic phase-field quasi-static brittle fracture propagation, CMAME 325 (2017), 577–611.
[115] Wick T, Goal functional evaluations for phase-field fracture using PU-based DWR mesh adaptivity, Computational Mechanics 57 (2016), 1017–1035.
[116] Wick T, Lee S and Wheeler MF, 3D Phase-Field for Pressurized Fracture Propagation in Heterogeneous Media, ECCOMAS and IACM Coupled Problems Proc., May 2015 at San Servolo, Venice, Italy, 2015.
[117] Wick T, Singh G and Wheeler MF, Fluid-filled fracture propagation using a phase-field approach and Coupling to a reservoir simulator, SPE Journal (2015), 1–19.
[118] Wilson ZA and Landis CM, Phase-field modeling of hydraulic fracture, Journal of the Mechanics and Physics of Solids 96 (2016), 264–290.
[119] Wloka J, Partielle Differentialgleichungen, B. G. Teubner Verlag, Stuttgart, 1982.
[120] Xu XP and Needleman A, Numerical simulations of fast crack growth in brittle solids, J. Mech. Phys. Solids 42 (1994), 1397–1434.
[121] Yogaraj S, An embedded interface finite element method for fluid-structure-fracture interaction, Ph.D. thesis, Technische Universitaet München, 2015.
[122] van der Zee KG, van Brummelen EH, Akkerman I and de Borst R, Goal-oriented error estimation and adaptivity for fluid-structure interaction using exact linearized adjoints, CMAME 200 (2011), 2738–2757.

Index

Radon Series on Computational and Applied Mathematics

Volume 19
Boris Khoromskij
Tensor Numerical Methods in Scientific Computing
ISBN: 978-3-11-037013-3, e-ISBN: 978-3-11-036591-7

Volume 18
Maïtine Bergounioux, Gabriel Peyré, Christoph Schnörr, Jean-Baptiste Caillau, Thomas Haberkorn (Eds.), 2016
Variational Methods. In Imaging and Geometric Control
ISBN: 978-3-11-043923-6, e-ISBN: 978-3-11-043039-4

Volume 17
Maïtine Bergounioux, Édouard Oudet, Martin Rumpf, Filippo Santambrogio, Guilaume Carlier, Thierry Champion (Eds.), 2016
Topological Optimization. Optimal Transport in the Applied Sciences
ISBN: 978-3-11-043926-7, e-ISBN: 978-3-11-043041-7, ISSN: 1865-3707

Volume 16
Harald Niederreiter, Alina Ostafe, Daniel Panario, Arne Winterhof (Eds.), 2014
Algebraic Curves and Finite Fields. Cryptography and Other Applications
ISBN: 978-3-11-031788-6, e-ISBN: 978-3-11-031791-6, ISSN: 1865-3707

Volume 15
Peter Kritzer, Harald Niederreiter, Friedrich Pillichshammer, Arne Winterhof (Eds.), 2014
Uniform Distribution and Quasi-Monte Carlo Methods. Discrepancy, Integration and Applications
ISBN: 978-3-11-031789-3, e-ISBN: 978-3-11-031793-0, ISSN: 1865-3707

Volume 14
Ivan G. Graham, Ulrich Langer, Jens M. Melenk, Mourad Sini (Eds.), 2013
Direct and Inverse Problems in Wave Propagation and Applications
ISBN: 978-3-11-028223-8, eISBN: 978-3-11-028228-3, ISSN: 1865-3707

Bei Fragen zur Produktsicherheit wenden Sie sich bitte an:
If you have any questions regarding product safety,
please contact:

Walter de Gruyter GmbH
Genthiner Straße 13
10785 Berlin
productsafety@degruyterbrill.com